Man Shapes the Earth

A TOPICAL GEOGRAPHY

Cartography by
HANS J. STOLLE
Western Michigan University

HARM J. DE BLIJ
University of Miami

Man Shapes the Earth

A TOPICAL GEOGRAPHY

HAMILTON PUBLISHING COMPANY
Santa Barbara, California

Published by **Hamilton Publishing Company,**
a Division of John Wiley & Sons, Inc.

Library of Congress Cataloging in Publication Data:

De Blij, Harm J.
Man shapes the earth.

1. Geography. I. Title.
G115.D35 910 73-20159
ISBN 0-471-20049-2

Printed in the United States of America

10 9 8 7 6 5 4 3 2 1

For Tanya

ABOUT THE AUTHOR

Harm J. de Blij received his early education in the Netherlands, took a bachelor's degree in Africa, and attained his doctoral degree at Northwestern University. From 1961 until 1969 he taught geography at Michigan State University, where he was also Associate Director of the African Studies Center. Since then he has been Chairman of the Department of Geography at the University of Miami, Florida. Dr. de Blij's systematic interests in geography are in political and physical fields, and his regional concentration has been in Africa. He has written *Africa South* and two books on East African cities, all published by Northwestern University Press; *A Geography of Subsaharan Africa*, published by Rand McNally; and *Systematic Political Geography* and *Geography: Regions and Concepts*, published by John Wiley. In addition to teaching geography at the University of Miami, Dr. de Blij is Editor of the *Journal of Geography* and Secretary of the Association of American Geographers.

CREDITS

The author is grateful for the kind cooperation of the many individuals and organizations who provided photographs to illustrate this text. Sources for these photographs are listed below. Photographs that are not otherwise credited were supplied by the author.

TITLE PAGE
NASA

CHAPTER ONE
Oregon State Highway Department photos 1–2, 34 NASA 4 Information Canada Photothèque 30 Geological Survey of Canada, Ottawa, G.S.C. 202183 31 F.A.O. photo 32 Geological Survey of Canada, Ottawa, G.S.C. 152353 46

CHAPTER TWO
Wide World Photos 57–58 Matthew Klein, Magnum Photos, Inc. 75 NOAA 77 Photo Department of Information, Papua New Guinea 86 F.A.O. photo 87 United Nations 91 International Society for Educational Information Tokyo, Inc. 93

CHAPTER THREE
F.A.O. photos 99–100, 113, 134 Brian Brake, copyright Rapho Guillumette Pictures 107 UNICEF photo by John Weisblat 111 Bruce Davidson, Magnum Photos, Inc. 115 UNICEF photo by Horst Cerni 130 DSCS photo 136 Photo courtesy Hong Kong Tourist Association 137 Bruce Roberts, copyright Rapho Guillumette Pictures 138 UNICEF photo by Jack Ling 139

CHAPTER FOUR
Wide World Photos 145–146 Henri Cartier-Bresson, Magnum Photos, Inc. 152 Information Canada Photothèque 174 German National Tourist Office (left); From Afghan Tourist Organization by Wakili Kabuli (right) 180 Dennis Stock, Magnum Photos, Inc. (top); Photo courtesy Hong Kong Tourist Association (bottom) 187

CHAPTER FIVE
F.A.O. photo 189–190 Ministry of Information, Iran (2); Minister of Tourism, Morocco (2); F.A.O. photo (1) 196 Irish Tourist Board (2); German National Tourist Board (2) 200 Homer Page, WHO photo (2); United Nations (1); D. Henrioud, WHO photo (1) 206 Photos courtesy Hong Kong Tourist Association (2); F.A.O. photo (1); United Nations (1); Tourism Bureau, Republic of China (1) 209 DSCS photos (3); Singapore Tourist Promotion Board (2) 214 WHO photo (1); F.A.O. photo (1); Pierre Pittet, WHO photo (1); Paul Almasy, WHO photo (1) 218 F.A.O. photo (1); El Salvador Institute of Tourism (1); Courtesy of the Panama

Government Tourist Bureau (2) 222 Department of Public Works, City of Chicago (1); Louisiana Tourist Commission (4); Photograph courtesy of Department of Natural Resources, State of Washington (1) 226 Australia Information Service (7); Photo Department of Information, Papua New Guinea (1) 229 Novosti Press Agency (2) 232 International Society for Educational Information Tokyo, Inc. (4) 236

CHAPTER SIX

Piedrabuena O'Sullivan, Magnum Photos, Inc. (top); Thai Forestry Department (bottom) 247 Paolo Koch, copyright Rapho Guillumette Pictures 256 German National Tourist Office 260 Rene Burri, Magnum Photos, Inc. (top); Henri Cartier-Bresson, Magnum Photos, Inc. (bottom) 264 Robert Janke 268 Georg Gerster, copyright Rapho Guillumette Pictures 277

CHAPTER SEVEN

F.A.O. photo 291–292 Lynn McLaren, copyright Rapho Guillumette Pictures 299, 306 UNICEF photo by Bernard Pierre Wolff 303 Marc Riboud, Magnum Photos, Inc. 313 Idaho Department of Commerce and Development 316 Rene Burri, Magnum Photos, Inc. 324 Novosti Press Agency 327 United Nations 334

CHAPTER EIGHT

Novosti Press Agency 337–338 UNICEF photo by Nagata 341 Information Canada Photothèque 349 International Society for Educational Information Tokyo, Inc. 354, 374 Novosti Press Agency 370 El Salvador Institute of Tourism 383

CHAPTER NINE

Wide World Photos 385–386 Paolo Koch, copyright Rapho Guillumette Pictures 389 BOAC 393 Louis Goldman, copyright Rapho Guillumette Pictures 396 Henri Cartier-Bresson, Magnum Photos, Inc. 399 Department of Public Works, City of Chicago 414 Wide World Photos 419 Hedrich-Blessing, Bertrand Goldberg Associates 421 Rene Burri, Magnum Photos, Inc. 422

CHAPTER TEN

Courtesy of the Panama Government Tourist Bureau 425–426 NOAA 438 U.S. Geological Survey 441 Rene Burri, Magnum Photos, Inc. 444 EPA 449 Unations 455

The homolosine interrupted projection used throughout the text is based on Goode Map No. 201HC World Homolosine, Goode Base Map Series. Courtesy of the Department of Geography, the University of Chicago. Copyright by The University of Chicago, Department of Geography.

PREFACE

The principal objective of this book is to present in easily readable, conversational language a series of ten subjects of geographic interest. Together these ten topics provide considerable coverage of the field and constitute a quite comprehensive introduction to physical, human, regional, and systematic contents of the discipline. I have intentionally kept the topics separate from each other. Each stands by itself as a learning unit and in such a way that if one is deleted or a substitution is made none of the others will be damaged and the continuity will not be seriously impaired.

In selecting the ten topics, I have tried to represent a rather wide range of geographic interests. My practice has been to spend a week, or slightly longer, on each of the subjects, using the text as "take-off" material and guiding the classroom discussion in slightly different directions. In a sense this particular selection of topics reflects both the demonstrated interests of the students and their self-realized lack of fundamentals. For example, it is obvious that very few students reach the college or university level with any sort of systematic background in physical geography. Yet they find themselves fascinated by the implications of continental drift and frustrated by their lack of personal knowledge and consequent inability to discuss, debate, and judge. I believe that general awareness is one of the great assets of geographic orientation, and that includes a capacity to look at landscapes in terms other than turnpike miles.

The major part of this book, however, deals with aspects of human geography. Obviously it has been necessary to focus somewhat on particular areas within each subject: the population section centers on growth problems; the culture section, on criteria; the regional chapter, on world realms; the political chapter, on the state; the agriculture chapter, on major product movements; the manufacturing chapter, on development; and the urban chapter, on evolutionary patterns. In the final chapter, which focuses upon the oceans, all sorts of issues are tied together: some physical geography in the morphology of coastlines, a bit of economic geography in connection with continental-shelf resources, a little demography in the context of the oceans as food-supply sources, and some political geography relating to territorial waters and the penetration of man's last earthly frontier. The ocean realm provides a good opportunity for a relevant summary to any introductory geography sequence.

Beginning students are not often attuned to checking footnotes, so I have integrated citations into the main text. Additionally, I have placed "asides" and definitions, among other items, in framed units separate from the continuous commentary. This improves the continuity of the major body of material and, through the book's design, provides an easy frame of reference.

I am most grateful to several colleagues who have read and commented on this volume. Dr. Clifford Tiedemann of the University of Illinois, Chicago Circle, gave an enormous amount of his time and energy to prepare a detailed review. Mr. Richard Dastyck of Fullerton Junior College, California, also provided a most useful commentary. Dean Theodore Kury of the State University College at Buffalo, New York, read much of the book and sent me a productive response. Dr. Stephen Birdsall of the University of North Carolina commented on the book's organization and on other aspects as well. Dr. John Hehr of the University of Miami did a detailed review of the chapters on physical geography and wrote "The Seventh Approximation," page 37. Mr. Alan F. Ryan, also of the University of Miami, greatly aided in the revision of Chapter Ten.

I am especially appreciative of the work done by cartographer Hans J. Stolle of Western Michigan University, whose attention to detail and sustained interest in the whole project were matters of pleasure and stimulation. I am proud that Mr. Stolle's work appears once again in a book of mine.

In addition I should like to thank the staff of Hamilton Publishing Company for the work done relating to this volume: Paul A. Lee, Principal Editor, Ronald Q. Lewton, Editor for Production, Janet Bollow, designer, and Judith Chaffin, copy editor and photo researcher. For the shortcomings that remain I am, of course, solely responsible.

HARM J. DE BLIJ

Coral Gables, Florida
July, 1973

Contents

CHAPTER ONE

The Finite Earth: A Heritage of Abundance

ONE

Ours is not a large planet. Its familiar dimensions—a diameter of under 8000 miles, a circumference of about 25,000 miles—seem tiny against the vastness of a universe in which distances are measured in millions of light years. And even in our particular corner of the universe, the solar system, the earth is no giant. The planet Jupiter has a diameter more than 11 times as large as Earth's. Saturn is much larger still. True, Earth is slightly larger than Venus and Mars, but more than 70 percent of our planet is covered by water, making our living space smaller still.

What the earth lacks in size, however, it makes up in the wealth and variety of its contents. More than 6 billion years ago, when it was born from a gaseous mass of material that slowly collapsed and congealed, the earth was endowed with an almost infinitely complex chemistry. Throughout the planet's life, that chemistry was modified continuously. The molten earth mass, turning and cooling, differentiated into several shell-like layers, of which the crust is but one. Gases escaped and began to collect in low-lying areas as liquids, signaling the formation of the oceans. Other volatiles created the earth's initial atmosphere. Eventually that atmosphere supported the beginnings of life, and late in the planet's history it attained levels of oxygen content that made possible the evolution of complex life forms. Man is the culmination of this process, and mankind became the beneficiary of those billions of years of accumulation of natural wealth.

STRUCTURE AND TIME

Man's impact upon this enormous legacy of riches, and humanity's interaction with its various environments, are central themes we shall pursue in this book. But before we consider man's habitat—the surface of the earth—we should briefly consider the earth as a whole. An understanding of the surface features of our globe begins with a knowledge of what lies beneath, for forces originating deep inside the planet have much effect on its outermost layer, the crust. For centuries physical geographers and other scientists have been trying to discover the nature of the processes that shape the landscapes with which we are so familiar. In recent years, those efforts have taken a new direction: for years we have been interested in the reasons why earthquakes occur, for example, but now the search is also on for a means to predict when and where the next quakes will take place. The process of volcanism is studied not only for its own sake, but also for finding ways to forewarn people

The earth as photographed from the *Apollo 17* spacecraft. Almost the entire coastline of the African continent is visible. The white area in the lower part of the globe is the Antarctic ice cap. Heavy cloud cover extends across the southern part of the Southern Hemisphere. The Red Sea and the Arabian Peninsula are clearly outlined in the upper center of the photograph.

who might be endangered that an eruption is due. Landslides, river floods, pollution threats of various kinds—all these and many more properties of the environment are studied for these purposes.

We have to be careful, of course, that in our search for the really relevant areas of physical geography (or those that seem relevant to us now) we do not lose sight of the greater framework within which all this occurs. You never know what might turn out to be of direct consequence. The history of the concept of drifting continents is a good case in point. In the 1950s the idea that our continents might be pieces of the earth's crust actually in motion, "drifting" in various directions, had little support in the United States. Now, in the 1970s, the principle of continental drift is so well accepted that it even has the sanction of *Time* Magazine! What will become clearer, as this decade progresses, is that practically all features of physical geography—coastlines, deserts, mountain ranges, river courses—are in some way related to the drift process. Take a look at a picture of the 1906 San Francisco earthquake and ask again how "relevant" continental drift could possibly be to you. North America is moving into the Pacific Basin, and the effects of that process will be felt from Alaska to Mexico and beyond. There will be more San Franciscos and Managuas.

The Earth's Interior

To learn about the interior of the earth we must turn to geology and geophysics. Scientists in these areas of research use delicate instruments, careful extrapolation, and considerable ingenuity to reach their findings, for we have very little in the way of direct observation of the earth's interior. Men have carried rocks back from the moon, but the deepest mines on our own planet penetrate less than 2 miles into the crust at most. Bore holes go deeper, but not enough to penetrate through the crust. In fact, we are still far away from seeing just one-thousandth of the total diameter of our own planet.

Geophysicists have discovered, nevertheless, that the earth's crust is very thin and that our planet has a twin core—an *inner core* with a radius of just under 800 miles and an *outer core* whose thickness is somewhat less than 1400 miles (Fig. 1.1). The inner core appears to be solid and made up of extremely heavy rock material, but the outer core is believed to be liquid, rather like molten steel. Between this outer core and the crust lies the earth's *mantle*, 1800 miles thick and made of rock material of high density and great rigidity.

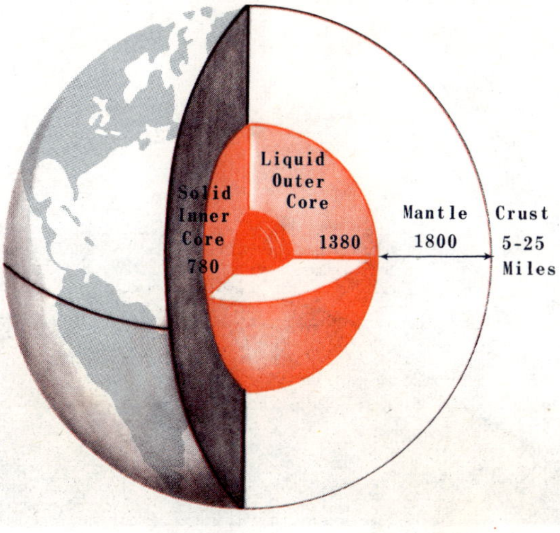

FIG. 1.1 Zones of the earth's interior.

The Crust

Naturally it is the crust of the earth that interests us most of all. So far as we know, this is the thinnest of all the earth's "shells"; there are places where it is a mere 5 miles thick. Elsewhere it may reach 25 miles, and the average appears to be about 10. The crust's main chemical components are silica and alumina. Here is a bit of that extrapolation the geophysicist uses: when geologists sample all of the crust's rock types and measure their specific gravity, the average turns out to be just under 3.0. But calculations of the total mass of the earth, based on its role in the balance of the solar system, produces an overall specific gravity of 5.5. Now, if the crust averages less than 3.0 and the whole planet 5.5, then the earth's interior must be much heavier than the crust. To compensate for the crust's lightness, the deep interior of the globe must have a specific gravity as high as 10 or even more. We know of only a few substances so heavy, including nickel and iron; hence we conclude that these are major elements of the inner and outer core.

The crust, then, is rather light; rocks such as sandstone and granite weigh little compared to the deep interior. Because this is the rock layer of the earth, the crust is also referred to as the *lithosphere*. The bottom of the crust is of special interest, for it does not merge into the mantle below, but is sharply demarcated from it. This surface, where the crust lies upon the mantle below, was identified (from earthquake records) by a geophysicist named Mohorovičić, and understandably it became known as the *Moho discontinuity* (Fig. 1.2).

The crust may, in general terms, be divided into two distinct zones, or layers, whose names are based upon their chemical composition. At the surface

FIG. 1.2 The earth's crust.

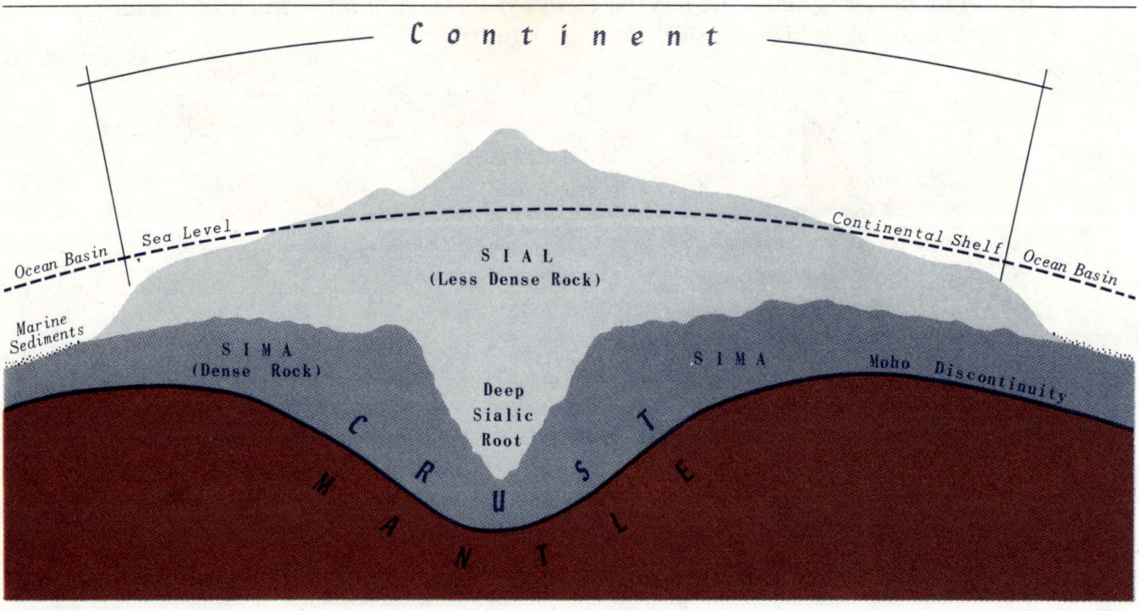

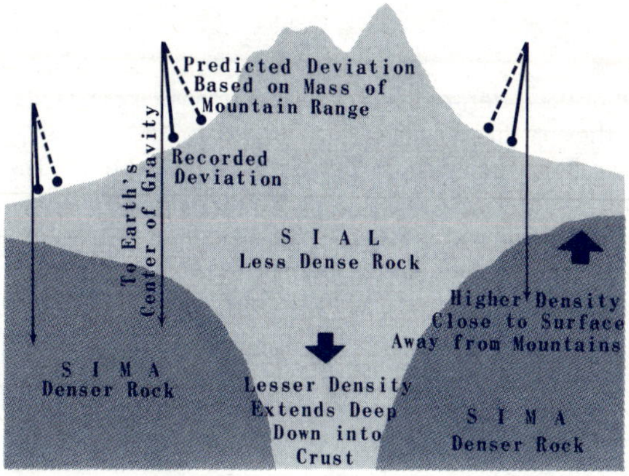

FIG. 1.3 Plumb-line deviations.

and in the outer layer, granitic-type rocks predominate and aluminum-rich silicates prevail. Deeper down, darker, more basaltic material occurs, and magnesium, the substance also occurring in the mantle, is more common. These two layers have been given distinctive names: the *si*lica, *al*uminum-rich upper crust is known as the *sial,* and the *si*lica, *ma*gnesium-rich lower layer is called the *sima.*

Spatially, the crust also shows variation. Under the continents, which are basically sialic, granitic masses protruding above sea level, the crust is much thicker than under the low-lying ocean basins. In fact the sial is noncontinuous, for it is absent under much of the oceanic realm of the globe, where sima, overlain by marine sediments of course, comes to the surface. The continents have "roots" of sialic material sticking down into the sima, and the Moho discontinuity (marking the base of the crust) normally lies far deeper below the surface under the continents than under the ocean bed (Fig. 1.2).

Isostasy

That the continents have sialic "roots" has been surmised for more than a century. When measurements involving a plumb line were being made near the Himalaya Mountains, an astounding result was obtained: the giant, vast Himalayas failed to attract the plumb bob nearly as much as their enormous mass ought to. There could be only one conclusion: the rocks beneath the Himalayas did not have sufficient mass to attract the plumb line. Thus the relatively light, sialic rocks of the mountains must extend deep down, displacing the heavier, denser sima that would have attracted the plumb line to a greater degree. The same results were later recorded in measurements on the flanks of the South American Andes. In one such measurement, on the western side of the Andes, the plumb line actually stood *away* from the mountains!

This apparent anomaly was subsequently confirmed by gravity measurements, which again revealed that the greater mass of the mountains is compensated by the lesser density of the rocks making them up, a low density that extends deep down into the crust (Fig. 1.3; see also Fig. 1.2). The situation can be illustrated

THE EARTH AND TIME

When geologists began to realize that there is a degree of order in the massive accumulations of rocks that make up the crust, they realized that it was necessary to establish a timetable according to which newly identified rocks could be permanently labeled. The geologic time scale was first developed more than a century and a half ago by British geologists. They were able to see how a geologic cross section of England involved several discrete phases of accumulation of rocks and deformation by bending and breaking of the crust. They called the oldest sequence of rocks they were able to recognize the *Primary* and the next two, the *Secondary* and *Tertiary.* Later the very youngest rocks, such as sediments now accumulating in river deltas, were separately identified as *Quaternary.* If we used these original terms, our geologic time scale would logically look like this (with the oldest rocks at the bottom, of course):

4. Quaternary (Recent)
3. Tertiary (Young)
2. Secondary (Intermediate)
1. Primary (Oldest)

Field research in the decades after the establishment of the first geologic time scale produced such a vast quantity of detail that it soon became necessary to subdivide these original major units. Bit by bit the geologic time scale became more exact, and more complicated. More was learned about fossils, and correlations were made with discoveries in distant countries. How important the discoveries pertaining to the evolution of life were can be seen from the names in the geologic time scale now in use: Primary has become *Paleozoic* (era of ancient life); Secondary is now *Mesozoic* (era of medieval life). We can still find Tertiary and Quaternary in the modern time scale, but they are now subdivisions of a third great era, the *Cenozoic* (era of recent life). It has also been realized that there are rocks older than the Paleozoic (older than about 600 million years); surprisingly, this era is not called the Prepaleozoic, as would seem reasonable, but *Precambrian,* which means that it is older than the oldest period of the Paleozoic.

A simplified geologic time scale is given on page 9. It shows that more is known about relatively recent geologic times than older periods; subdivisions at the top of the table are more numerous, and they cover shorter time intervals. This is not surprising, for the youngest rocks tend to be closest to the surface, some of them still being deposited, and others, such as those layers left in the northern United States and Canada during a recent glaciation, are fresh and have hardly changed, geologically speaking.

The time scale shows the three major *eras* divided into *periods,* which, in the case of the Tertiary and Quaternary, are shown further subdivided into *epochs.* Even so, the time scale can be seen to cover only the last 600 million years of the earth's life in any detail. This may be less than 10 percent of the total life span of this planet. But in those 600 million years momentous developments occurred that shaped the landscapes of the world as we know it and witnessed the rise of mammalian life and man.

THE GEOLOGIC TIME SCALE

ERA	PERIOD	EPOCH	MIL YRS	EVENTS	FEATURES	AGE OF
CENOZOIC	QUATERNARY	Recent			Civilizations	Man
		Pleistocene	1	Glaciers overspread large areas in four stages	Rise of man	
	TERTIARY	Pliocene	12	Cascades and Sierra Nevadas formed	Large mammals	Mammals
		Miocene	25	Large deposition of sediments on coastal plain of Gulf of Mexico and Atlantic Ocean; also in Pacific coastal areas	Monkeys Horses	
		Oligocene	40		Grasses spread; grazing animals	
		Eocene	60			
		Paleocene	70		Mammals develop	
MESOZOIC	CRETACEOUS		130	Rocky Mountains formed	Dinosaurs extinct	Reptiles
	JURASSIC		190	Mountain-building movements and great granitic intrusions in Sierra Nevada region	Flowering plants	
	TRIASSIC		220		Dinosaurs Birds	
PALEOZOIC	PERMIAN		270	Appalachian Mountains formed	First dinosaurs Primitive mammals	
	PENNSYLVANIAN		310	Coal-bearing layers deposited	Conifer forests	Amphibians
	MISSISSIPPIAN		350		Reptiles spread	
	DEVONIAN		400	Much crustal activity and granitic intrusion in New England	Sharks Amphibians appear	
	SILURIAN		440	Much deposition of marine sediments; crustal folding in eastern and western North America	First forests	
	ORDOVICIAN		500		Fishes develop	Fishes
	CAMBRIAN		625	Great sequences of deposition and crustal deformation; not well known	Marine life First vertebrates	Marine invertebrates
	PRECAMBRIAN				First life forms	

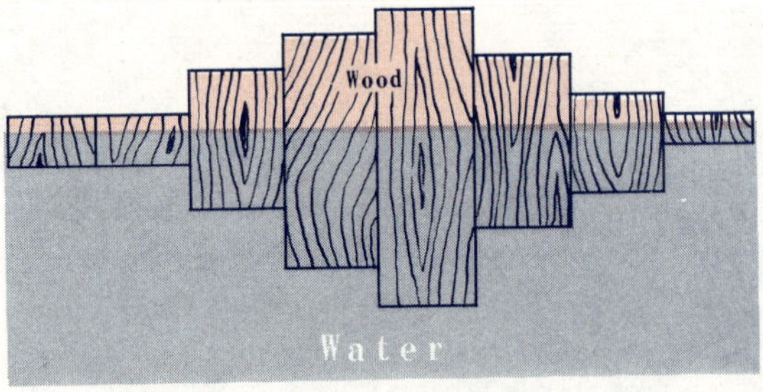

FIG. 1.4 Isostasy model.

by placing a set of wooden blocks in a basin of water (Fig. 1.4). Each block of wood simulates a column of the earth's crust. The taller block protrudes farther above the water, but it also extends deeper into it. If we simulate erosion by sawing off the top of a block, we will note that the block rises slightly in adjustment and water takes the place of a section of its underwater support. If we place the sawed-off piece on top of another block, it will sink slightly, displacing some water below.

This principle of balance within the earth's crust is the principle of *isostasy*, although isostatic equilibrium is much more complicated, of course, than here described. The crust does not consist of a set of convenient columns capable of moving independently. But we do know that erosion causes the slow accumulation of vast quantities of sediment of great weight and that the crust sinks under that deposition. We also know that eroded areas (or areas relieved of a great mass of ice after a recent glaciation) rise gently. It is reasonable to conclude that slow compensatory flowage of rock in the sima does take place to adjust for this surface transfer.

SURFACE AND SPACE

In this book we will take a particular interest in the surface of the earth as the human habitat, in the environments it holds, and in man's interrelationships with these environments. As geographers, we will take note of the spatial distribution of phenomena, the shape (literally) of things, and the processes shaping them. Take, for example, a globe showing the continents and oceans and consider it thoughtfully. There are some striking distributional features and features of shape that quickly lead us to ask *why*. Why do three of the continents have roughly triangular shapes, with the apexes pointing in the same southern direction? Is it really an accident or does this reveal something basic about the structure of the crust? And why, when the earth is a sphere and everything on it so mobile, are the continents clustered on one half of it (on what we call the land hemisphere), with a sea hemisphere that is virtually all ocean? And looking somewhat more carefully, why can we see long, linear mountain ranges on all the major continents except one, Africa? There is an Andes in South America, a Rocky Mountains in North America, an Alps in Europe, a Himalayas in Asia, a Great Dividing Range in Australia. Why is Africa's equivalent missing? And there is a question of location. Those elongated mountain chains seem to have a habit of lying on the Pacific side of the continents. South and North America's major mountains lie in the west, on or near Pacific shores; Australia's ranges lie near the eastern margin of that continent, again on the Pacific side. Do the continents share a relationship with the Pacific Basin that generates this situation?

It may be, of course, that the answers to such questions must be found by scientists in disciplines other than geography. That does not matter. The important thing is to be alert enough to recognize the need for a solution in the first place. The questions we just posed have been attacked by geologists, geophysicists, paleontologists, physical geographers, and others. The following paragraphs report on some of the things they have learned.

Continents and Oceans

The continental landmasses that protrude above the ocean waters constitute just 30 percent of the total surface of the globe. Actually, sea level has varied through geologic time, and if the water level dropped just 600 feet, it would expose large, adjacent *continental shelves.* As Fig. 1.5 shows, these continental shelf areas are especially extensive around Eurasia and northern and eastern North America. If their area were added to that of the continents, the continental landmasses would amount to 35 percent of the total. Since the sialic material extends to the edge of the continental shelf (called the *continental slope*), it is reasonable, in geologic terms, to view the continents this way. From other points of view it is appropriate also: not only are the continental shelves among the world's richest fishing grounds, but also more and more of the resources below the shelf, in the *subsoil,* are being brought within man's reach by advancing technology. There has even been talk of permanent underwater settlements. The continental shelves are part of the continents in more than one way.

SHIELDS When we consider the continents, then, we should *not* use as our reference the familiar coastline we see on world maps; instead we should use the margins of the continental shelves as shown on Fig. 1.5. This is important in connection with those questions we raised about the distribution of landmasses and their spatial qualities. Figure 1.5 tells us that the continents are mainly constituted by two kinds of landforms, shields and mountain belts. The *shield* areas form the cores of the continents. These vast shields are made of very old rocks. On the time scale (page 9) they would lie in the Precambrian and thus be older than 600 million years, but most are well over 1 billion years old, and some shield rocks are thought to be between 3 and 4 billion years old. These rocks are mostly of the granitic type, though some have been altered by subsequent volcanic activity. Apparently the shield areas represent the original heartlands of the continents. Today they are typically low, undulating plains bearing witness to the lengthy erosion and degradation they have undergone. North America's core area is the *Canadian Shield;* Asia's is the *Russian-Baltic Shield.* In the Southern Hemisphere,

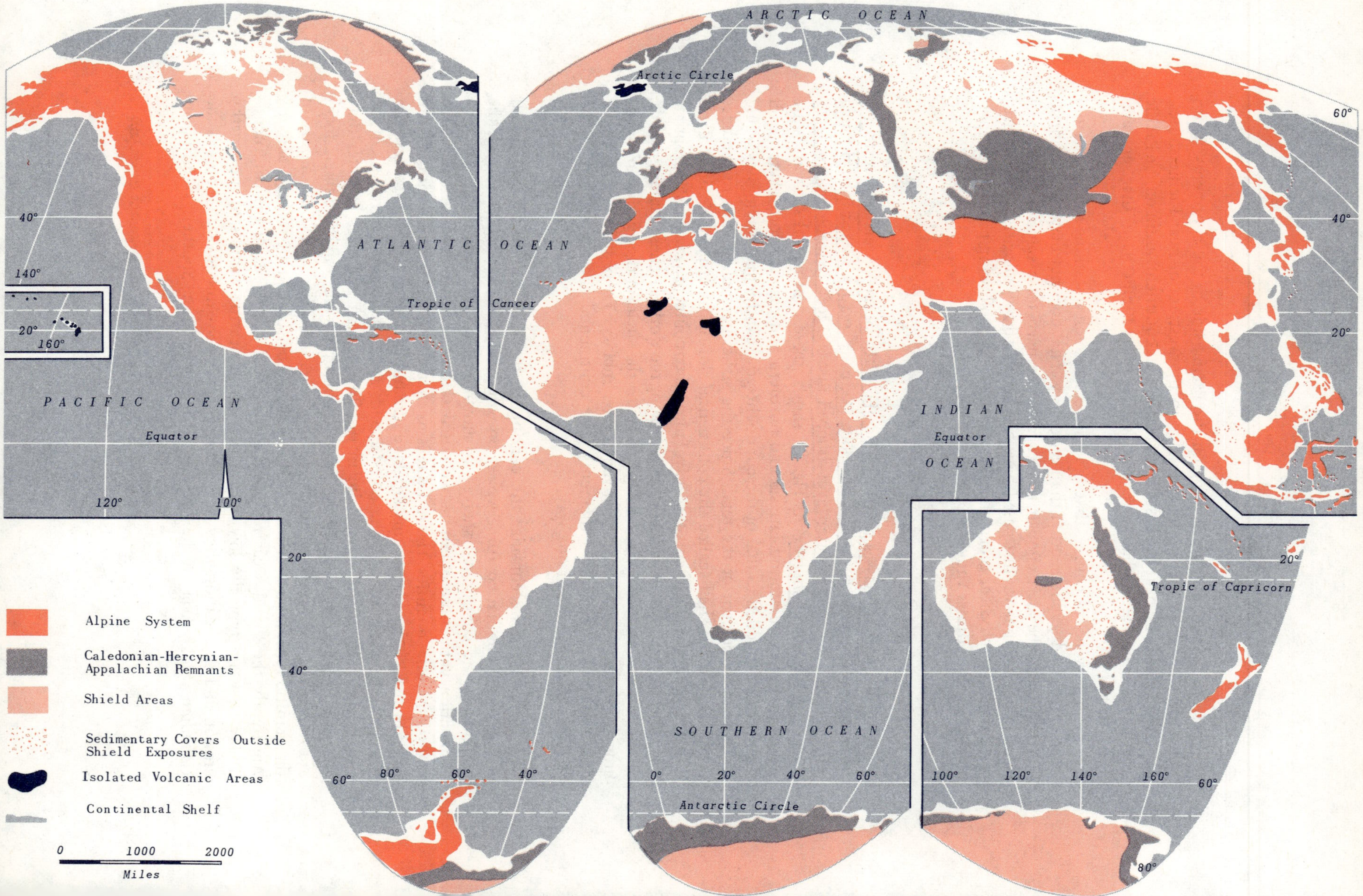

FIG. 1.5 Landforms of the world. (Modified after Richard E. Murphy.)

all three of the continents have huge shield areas, especially Africa. Note that the shields probably extend considerably under the sedimentary material laid down upon their margins. Figure 1.5 shows only the exposed portions of the continental shields.

MOUNTAINS Adjacent to the shields lie mountain ranges, also known as *orogenic belts*. Whereas the shield areas are the stable zones of the earth, the orogenic belts are places where the crust is deformed, where earthquakes occur frequently, and where volcanic mountains exist. A British geographer, Marion Newbigin, suggested that the contrasting environments presented by the flat, stable shields and the rugged, changeable mountains produced differences in the character of the peoples inhabiting these regions. The peoples of the plateaus and plains were likely to be complacent, uninventive, and rather dull. Those used to the varied mountainous landscapes, on the other hand, were innovative, aggressive, and energetic. Such generalizations could never be proved, no matter how many examples one could gather, but there is no doubt that shield areas and mountainous zones constitute very different environments.

One such area of contrast lies in the rocks that underlie shield areas and mountain belts. Shields are crystalline, as we noted, but mountain ranges are in large part made of sedimentary rocks, which, along with other rocks, have been crushed and folded into gigantic ranges. This happens during certain periods of earth history, and the term *orogeny* is used to identify these. For example, on the geologic time scale, where the formation of the Rocky Mountains is indicated the *Laramide Orogeny* occurred. Just why orogenies have taken place throughout earth history during certain periods but not during others is not clear; neither is the mechanism by which all this happens. We know that rocks that have been laid down over long periods of time are eventually bent, folded up, crushed, and pushed into huge, linear flexures. We can see the evidence in the landscape, in road cuts and in other places where the structures are exposed. But why it happens exactly where it does and how it takes place are still not known for certain.

What we do know is that erosion attacks the mountains immediately, in fact, while they are still being formed. Thus younger mountains stand out more prominently than older ones, time being the great ally of weathering and erosion. In our country, the Appalachians are much older than the Rocky Mountains, and if you have flown over those two physiographic regions, you will know the difference. The Appalachians are rounded, rather smooth even, vegetation-covered, comparatively low. The Rockies are much higher, more rugged, still bare in many places. Eventually they, too, will be worn down to low relief.

Not only are shield areas and mountain ranges made of different kinds of rocks, but they also contain different kinds of mineral associations. Obviously there is no point in looking for coal in granitic rocks. Things get molten and assimilated when granite forms, not covered and compacted. Between the sedimentary rock strata of which mountains are often made lie layers of compressed vegetation, and when the circumstances are right, the process of folding can produce excellent coal. Pennsylvania is known for the high-quality coal called *anthracite*, so pure that most of the noncombustible material has been squeezed out. But we shall return to the topic of mineral associations shortly.

ENVIRONMENTS It is not difficult to imagine how different the living environments of the world's shield areas and mountain ranges are. The landscapes of the shield areas are not only flatter (that is, of lower relief) than mountain belts, but they are also much more stable and quiet. If we made a map of the incidence of earthquakes around the world, we would find that the majority of them occur in and along the world's mountain ranges; the shield areas are comparatively free of them. Those orogenies we just spoke of are not something of the past; they still continue, and they produce a steady rumble of tremors, shakes, and quakes in the crust. Many of these earthquakes occur away from human settlements, but, all too often, towns and cities are victims of their violent forces. The whole western margin of the Americas, from Alaska to southern Chile, is an orogenically active zone, and if you have been following the news over the past several years, you are aware that severe quakes have struck Alaska, California, Nicaragua, and Peru quite recently. Parts of Los Angeles were severely damaged in the 1971 earthquake that shook southern California; but the results were nothing compared with the earthquake that hit Peru on May 31, 1970, where the official death toll was 66,794. Just before Christmas, 1972, a series of quakes destroyed practically all of Managua, capital city of Nicaragua in Central America.

Earthquakes are not the only hazard in mountain zones: landslides, avalanches, and fires that can engulf vegetation-clad slopes in a matter of minutes all contribute to a high danger factor. Small creeks in narrow valleys can turn into raging, destructive torrents during a concentrated and prolonged rain. In spite of this, people live on the steepest of slopes, on the sides of volcanoes known to be capable of eruption, and in areas repeatedly struck by earthquakes. The greatest danger is in the younger mountain ranges, whose life cycle is still active. On Fig. 1.5, two major world-scale orogenic belts are shown: the older Caledonian and Hercynian belt (the Appalachians form part of the latter) and the so-called Alpine system. The older Caledonian-Hercynian Orogeny is quiescent now, a remnant of its earlier prominence and activity. Erosion is wearing these mountains down, and the geologic forces that once sustained and built them have moved elsewhere. Today the Alpine system dominates the areas of high relief of Eurasia and the Americas. It includes Europe's Alps and its neighboring ranges, the mountains of southern Asia all the way through the Himalayas, and all of eastern Asia. The Alpine Orogeny even extends into the coastal waters of eastern Asia. The archipelagoes of Japan and the Philippines are huge folds in the crust, parts of which emerge above the waters of the Pacific Ocean. No wonder that Japan (among other areas in East Asia) has a high frequency of earthquakes!

Geographers have taken a considerable interest, notably in recent years, in the reasons behind people's decisions to continue to live in areas threatened by natural disasters such as earthquakes. This has to do with the way they perceive the threats that natural disasters pose and the value they attach to the resources that form the basis for their lives in the threatened area. People have a peculiar capacity to dismiss danger and ignore possible avenues of protection. This shows in the failure of so many people to fasten seat belts despite 50,000 highway fatalities in the United States annually, and it also shows in their neglect of options that might reduce the impact of floods, earthquakes, and other disasters.

Rocks

The landscapes of the continents are carved out of rocks. Although the vegetative cover and the mantle of soil often hide the rock layers that sustain the landscape, rocks exist under every part of the earth's surface, under ocean waters as well as polar ice. Most simply, a rock can be defined as a natural mass forming part of the earth's crust. This means that such very different materials as hard granite and loose desert sand are technically both rocks.

Rocks are aggregates of minerals, and they form in a variety of ways. Some are created when a liquid mass of molten earth material cools and solidifies, or "freezes." The continental shield areas first formed in this way. Other rocks form when oceans and rivers deposit layers of sediment, which later become hard and dry. Still others are created when hardened or deposited rock is subjected for some reason to high pressures or high temperatures or both. This is not difficult to visualize: imagine an outpouring of molten lava over a sandy beach. The lava itself simply hardens, but the sand over which it flowed is heated up by hundreds of degrees, possibly to several feet deep. The sand is partly melted, and it will cake and become hard itself because of the high temperatures and (under the weight of the lava) increased pressure. Thus the sand may be made into quartzite.

TYPES The three processes of rock formation, hardening, accumulation, and alteration, produce three different types of rocks. *Igneous* rocks are those that result directly from the solidification of molten earth material. Actually, it may be better to speak of crystallization because the cooling of a molten, liquid mass may produce sets of crystals as the temperature goes down through various levels. This process depends somewhat upon the rate of cooling. The slower the rate, the greater the opportunity for large, well-formed crystals to develop. A piece of granite with large crystals of quartz, feldspar, and flakes of mica probably cooled slowly somewhere deep in the crust and was exhumed by subsequent erosion.

Weathering, erosion, and deposition result in the accumulation of *sedimentary* rocks. Sedimentary rocks go through another stage, a stage of cementation, or compaction. Although it is true that river alluvium and desert sand are technically sedimentary rocks, we consider sandstone to be representative of this type. Sandstone was loose sand at one time, but some agent caused the loose grains to stick together. This agent may have been filtering water that deposited a cement (such as calcium carbonate or an iron compound) between the grains. Or perhaps the gathering weight of overlying sediments aided the compaction process.

When preexisting igneous or sedimentary rocks are altered through increases in temperature or pressure or both and by the invasion of hot, permeating fluids, *metamorphic* rocks result. This process can occur repeatedly, with rocks being metamorphosed several times. As a result metamorphic rocks often are very complex, bearing the effects of successive alterations. Some familiar metamorphic rocks are slate (compacted shale) and quartzite (sandstone baked into a harder rock). When granite is metamorphosed, *gneiss* may result.

Continental Surfaces: Landscapes and Landforms

Physical geographers are especially interested in the processes that create landscapes out of the rocks of which the earth's crust is made. The work of weathering and erosion carves valleys and scarps from granites and sandstones, and rivers carry ground-up sediments to coasts where deltas may form. Slowly the upper layer of rocks is transformed into soil, and vegetation clothes the countryside. How does all this happen? Are there laws and rules of nature that govern the processes whereby the world's varied scenery is created?

First, let us differentiate between the terms *landscape* and *landform*. A land*form* is a single, discrete, physical feature: a dune is a landform, as is a volcano, a river valley, and a scarp. The land*scape*, on the other hand, is the total natural appearance of an area. A landscape is constituted by all the landforms in the area, by the relief, the soils, the vegetation. A landscape, thus, is an aggregate of features; a landform is a single feature.

When we try to reconstruct the sequence of events in the formation of a certain landscape, individual landforms often provide key bits of information. For instance, when an area is glaciated and the glaciers retreat, they leave behind mounds, hills, small ridges, and other landforms so characteristic that there can be no doubt about that landscape's genesis. A trained physical geographer can tell almost at a glance that the landscape owes its origin to glacial conditions. But there are other occasions when landscapes are difficult to explain and the search for answers leads only to ever larger questions.

THE MAP OF AFRICA A marvelous example of the lessons landscapes can teach us relates to our earlier observation about the distribution of shield areas and orogenic belts. Among the continents, we noted that Africa has much shield and very little evidence of large-scale orogenic activity. Only in the Atlas Mountains in the north and in the Cape Ranges in the far south are there landscapes comparable, say, to the Andes and Appalachians. From Fig. 1.5 we conclude that most of the continent is relatively flat or undulating country.

When we examine the map of Africa in greater detail, we find that conclusion to be essentially correct, but with a qualification: Africa's shield areas are not low-lying as are most shields in the Northern Hemisphere, but plateaulike in character. In fact, much of the continent is rimmed by a scarp that leads rather suddenly (sometimes in a series of steps) from this plateau to a very narrow coastal zone (Fig. 1.6). In Africa this is called the Great Escarpment, and in various degrees

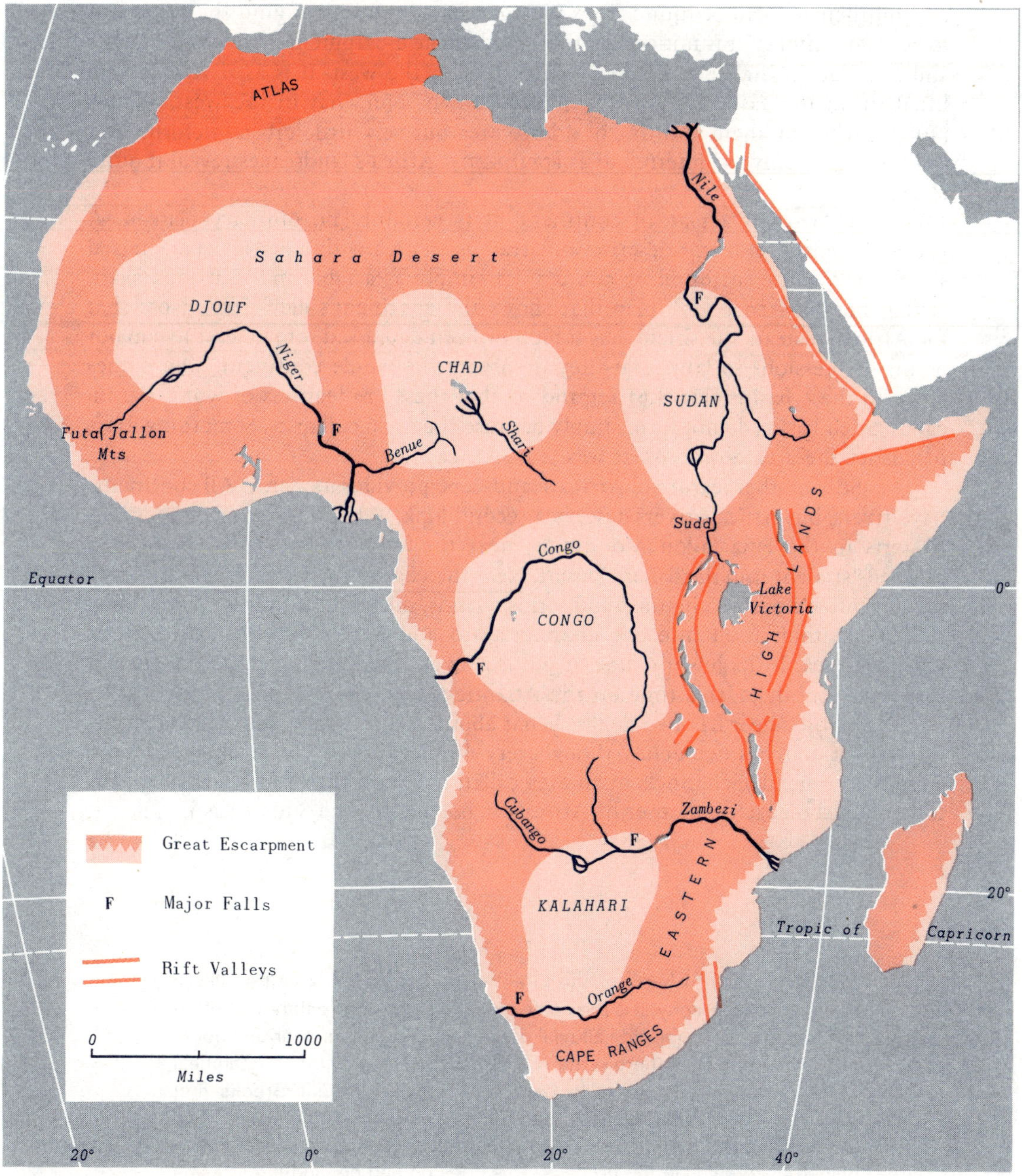

FIG. 1.6 African landscapes.

of prominence it can be found from West Africa all the way around the continent to beyond Ethiopia. It is noticeable, though, that the escarpment is generally steeper and the plateau higher in eastern Africa than in the west. It is as if Africa were lifted up in the east and, perhaps, pushed down somewhat in the west. Such a process, whereby there is tilting of a large area but very little crustal deformation (and so no orogeny), is referred to as *epeirogeny.* Africa's landscapes reveal it quite clearly.

If Africa is a plateau continent, it is certainly no ordinary plateau. A textbook definition might describe a plateau as an area with a level surface, raised sharply above the surrounding country, or simply as a tableland. But look again at that map of Africa, *any* map that shows the continent's elevations. Note that the African plateau, the shield, has somehow been depressed into at least five major basins, depressions so large that they dominate the whole continent. The names given to these basins (Fig. 1.6) remind us that these are features of world significance. Such basins do not form simply because the crust collapses. Something drags or pushes the area down. What was it in Africa?

Still another feature of Africa's landscape provides us a clue. All continents have rivers, but Africa's rivers deserve a second look. Take, for example, the Niger. It starts in the Futa Jallon Mountains, near the West African coast, and then proceeds straight into the Sahara Desert. Near the edge of the Djouf Basin, the river divides into numerous channels and forms what is, in effect, an interior delta. Then the river gathers itself into one channel again and, in crossing the basin, turns sharply to sea. In its lower course it plunges over several major falls, and then it forms another delta, this time on the Atlantic Ocean shore.

So the Niger has two deltas. What about that most famous of delta rivers, the Nile? It, too, has two deltas, though very different ones. The river spreads into almost innumerable channels in an area called the Sudd (Fig. 1.6), which happens to be near the edge of another of Africa's major basins, the Sudan Basin. Then it

---

DELTA

Delta is the fourth letter of the Greek alphabet, and it is written like a triangle. In geography, it is the name given to a physical feature that often has a triangular shape: the alluvial deposition that occurs near the mouth of many rivers. Thus there is the Nile *Delta*, the Niger *Delta*, the Zambezi *Delta*. It was, in fact, the Nile Delta to which the ancient Greeks first gave this name. What happens is that the river deposits the alluvial load it is carrying and then flows over its own deposited material, dividing into numerous streams in the process. This goes on and on, and the streams divide time and again; a delta changes and grows all the time. Deltas often are very fertile areas for agriculture, but for obvious reasons they also present big problems to their inhabitants in times of flood.

---

travels across that basin, plunges over a series of falls and forms its more famous delta on the Mediterranean.

And note the Zambezi-Cubango river system. It, again, forms an interior delta near the margin of the Kalahari Basin, then plunges over the famed Victoria Falls and forms another delta on the coast of Moçambique. Note, too, the Congo. It traverses the Congo Basin and in so doing forms almost a full circle. There is hardly a major river in Africa whose course and character does not demand explanation.

Those rivers, including the Shari, which still flows into its Chad Basin delta, filled the basins with water and sediments. As they did so, the weight on the crust caused it to bend down. Today the water is mostly gone, but the sediments remain as desert sand in the Sahara and the Kalahari and as deep, tropical soil in the Congo Basin. Those sediments contain fossil evidence that the basins were once filled with standing water. They were, in effect, interior seas.

RIVERS Physical geographers can learn much about the history of a landscape from the valleys of rivers. Rivers, after all, are the removers of material loosened by weathering; they are constantly carrying boulders, stones, sediment, sand, and silt from higher areas downstream to lower reaches and, eventually, to the deltas and out to sea. Thus the valley of a river changes character as it leads from the interior of a continent to the coast. Ideally, a river starts in a mountainous area, where it rushes down steep slopes and tumbles over numerous falls. Here the river is *young*. Its valley is deep and narrow, and its "load" consists of boulders and large stones. But as the river proceeds coastward, its valley widens, its pace diminishes, the waters calm somewhat, and its load is made of finer material, some of it rolled along the valley bottom but much of it carried in suspension. Now the river is attaining *maturity,* and, if it mirrors the landscape through which it flows, it may begin to show signs of meandering. Nearby hills are more rounded than they were in the mountainous interior, and there is an impression that the river is rather well adjusted to its physical surroundings. Tributaries enter calmly, and, unless the area has been recently disturbed by orogenic or epeirogenic movements, there are no waterfalls. Finally, the river starts meandering very broadly, and its channel shows signs of dividing, or *braiding.* Eventually it does just that: its valley is now practically at the level of the landscape around it, sometimes even above it as natural *levees* contain the waters. It has reached *old age.* Then the delta forms, and the river reaches the sea.

The river does all this in response to the force of gravity and *baselevel,* the lowest level to which the river could erode and the level it has reached when it enters the sea. Ideally, then, a river would show a rapid drop in the far, mountainous interior, a more gentle flow through its mature stage, and very slow flowage when it is in old age, near its delta. Now to return to our African rivers. This sequence exists, not once, but *twice!* The Niger goes through the whole sequence, rushing down steep slopes in the Futa Jallon and then reaching a delta in the Sahara. Then it begins again, and it plunges over waterfalls before reaching another delta in

Nigeria. In other words, the Niger, and other African rivers with it, has a double genesis. When a whole continent reveals this kind of a situation, we must seek a total explanation.

What might have caused Africa's old rivers to have failed to reach the ocean and to have emptied into interior basins? One obvious answer, distance, is geographical. Could it be that Africa did not at one time have the coasts it has today and that its river system was so far from the shores then existing that it responded to *local* baselevels, that is, to interior seas rather than to the world's oceans? By this question we are postulating that other existing, sialic landmasses surrounded Africa, perhaps on all sides. Only when those landmasses broke away could the pent-up basins, filled with water, release their load. As they did so, the rivers rushed

The great Victoria Falls of the Zambezi River, in southern Africa. Here, near the town of Livingstone, the Zambezi plunges over a mile-wide, 300-foot-deep wall into a gorge that signals a rejuvenation of its course. Until it reaches this point, the river flows in adjustment with the landscape surface shown at the top of the photograph. Then it changes to a canyonlike trough, with numerous rapids. Above the falls the Zambezi carries mud and silt; below boulders, rocks, and other debris evince the river's new environment.

down toward the sea, plunging over falls and creating coastal deltas. Thus they were given their second life: their original, old courses filled the basins, and their new, young courses emptied them into the ocean. It works for the Niger, for the Zambezi, and for the Nile. But the hypothesis is far-reaching. Could such an event have occurred, and can the hypothesis be put to the test?

CONTINENTAL DRIFT

A South African geologist named A. L. du Toit was the first to observe African landscapes and to draw the conclusion I have just suggested. He was not the first to propose the idea that the earth's continents might not be permanently located where they are today, but he did produce the earliest well-researched, systematic study of African geology and landscapes in support of continental drift. Du Toit looked at more than just Africa's precipitous coastline, its basins, and its odd rivers. These gave the hint. Now what was needed was research into geology, the layering of the crust in Africa and the continents that once were supposed to have been adjacent to it, especially South America. Speaking of landscapes, who could escape noticing the amazing similarity between the bulges and indentations of the African and South American sides of the South Atlantic Ocean? Du Toit wrote two books, *A Geological Comparison Between Africa and South America* (1927) and his memorable work *Our Wandering Continents* (1937). He meticulously dissected and compared what he found on opposite sides of the Atlantic Ocean, finding that there were fossils that strongly indicated a former union, that the downfolds and upfolds in the Cape Ranges' rock strata (called *synclines* and *anticlines* respectively) were repeated in Argentina's Sierra de la Ventana, and that the proposed supercontinent was at one time glaciated.

These discoveries indeed provided tests for the hypothesis that the continents were at one time assembled approximately as shown in Fig. 1.7. Any continent, large or small, undergoes changes during its lifetime. Rivers attack it and seek to erode the higher parts down to lower levels. Ocean waves erode the coasts. Loose material is ground down into fine sand and deserts and dunes are created. And climates also change. Thus a supercontinent as large as Gondwana (Fig. 1.7) should, during its lifetime, have undergone glaciation. When the evidence for glaciation was found in southern Africa and similar evidence came from South America and India, the idea that drift might be reality received an enormous boost.

Certainly not everyone believed that continental drift could possibly be taking place. Even when Du Toit published his evidence for similar geological formations on opposite sides of the South Atlantic Ocean, for fossil distributions, and for glaciation, doubts remained. This was partly because of the way the hypothesis had been proposed several decades earlier in a book called *The Origin of Continents and Oceans* (1915) by Alfred Wegener. In that book, Wegener had combined hunches, inadequate evidence, insufficient data, and a proposal for a mechanism that could easily be disproved, and he had been roundly criticized. Although his book was the first modern statement of the hypothesis, it probably

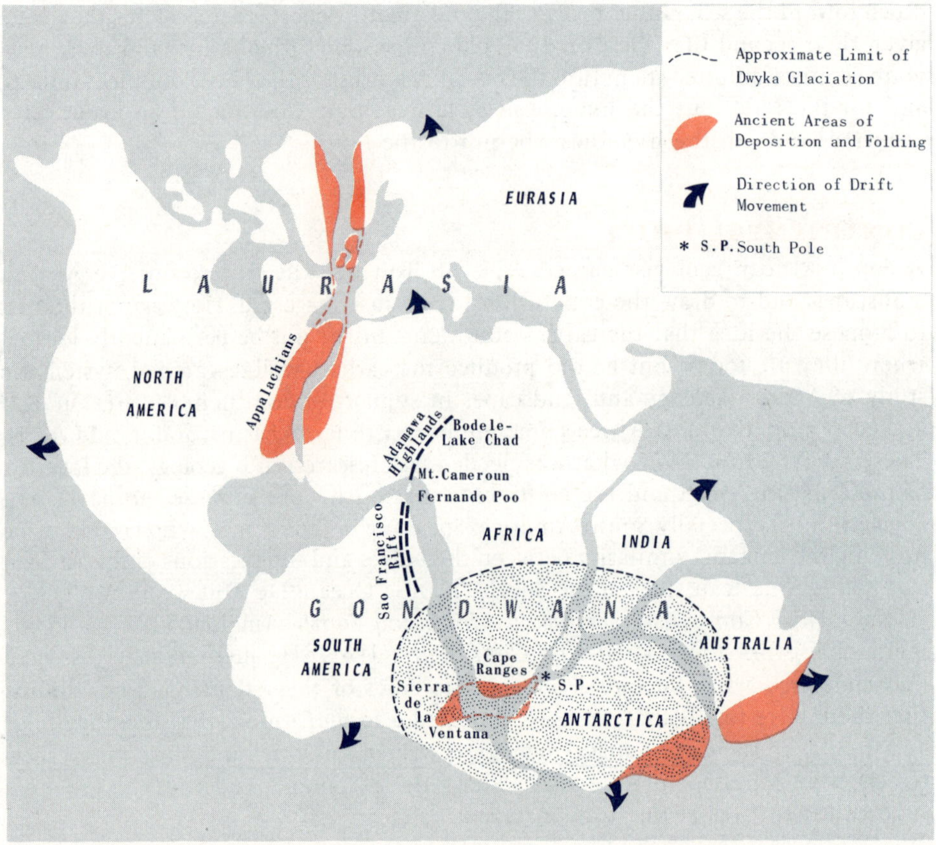

FIG. 1.7 The continents reassembled.

set the idea back. But Du Toit was not discouraged. In articles and monographs he painstakingly built the geological and geographical case for drift, reminding his readers that the world distribution of mountain ranges suggested that the continents had drifted away from Africa, crumpling and folding their leading edges into the great Pacific Coast mountains. As they drifted away, the African continent attained its escarpment, the great basins were drained of their water by newly formed rivers (the enormous volume of water carved great waterfalls below the basins), and the great rift valleys were formed.

Rift Valleys

Earlier we noted that when rock strata are compressed, they can fold and bend into huge linear, mountainous orogenic belts. But what happens to shield areas such as Africa when stresses are placed on them? Their rocks are not layered and bendable; they are hard and crystalline, rather like concrete. Another look at Fig. 1.6 suggests what may happen when a shield area is affected this way. It fractures. The plane along which the fracture occurs is called a *fault.*

In eastern Africa, from the Red Sea through Ethiopia and all the way to South Africa, lies a series of these faults. But these are not ordinary faults. There are sets of faults, parallel to each other, and the strips of crust between them have collapsed or have been pushed down to a lower elevation. In many areas the East African Plateau lies thousands of feet higher than those depressed strips of land, and it looks as though the continent was pulled apart and those strips, lacking support, simply dropped down.

That explanation is too simple, of course, but there can be no doubt that the rift valleys shown on Fig. 1.6 are related to the drift process. Investigations have shown that although the rift floors have in some cases indeed collapsed (note that several have filled up with water, as in Lakes Rudolf, Tanganyika, and Nyasa), the "blocks" between the faults in other places are being held down by the adjacent plateau (Fig. 1.8). This suggests that when a continent is affected by the drift process, it undergoes pull-apart forces as well as push-back stresses, with the pull-apart forces eventually stronger. To look at the East African rift system spatially: is it a coincidence that these rifts are approximately antipodal to the middle of the Pacific Ocean, the ocean into which the continents are apparently drifting?

The Mechanism

Much of the evidence for continental drift, then, comes from the earth's surface, the arena in which the geographer operates. For a very long time, the problem of a mechanism caused many scientists to doubt that drift could be taking place. But in recent years the recognition of convection cells in the earth's mantle produced a plausible possibility, and the idea of *plate tectonics* and *sea-floor spreading* was raised. Geologists noted that although rocks on the sialic continents are often very old, the ocean floors yield only relatively young samples, nothing much older than the Cretaceous (see the time scale, page 9). And they recalled that the ocean floors have major, sharply defined ridges that are earthquake-active. (The most famous one is the Mid-Atlantic Ridge.) Could it be that the ocean floors form near those

Rift valleys extend across East Africa from beyond Ethiopia in the north to Natal, South Africa. The widths of these troughs and the heights of their walls vary considerably. This photograph shows the rift valley wall in Swaziland, near the border with Natal. Viewed from a distance of about 7 miles, the scarp appears to be quite sheer and well defined. The surface on which the photographer stood when he took the picture has sunk from the level at the top of the fault scarp, about 2000 feet. If the photographer had turned around, he would have seen a nearly identical sight about 13 miles away. The Swaziland Rift Valley averages about 20 miles in width.

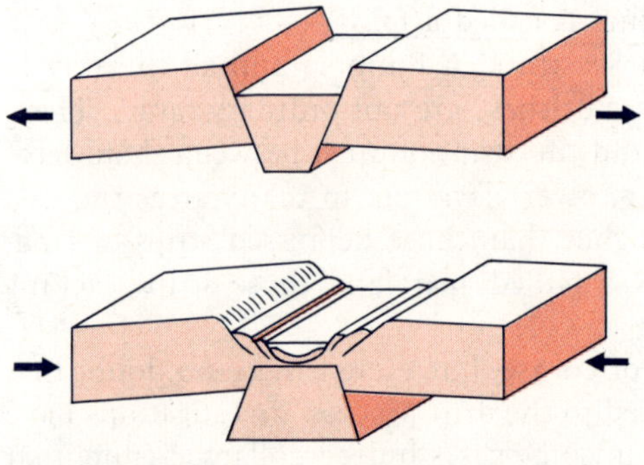

FIG. 1.8 Rift valleys.

centrally positioned ridges, spreading outward and pushing the continents away as they do so? It appears to be the case.

The present drift-redistribution of the continents appears to have begun sometime during the Cretaceous, say 100 million years ago, and the continents are still moving. Around the Pacific Ocean's margins, the force of the process is evident: volcanoes mark the mountains from South America through Alaska to New Zealand. Earthquakes rumble continuously. Great folds of crust create island arcs such as that of Japan; off eastern Asia a great new mountain belt may be forming. In those 100 million years, South America has moved about 4000 miles away from Africa, or over 20 million feet. That works out to about 1 foot every 5 years! But it is probably a mistake to believe that this drift process *began* during the Cretaceous. What we are seeing is probably the most recent cycle of drift, but not the only one. Is it reasonable to assume that the continents began to drift in the Cretaceous, but had been in place for billions of years before then? Of course not. The continents will presumably coalesce in some area of the globe, and then they will start to separate again in other directions. The continents on which we live are thin and, in places, fragile, sialic rafts.

GEOMORPHOLOGY

When we study landscapes, then, we should always be aware that every landscape and, in fact, practically every landform is in some way related to the drift process. Mountains are folded into place, volcanoes are activated, plateaus and plains are tilted, and river courses are altered because continents are always under the stress of mobility. One foot every five years is by no means slow; geologically it is very fast.

The study of physical landscapes is called *geomorphology*. Here we are interested in the genetics, the evolution, of landscape. Earlier we noted that a consideration of African continental landscapes could lead to some important conclusions about that continent's long-term history. But landscapes can also be studied on a smaller scale, and this is what most geomorphologists have done. For example, numerous studies have been made of the development of the landscape of the Appalachians. Some geomorphologists concentrate their work on the landforms and landscapes created by advancing and retreating ice sheets. Others interest themselves in deserts and dune types or in coastlines. Still others have made a life's work of research into *karst* topography, the topography of sinkholes and caves in limestone areas.

Another term sometimes used in conjunction with geomorphology is *physiography*. By physiography, however, we mean the more general study of the total natural environment. It includes the identification and description of landforms, the morphological study of relief, soils, natural vegetation, even the climate. A dominant concept that has come out of this work is the *physiographic province.* A physiographic province is a part of the earth's surface that displays a certain degree of physical homogeneity. In the United States, for example, one could say that the eastern coastal plain, the Great Plains, the Colorado Plateau, and the central Rocky Mountains constitute physiographic provinces. Wallace W. Atwood, famous American geographer, wrote a book entitled *The Physiographic Provinces of North America* (1940), basing his whole study on just such a subdivision of this continent.

Geomorphologists, on the other hand, are less interested in regions than in processes. When, for example, they see a nearly flat surface that has obviously been eroded down over a long period of time, they want to know the age of that surface, whether it can be correlated with other surfaces elsewhere (even on other continents), what happened to the sediments that were removed, and why certain remnants still rise above the otherwise even plain.

Those are the kinds of questions that interested one of the founders of the discipline in the United States, William Morris Davis. As long ago as 1889, Davis defined some of the world's extensive, nearly flat surfaces as *peneplains.* As he envisioned it, mountains and hills would be lowered into very low, rounded, almost imperceptible elevations in the landscape, merging into equally faint valleys. Here and there, a hill of especially hard rock (or one somehow protected from erosion) might still rise as a remnant above the peneplain. These were called *monadnocks,* after the prominent mountain in New England. The peneplains, he thought, would be created by long periods of erosion, toward the end of which almost all the higher areas would have been worn down by weathering and erosion. Davis was always concerned with processes, and he realized that such a plain could never become completely flat; rivers would stop flowing before that could happen. Thus he called those areas of gentle, broad slopes and wide valley floors *pene*plains, *near*-plains.

This concept of peneplanation seems so obvious and easily understood that no one would dispute it. Certainly Davis thought so. As he pointed out, so long as areas are not disturbed by orogeny or by epeirogeny (which would interrupt the cycle), erosion has free reign and the peneplain will become ever more nearly

WILLIAM MORRIS DAVIS

W. M. Davis, one of the founders of modern American geography, was born in Philadelphia in 1850. After attaining a bachelor's degree from Harvard in 1869, he spent three years with the Argentine Meteorological Observatory in Cordoba, Argentina. In 1876 he returned to Harvard on an assistantship, and in 1878 he was appointed instructor of physical geography. Initially he wrote prodigiously in meteorology, but in the late 1870s he began to become interested in geomorphology. The study of landforms and landscapes occupied him for the rest of his life. He rose to the rank of professor at Harvard (he retired in 1912) while becoming known for his penetrating field observations, rigorous analyses, and effective, powerful teaching. His celebrated debate with Penck took him to Germany where, in 1909, he was visiting professor at the University of Berlin. He was one of the founders and first president of the Association of American Geographers.

Although Davis received a master's degree (in engineering!), he never achieved the doctorate. Even so, numerous universities recognized Davis's enormous contribution to the discipline of geography by awarding him honorary doctoral degrees.

perfect. And he noted that some very old peneplains had been preserved when lava flows covered them or ice sheets and their deposits buried them. Now time was exhuming them so all could see that peneplanation had occurred throughout geologic eras.

But Davis was challenged by a powerful opponent, the German geomorphologist Walther Penck. Peneplanation presupposes that the slopes of a mountain or plateau wear *down*ward and become lower and lower as the forces of weathering and erosion attack. A landscape's slopes will, therefore, become convex, until finally the whole peneplain consists of convex interstream tracts. Penck argued that the landscapes he was studying in Germany had many concave rather than convex slopes, and this suggested to him that slopes do not wear downward but *back*ward. Think, for example, of an extensive mesa under erosion. Davis would suggest that it would become more and more rounded and that the flat upper surface would soon be worn away. Penck postulated that the mesa's sides would retreat under erosion until only a small butte was left; when this was ultimately destroyed, only the slight rise around its base would bear witness to the sequence (Fig. 1.9). This was the *pediment*. The whole issue became one of the most celebrated debates in the history of physical geography, as the two scientists exchanged articles and books written to strengthen their respective positions.

In the 1960s, Lester C. King developed Penck's views further, essentially on the basis of the point that surfaces (such as the top of that mesa) survive far longer than Davis had suggested. If Penck were correct, some very old plains (King

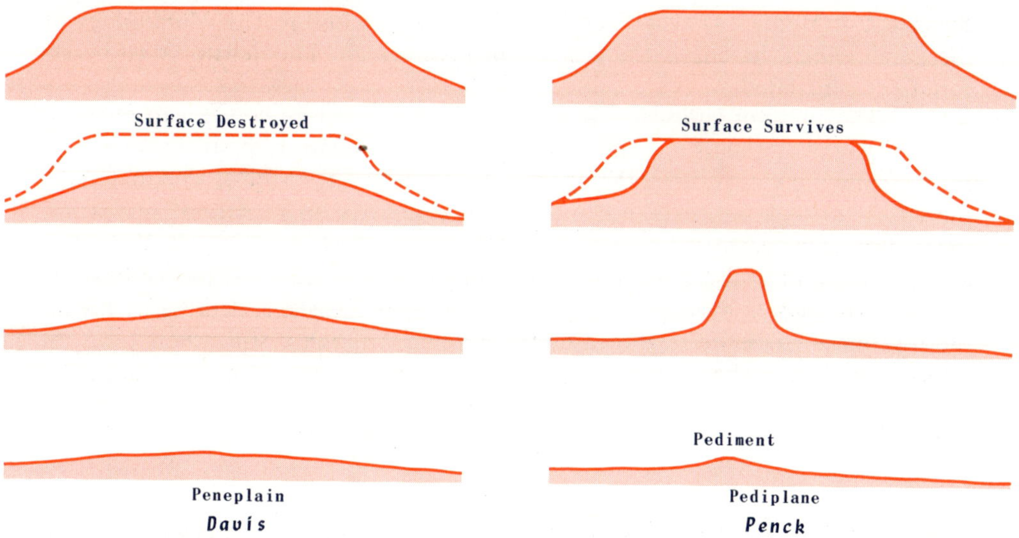

FIG. 1.9 Erosion-cycle model.

called them *pediplanes*) would survive. After all, they would not be destroyed until slopes retreating under erosion met. Now, asked King, could some erosion surfaces formed *before* Gondwana's supercontinent broke up still exist, even if only partially? He thought so, and he set about studying erosion surfaces in Africa, South America, Australia, and elsewhere. His conclusions were that remnants of erosion surfaces formed across Gondwana did indeed still exist and they could be mapped and

---

THE PENCKS

When W. M. Davis got involved in a debate over the development of landforms and landscapes, he faced a double-barreled attack: the father-and-son team of Albrecht and Walther Penck. Albrecht (1858–1945) was professor of geography at the University of Vienna from 1885 to 1906, when he went to the University of Berlin. In 1894 he published a two-volume work entitled *Morphology of the Earth's Surface*. But it was his son Walther (1888–1923) who posed the real challenge to Davis. Davis had published a work in German under the title *Die Erklärende Beschriebung der Landformen* (The explanatory description of landforms) in 1912, and Walther Penck replied with *Die Morphologische Analyse* (Morphological analysis) in 1922, a year before his untimely death. Penck's book was not translated into English until 1953, by which time L. C. King had taken up his side of the argument.

---

correlated. In numerous articles and several books, including *The Morphology of the Earth* (1962), he charts the progress of his research. The debate Davis began is not yet concluded.

The study of landscape planation has been a central concern in physical geography, and it entails a wide variety of topics: the nature of stream flow and the formation of river valleys, the deposition (aggradation) of the eroded materials, the effects of rock hardness and underlying structures on planation, and much more. But some geomorphologists have concentrated their work on altogether different kinds of landscapes: those created or affected by ice, those related to wind action, those carved by waves and tidal action along coasts, and those involving the solution of rocks as in the case of caves and sinkholes. Still others have researched such special features as volcanic landscapes, meteorite craters, and even ocean floors and deeps.

Landscapes of Glaciation

During its existence, the old supercontinent of Gondwana was glaciated at least once, and the landmasses that drifted apart all carry evidence of that glaciation. Thick sheets of ice formed in high-latitude (that is, near-polar) areas and spread over the countryside, scraping and scouring the rocks below and piling up the rubble along their leading edges. Later, the ice age waned and only the ground-up material remained to bear witness to its occurrence. Lava flows and sediments were piled on top of the glacial rubble (called *till*) and buried it deeply, but an occasional exposure through erosion and a great deal of bore-hole evidence have confirmed the Dwyka ice age (Fig. 1.7) beyond a doubt.

That ancient glaciation was not the last of its kind the earth has experienced. In fact, we may be in the middle of another one right now. About 1 million years ago, during the Pleistocene, a new ice age occurred, which generated enormous ice sheets. These continental ice sheets pushed from Canada across the Great Lakes and into the Midwest as far south as the Ohio River (Fig. 1.10); in Europe they covered nearly all of Great Britain, Scandinavia, the Netherlands, and most of Germany, as well as Poland and the whole northern part of eastern Europe. After much study of the till these glaciers left behind, the scratches, or *striations,* they made in the rocks over which they moved, and the topography that remained after they disappeared, it was clear that the Pleistocene glaciation has involved not just one, but four advances of ice, the last one ending just a few thousands of years ago. Between these glacial advances (each named after a state where representative till occurs: Nebraskan, Kansan, Illinoian, Wisconsin) came periods when the ice retreated and the climate warmed somewhat. These periods are called *interglacials,* and we may well be living in another such interglacial. The ice may be back.

A glaciation, of course, is caused by a climatic change involving a drop in temperature. Exactly what causes this drop and what determines the timing when it occurs are still not known for certain. But there can be no mistaking the impact of glaciation on the earth's landscapes, and this is what interests geomorphologists.

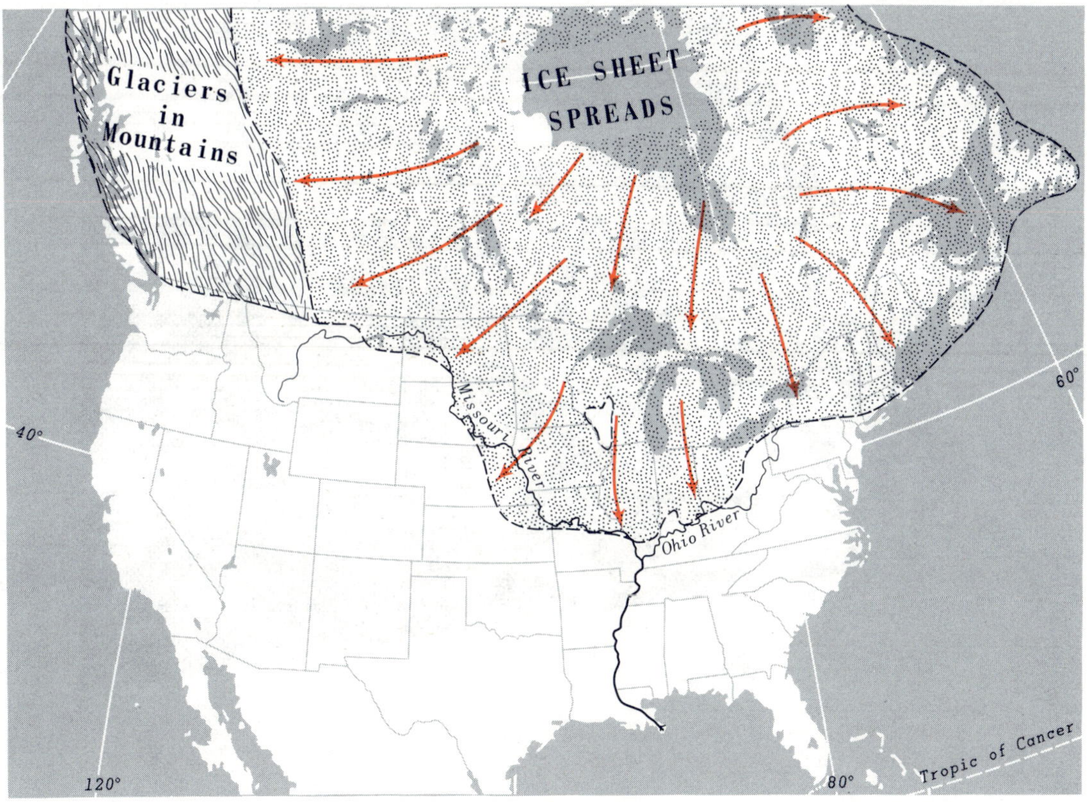

FIG. 1.10 Pleistocene ice sheet in North America.

CONTINENTAL GLACIATION It is practical to distinguish between continental glaciation, in which large areas are covered by major ice sheets, and *mountain* glaciation. When the climate cools, mountainous areas also acquire large glaciers, but these then flow outward through valleys and are generally more confined by the underlying relief. Continental glaciers, on the other hand, are independent and larger. Fed by near-continuous snowfalls piling on enormous quantities of ice, these sheets of ice expand outward, weighing heavily on the underlying bedrock. Not surprisingly, the bedrock below is scraped, shattered, and scoured; pieces of the rock become mixed with the lower layers of the ice. The ice sheets are huge; the quantities of rock dislodged and carried along are similarly great. A look at a section of the Canadian Shield shows hollowed-out areas now filled by water, forming the regions's numerous lakes, and flattened, scraped-down areas between them.

What happens to the rock debris carried along the ice sheet? Ice sheets erode; they also deposit. Glacial deposits, or *drift,* consist of all kinds of rock debris. Sometimes when melting began and flowing water became involved in the process,

Continental glaciation scours the countryside, leaving bare rock surfaces, hollowed-out areas, irregular drainage patterns, and other scars. The photograph is an overview of the area near Atikika, Ontario.

the rubble was sorted. The resulting material is called *stratified drift.* More often, however, the glacier simply dropped its load of debris in an unsorted mass called *till.* An idea of the amount of rock carried along can be gained from the thicknesses of till that have been found in the Midwest. In some places, parts of Iowa, Illinois, and Ohio, for instance, it measures *hundreds* of feet in depth, and the average is probably about 50 feet. Along their leading edges, the ice sheets left belts of piled-up debris known as *terminal moraines,* which can be seen in many parts of the Midwest, diversifying an otherwise rather monotonous landscape. Speaking of landscapes, what happened to the landscape of areas covered by the great ice sheets? It was buried by ice and covered by till brought from elsewhere. In one area, the so-called driftless area of southwestern Wisconsin, the glaciers for some reason (still uncertain) failed to cover the scenery. There we can get a glimpse of what this part of the Midwest may have looked like.

MOUNTAIN GLACIATION When large glaciers develop in and over areas whose relief is already mountainous, the results are somewhat different. Glaciers fill the existing valleys and flow slowly downward, hollowing out the valley and giving it a U shape, cutting off spurs (or *truncating* them, as the term is used), and leaving the valleys of former river tributaries hanging high up the valley wall. The weight and force of the valley glacier, like the sheet ice, cause scraping and scouring of the rocks below; and the rubble can be seen to streak the glacier with dark bands in the middle (the *medial moraine*) and near the sides (the *lateral moraines*). The valley glaciers also push material ahead, which is left as *terminal moraines* when the glacier yields to warmer temperatures.

The topography of an area that has undergone mountain glaciation is unmistakable: the mountain tops have become angular, jagged peaks, ridges are sharp, and the sides of hills have been hollowed out to accommodate masses of ice. In the U-shaped valleys the glaciers have vacated there are ridges and mounds of debris left behind by the melting ice, and lakes fill parts of those valleys, dammed up by such material. Long, narrow lakes are an important indicator of glaciation. New York's Finger Lakes are one example of this.

It was during the Pleistocene Ice Age that man's modern development and dissemination occurred, and we will refer later to the role of environmental change in that process. But it is worth noting that Europe, during the last glacial advance (there called the Würm Glacial), was covered to its very heart by sheet ice, that great mountain glaciers lay over the Alps and the Pyrenees, and that such comfortable country as the Riviera, Bordeaux's wine region, and the area of Venice and the northern Adriatic were under tundra conditions. On the map of world vegetation (page 43) note where the tundra lies today! And in Spain and Italy pine trees grew in stands like those we find today in the Canadian and Siberian high latitudes. The Würm glaciers withdrew, and man entered. What he found will be among our future topics.

A glacier emerges from the flanks of Mount Athabasca, Canada. Known as the Athabasca glacier, this enormous tongue of ice and snow comes out of the Columbia ice field and flows slowly toward the Sunwapta River. Note the lateral moraines on both sides of the glacier. The trees in the left foreground provide an impression of scale.

Landscapes of Wind Action

When we think of landscapes where wind action prevails, deserts come immediately to mind. But even in deserts, running water from rainstorms, however infrequent, has a great impact on the development of the landscape. Physical geographers have found it difficult to find landforms created exclusively by wind erosion, but there is no doubt that wind plays a significant role in the deposition and accumulation of loose material.

When wind acts as an erosive agent, it does so by moving particles (from the tiniest grains to sizable pebbles) from one place to another and by hurling small grains against obstacles, wearing them down. Most of us have experienced the latter effect by being on a beach on a windy, dry day. Sand grains may be swept along in sheets, hitting ankles and legs. Anything permanently in the way of such abrasion will eventually erode away.

The landform feature most closely associated with wind action, and deserts, is the dune. Dunes are given their shape and form by prevailing wind activity, and fields of sand dunes are characteristics of many (but not all) deserts. Several

Dunes in the Algerian Sahara, north of the Ahaggar Mountains. The oasis at Zaouia el Kahla lies surrounded by ripple-marked longitudinal dunes. The sand is nearly always moving, and the dune landscape is constantly changing. Natural springs supplemented by artesian wells provide the oasis settlement with its water. In the far distance, the sand-covered Sahara gives way to a hard-rock plateau, which is still a desert area although it is not covered by dunes.

kinds of dunes have been identified. Among the most typical is the *barchan,* a crescent-shaped dune whose ends are pointed downwind. When an area is completely and thickly covered by loose sand, the wind may throw the sand into a series of parallel ridges called *transverse dunes.* These transverse dunes lie at right angles to the wind direction. There are other places where the dune chains lie parallel to the wind. These are *longitudinal dunes.*

LOESS A very important feature associated with wind action is a very fine-grained, wind-carried silt deposit known as *loess.* Loess occurs in a large area in China, in extensive sections of the middle Mississippi basin, in Europe, South America, and elsewhere. One-hundred-feet depths of loess are not uncommon, and it is believed that this material originated during dry periods of the Pleistocene glaciation. Strong winds probably swept the margins of the ice sheets, where streams emerged to carry loads of silty sediment away from the glaciers. They dried out the river beds and picked up the finest silt and carried it to nearby areas, where it accumulated in quantity. For man, it turned out to be a most fortunate sequence of events. In China, the Midwest, Europe, and Argentina's Pampa, among other places, this loess has developed highly fertile soils capable of sustained production.

Coastal Landscapes

A whole host of factors enters into the interpretation of coastal landscapes. First, we should think about the nature of wave action. Waves behave differently in deep and in shallow water. Breakers and backwash develop when deep ocean waves are swept up a beach. Second, weather conditions must be considered. When there is little wind, waves are rather gentle and bring sediments toward the beach, enlarging it. But during storms, waves are hurled up high on the beach, pieces of rock are broken loose, and these in turn attack the whole shore. Much erosion can take place in an incredibly short time under such conditions. Third, it is important to know whether the coastline being studied is undergoing any epeirogenic change. Coastlines that are being lifted up show characteristics quite different from those being depressed. Fourth, the types of rock that make up the coastline must be known as well as the nature of the geologic structures involved.

Among the most familiar landforms created by marine erosion are *cliffs.* Who has not heard of the Cliffs of Dover? When a part of the shoreline is open to wave attack, the water is deep, and the rocks are of suitable composition, the waves will carve out a steep wall, even cutting under the rocks to form caves and arches. Whenever we see a cliff, we should look for other evidence that this area is being depressed. Cliffs may be evidence of comparatively recent submergence; gentle, sloping *beaches,* on the other hand, are depositional features of the marine geomorphic process. Sometimes a coastline has two parallel beaches, one on the mainland and another some distance offshore, where an accumulation of sand may be forming an *offshore bar.* A lagoon may form between the mainland and this offshore bar, and such a situation suggests that the land here is rising or emerging.

A section of the coastline of southern Oregon. Cliffs, stacks, and arches (an arch is forming in the left foreground) are evidence of deep water and coastal submergence.

An offshore bar can create serious problems to navigation. It may form across the mouth of a river or an estuary, and an opening may have to be maintained artificially by dredging or some other means. On occasion, a bar may build from the tip of a peninsula, terminating some distance away. Such a partial bar is known as a *spit*, and, like the offshore bar, it lies parallel to the coast except where strong currents or tides cause it to grow outward at an angle.

The study of coastal landscapes is a complicated business, for it is along coastlines that the continental landmasses undergo much structural change. Offshore bars reflect the process of emergence in Florida and the southeast coast of the United States; but northward, estuaries and deep bays reveal that drowning, or submergence, has taken place. But in the Northeast, some of that submergence was related to the weight of the ice sheets that covered this area during the Pleistocene glaciation. The weight is now removed, and an apparently drowned coastline, full of islands and estuaries, is actually undergoing emergence. Local uplift has raised marine, wave-cut terraces well above sea level; elsewhere, local submergence of such terraces makes correlation difficult. And speaking of sea level, that changes, too. The identification of individual landforms created by wave action is one thing; the classification and correlation of whole shorelines is quite another.

Other Landscapes

Streams, ice, wind, and ocean water thus create their own landscapes. These are the major erosive agents acting upon the crust of the earth. They remove (degrade) and accumulate (aggrade), and they do so while the surface they modify is itself being changed by internal earth forces. Only very few landscapes cannot be immediately associated with one of these agents. Volcanic landscapes exist upon which stream erosion has barely begun to have an impact. Landforms apparently formed by the impact of meteorites also stand apart. And *karst* topography is also a special case. Here, rocks attacked by stream erosion yield, not by fragmentation but by dissolving. Limestone, a rock made of calcium and magnesium carbonate, sometimes occurs in layers between other sedimentary rocks such as sandstone and shale. Imagine a landscape that has developed upon sandstone, which is in turn supported by limestone. When the streams on the sandstone erode down to the limestone, some of the water goes down into cracks in the limestone; it seeps along the strata to a distant outlet point. The cracks become bigger; so do the channels along which the water flows away. Eventually the cracks develop into wide, gaping openings called *sinkholes,* and the channels along which the water moved away underground grow into caverns. From the air, karst topography sometimes looks pockmarked, like an area full of bomb craters.

There are several properties of the earth's crust and surface to which we have not yet referred, but which have been crucial to human development on this globe. These include the development on the earth's rocky crust of a mantle of soils, the emergence of a cover of natural vegetation, and the accumulation of a wealth of minerals and fuels below.

SOILS

The processes of change we have discussed so far—mountain building, glaciation, wave action, and so forth—suggest major, far-reaching alteration of the earth's crust and its upper surface. But another important process, much less perceptible but no less important, goes on all the time. This is the formation of soils. Much (though not all) of the earth's land surface is covered by a cloak of soils, soils of many types. These soils are continuously being modified. As time goes on, they become more mature, thicker, and often richer in plant nutrients. While they are developing down below, they may be eroded away at the surface. Water passes through the spaces between the soil's particles; it mixes with chemical components in the soil and reacts chemically with the solids. Plant roots penetrate the soil's layers and interact biologically with it. The soil mantle may seem static and unchanging, but, in fact, it is dynamic, full of life and change.

How do soils develop from bedrock? Several factors and conditions play a role in the process of soil formation. The first of these is the *parent material,* the bedrock from which the soil is derived. The influence of the bedrock is greater when the soil is young. As the soil grows older, the bedrock matters less. Of course, the parent material may be something *other* than bedrock. Soils also develop on glacial till, on river alluvium, even on dune sand. A second important factor is *slope.* Obviously, steep slopes inhibit the development of deep soils. Erosion is greater, rainwater tends to run off quickly rather than seeping in to feed the soil, and vegetation has a harder time holding on. Thus flat valley land and relatively flat upland areas tend to have thicker, better-developed soils. A third factor in soil formation seems obvious enough. That is *time.* Rock fragments are naturally not transformed into soil particles overnight. Although soils mature at different rates, most require several thousand years to develop fully. Some soils are still very young, those developing on recent lavas or on river alluvium, for example, while others are very old, like those on peneplains that have existed for a million years or more. During all this time, the fourth factor, *climate,* acts upon and interacts with the soil. The moisture and precipitation that affect soils assist the many chemical and biological changes the soil must undergo to mature. But there are extremes. Soils can suffer from too much water, which causes the leaching away of important ingredients from the upper layer of the soil; they can also suffer from too little

Severe soil erosion in eastern Nigeria, West Africa. Deep soils, a moist climate, and overuse by humans (with destruction of the natural vegetation) combine to produce enormous scars in the landscape. Here a valley side, once covered with a dense mantle of tropical forest, is being transformed into a huge gulley. During the rainy season, the forest retreats almost daily. Villages must move, and life is disrupted. What remains is useless and unproductive—an ugly wound.

THE SEVENTH APPROXIMATION

The concept of a soil has gradually evolved as improved technologies have provided more detailed information concerning the various properties of the soil body. Until the 1930s, soils were studied and classified essentially as single points on the landscape. However, during that decade the concept of the profile was refined, and soils were thought of thereafter as three-dimensional bodies at the earth's surface. This and other new approaches to the study of soils led to the U.S. Department of Agriculture Yearbook, *Soils and Man,* published in 1938, which included a new classification of soils for the United States.

When the 1938 classification was used in the field, it was found to have many serious shortcomings. So, in 1951, Dr. Guy D. Smith and the soil survey staff of the U.S. Soil Conservation Service initiated work on a new system that was introduced through a number of drafts, or "approximations." Each draft was circulated among a number of soil scientists for field testing so that deficiencies could be uncovered and corrected. In 1960 the seventh draft or "approximation"* was introduced by the soil survey staff. It has been amended several times but still remains the basis of the new classification.

The Seventh Approximation differs from the older classifications in two important ways: (1) the nomenclature of the higher categories (order, suborder, great group, subgroup, and family) is entirely new; and (2) the criteria used to define the various classes are considerably more quantitative and specific. The nomenclature in the new classification is based on Greek and Latin roots and is systematic so that it connotes the most significant soil characteristics. The table below illustrates the 10 new orders of the Seventh Approximation and their approximate equivalents in the classification of 1938.

ORDER	MEANING	APPROXIMATE EQUIVALENTS
1. Entisols	Recent soil	Azonal soils and some low humic gley soils
2. Vertisols	Inversion soil	Grumusols
3. Inceptisols	Inception soil	Ando, sol brun acide, some brown forest, low humic gley, and humic gley soils
4. Aridisols	Arid soil	Desert, reddish desert, sierozem, solonchak, some brown and reddish-brown soils, and associated solonetz
5. Mollisols	Mollic epipedon	Chestnut, chernozem, brunizem (prairie), rendzinas, some brown, brown forest, and associated solonetz and humic gley soils
6. Spodosols	Podzol	Podzols, brown podzolic soils, and ground-water podzols
7. Alfisols	Pedalfer soil	Gray-brown podzolic, gray wooded soils, noncalcic brown soils, degraded chernozem, and associated planosols, and some half-bog soils
8. Ultisols	Ultimate soil	Red-yellow podzolic soils, reddish-brown lateritic soils of the U.S., and associated planosols and half-bog soils
9. Oxisols	Oxide soil	Laterite soils, latosols
10. Histosols	Tissue soil	Bog soils

*For more information concerning the Seventh Approximation see: "Soil Classification, A Comprehensive System, 7th Approximation," U.S. Department of Agriculture, Soil Conservation Service, 1960, 1967.

Source: Richard Handy, "Screenings from the Soil Research Lab," Iowa State University, 5, 3, 1969.

water, which leads to excessive evaporation and a drying out of the upper soil layer. Temperature, too, influences soil development. Warmer temperatures promote chemical change, and the thickness of some tropical soils attests to this. In cold areas, the mechanical freeze-thaw process plays a major role in the breakdown of parent material, and chemical change is much less prevalent. But there is another side to this. Those thick tropical soils also have very little *humus*, or decomposed plant matter, so important in vegetative growth. Why? Because warm temperatures also accelerate bacterial action, and bacteria thrive and consume almost all the dead plant matter lying on the ground. Thick as they sometimes are, tropical soils are generally not very fertile. And this brings us to the last of the soil-forming factors, *biological activity.* Such activity goes on in all soils to a degree, not just in tropical soils. Tree and plant roots, bacteria, and fungi are all at work, consuming soil nutrients, altering them, and returning them in some form to the soil. Roots penetrate the soil and absorb such elements as calcium, magnesium, and potassium. These are modified and transmitted to the above-ground portion of the growing plant, in the stems and leaves. Then the plant dies, and its contents are returned by the decomposing, decaying foliage.

A map of the major soil groups of the world (Fig. 1.11) shows the major distributions. Between the desert soils of the southwestern United States, Sahara-Arabia, and interior Asia, on the one hand, and the Arctic and tundra soils of the near-polar areas, on the other, lies a belt of soils identified by such terms as *podzolic, chernozem,* or *prairie.* Here lie areas where the temperatures are neither too hot nor too cold and where the rainfall is adequate, the parent material is suitable, and the topography is not too rough to permit the maturing of rich soils. Given the size of the continents, these areas seem small on our map, but they are of great significance. These soils share the characteristic that comes from limited bacterial action, the accumulation of nutritious humus. The Russian word *chernozem* means "black earth," black with the dark, organic matter on which new plants thrive. And, as the map shows, the Soviet Union has a large area of such soils, as do the United States and Canada.

The soils of tropical areas cover much of South America, Africa between the Sahara and Kalahari deserts, southern and southeastern Asia, and northern Australia (Fig. 1.11). These soils are *laterized* rather than podzolized. Ample or adequate rainfall and relatively high temperatures promote bacterial action to such a degree that all or nearly all humus is destroyed, and what remain behind after the leaching process are the insoluble iron compounds that color tropical and subtropical soils their characteristic red. These, then, are the *latosols.*

An important soil category has not yet been mentioned, although it appears at the top of the legend for Fig. 1.11. This is *alluvial* soil, which occurs on the often richly fertile deposits of rivers such as the Nile, Ganges, Hwang (China), Mississippi, and others. It is hardly necessary to stress the significance of these soils. In areas where the climate is suitable, their fertility is almost legendary, and they produce crops that range from corn to cotton, wheat to jute, and rice to rubber trees. It was perhaps on the alluvial soils of the Nile Valley and the lowland of

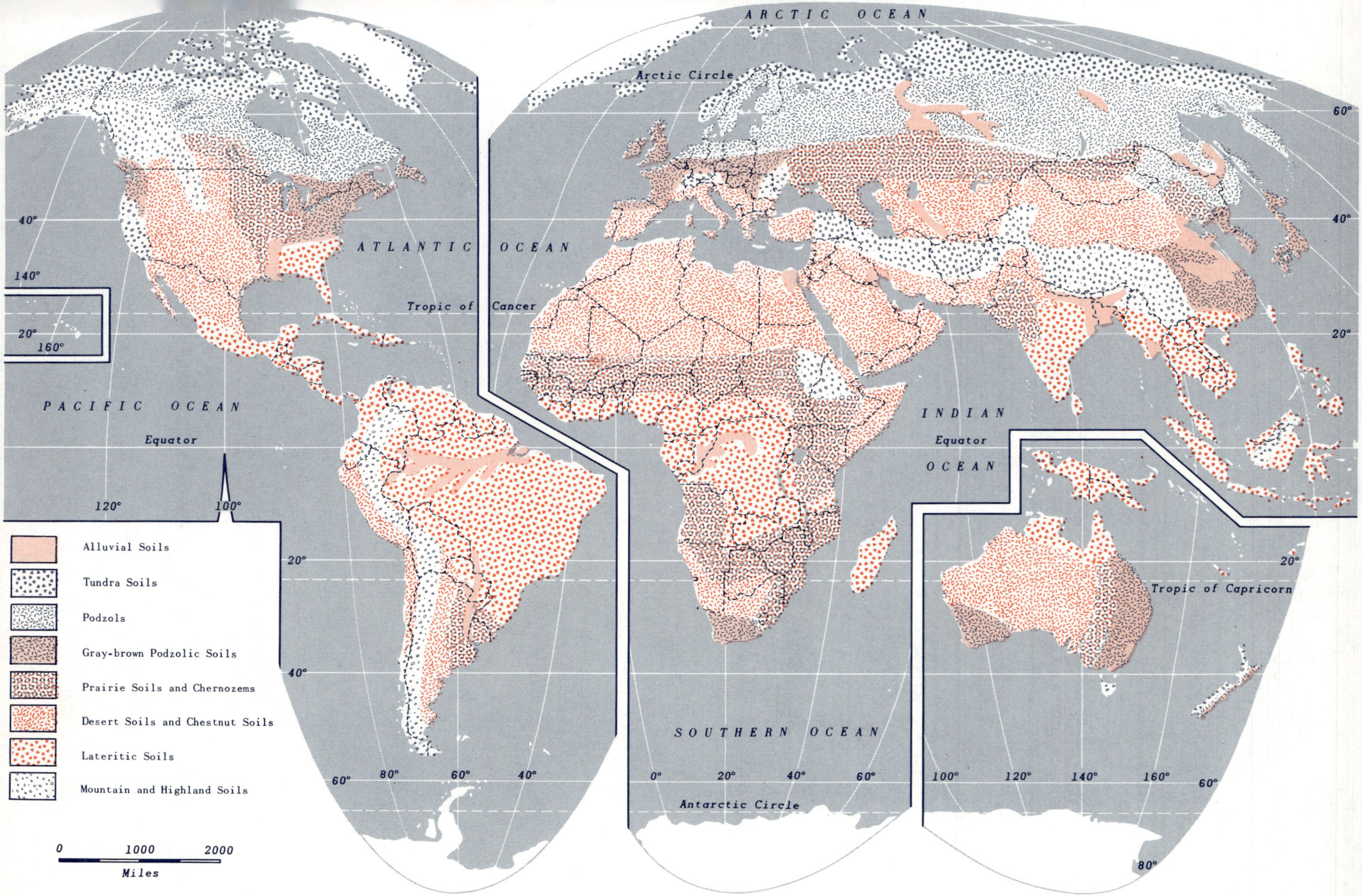

FIG. 1.11 Major soil groups of the world. (Modified from J. F. Lounsbury and L. Ogden, *Earth Science*, Harper & Row, New York, 1973, p. 372.)

the Tigris-Euphrates that man began his experiments in plant domestication. Still today, hundreds of millions of the world's people cluster on or near these immensely productive river lowlands.

Again it is worth observing how small the area of the alluvial regions is and that not all alluvial lowlands are very productive. For example, the Amazon and Congo (Zaïre) basins have alluvium, but the climatic regime is not conducive to the maintenance of fertility. Not even all of the chernozem areas are intensely productive. Substantial areas of fertile soil lie amid much larger regions of unproductive or marginally productive soils; the conservation of this asset is a critical responsibility.

VEGETATION

The soils of our earth are an invaluable resource. Our natural vegetation is no less an asset. The earth's flora and fauna were man's earliest resources, and, although man has destroyed and replaced much of the natural vegetation he found on this globe, what remains is still a vital element of his environment.

The vegetative cover of the earth consists of trees, shrubs, grasses, mosses, and an enormous variety of other plants. Geographers who study this aspect of our environment (biogeographers, or, more specifically, phytogeographers) recognize four subdivisions, or *biochores*, out of this mass of vegetation. These are the forest, the savanna, the grassland, and the desert (Fig. 1.12). As the figure shows, the forest is an association of plants in which trees dominate. The trees stand so close together that their foliage shades the ground below almost entirely. This stunts the flora underneath, and a secondary layer often develops, whose smaller trees await the death of the larger ones to shoot upward to the patch of sunshine. The savanna is marked by much more open space, with trees standing alone or in clusters separated by areas of grassland. The appearance is parklike, and parts of the savanna are, in fact, identified as parkland savanna. Here the rainfall is generally less than over the forests. Where there is still less precipitation, the grassland biochore develops. Here the vegetation consists almost entirely of varieties of herbs. These grasses may be dense and cover the ground solidly, they may be sparse, and they may be mere tufts scattered over dry, steppe country. A few trees may exist in more moist patches and along stream courses, but not in the uplands. Finally, the desert biochore is treeless, arid, quite barren, with scattered small plants such as herbs and lichens.

As a generalized map of the world's natural vegetation regions shows, these four biochores exist at several latitudes (Fig. 1.13). There are tropical, temperate, and cold-climate forests, and there are deserts in low as well as high latitudes. The prairie is part of the grassland biochore; so is the tundra. The internal differences within biochores (that is, between equatorial rain forest, temperate rain forest, deciduous forest, and needle-leaf forest) depend upon the availability of moisture and the prevailing conditions of temperature, which, as we know, also affect soil formation. For example, a difference between the equatorial rain forest (see Fig.

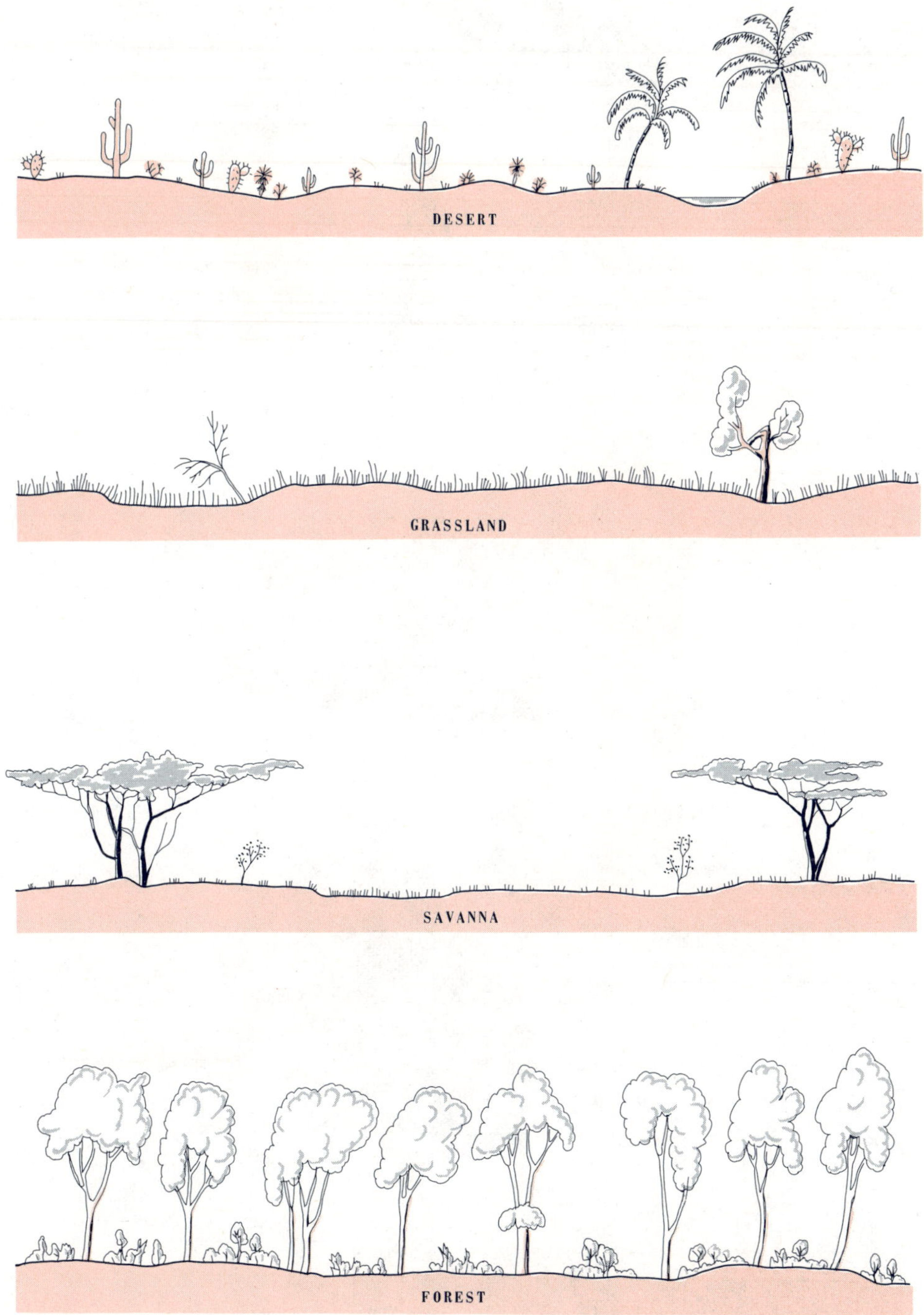

FIG. 1.12 The biochores.

FIG. 1.13 World natural vegetation regions. (After a map by H. Brockmann Jerosh, 1951.)

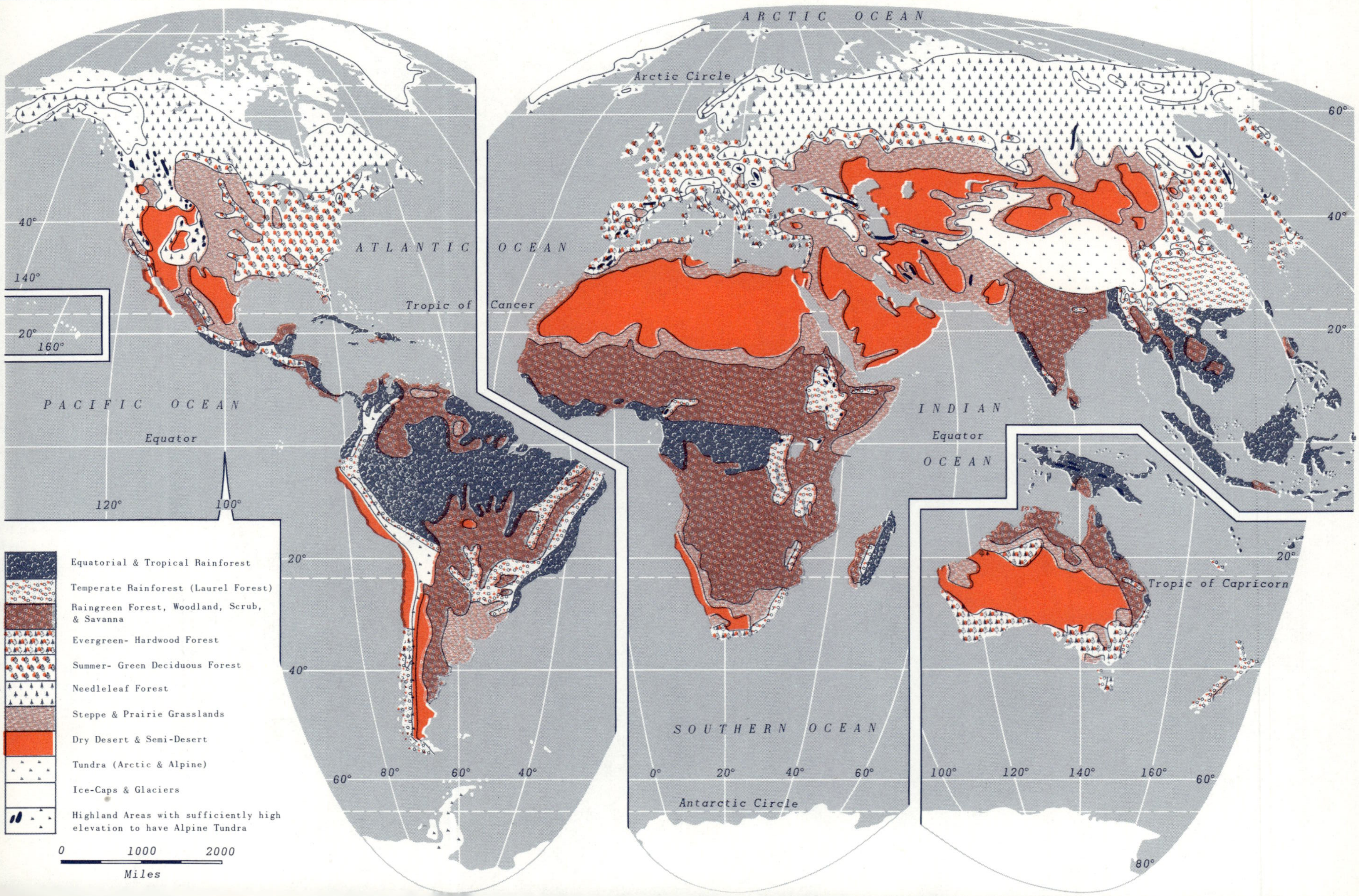

1.13) and the deciduous forest (shown on the same map) is the source of the nutrients for the trees. In the tropical rain forest, these nutrients come essentially from the vegetation itself, that is, from the foliage, stems, and branches that have fallen on the ground. In the deciduous forest, the nutrients come from the soil, where the dead leaves and branches have been transformed into humus by just the right amount of bacterial activity. One consequence of this difference is that if the tropical forest is cut down, the growth cycle is interrupted. Vegetation planted in place of the rain forest may grow a year or two, but then the nutrients are exhausted, and the soil cannot offer much. Unless major fertilizer additives are put in the soil, the plants will die, as many a tropical plantation owner can attest. But cut down the deciduous forest, and the supply of humus will still be there to support newly planted crops.

And that is exactly what man has done. He has cut down the forests. Our map of world vegetation regions shows where these regions ought to be, and it was drawn from our knowledge of what the prevailing conditions would generate as *climax vegetation*, the vegetative association that develops under given circumstances over a long period of competition and adjustment. But look at the map closely. Is the whole eastern United States still under deciduous forest? Is the Spanish plateau and the Italian peninsula really covered with hardwood evergreens? Of course not. From the Mediterranean to Mexico, from New York to New Zealand, man has cleared the forests to build houses and ships from the lumber, to burn the wood for fuel, and to cultivate the cutover land. It is estimated that one-fourth of the land areas of this earth were at one time covered by forests and that the figure is now under 15 percent. And although reforestation programs exist and conservation-conscious organizations are using recycled paper to save living trees, the forests are still losing to man.

Where was it that man first began to use the forest as a resource? Consider again Fig. 1.13. Those lines that separate vegetative regions do not, of course, exist as sharp divisions in nature. A forest does not give way to savanna over a distance of a few hundred yards. The lines represent transition zones. Along the margins of the rain forest, the trees begin to thin out, grasses begin to appear, and eventually the savanna biochore takes over. It seems likely that earliest man had his greatest opportunity here, in this transition zone, to benefit from the resources offered by both the forest and the savanna plains. These resources differ considerably, and in some ways they complement each other. The animal life of the forest (the fauna) differs substantially from that of the savanna. Early man was a hunter. On the plains he found the herds of buffalo, zebra, and other herbivores that could ensure his survival. But early man was also a gatherer, and in the forest he could find roots, fruits, and other essentials of life.

The natural vegetation, wildlife, and the soil mantle of the earth constituted man's earliest resources. They are still crucial today, but things have changed. The natural vegetation cover has been rolled back and man has inserted his domesticated plants into the growth cycle. The world's wildlife has been enormously depleted, and herds of cattle graze where buffalo once roamed. And from the soil man demands more every year, even though overgrazing and overpopulation have turned large portions of the soil cover into fields of eroded gulleys. Still the natural vegetation, through its interaction with soil and atmosphere and its various products, constitutes a vital element in our life support system. Man now seeks to protect and conserve some the earth's remaining wildlife, recognizing it as an irreplaceable part of this planet's evolutionary heritage. And while man strains the soil, he also seeks to sustain it—through hillside terracing, drainage control, fallowing, crop rotation, and fertilizing.

MINERALS AND FUELS

Even as our earliest human ancestors led a hunting and gathering existence, they began to make use of the earth's rocks and minerals. From those beginnings, when spear- and arrowheads were chipped from fine-grained flint (a kind of quartz), man's search for this planet's mineral wealth has continuously intensified. One of the first metallic minerals to be used was copper, which occurs pure in the native state, and which can, in that state, be fashioned rather easily into any particular shape. Historical geographers say that gold and copper were sought in the Middle East more than 10,000 years ago. But it was not until bronze was made by combining copper and tin that the real usefulness of metal for toolmaking was understood. This occurred between 6000 and 7000 years ago, and it heralded the beginning of the Bronze Age.

The people of those ancient civilizations in Southwest Asia and North Africa found uses for an ever larger number of such *nonferrous* minerals as copper, tin, zinc, and lead. They devised smelting methods to purify ores that were impure, and they began systematic searches for deposits. For thousands of years they had been taking gold from the river waters of their region, and their first searches for the ores took them upstream into the mountains overlooking their lowland realms. Indeed they found metallic ores, and they mined them systematically, beginning to learn some geology in the process. They did not find iron until later; the only iron known during much of the Bronze Age seems to have been that found in meteorites. Meteorites are not especially common, and it is not surprising that iron was valued even more highly than gold in ancient times.

Eventually, iron ores were found and the smelting of iron was learned, and the Iron Age superseded the Bronze Age. Iron tools and weapons were harder still than those made of bronze, and the knowledge and practice of iron metallurgy was carefully monopolized by those in power because it gave them an insurmountable advantage over adversaries. But such knowledge did spread to other regions, and the Iron Age reached different parts of the world at different times (Europe was behind the Middle East and Africa, for example). Wherever it arrived, it stimulated the search for iron ores.

Iron, the *ferrous* metal, happens to be one of the most common among the minerals in the crust of the earth. There are places, such as at Iron Mountain, Wyoming, in New York's Adirondacks, and at Kiruna, Sweden, where an iron-rich igneous mass intruded to very near the surface of the crust. Elsewhere, iron ores were formed by metamorphic activity, and there are even places where sediments accumulated in standing water so rich in iron that the resulting rocks can be profitably mined today. The earth's rocks, on an average, contain more than 5 percent iron. The planet proved to have in abundance the mineral that became fundamental to industrialization.

Those ancient Middle Easterners, then, used *metallic* minerals as resources (ferrous as well as nonferrous); but they also used *nonmetallic* minerals, and in considerable quantities. Clay was among the first. It was used for bricks in building and in art works some of which have been dated as over 20,000 years old. Salt was another, and it appears in the very oldest records of commerce. Stone was quarried for the construction of huge buildings. Gemstones were coveted. Like the ancient Greeks, the Babylonians and Egyptians were fond of bright colors, and they painted their buildings and statues with ocher and other pigments.

Resources

This raises a question about resources. What is a resource and what is not? To the earliest, forest-dwelling, hunting man, copper and zinc were not resources. Copper did not become a resource until it was deemed to have value as a malleable substance for the creation of various implements; its worth was heightened when, in combination with tin, it produced bronze. To the ancient Egyptians, petroleum was not a resource; our incredibly fast consumption of mineral fuels attended the emergence of the industrial age. Uranium did not become a resource until the twentieth century. There may be substances in the earth that will become resources in the future, but that are not resources today. In other words, human development determines and defines the nature of resources; what is and what is not a resource is a cultural matter. Each civilization, in the history of man, has depended on its own set or complex of resources. In ours, at present, the power resources dominate, and the earth's mineral fuels (coal, petroleum, natural gas) are the critical elements of the complex. Although it is true that iron is also fundamental to our industrial age, numerous substitutes for iron have already been developed, aluminum, for instance. Substituting for our current power supply is quite another matter, and although scientists are searching for alternatives (nuclear fission is one such possibility), no one is sure where the energy the world will require, say, 100 years from now will be coming from.

And this brings us to another aspect of resource use: changes brought about by necessity, technological progress, or both. A prominent example is that of iron ore. One of the world's largest sources of iron for a long time was the Mesabi Range, not far from the western end of Lake Superior. The Mesabi deposit is the largest of six such deposits extending from Minnesota and Wisconsin into the Upper Peninsula of Michigan. These deposits are of sedimentary origin, and circulating waters for hundreds of millions of years have been removing silica and other impurities from it. The iron content, when man began mining these ores, was as

The extraction of iron ores, coal deposits, and certain other resources often requires that comparatively thin overlying soil and rock be scraped away before the mineral can be excavated. This process has denuded and scarred large areas of the landscape, as here at the Marmora iron pit, Ontario, Canada.

high as 56 percent on the average and even went to 62 percent in places. The rock, called *hematite,* became the leading source for iron ores used in the steel mills on the shores of the Great Lakes and beyond. But it could not last long in the face of the massive mining operations, and as it became exhausted, steel makers had to look for other iron ores. Since major ores were located in West Africa and South America, it was sensible to build new plants on the East Coast, so that the cost of importing those overseas ores could be minimized. Certainly there were other ores around Mesabi, but they were so low grade that it was not practical to exploit them. One of these was a rock called *taconite,* just a rock with 30 percent or so iron. Taconite was no more a resource than sandstone.

But then technology changed things. It was no accident, and indeed a matter of necessity, that scientists had been working on methods to extract iron from low-grade ores. In the 1950s they succeeded, and suddenly it became feasible to refine taconite's iron at costs that were competitive. Just as suddenly, taconite, which had been merely another rock, became a valued resource, and the iron-workings at Mesabi and its surroundings had a new lease on life.

The taconite story is not unique. In South Africa, where gold has been mined on a large scale for nearly a century, the waste from the mining operations was piled in big heaps, the so-called mine dumps that dominate the urban scene of parts of Johannesburg and nearby towns. Gold-extracting techniques have improved to the point where the gold left behind in the waste material of the old days can be profitably extracted. The mine dumps themselves have become a resource. The same thing has happened with copper. Not only can ores long considered to be too low in copper content be mined, but also the mining wastes of decades ago contain extractable amounts of copper.

Thus it is reasonable to identify *potential* resources, deposits that cannot under present conditions of technology and economics be exploited, but that await the necessary developments. In this respect our modern age of science differs from preceding periods of human history. We *seek* ways to make resources out of presently unproductive sources whereas for centuries our predecessors relied chiefly on accident, trial, and error in this context. People were smelting iron in pre-Christian times; they were doing essentially the same thing as late as the nineteenth century. The industrial revolution brought with it changed (and still changing) perceptions of the earth's resource base.

Mineral Fuels

One of these changes has been our realization of the possibilities of another set of nonmetallic minerals, the *mineral fuels*. Although there is evidence that man has used petroleum for some purposes for many hundreds of years, the systematic exploitation of oil is just over a century old. In some places in the Middle East and North America and elsewhere, oil seeps to the surface from underground sources, and the ancient Mediterranean seafarers used it for stopping leaks in their boats. An associated product, natural asphalt, was used by the builders of cities to cover streets and walls. American Indians sometimes used oil for fuel. But petroleum became a full-scale resource only in the second half of the nineteenth century when the first wells were drilled. From those beginnings, a single well producing a few barrels a day in Pennsylvania, production has risen to 50 million barrels a *day* in 1972. But if this seems an enormous production, the world's consumption of petroleum, now more than 18 billion barrels annually, is rising at the rate of about 9 percent per year. Since known reserves total about 621 billion barrels (1971), the world has an oil supply that is guaranteed for only 30 years, although the picture is probably not that dark. New reserves are discovered every year, and it is expected that the continental shelves will yield perhaps as much as one-quarter of all the oil produced.

A map of present world petroleum producing areas (Fig. 1.14) shows four major regions and a number of smaller areas of production. A comparison to a map of the world's shield areas will indicate that petroleum occurs in sedimentary rocks. In fact, petroleum appears to have originated in the ocean, where, scientists believe, microscopic plant forms (such as diatoms) contain it. Upon death, these tiny plants release a very small amount of this substance, which then drifts to the bottom and becomes part of the sediments accumulating there. Later, perhaps

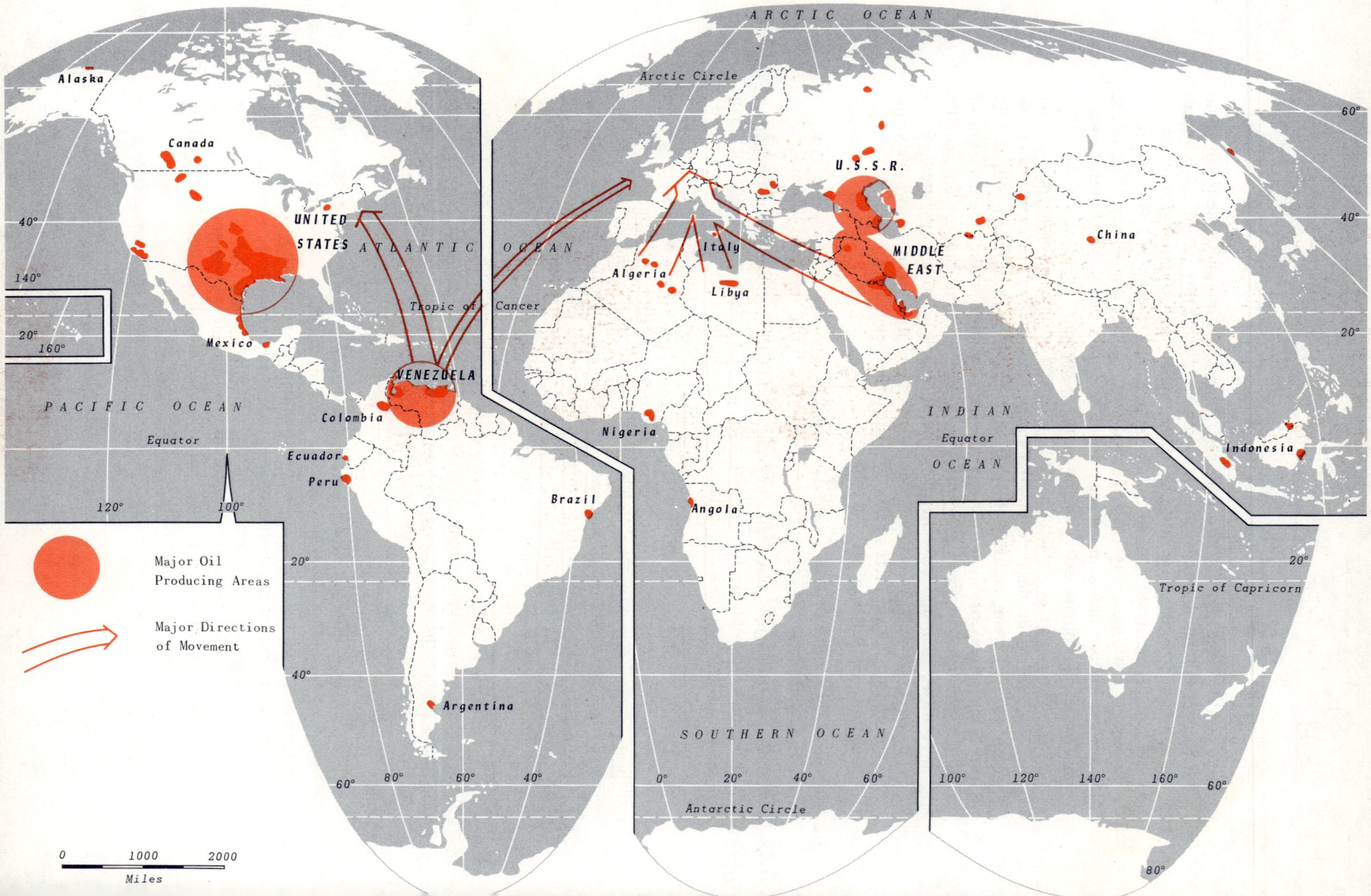

FIG. 1.14 World petroleum producing areas.

through the weight of overlying sediments, the accumulated oil is squeezed into a reservoir. Such a reservoir may be an upfold in the rock (an anticline), or a dome capped by a hard, impermeable layer. There it is under pressure, until nature creates an outlet or man drills through the caprock.

Another aspect of the distribution of petroleum areas suggested by the map is the coastal and near-coastal location of many of the deposits. The map practically demands that the continental shelf be examined for the continuation of these oil fields, and indeed, improved technology is bringing more and more of the continental shelves' oil reserves into production. And improved methods of exploration suggest that the continental shelves have only begun to yield their content.

Our map also suggests where the majority of the world's petroleum is consumed. In 1970, the United States used 15 million barrels a day, or 30 percent of the whole world's consumption; Western Europe needed more than 13 million barrels, or nearly 28 percent. But the United States produces only about 20 percent of the world total, and Western Europe still very little. It is not difficult to see what the Caribbean production means to the United States, and that of the Middle East to Europe. And when the proven reserves are brought into the picture, another serious aspect of the situation is emphasized: United States reserves amount to less than 10 percent of the world total. The Middle East, on the other hand, has some 55 percent.

Two factors are relevant in this connection, apart from the ever-present possibility of major new finds (as happened recently in Alaska). One has to do with the rejuvenation of oil fields. In a development almost paralleling that of the discard piles from gold and copper mining, it turns out that old oil wells are not useless after all. Capped and left untapped for a time, the underground reservoirs are partially refilled as more oil seeps into them, and the wells can produce again for a time. Oil-field rejuvenation has altered the reserve picture to the positive side.

Another positive feature of oil fields, and sometimes reservoirs without appreciable oil accumulation, is the presence of another, alternative source of energy, *natural gas.* As Fig. 1.15 shows, natural gas can be trapped by the same structures as petroleum, and it is most often found in association with oil fields. The use of natural gas for the creation of energy has increased enormously since the Second World War, especially in the United States. In 1971, the United States was consuming about one-half of all the natural gas used in the world, and even then the demand was somewhat greater than the supply. Discoveries of natural-gas reserves in the Netherlands, Germany, and under the North Sea stimulated the industry in Europe, and a growing network of pipelines began distributing the fuel across the continent. In the United States there were more than 2 *million* miles of pipeline carrying natural gas by 1970, with more under construction; in 1972 alternatives were under study to carry gas from newly discovered Alaskan reserves to the major consuming areas.

Natural gas is a clean and efficient source of energy, and although this fuel has reduced somewhat the pressure on petroleum production, we do not know how long the supply will last. In the early 1970s, proven reserves in the United States

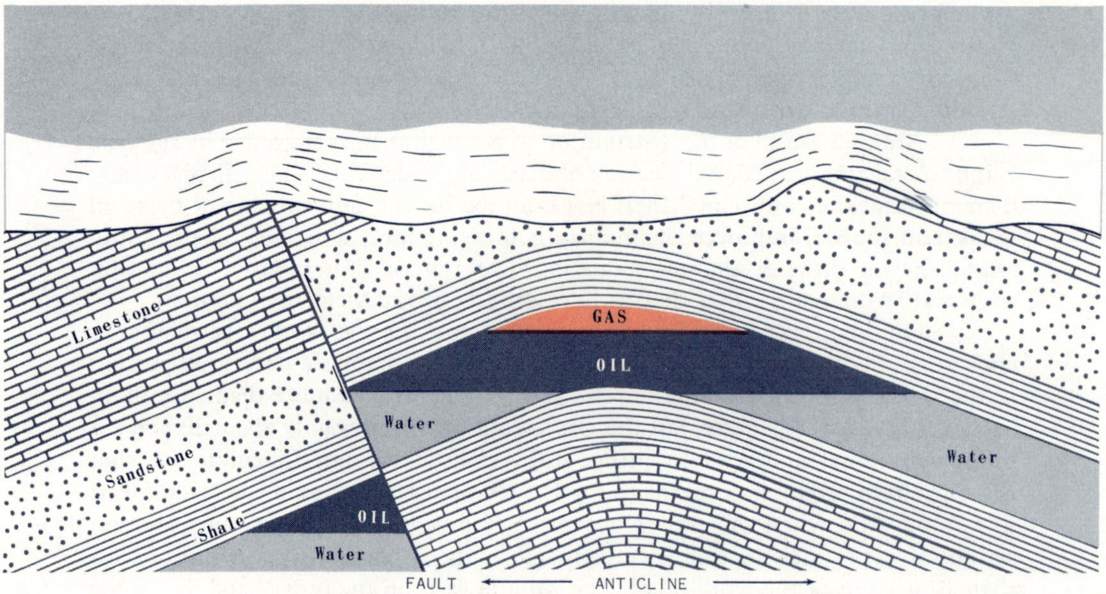

FIG. 1.15 Two oil traps.

were about 250 trillion cubic feet, and consumption about 25 trillion cubic feet, figures not calculated to stimulate long-term confidence. Already United States gas companies are negotiating to buy foreign supplies. But elsewhere the consumption of gas is rising too—in the United Kingdom, the U.S.S.R., and Japan.

If it is clear that there remain oil reserves to be discovered, it is even more obvious that the search for natural gas will yield major rewards in the future. The reserve picture prevailing in the early 1970s will probably change, therefore, but a pattern is emerging. Best endowed, apparently, is the U.S.S.R., with nearly 30 percent of the world's known reserves. The United States has about 15 percent, and Iran, Algeria, the Netherlands, and Canada follow. Using as much natural gas annually as the rest of the world's countries combined, the United States clearly has the biggest stake in the exploration for more reserves.

Finally, we should consider what was once the most important source of energy, a commodity now being replaced by petroleum and natural gas, *coal.* Sporadic use of coal occurred as early as Roman times, and when Europe's forests began to become exhausted in some areas during the Middle Ages, the commercial exploitation of easily accessible coal deposits commenced. But it was not until coal started to play its role in the smelting of iron and in the driving of the machinery of the industrial revolution in general that this resource was fully taxed.

Coal has its origins in swamp or bog environments, usually associated with forests. Under such conditions of stagnant water, the bacterial activity that destroys plant matter is slowed down, and the vegetation accumulates faster than it decays.

The initial product of this process is *peat,* a soft mass of vegetative tissue used in some parts of the world as a household fuel. During the Mississippian and Pennsylvanian periods (see again the geologic time scale on page 9) the continents were eroded down to very low relief, and vast expanses of plains existed, with enormous swamps. Great accumulations of peat developed, hundreds of feet thick in many places. Then these peat layers were buried when new sediments were laid down on top, and the sediments began to weigh down on the fibrous mass below. This caused the peat to change into a somewhat higher grade of fuel known as *lignite.* Although there are large lignite deposits in many parts of the world, this fuel is not very useful because there is too much water in it and too great a quantity of volatiles (such as hydrogen and oxygen). But compaction continued as more sediments were laid down over the original peat areas, and the weight of the sediments started to drive off almost all the water. The coal became harder and darker in color, a much better quality fuel. This, *bituminous* coal, is the variety of coal mined in most of the areas shown on Fig. 1.16. It is the coal that proved usable in the smelting of iron and the production of steel. There is a still higher grade of coal. When layers of coal between sedimentary rocks became caught up in orogenic movements, complete with intense folding, enormous pressures, and rises in temperature, practically all the water and all the volatiles were driven out and what remained was the purest coal of all, consisting of over 90 percent fixed carbon, *anthracite.* This, the cleanest of the coals, long was a fuel in great demand for heating homes.

The map of known major coal deposits (Fig. 1.16) indicates that coal-bearing sedimentary strata are widely distributed across the earth; lignite reserves are not even shown. The paucity of coal in Africa suggests the absence of this fuel (along with the others) from shield areas; the band of coals across North America and Eurasia from the Midwestern United States through Europe and Siberia into China suggests the shield-sedimentary contact zone as a favorable area for exploitation. Undoubtedly there are other coals, buried more deeply under the sediments; presently reserves are estimated to be around 2 trillion tons. One of the countries that found itself well endowed with good, accessible coal was the United Kingdom, source of the industrial revolution. Continental Europe, too, has major deposits across northern France, Belgium, the Netherlands, and Germany. Here, also, the fuel was available when the impact of the industrial revolution came. The United States, too, has very large reserves, and there are major fields in the Soviet Union from west to east. Clustered about the Hwang Ho in China lie coal reserves that sustain China as one of the world's major producers.

In terms of world production, the United States, Soviet Union, Europe (especially West Germany and the United Kingdom), and China extract about three-quarters of all the coal mined annually. In terms of total tonnage, the Soviet Union leads, but its 1970 production of 615 million tons included 147 million tons of lignite; the United States produced 500 million tons, almost all of it bituminous. Since world production in 1970 was nearly 2.2 billion tons, the U.S.S.R. and United States each accounted for about 23 percent of the total. Europe's total volume was

FIG. 1.16 Known world coal deposits.

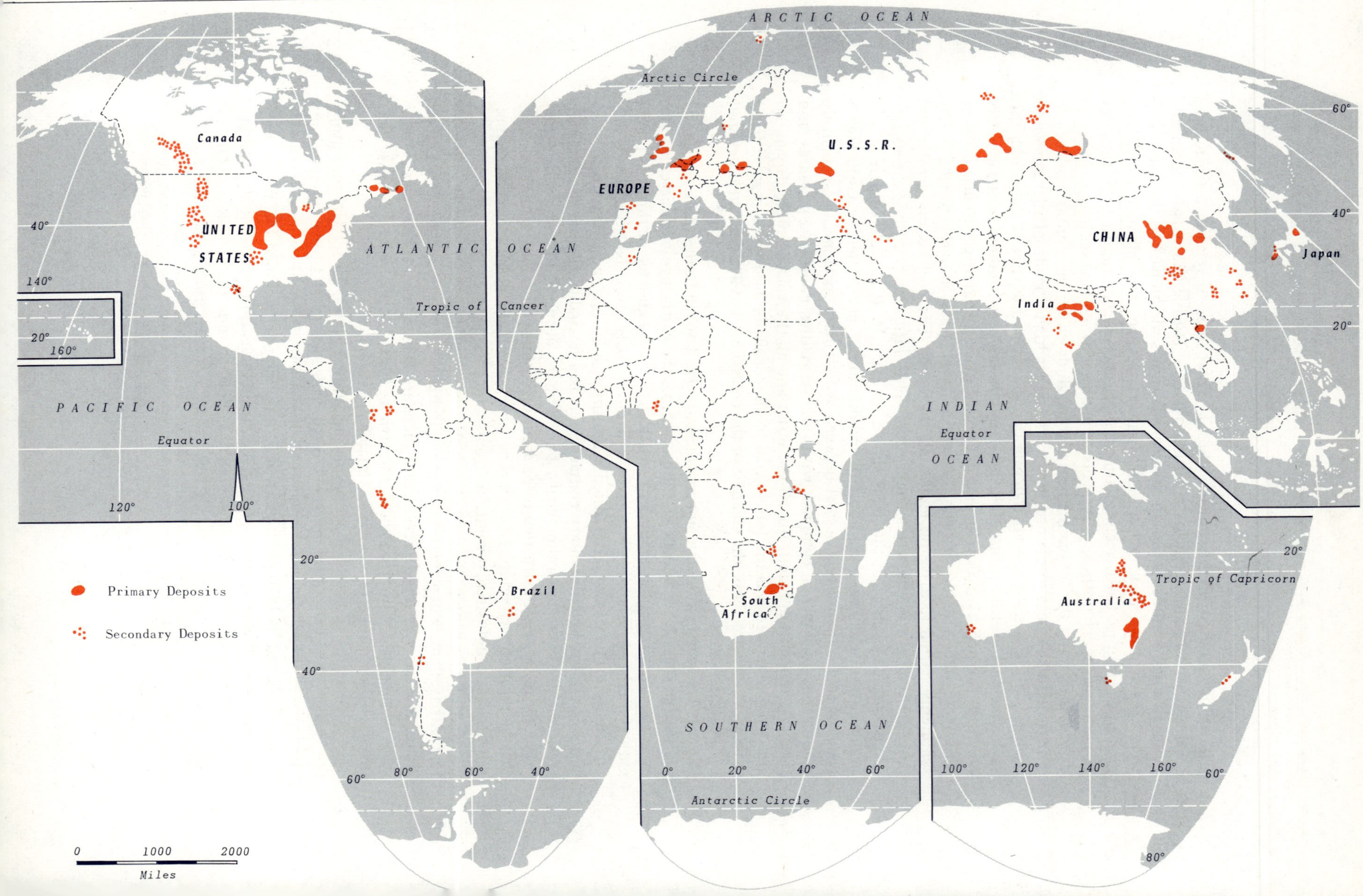

of the same magnitude, and China, for which there still are only estimates, probably produced about 12 percent of the world total. These figures of millions of tons seem staggering, but against the vastness of the earth's coal reserves they diminish considerably. When it comes to coal, we can confidently speak of adequate reserves for hundreds of years to come.

A HERITAGE OF ABUNDANCE

This brings us back to our theme of the earth as a storehouse of accumulated natural wealth from which man has, from his beginnings, chosen his resources. It is not true that the specter of short supply is now confronting us for the first time: man in his former dependence on wood for fuel, building material, and other purposes in various parts of the world faced crises when the forests were exhausted. It is not true that the issue of pollution is being faced for the first time by our generation: King Edward I of Britain, reflecting public opinion about the danger of coal burning, forbade the burning of coal in London during meetings of the Parliament. King Edward ruled during the thirteenth century. But it is true that man, in larger numbers on this earth today than ever before, is straining the earth's resource reserves more than he has ever done and is altering his environment in unprecedented ways. In July, 1972, four drums of water buried since the Second World War under the Sahara's sand were recovered, and the recovery created great excitement. They contained the only water unaffected by radiation pollution, according to scientists, and they were treated with the kind of care usually reserved for moon rocks. Are we damaging the environment bequeathed to us in an irreparable way? Are we, in this age of energy, rushing toward an energy crisis in which, for the first time, man will be unable to switch to alternative resources the earth may have to offer?

The two questions are obviously related, but we shall deal with the environment in greater detail in the next chapter. The prospect of an energy crisis, however, directly involves our consumption of the *nonrenewable* resources we have discussed in the previous paragraphs. It is important to draw this distinction between nonrenewable and *renewable* resources, for some of the resources we use do indeed renew themselves and others do not. The soil, for example, can regenerate after it has been exhausted or overused unless it is totally destroyed, as man-induced erosion has done in places too numerous to mention. And the natural vegetation, too, can recover. Much of the forest we see in the eastern half of the United States is second growth, having replaced the cutover, original trees. Water supplies, also, constitute a renewable resource. A well that runs dry during a drought period will flow again in future months and years, after the rains have come to replenish the water-bearing rocks being tapped below.

But coal, natural gas, petroleum, iron ore, and other mineral resources exist in the earth in finite quantity and can be used up. Over the hundreds of millions of years of geologic time that preceded the emergence of man on this globe, vast amounts of these resources developed, but man has an enormous appetite. Some experts estimate that presently known reserves of zinc, lead, and tin ores will be exhausted early in the twenty-first century, that the world's usable forests will not last much longer, and that even iron ore, the taconite experience and its implications notwithstanding, will soon follow. Still, the history of man is one of technological innovation, and somehow the prospect of alloy shortages does not seem awfully alarming. As for iron ore, other metals have already been substituted for iron in many instances, including automobile engines. In any case, low-grade iron ore, lower even than taconite, exists in vast quantities.

This, however, is an age of energy, and it is the energy picture that has resource specialists worried. It is not that we will run out of oil, gas, and coal suddenly some summer (although some observers have even forecast such a doomsday); the problem is that demand is skyrocketing and supply looks to be on the verge of falling seriously short. The most accessible reserves have been worked out, costs are rising, and technological developments are not coming as fast as they are needed. The effect of all this is felt especially strongly on electricity, the form of power for which demand is rising most rapidly of all.

Electric power is manufactured from oil, coal, and gas; it is produced by hydroelectric plants; and increasingly it comes from the alternative of the future, nuclear fission. Today, the United States still uses 25 percent of its fossil fuels for the generation of electricity. Industry uses most of the output (more than 40 percent), and private homes and commercial establishments consume most of the rest. The problem with the conversion of these fuels into electric power is that it is a very dirty process. Just over one-third of the fuel is made into electricity, and the rest becomes wasted smoke and heat. An increase in electricity supply, therefore, requires a disproportional increase in fuel supply and consumption, which means more landscape-destroying strip mining, ocean-polluting oil spills, and atmosphere-befouling soot and noxious gases. Is man's technology finally failing him?

Not yet, but it looks like a close call is at hand. Even while industrial countries including the United States, the United Kingdom, West Germany, and others are converting much of their power-generating equipment to oil and natural gas, preparations are under way to switch to nuclear reactors, the apparent alternative when it becomes uneconomical to exploit oil and gas. Such nuclear plants already exist and are operating, but today they supply a mere 3 percent or so of United States electricity requirements. Some specialists predict that by 1990 the figure will be 50 percent and perhaps more.

If we already know the source of future power, why not give up on oil and gas (as we are giving up on coal before it is exhausted) and convert entirely to nuclear power? The answer lies in the exponential growth of man's numbers and needs and, to some extent, in his growing concern over the environment. On the one hand, technologists are having trouble planning, designing, and locating nuclear plants as fast as desired. On the other, individuals and groups of people are blocking construction because they fear that lethal radiation pollution will occur just as surely as soot emanates from coal-using plants. And certainly they have reasons. The uranium-powered nuclear fission process, apart from dangers of pollution or explosion, produces radioactive wastes whose storage and disposal constitute problems not only for today, but for future generations as well. This is one reason why the so-called breeder reactors, now emerging as the likely future generation of nuclear plants, are opposed by conservationists. And so the search is on for still other alternatives for the generation of electric power. At present, the most attractive prospect seems to lie in nuclear *fusion*. In this process, the joining of atomic nuclei sets off an enormous amount of energy. The high temperatures involved (this, in fact, is the process that powers the hydrogen bomb) and problems of storage will delay the appearance of fusion reactors, but when they emerge, the prospect of an energy crisis may disappear. Hydrogen, the fuel, is so common on this planet (in sea water, for example) that the development of the nuclear fusion reactor will guarantee perhaps millions of years of power. And the process is clean. If it can also be made safe, it will solve the problems fission poses. Another potential advantage may be the greatest of all: it may be possible to develop "cells" that convert fusion power directly into electricity. This may mean the end of huge power plants for many purposes. Cells may be installed in private homes, and when you buy a television set, it will contain its own power source. No more wires and plugs.

Notwithstanding all this long-term optimism, however, man is clearly overstraining his current energy sources. His capacity to overcome problems of resource-exhaustion through technological innovation is being overtaxed by a demand that is rising faster than ever. The ultimate solution is apparently in sight, but the need is here and now. In these terms, our age of energy is clearly on a crisis course.

CHAPTER TWO

Life Layer Under Stress

TWO

The earth has six continents and seven seas—but only one atmosphere. On this atmosphere we all depend for our survival. We breathe its oxygen. It moderates temperatures. It carries moisture from the oceans over the land, sustaining the vegetation and replenishing soils and aquifers. It is the essence of life on this planet.

The atmosphere is another of the shells of which this earth is made, a layer of air that lies upon lands and oceans. But the breathable layer is not thick. When mountaineers climb the higher peaks, they need oxygen masks because the air becomes too thin. Athletes find their endurance lessened in the rarified air of places like Mexico City and Johannesburg. The layer of atmosphere upon which we directly depend is thinner even than the earth's crust.

Like the solid earth itself, the atmosphere consists of several layers. The lowest layer, the *troposphere,* is the one whose air we breathe, and it is only 8 to 9 miles thick (Fig. 2.1). Above the troposphere is the *stratosphere,* about 20 miles thick. Beyond the stratosphere lies the *mesosphere,* which extends to about 50 miles from the earth's surface. There are traces of atmosphere to some 6000 miles out into space, but there the "air" is so thin that what remains of the atmosphere is practically a vacuum. The boundaries between the troposphere, stratosphere, mesosphere, and the outermost *thermosphere* are marked by rather abrupt shifts in the rate of temperature change. Down here in the troposphere, we might assume that the farther upward one goes, the lower the temperature; after all, the highest mountains, even on the equator, carry snow. But at the tropopause, this drop in temperature (known as the *normal temperature lapse rate,* $3\frac{1}{2}$°F per 1000 feet) suddenly changes. In the stratosphere temperatures actually *rise* slightly, and in the mesosphere they first rise sharply, then begin to drop off.

THE TROPOSPHERE

Our greatest interest is in the troposphere, for it is in this layer of atmosphere that we live. The clouds, haze, and smog we observe all lie in the troposphere. Winds and hurricanes, snowstorms and monsoons—all are products of the troposphere.

The air of the troposphere has some properties familiar to all of us: it consists of oxygen (nearly 21 percent), nitrogen (over 78 percent), and less than 1 percent of other gases, among which carbon dioxide is the most important. Air has weight, and it exerts *atmospheric pressure* on every part of the earth's surface,

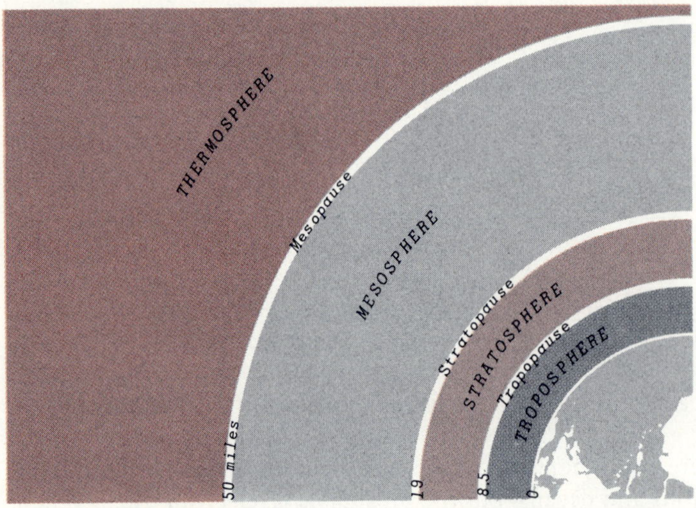

FIG. 2.1 Layers of the atmosphere.

continents as well as oceans. Air also has *temperature*, as anyone who has seen balmy summer days yield to raw winter winds will testify. And air has moisture. Water vapor, which is water in a gaseous state, enters the atmosphere to give it a degree of *humidity*. When that water vapor is condensed, through cooling of the air, the result is rain, snow, or some other form of precipitation.

Temperature in the Troposphere

To us, the most noticeable quality of the atmosphere is probably its temperature, the amount of heat energy it contains. In any weather report, this is one of the first items accounted for. Frequently, the *sensible temperature* is also given. This is an index that takes account not only of the recorded temperature, but also of conditions of wind and humidity, among others. On a winter day, a temperature of 10°F may be comparatively comfortable if there is no wind and the sky is blue. But if it is 10°F and there is a north wind blowing at 30 miles per hour, bringing dust and haze, it may feel like the temperature is minus 20°F. In such a situation, 10°F is the recorded temperature, and minus 20°F is reported as the sensible temperature. At the other end of the scale, we all know the difference between 80°F on a clear, dry day and 80°F when it is humid and muggy.

The earth's surface receives its heat energy from the sun. This heat, or *solar radiation*, comes chiefly in two forms: shortwave, consisting of light rays we are able to see, and infrared rays, which are thermal rays whose wavelength is beyond the visible spectrum. As the earth orbits around the sun, more than 90 million miles away, it receives a tiny fraction of all the heat energy the sun radiates in all directions. But that tiny fraction is enough to penetrate the protective atmosphere and heat it as well as the land and water of this planet.

When the sun's radiation reaches the earth's outer atmosphere, several processes begin. Molecules of nitrogen and oxygen, dust particles, water vapor, and other contents of the atmosphere interfere in various ways with the rays. Some of the radiation is reflected back into space. Some of it is scattered. Some of the infrared rays are absorbed by water vapor or carbon dioxide in the atmosphere. Clouds play a considerable role in reflecting radiation back into space. On a cloudy day, with lots of moisture and a great deal of reflection, as much as two-thirds of the radiation headed toward the surface of the earth may be lost through these processes. There is even some small loss at the globe's surface itself, where land and water reflect small percentages of radiation back out. Meteorologists, specialists who study the physics of the atmosphere, estimate that more than half of the arriving radiation is lost in any given year.

What happens to the radiation that does penetrate all the way through the troposphere to reach the surface of the earth? It is turned into heat. When shortwave rays reach an exposed surface, a process called *insolation*, they are converted to long-wave rays. Most of this long-wave radiation is absorbed by particles of the atmosphere, and the air temperature rises. Anyone who has seen a distant section of asphalt road shimmer in the midday heat has seen this process going on. Not only is the air heated by ground radiation, but conduction, too, raises its temperature. Thus our troposphere is actually heated from *below*, not directly by the sun. The atmosphere allows shortwave radiation to pass through, and it is the earth's surface (water as well as land) that transforms it into heat, which in turn warms the air. This is often referred to as the *greenhouse effect*, because greenhouses function on the same principle: shortwave radiation passes through the glass, the interior surface is insolated, long-wave radiation heats the air inside the greenhouse, and the same glass that let the radiation in prevents the heat from being lost.

Different parts of the earth receive different amounts of the sun's solar energy, and herein lies the root of a ceaseless system of circulation that keeps the troposphere in motion all the time. The ultimate effect of this contrast is familiar enough: equatorial regions are associated with heat, and polar areas display the effects of low-intensity insolation. What brings it about is a combination of factors: (1) the earth's spheroidal shape, and (2) the inclination of the earth's axis of rotation.

The earth's spheroidal shape has the effect of reducing the intensity of radiation received at greater distances from equatorial areas. Figure 2.2 is a cross section of the earth, showing only its surface and two parallel rays arriving from the sun. Note that rays *X* and *Y* have equal width. Ray *X* strikes the surface of the earth at right angles, and the sun angle where it is marked on Fig. 2.2 would be 90 degrees. The area this ray insolates is represented by *A-B*. Ray *Y*, on the other hand, strikes the surface at a low angle. Its area of insolation is represented by line *C-D*. Now, although the rays are drawn to equal width, *C-D* is much longer than *A-B* and therefore represents a much larger area. Thus the same amount of radiation received by a rather small area at the equator is spread over a considerably larger area at higher latitudes. The larger area will, therefore, be insolated with less

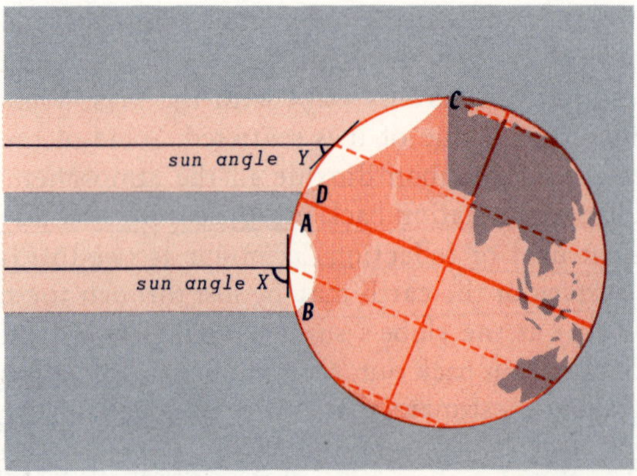

FIG. 2.2 Angle of the sun and intensity of insolation.

intensity. The earth's spheroidal shape has this effect: where the angle of the sun's rays is high, the intensity of insolation is greatest.

This brings us to the reasons why the equatorial areas happen to be the part of the globe to receive this maximum insolation. Our planet orbits around the sun at a distance averaging about 93 million miles. While it does so, it rotates around its own axis, so that for about half of each 24-hour period we are on the sunward side. For the other half we are in darkness. If we visualize the earth's path of orbit and the sun as a surface, we thereby conceptualize the *plane of the ecliptic*, the name given to it by astronomers. Importantly, the earth's axis of rotation is not vertical to this plane: it sits at an angle of 23½ degrees (Fig. 2.3). If indeed the axis of rotation were vertical to the plane of the ecliptic, then the sun's rays would always fall vertically on the equator alone. But because of the 23½-degree angle, the earth during its 365-day orbit is swept by the vertical ray between latitudes 23½ degrees North and 23½ degrees South.

Figures 2.3 and 2.4 help explain this. Figure 2.3 shows the earth in four positions during its orbit. Figure 2.4 shows in detail how the sun will be vertically above the Tropic of Cancer when the earth is on one "side" of it and how the ray falls vertically on the Tropic of Capricorn when the earth has moved to the opposite side. These are the two *solstices* that occur during each orbit, during every year. Because of the inclination of the earth's axis, this is the farthest north and south the vertical ray reaches. Midway between the solstices, the vertical ray falls on the equator. That situation can be simulated by pointing a pencil straight down to the center of the sphere on Fig. 2.4. It happens twice, as Fig. 2.3 shows, once in late March and once in late September.

We have focused in some detail on this matter of the inclination of the earth's axis of rotation, but with good reason. It is perhaps the critical element in the whole system of heat reception and distribution over this planet. From our vantage point here in the Northern Hemisphere most of us see the sun angle rise each spring, reach a maximum on June 21 (the longest day), and then decline again.

The warmest weather is still to come, during July and August. There is a lag because the earth and its atmosphere need time to convert and distribute all the energy received in such enormous quantities in tropical areas. Such a lag, incidentally, can also be observed to occur each day. On a hot summer day the hottest time will be about 2 P.M., although the highest the sun rises in the sky is as early as 12 noon. Again, the conversion of heat and its accumulation have the greatest effect sometime *after* maximum insolation. And there is another reason to look into the earth's axis of rotation and its effects. Anyone who looks at the globe will notice that maximum radiation areas, the tropical areas, consist of much water and relatively little land. Scientists have already wondered aloud about the effects of moving the axis and artificially relocating the equator in a more advantageous position. Almost no one takes such a suggestion seriously, but do not be surprised if it becomes practical in the future.

One additional factor relating to the distribution of temperature on our globe becomes apparent from Fig. 2.3. Again, if the earth's axis were vertical to the plane of the ecliptic and thus to the sun, all our days everywhere on this earth

FIG. 2.3 Sun-earth relationships and the seasons.

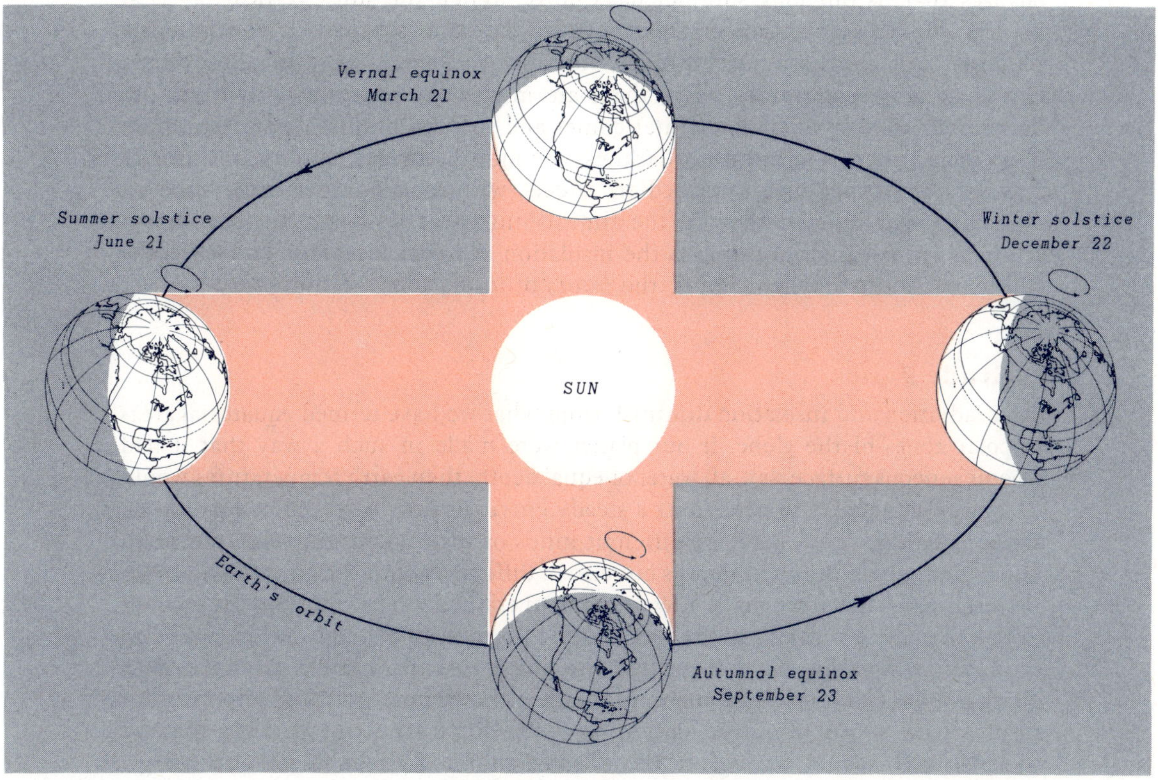

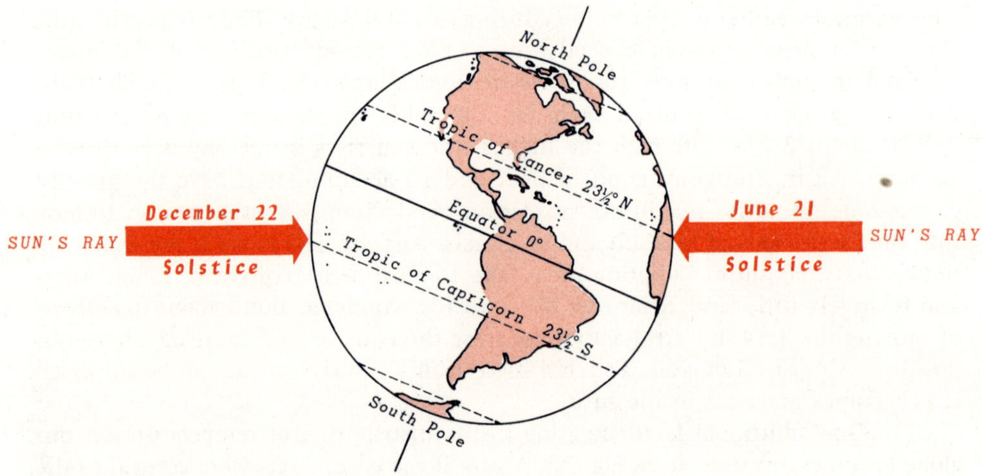

FIG. 2.4 Summer and winter solstices.

would consist of 12 hours of daylight and 12 hours of night. But our inclined axis changes that, as our long summer days prove. When the sun's vertical ray is at or near the Tropic of Cancer (that is, near the June solstice), it continuously illuminates the northernmost latitudes, the North Pole and as far south as the Arctic Circle. In these polar areas, then, there are months of continuous light and no darkness at all. But when the winter comes and the sun illuminates the southern polar areas, there is continuous darkness for months. Areas between the tropics and the poles experience intermediate effects, with long summer days but short winter daylight hours. In this way the diminishing effect of low sun angles at higher latitudes is increased: not only is the insolation of *lower intensity,* but when the sun is lowest, the daylight hours, the *hours of insolation,* are shorter, too.

Latitude Zones

Solar radiation and insolation diminish from what we have termed equatorial areas to polar zones of the globe. If our planet were made in such a way that it had a homogeneous surface, say, all water of equal depth, then earth temperatures would decline from equator to pole along a steady curve. But the earth has a very varied surface, with the land made of different kinds of rocks at varying elevations and waters not only of different depths but also of different composition (salt and fresh, for instance). Thus if we made a record of temperature conditions from the equator to the pole, say from Africa through the Mediterranean Sea and Europe, or from the interior Amazon Basin in South America across the Caribbean and eastern North America, our curve of poleward temperature decline would be anything but steady. Even so, we would be able to recognize some recurring patterns that are so pervasive that they have given rise to some common geographical terminology.

At the lowest latitudes lies the equatorial zone, known for its heat and (with so much water and so little land) its high humidity (Fig. 2.5). This equatorial zone is not bounded by any sharp line (none of the other latitudinal zones we will identify is either), but its width averages about 20 degrees, 10 degrees on either side of the equator. Here the sun angle is always very high, and there is not much seasonal variation in the length of days and nights, which are about equal. Beyond the equatorial zone lie the tropics. Here we can recognize the beginning of seasonal variations, although insolation is still very intense. As we will observe, one feature of this seasonal cycle is reflected by the precipitation regime as the steady, year-round rainfall of the equatorial zone gives way to a wet and dry season. Technically the tropical zone extends to the Tropics of Cancer and Capricorn, but again, we are dealing with generalized zones and not with specific latitudes. The tropical zone gives way to what is normally called the subtropical zone, where tropical conditions prevail during the high-sun season, but where the moderations of higher latitudes are really beginning to be felt. In North America, Florida lies in this zone. Adjacent

---

LATITUDE AND LONGITUDE

In order to make it possible to identify precisely the location of places and areas on the surface of the globe, a grid system has been devised. This system consists of a set of lines running east-west, parallel to the equator, and a set of lines running north-south, connecting the poles.

The lines running east-west are called *parallels*, and they are measured by the angle from the plane of the equator, as shown below. The poles, then, are at 90 degrees. Positions of latitude are given as *x* degrees North or South latitude. For greater precision, the degrees are in turn divided into minutes and seconds. One degree of latitude is about 69 miles on the surface of the earth.

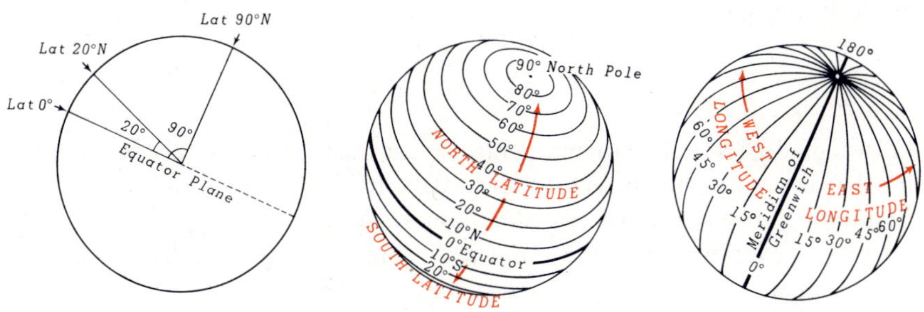

The lines connecting the poles are *meridians*. In this case there is no convenient equator from which to measure, so a *prime meridian* was established to run through Greenwich, England. From there, 180 degrees East and West longitude are measured. In the sketch above, the meridans are shown at 15-degree intervals, the parallels at 10-degree intervals.

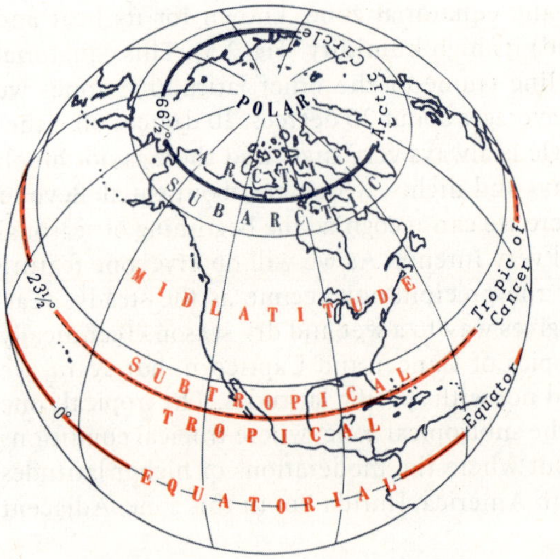

FIG. 2.5 Latitude zones of the Northern Hemisphere.

to the subtropical zone is the mid-latitude zone, an appropriate term because it lies astride the middle parallel, 45 degrees. It extends about 10 degrees in either direction, and in this zone the seasonal contrasts brought by the movement of the sun's rays are strong. Summer days lengthen and winter days are short. Summers can be balmy, tropiclike; winters are laced with polar air.

Beyond the northern margins of the mid-latitudes (about 55 degrees) lie the subarctic (and subantarctic, in the Southern Hemisphere), arctic (antarctic), and polar zones. Here we are at the opposite end of the curve we began at the equator. Variations in the length of days and nights are enormous, climaxing in the polar six-month day and six-month night, and seasonal contrasts in insolation are similarly immense.

Contrasts on the Surface

Earlier we made reference to temperature contrasts induced by the variety of materials making up the surface of the earth. It is important to realize that the conversion of shortwave radiation to the long waves that heat the troposphere does not take place everywhere at the same rate. This is obvious really: on a hot day

---

ISOTHERM

An *isotherm* is a line connecting points of equal temperature. On a map, the temperature along such a line is the same. It will be warmer on one side of the line and cooler on the other.

---

the sand of a beach may become too hot to walk on, but the water will never become that warm. Neither by ground radiation nor by conduction could the water ever heat the air above it as fast as the adjacent land could. This is true on a large scale as well. Under a hot sun of summer whole continental areas heat up much faster and to higher temperatures than the oceans of the same latitudes.

Why should this be so? In the first place, the water is constantly in motion while the land is not. No sooner is the top layer of water heated up than a wave pushes it several feet down and cool water replaces it. The heat is redistributed. In contrast, the radiation received by land surfaces builds up in the stationary upper layer. And water is transparent, which allows the sun's rays to penetrate to a considerable depth. Rocks, of course, are opaque. And although evaporation takes place both on sea and on land, the rate involved, also a cooling process, is much greater at sea than on land. Altogether, then, it takes nearly five times as much heat to increase the temperatures of water and rock the same amount.

But there is another side to this situation. Although the land surface heats up very rapidly, it also loses its heat much faster than the water does. Returning to that beach late at night, you may find that the sand is unpleasantly cold, but the water feels comfortably warm. When the solar radiation is ended by the coming of night, the rock and soil quickly lose much of their heat. This means, again on a larger scale, that temperature contrasts over land areas (seasonal as well as daily) are much greater than over water areas. Water tends to retain its warmth longer, partly for the reasons we identified before, the depth of heat distribution and mobility. Shortly, we will see that this land-water contrast has great implications for the world's distribution of climates.

Air Pressure and Winds in the Troposphere

When, in 1783, two Frenchmen filled a large bag with air from a wood fire and watched it climb several hundred feet into the sky, they discovered something that is fundamental to the study of the atmosphere. What they realized is that when air is warmed, it expands. It becomes lighter than colder air and will rise.

On this principle the entire atmosphere of the earth, especially the troposphere, is kept in constant motion. We know that different parts of the earth receive different amounts of solar energy and that differences in surface material, such as between land and water, create differences in rates of ground radiation. Let us reconsider for a moment that day at the beach when the sand became so hot. What happened to the air there? It was heated by intense ground radiation and by conduction, and it began to rise. All afternoon, the air over the land rose like the Frenchmen's balloon. What replaced the rising air? Air from the water, where it had remained relatively cool. Thus the bathers enjoyed a *sea breeze,* the cool air from over the water, blowing landward to fill the place of the hot, rising air over the land. But that late night, when we went back to find the sand rather cool but the water pleasantly warm, we could observe the reverse. Now the air over the water was warmest and rising, though not as rapidly as the air over the land had earlier.

Nevertheless, smoke from a beach fire would drift out over the water. Now air from the cooler land was replacing the rising air over the water, and a *land breeze* was blowing.

The buoyancy of air is translated to maps and charts by measurement of the amount of *pressure* it exerts. Air heated to a high temperature (and rising as a result) exerts less pressure on the surface than air that is cold. Thus the differential heating of the earth we discussed before has the effect of creating not only temperature contrasts but also pressure differences. And pressure differences lead to constant readjustment as the air moves to replace ascending (or descending) air.

This would lead us to assume that the equatorial zone, where so much heat is received, should be a zone of rather low pressure. And so it is. Here the process of *convection* prevails. On either side of this equatorial zone of low pressure, at about 25 to 30 degrees North and South latitude, is the reverse, a zone of subtropical high-pressure belts. The *equatorial trough* (as the low-pressure zone is called) and the *subtropical high* (sometimes called the high *horse latitudes*) are part of a global system of pressure zones and wind belts we shall shortly dissect.

Before we do so, however, we must consider one important factor. Our earth rotates—and not slowly. Consider the circumference of the globe, about 25,000 miles. It turns on its axis one complete rotation in 24 hours. If you live on or near the equator, you travel 25,000 miles each 24 hours (not to mention your progress in orbit), or more than 1000 miles per hour! Of course, we barely notice it. The atmosphere moves with us, and gravity keeps us firmly on the ground. But this does not mean that the rotation of our planet has no effect on us.

Indeed, the earth's rotation in some way affects all objects (gaseous, liquid, or solid) moving in any direction on its surface. This force of deflection, named after Coriolis, causes all moving objects in the Northern Hemisphere to be deflected to the right of their direction of motion and all objects in the Southern Hemisphere to be deflected to the left of their direction of motion. Only on the equator itself is this force absent. This is not just some abstract bit of theoretical physics. Rivers in the Northern Hemisphere tend to erode their right banks more strongly; rivers in the Southern Hemisphere their left banks. Long-range missiles are deflected over their intended course. In the *London Times* some years ago there was a letter from an irate pedestrian who asked why people on sidewalks would not keep to the left, as motorists in Britain do. Said a geographer: could it be that crowds, like water

---

THE NAME OF A WIND

A prevailing wind is named after the direction *from which it comes*. Thus a westerly wind comes from the west and blows in an easterly direction. The southeast trades come from the southeast and blow toward the northwest. A sea breeze comes from the sea; a land wind blows outward from land to water.

---

and wind, tend to be pushed to the right by Coriolis force and that walking on the left is a matter of conscious effort in London?

Whether or not crowds can be affected this way, winds certainly are. If Coriolis force did not exist, air would simply flow from the subtropical high to the equatorial trough in a straight north-south direction. But deflect a north-south wind to the right, and a wind will blow from northeast to southwest. In the Southern Hemisphere the equatorward wind would blow due north, but Coriolis force deflects it to the left, and the southeast trades develop. Figure 2.6 shows the approximate relationships involved, along with a vertical view of circulation in the lower atmosphere, but it is important to note that this is a highly generalized picture. For example, the subtropical high is pictured as a continuous belt; in fact, it is a set of cells of high pressure with intervening areas of variable character. North and south of the subtropical high lies the belt of westerly winds with which most of us are directly familiar. This is the wind belt influencing most of the heart of North America. But we all know that the winds in, say, Chicago or Memphis do not always blow from the southwest; they sometimes blow from the north, from the southeast, and from other directions. What is meant by the belt of westerlies on Fig. 2.6 is that winds have a stronger westerly component than any other. On weather maps

FIG. 2.6 Global pressure and wind belts.

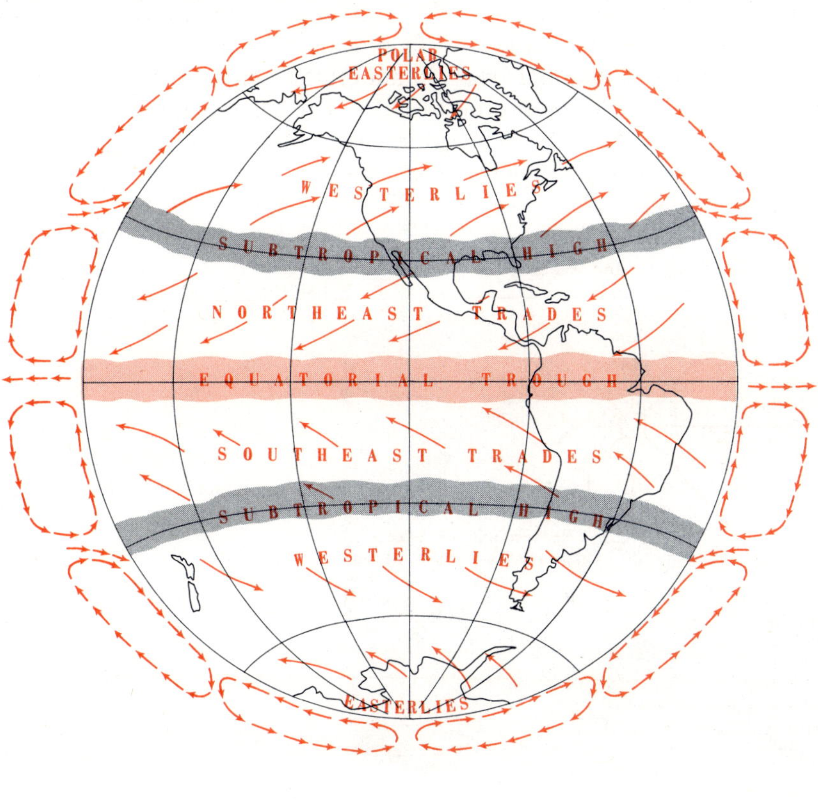

this is illustrated by a *wind rose,* in which the line length indicates at the same time the prevailing direction of the wind, its average strength, and the percentage of time it blew (Fig. 2.7). On Fig. 2.7, which shows a hypothetical station in the westerly wind belt, there was a southwesterly wind 40 percent of the time and a westerly wind 20 percent of the time.

Another reality not reflected by Fig. 2.6 relates to the annual migration of the sun's vertical ray. The situation as pictured by Fig. 2.6 obtains shortly after the two equinoxes, but the whole system is pulled to the north during the northern summer, after the June solstice, and to the south after the December solstice, during the southern summer. This adds a whole new dimension: for example, the southeast trades in the Atlantic Ocean, after crossing the equator during the northern summer, are affected by the Northern Hemisphere's Coriolis force. In the Southern Hemisphere they had been deflected to the left of their path of motion, but upon crossing the equator they tend to be deflected in the opposite direction. Thus the trade winds actually bend about 90 degrees and change direction from southeast to southwest!

Figure 2.6, then, is a highly generalized average of countless recorded observations. Despite these disclaimers it does have some relationship to reality. In the equatorial trough, where heat is intense and vertical movement of air prevails, there is relatively little horizontal movement of air. Small wonder that the ancient mariners came to call this the *"doldrums."* Their sailing ships were becalmed, sometimes for months on end. And the winds flowing from the subtropical high to the low-pressure zone in equatorial areas did not get their name, the *trade winds,* for nothing: these predictable, prevailing easterlies carried the intercontinental commerce of the world for centuries. The subtropical high was probably called the high *horse latitudes* because it sometimes became necessary to throw animals

FIG. 2.7 Wind rose for station in the westerly wind belt.

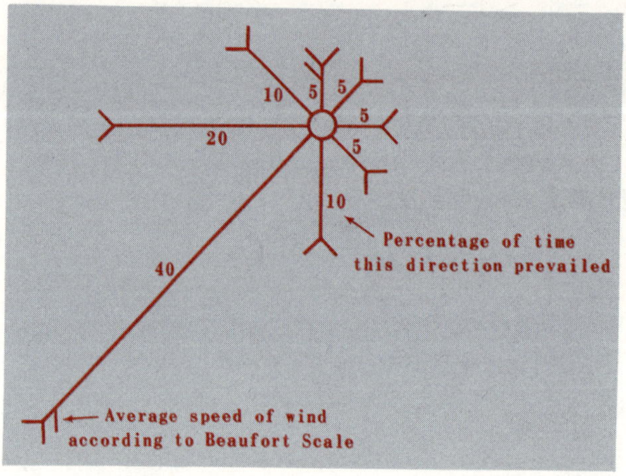

overboard to conserve water and food on ships becalmed in the quiet air characteristic of those zones. The high-pressure, descending-air quality of the high horse latitudes also is expressed on land areas. Note on the map of climatic regions of the world (Fig. 2.12) that there is a remarkable similarity between the position of this belt and some of the world's largest deserts. In the *westerlies,* winds as well as the weather generally are highly variable, but, as many of us know from experience, weather systems tend indeed to move from west to east. Because there is much more land area in the Northern Hemisphere than in the Southern Hemisphere, the westerlies are rather moderate in strength here. In the Southern Hemisphere, as a look at a globe shows, there is very little land in the latitudes of the westerlies, and they develop unimpeded into the howling storms that sailors have feared for centuries.

PRESSURE CELLS So far, we have discussed pressure and wind zones on a global scale. But when we listen to weather reports, we are told that there is a "a large high" over central Canada or "a low-pressure cell" over the southern Great Plains. What this means is that centers of high and low pressure develop within the system of global zones described earlier.

How does this happen? Several factors are involved. One phenomenon accurately suggested by Fig. 2.6 relates to the persistence of air in specific parts of the troposphere. Note how air flows from the subtropical high through the tropics to equatorial regions at the surface and back again at upper levels; note also (on the vertical sketch) how polar air tends to get caught up in polar-arctic circulations.

CORIOLIS

Who was Coriolis, after whom the famous *force* is named? Gaspard Gustave de Coriolis was born in Paris in 1792. He studied engineering and mechanics and became a professor teaching at Paris's Polytechnical Institute.

Even though Coriolis was plagued by illness through much of his life, he managed to do some significant research. In 1835 he published a paper in which he showed that existing laws of the motion of bodies are applicable to a rotating condition when an inertia force is inserted in the formula, a force acting to the right of the direction of motion for counterclockwise rotation, and to the left under conditions of clockwise rotation. The significance of that article in studies of meteorology and oceanography was immediately obvious, and the force factor in equations of motion came to be known as the Coriolis factor. Coriolis's name is associated even today with calculations in the field of rocketry. Missile paths require Coriolis adjustment for accurate prediction. Coriolis died in Paris in 1843, but not before he achieved recognition by election to the Academy of Sciences.

Thus air *masses* develop certain semipermanent characteristics, and we can recognize tropical air masses and polar air masses. They develop these characteristics by prolonged exposure to certain conditions (heat, humidity; cold, dryness) prevailing in a certain area, referred to appropriately as a *source area.* Northern Canada is such a source area. The cold, dry air that sweeps south into the heart of North America during the winter develops there. The Gulf of Mexico is another source area, in this case for warm, moist, tropical air well known by anyone who has spent the muggy days of a Midwestern summer in a sweltering city.

There is more involved. We know that cold air is heavier than warm air and exerts more pressure on the surface. That polar-arctic air, therefore, develops high-pressure characteristics. This means that it is, in effect, a heavy pile or bulge of air and that the air is busy flowing outward from it in all directions to places where it can push lighter air away. But now Coriolis force comes into play. Imagine the outflowing air by drawing a series of arrows fanning straight out from the center of the polar air mass. Coriolis force begins to affect those arrows, and (in the Northern Hemisphere) they are deflected to the right. As Fig. 2.8 shows, that results in a closed cell of circulation in which the air moves clockwise. This is called an *anticyclone.*

Low-pressure tropical air can produce the reverse effect. Imagine a mass of such warm air has flowed into the southern Great Plains, where it is additionally heated by the high-angle July sun. Air in the center of the mass rises, heated by ground radiation, and air from adjacent parts of the air mass flows into the center, replacing the risen air. Again, the arrows would point straight along the pressure gradient (Fig. 2.8, right top), but Coriolis force deflects their movement, again to the right. The result of this is a counterclockwise circulating system known as a *cyclone.* (In the Southern Hemisphere, where Coriolis force deflects to the left, the cyclone's circulation is clockwise and the anticyclone counterclockwise).

The weather map that is part of a good daily weather report, thus, has lines connecting points of equal pressure (*isobars*) closing around cyclones and anticyclones, and it shows wind directions associated with these systems. Weather forecasters look for changes in these systems as they traverse the country, and some of these changes will shortly be discussed. In the meantime, let us not lose sight of this important reality: even though a cyclone may control the whole central United States and winds in this cyclone blow from all directions (as Fig. 2.8 shows), the whole cyclone lies in the westerly wind belt (Fig. 2.6), and it will slowly travel an eastward path. This is why, if your home were located at point *X* on Fig. 2.8, you could predict with fair certainty that winds in the next few days would change from south to southwest, then to west, and eventually to northwest. All you have to do is move the cyclone across point *X* and read the arrows!

THE JET STREAM Cyclones and anticyclones are phenomena of the lower troposphere, that is, of the surface and of comparatively low elevations, and one problem weather forecasters face is that things happen at high altitudes that affect the weather on the ground, altitudes at which very few data are readily available. One

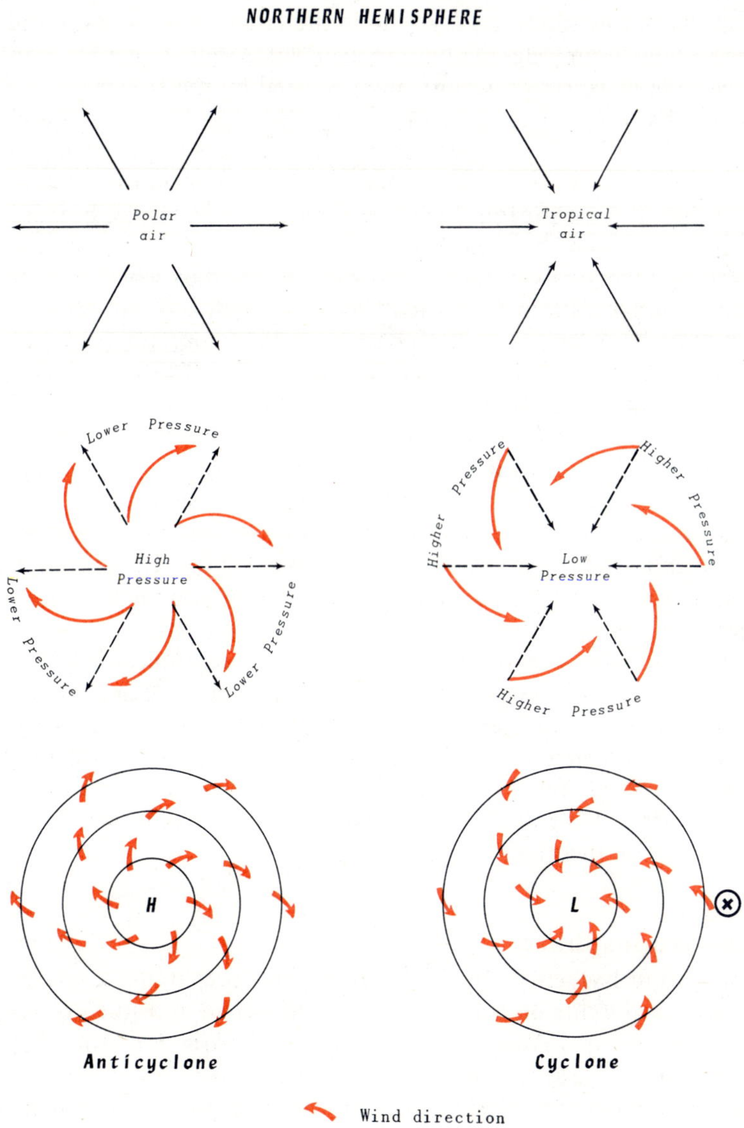

FIG. 2.8 High- and low-pressure cells.

of these high-altitude phenomena is the *jet stream*. Actually, it would be more correct to say jet streams, for more than one of these globe-circling, snakelike streams of air exists. The most important one to us is the mid-latitude (polar-front) jet stream. It lies between 30 and 45 degrees North and South latitude and at an altitude of 30,000 to 45,000 feet, at the top of the troposphere, near the tropopause. It is a westerly jet (the air moves eastward), and it is made of a set of concentric layers of air, among which the innermost layer flows the fastest. Air speeds in the jet stream vary, from a few dozen miles per hour to over 300. Airplanes have ridden the jets and covered long distances in record time, boosted by the stream.

The jet stream has a significant effect on the paths taken by cyclones and anticyclones at lower altitudes. Since jet-stream paths, once established, tend to remain stable for some time, they assist in the determination of long-range (30-day) weather predictions. But the jet streams also pose forecasting problems, because (1) the jet stream is not regular, but twists and bends like a giant, meandering stream, and (2) it moves suddenly from one latitude to another when it *does* relocate. When the jet moves, storm tracks tend to move with it. One example of the impact of jet-stream movement can be seen in the Indian monsoon. For years, climatologists have wondered why the enormous increase in rainfall associated with the monsoon starts so suddenly. Now it is believed that this increase is heralded by the movement of the jet stream. Before the rains come, the jet lies south of the Himalaya mountain range. There it is associated with an anticyclone, a high-pressure area that brings comparatively dry weather. Then, quite suddenly, the jet stream shifts to its summer position to the north of the Himalayas. The high-pressure area moves with it, and low-pressure conditions begin to prevail over much of peninsular India. As a result, warm, moist ocean air is drawn into the landmass, and the rains begin.

Analysis of weather and climate must depend on long-term recorded data. Over much of the land area of the globe (let alone the oceans), there are not enough weather stations, and data where they exist are still short-term. Upper-air information is even less complete, and meteorologists depend on data sent down by weather balloons, aircraft, and satellites. Yet the upper air greatly influences the course of our weather on the ground. Most of the time weathermen are not to blame for forecasts that turn out to be incorrect.

Moisture in the Troposphere

The air of the troposphere has temperature and exerts pressure; it also contains an ingredient that is crucial to life on this planet. This ingredient, moisture, exists in the atmosphere as water vapor. Water can enter the air through the process of evaporation, and moisture in the air can return to the liquid state by condensation. When we read weather reports, the *relative humidity* of the outside air is usually given. This index denotes the amount of water vapor the air is carrying, and it is given as a percentage of the maximum amount the air can contain at the prevailing temperature. The warmer air becomes, the more water vapor it can contain. Thus when the outside temperature is 85°F and the relative humidity is reported as 90 percent, then the air holds an enormous quantity of moisture. But if the temperature of the air is only 20°F, 90 percent relative humidity represents far less actual water vapor present.

Air also has a temperature at which the amount of water vapor present fully saturates it. To use our weather report again: that air at 85°F with a relative humidity of 90 percent need only be cooled slightly to reach a relative humidity of 100 percent, since cooler air can contain less water vapor. When the air is cooled below that point, some of the water vapor will have to go, and it goes by condensation into rain drops. That critical temperature, where condensation begins, is the *dew point.*

What can cause the air to be cooled in this way? Mere loss of temperature at night, when ground radiation slows down and the air loses heat, is not sufficient—not enough, anyway, to cause precipitation. Droplets of water do form on the ground and on grass and other vegetation when the air layer adjacent to the surface is cooled below dew point. But rainfall and other forms of precipitation can only occur when large quantities of air are forced to rise. This can happen in three ways: first, when a mass of air is warmed by ground radiation on a hot day and begins to rise; second, when air is pushed up against a mountainside; and third, when air masses of different character (such as polar and tropical) collide and the cold, heavy air pushes the warm, buoyant air upward to higher altitudes.

CONVECTIONAL PRECIPITATION Everyone at some time or another has seen a huge, cauliflower-shaped cloud billowing high into the summer sky. As the day wears on, the base of the cloud grows darker, and late in the afternoon, rain can be seen falling from it. This is a manifestation of *convection.* The air over a particular, exposed area has been heated, has begun to rise, and has continued to rise in a column of updraft that may exceed 40,000 feet. The air is moist, and condensation

A rainstorm falls from a convection column in Wyoming. The time is late in the afternoon, the ground surface has been heated, and the rising air has developed cumulus and dark-based cumulonimbus clouds. Isolated storms like this are widely distributed over the landscape. Such a statement in the weather forecast as "Rainfall probability is 30 percent" means that during the period of the forecast 30 percent of the area covered by the forecast is expected to receive rainfall.

first produces droplets of water (and, at the highest altitudes, even ice) so tiny that the updraft is sufficiently strong to keep them airborne. But as the air continues to ascend, larger-scale condensation begins and sizable drops of water form. Precipitation begins when the drops reach the ground, and in the course of the late afternoon and evening the convection cell may produce tons of rainwater. If the updrafts in the cell are strong and sustained, drops may be carried to high altitudes where temperatures are well below freezing. There they become ice, and before they eventually reach the ground they may be coated with water, taken upward, and frozen anew. This can happen again and again. The result is precipitation that can be quite destructive, hail.

OROGRAPHIC PRECIPITATION When air is forced upward because a mountain range blocks its path, it is cooled. If the mountainous obstacle is high enough, the air temperature will drop below the level at which condensation occurs, and precipitation takes place. When the air first begins to rise, the rate at which it cools is 5.5°F per 1000 feet. This called the *dry adiabatic rate,* and it holds until the air becomes saturated. Then the rate changes to 3.2°F per 1000 feet of ascent, the *wet adiabatic rate.* This change occurs because the condensation process generates some latent heat, which counteracts the temperature loss sustained as the air continues to rise. It follows, then, that air that has crossed the mountain range and descends on the other side will warm up. Since it does so, and no rain is

CLOUDS

Clouds form when tiny droplets of water (or ice) collect around extremely small dust particles in the atmosphere. These small dust particles are essential, because they serve as the nuclei of condensation. This is true of fog also. A dense city fog is generated by the high pollutant level in the lower air layer and by radiation cooling.

Cloud types are classified according to their appearance and the altitude at which they occur. *Stratiform* clouds are sheetlike, of wide areal extent, but thin. *Cumuliform* clouds are billowing, tall clouds.

Altitude is signified by prefixes. The lowest stratiform cloud is simply known as *stratus.* Above 6500 feet or so, a stratiform cloud is known as *alto*stratus. Above 20,000 feet, it is *cirro*stratus. The lowest cumuliform cloud is *strato*cumulus, or simply *cumulus,* the middle form is *alto*cumulus, and the highest, *cirro*cumulus. Very high altitude streaks of thin cloud are called *cirrus.*

Cumuliform clouds tend to develop to great height. Cumulus clouds, caught up in a convection cell, may develop a base at 1000 feet and a head, or *anvil,* at 35,000 feet or even higher. When this happens the cloud is identified as a *cumulonimbus,* signifying that it produces rainstorms and often thunder and lightning, along with squalls of high wind.

Four cloud types. A low *stratus* cloud lies near the base of an escarpment (upper left). The beginnings of cumulus development at middle altitudes is *altocumulus* (upper right). Vertical, convectional air movement results in *cumulus* (lower left). Further convectional updrafts produce *cumulonimbus* (lower right); rain can be seen falling from the base of the cloud.

produced, the rate of heating is the same as the dry adiabatic rate, all the way down. This other side of the mountain range lies in the *rain shadow*; it is dry, and the rapidly heating, downward surging air generates winds that are sometimes hot and uncomfortable. Next winter, keep a check on temperatures on the lee (east) sides of our northwestern mountain ranges, in Montana, Wyoming, and the Dakotas. The winter cold suddenly breaks as the *chinook* brings unseasonably warm temperatures in just the way we have described. In a matter of hours, the temperature may go from the 20s to the 70s.

CYCLONIC PRECIPITATION The third process that generates substantial precipitation involves air masses in low-pressure cells, or cyclones. Earlier we discussed the development of these systems, which, in the middle latitudes, literally become arenas of conflict between contrasting air masses.

Consider as we did previously a low-pressure cell that occupies the central part of the continent, lying between the Rocky Mountains and the Appalachians and extending from southern Canada to northern Texas. We know that this cyclone has winds converging on the center and that these winds emanate from central Canada as well as southern Texas and the Gulf of Mexico. Imagine that this cyclone prevails during the fall, say, in late October. The air from the north, with its near-Arctic source area, is cold, dry, heavy, stable. It flows into the system's northern half and is caught up in the circulation. (On the weather map this air may be identified as cP, continental polar in character.) From the south comes air from the Gulf, warm, moist, buoyant, unstable. It flows into the southern half of the system. (The map shows this as mT air, maritime tropical in nature.) Now our low-pressure cell has two air masses in it, air masses with very different properties.

How do these air masses interact? Do they mix and eliminate their differences? No. Rather, the air masses make contact with each other along a surface called a *front*. If this term suggests wartime combat, that impression is correct. Just as ground armies face each other along a front and seek to push each other back, so the front between air masses is a zone of contact and combat. Figure 2.9 indicates where these fronts might be positioned in our hypothetical low-pressure cell. In the southwestern quadrant, the cP air is pushing back the warm air along a front that looks like a line on our map, but is in reality a plane of contact, as the cross section A shows. In the southeastern quadrant, warm mT air is pushing back the cool air (cross section B). The small arrows indicate the wind directions in this particular system.

Here, then, we have the third process whereby air can be forced to rise to higher altitudes, generating rainfall or other forms of precipitation. Along the cold front, the heavy cP air wedges beneath the more buoyant mT air, forces it to rise quite rapidly, and lowers its temperature below the dew point. Such a cold front is marked by billowing cumulus-type clouds, intense rain storms, and, often, lightning and thunder. Along the warm front, the updrafts are less rapid and the clouds more likely to be of the stratus variety, with rather persistent, gentle rain. Here the warm mT air overrides the cooler air (cross section B) and cools below condensation temperatures; in the cool air, too, there is some updrafting. Along the warm front, mixing of air is more pronounced than along the cold front.

In our middle latitudes, these frontal cyclones traverse parts of the North American continent almost daily, traveling eastward at rates varying from a few to several dozen miles per hour. Their paths are difficult to predict, for reasons we have discussed. They may move to the northeast, into Canada, stay on a due easterly track, or turn to the southeast. In our example, much of North America between the Rockies and the Appalachians was dominated by the cyclone pictured on Fig. 2.9, but it will move eastward, and soon another system will take its place. Not all systems that cross the continent are cyclones. Sometimes, especially in the fall, winter, and early spring, an uncontested high-pressure area will overspread the region, bringing stable air, dryness, and blue skies. Fronts do not develop in such anticyclones as they move eastward; they do not draw in air, but rather disperse air outward. A careful look at tonight's weather report might reveal some high- as well as low-pressure areas in control of parts of the country.

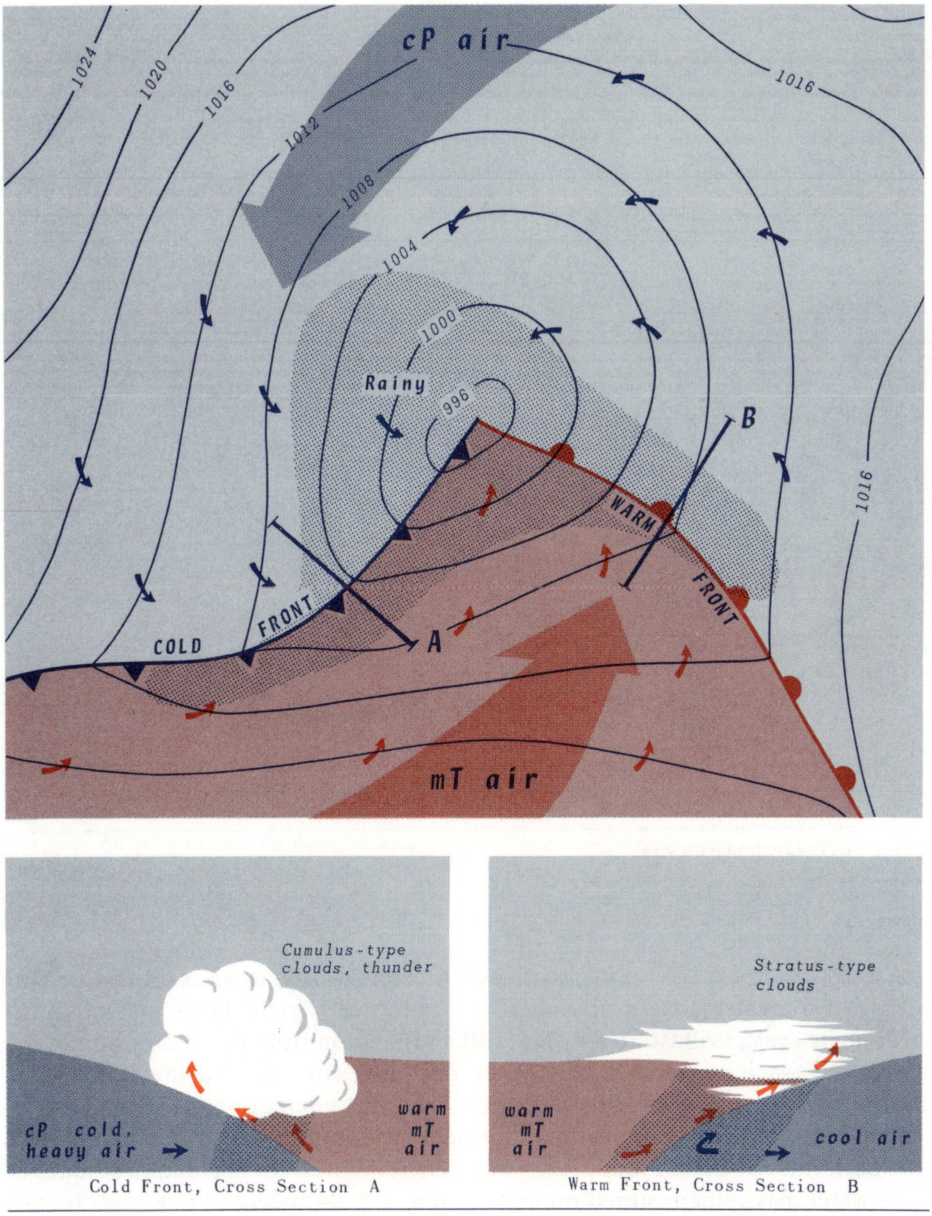

FIG. 2.9 Cyclonic system and fronts.

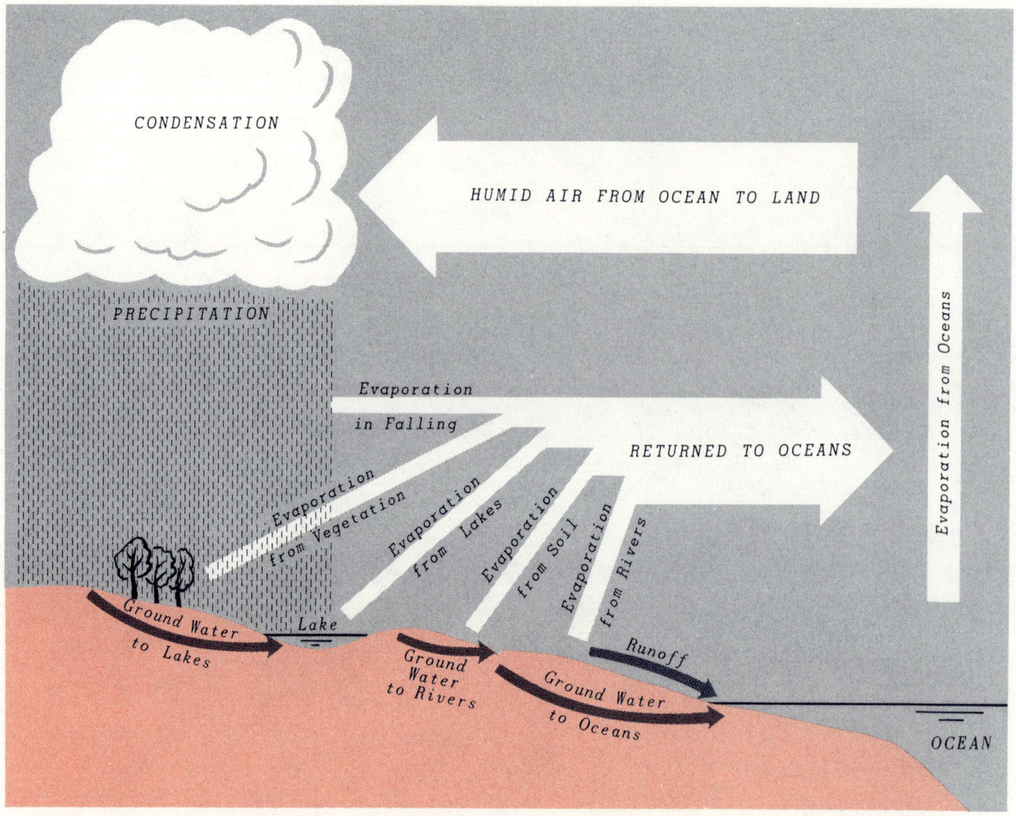

FIG. 2.10 The hydrologic cycle.

THE HYDROLOGIC CYCLE The processes of precipitation we have examined are only a part of a greater system of global water transfer. This system, known as the *hydrologic cycle*, is illustrated by Fig. 2.10. Evaporation from the oceans humidifies air masses (such as the mT air mass in our last example), and these air masses are then carried over the earth's land areas. There, condensation occurs and precipitation falls. Some of the moisture that reaches the ground runs directly into rivers and back into the oceans, but other water penetrates into the soil and into aquifers. This water may remain stored for a long time, but ultimately it, too, finds its way back to the oceans by way of evaporation from vegetation (which transpires water), from soil, and from lakes and rivers. Back in the oceans, the water can evaporate into the air all over again. Figure 2.10 is a schematic diagram, of course, and it does not represent the relative quantities of moisture carried through the cycle. It does, however, represent the cycle's chief elements.

Precipitation: A World Perspective

Now that we have examined the way that precipitation comes about, let us determine to what degree various parts of the world are endowed with moisture by these processes. Figure 2.11 displays, in generalized form, the distribution of precipitation across the land areas of the globe during an average year. The equatorial and tropical areas immediately draw attention because they receive comparatively high amounts of rainfall. South and Southeast Asia, equatorial Africa, and northern South America generally receive more than 40 inches of rain, and large areas in these regions record over 80 inches annually. Some locales (West Africa, coastal southwest India, part of Indonesia and New Guinea, for example) get over 120 inches. The significant exception is East Africa, which appears to lie in the path of Indian Ocean trade winds and should have far more rainfall than it does. We do not know exactly what interferes with rain-bringing systems here, although there are theories that the trade winds are broken (by Madagascar from the southeast), that the layer of moisture-producing air is thin and the upper air is dry, and that East Africa gets leftover rainfall from the west, from equatorial Africa, rather than from the Indian Ocean. Even on a scale like this, the world map still poses unsolved problems.

It is clear, however, that the abundance of rainfall in the equatorial and tropical areas relates to the limited size of the land areas, the warmth of the air, and its capacity to carry huge amounts of moisture. Most of the rainfall comes by convection, though orographic processes also play a role, as in the case of the Indian monsoon and the escarpment of eastern Madagascar. And the apparently anomalous dryness of the west coast of South America and western Africa south of the equator also can be explained. Here subtropical high-pressure systems extend somewhat inland and since high pressure means subsidence and warming air, there is little or no rainfall.

Poleward from the equatorial and tropical zone of abundance, rainfall declines, a precipitous decline in Asia north of India and in Africa north and south of the Congo (Zaïre). Here we are in the belt of subtropical highs. Australia, southern Africa, and South America all have desert and near-desert (steppe) areas at these latitudes, and in the Northern Hemisphere the vast Sahara-Arabian Desert lies astride the Tropic of Cancer. In Asia, the zone of persistent high pressure shifts somewhat northward and lies to the north of the Himalaya Mountains. Note how dry much of interior Asia is. Remoteness from the moisture-bringing air masses and mountainous obstacles to air flow have much to do with this.

In the middle latitudes, the effect of the westerlies is shown clearly in North America from about 30 degrees northward, in South America from the same latitude southward, and in Europe. In North and South America, the areal extent of orographically induced precipitation is rather limited because coastal ranges (the Rocky Mountains and the Andes) confine the rainy areas to coastal strips, but in Europe there is less impediment to the flow of the moist air. Note how, in North America, the trend of the Rocky Mountains coincides with the zone of annual precipitation below 20 inches and how, in Europe, the moist coastal zone extends as far eastward

FIG. 2.11 Mean annual precipitation of the world.

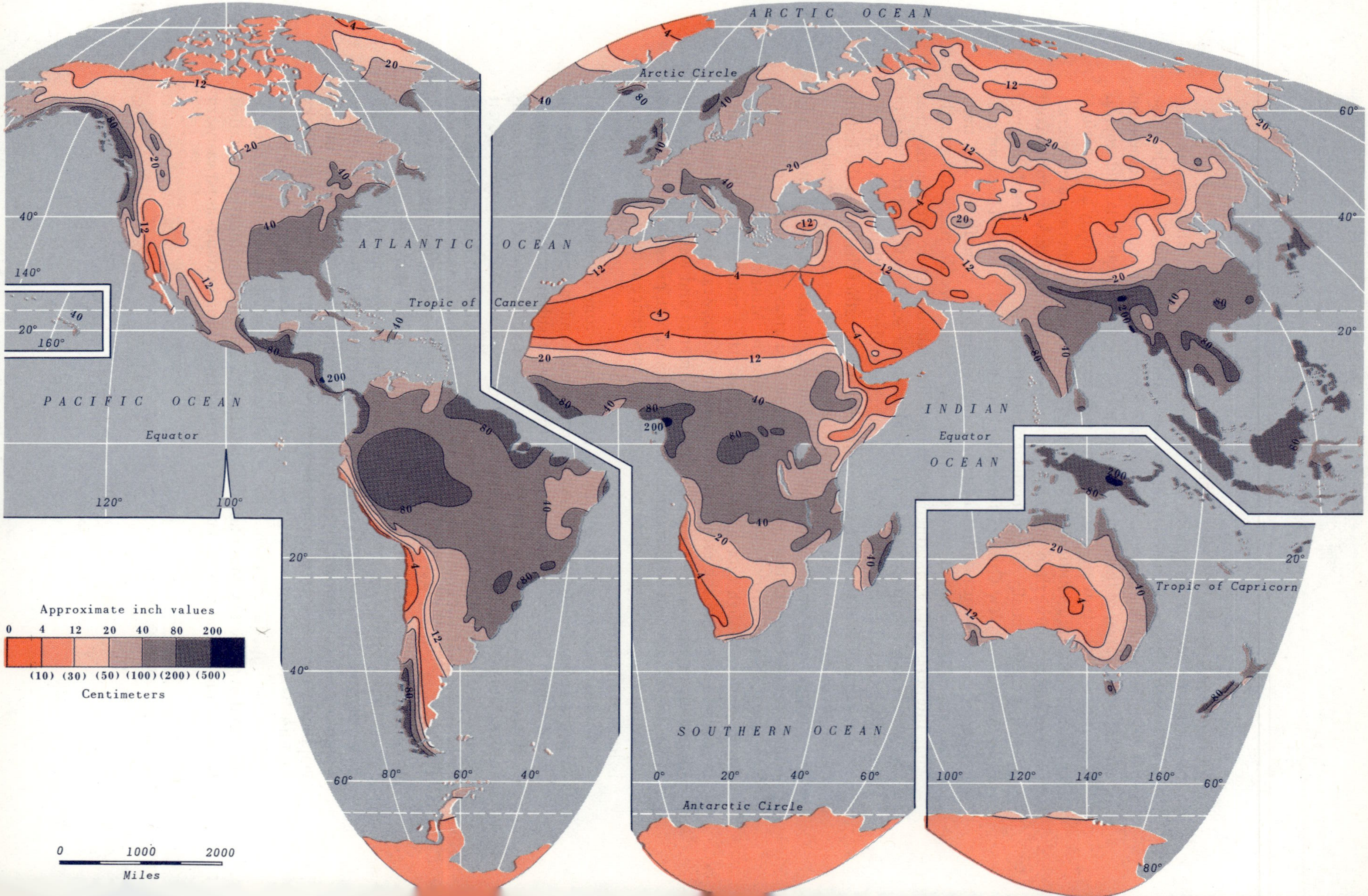

as the Soviet Union. East of the Rockies, much of the rainfall comes from the south, from the Gulf of Mexico.

North of the middle latitudes, the arctic and polar areas receive relatively little precipitation, as Fig. 2.11 shows. There, low air temperatures reduce the moisture-carrying capacity of the atmosphere.

EVAPOTRANSPIRATION Before leaving the topic of world precipitation, we should consider the relationships between precipitation and the world distribution of temperature. From Fig. 2.11 we might conclude that all the areas receiving more than 40 inches of rainfall are thereby amply supplied with moisture. But there are equatorial areas with temperatures averaging over 70°F receiving 40 inches, and there are much cooler areas, for example in Europe and New Zealand, that also receive 40 inches of rain annually. Obviously evaporation goes on much more rapidly in the equatorial and tropical areas than in those mid-latitude zones. When we look again at the diagram of the hydrologic cycle (Fig. 2.10), we notice that evaporation from the soil and from vegetation constitutes an important part of the return phase of the system. But the contribution by evaporation in hotter parts of the world is greater than elsewhere.

Similarly, evaporation from vegetation goes on very rapidly in equatorial regions. This evaporation from vegetation is actually a three-stage process. Roots of plants absorb water in the soil. It is transmitted through the organism and reaches the leafy parts, which transpire in warm weather much in the way we perspire. And from the surface of the leaves, the moisture evaporates. Thus a plant acts like a pump, and the process we described as "evaporation from vegetation" is actually a process of transpiration and evaporation, or *evapotranspiration*.

Now it becomes clear that 40 inches of rainfall in a tropical area might very well be insufficient and inadequate. Not only does the ground lose moisture at a rapid rate, but also plants transmit enormous amounts of moisture to the air. It may very well be that if we calculate the amount thus lost, it exceeds the 40 inches received. Some of those "moist" tropical areas, even those with over 60 inches of rainfall annually, can be shown to be moisture deficient, that is, they would use more rainfall if it were available. On the other hand, in cooler parts of the world 30 inches may be enough to keep the soil moist and the vegetation at optimum growth. Figure 2.11 shows us much, but it also conceals factors.

CLIMATES OF THE WORLD

It is tempting to seek some form of order in the earth's varied patterns of weather, to take all the records of temperature and rainfall, to make maps of soils and vegetation (both related to prevailing climate), and to arrive at a fairly simple classification of the world's climate. Many geographers have tried it, but every scheme they have devised has been criticized. Still, those efforts have been worthwhile, for each provided some additional insight, and each stimulated the search for still more adequate classifications.

One problem relating to any map representing the classification of world climates will already be obvious to us. We have seen how weather and climate change from place to place, almost everywhere gradually. Yet the map demands that we draw lines, which put us in one climate on one side of the line and in another climate on the other side of the line. There are not many places on this globe where you can walk from one climate to another in 50 feet or so!

Another problem has to do with the criteria we ought to use in classifying world climates. Some geographers have preferred to base their systems on known temperature and precipitation records and on world patterns of soil and climate, in a sense letting nature record the dividing lines (for example, between tropical rain forest, or "jungle," and savanna). Others have argued that it is pointless to bother with classifications that do not account for such dynamics as the evapotranspiration process.

In the present discussion, we shall base our classification on a system devised by Köppen and modified by Geiger and Pohl (Fig. 2.12). The system is clear, but considerably generalized. You may well be confronted by alternatives, and the comparison should be instructive.

With these cautions in mind, let us examine Fig. 2.12 in some detail. On this map, each of the climates is identified by a set of letters. The letter code is described in the legend, and we will not dissect it here. Suffice it to note that the first (capital) letter is the critical one. Thus the A climates are equatorial-tropical climes, marked by year-round warmth, the B climates are characterized by dryness, the C climates feature mid-latitudinal properties, the D climates have cold, snowy winters, and the E climates have no really warm summer month.

Equatorial-Tropical Climates (A)

True equatorial climates, with very little deviation from a high monthly mean temperature, occur in the inner Amazon Basin in South America, in the Congo (Zaïre) Basin in Africa, and in areas of south and Southeast Asia. In addition to the prevailing high temperatures, there is also a steady (if not as even) supply of precipitation, and these equatorial climes are both hot *and* moist. Substantial rainfall is recorded every month, but in some areas the rainfall record shows a sudden, enormous bulge due to the arrival of the annual monsoon. Coastal India and the strip along farthest coastal West Africa experience this pattern. Where the rainfall is spread fairly evenly throughout the year, the climate has generated the vegetation association we call the *equatorial rain forest,* and the prevailing climate has been given this name as well. It is marked Af on the map. The *monsoon*-dominated climes are identified as Am. The third variety of A climate covers large areas in Brazil, in Africa from the Atlantic to the Indian Ocean, and in India, Southeast Asia, and northern Australia. Again a reference to Fig. 1.13, the map of world natural vegetation regions, shows the correspondence between a particular vegetation association and this climatic region, the *savanna.* The savanna climatic type is marked by greater extremes in temperature and by the occurrence of wet and dry

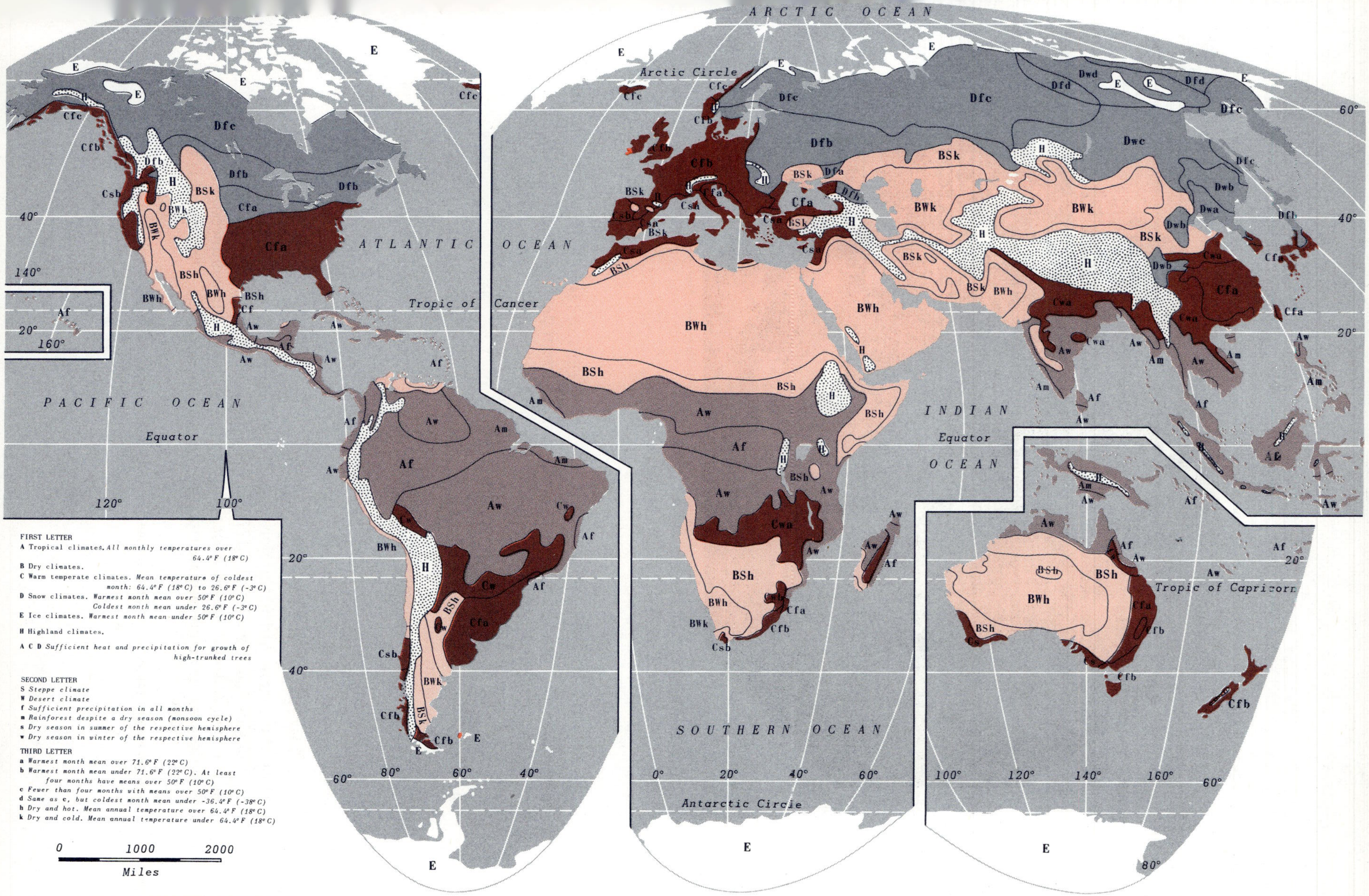

FIG. 2.12 Köppen-Geiger climate classification system.

Areas of tropical rain forest exist in Middle and South America, Africa, and South and Southeast Asia. This is a section of the dense rain forest at East New Britain in Papua (New Guinea). Note the enormous variety of plants, from ferns and small trees to the largest trees, all competing for space and sunlight.

seasons. Here, affected by greater distance from the equator (as in India and Brazil) or by elevation (as in East Africa's highlands), a pattern develops with which we are quite familiar in our own latitude. The high-sun season brings rainfall; the low-sun season brings drought. In the savanna, where the sun's vertical ray passes an area twice each year, *two* rainy seasons may develop and two dry seasons. In Africa one frequently hears references to the coming of "the long rains" and the end of "the short rains." At the same time, the total amount of rainfall received in savanna areas is generally less than in equatorial regions, and the dry seasons are often periods of grayness, baked soils, and parched landscapes. For animals and sometimes for people, the rains break too late.

Dry Climates (B)

Most of northern Africa, interior Australia, and inner Asia have climates marked by droughts and low annual precipitation. The heart of these regions (found also in southern and west coastal Africa and South America and in southwestern North America) is true *desert* climate (identified as BW on Fig. 2.12). The rainfall is very low (see Fig. 2.11: the central Sahara gets less than 4 inches annually), and daily temperature extremes can be enormous. In the daytime the sun's rays fall unobstructed on the surface, and afternoon temperatures, even under a canvas shade, may exceed 120°F. Then, after sunset, the loss of heat from the ground is rapid and by early morning the temperature may be in the 40s. Daily temperature ranges of 60 degrees are not at all uncommon in low-latitude desert areas. Monthly ranges are less and reflect the latitudinal position of the desert regions.

The desert margins, transition areas to adjacent, moister regimes, are themselves somewhat better endowed with precipitation. Note, for example, that in Africa there is a narrow belt of BS climate between the savanna and the desert proper. This is the *steppe,* where the meager rainfall is nevertheless enough to sustain some hardy, if sparse, vegetation. Our precipitation map suggests that we are dealing here with areas having a rainfall between 10 and 20 inches.

Open savanna country in Tanzania, East Africa. The xerophytic (moisture-preserving) vegetation in the foreground reflects the comparative dryness of this area, as do the rather sparse, short grasses. In the distance, where the country is somewhat lower, slightly more moisture supports a somewhat denser growth of trees and bush. The soil is stony, and there are numerous boulders; this is at best grazing country. Note also the slight cumulus development. On this day these clouds will not produce rain.

Mid-Latitude Climates (C)

These are the climates identified as *warm temperate* in the legend of Fig. 2.12, and a look at their distribution will suggest why. This is the prevailing climate in the southeastern United States, from Kentucky to Florida, in California, in Western Europe and the Mediterranean, in southern Brazil and northern Argentina, in coastal South Africa, in parts of coastal Australia, in New Zealand, and in southern China and Japan. We would associate none of these areas with climatic severity and extremes, although winter months can be fairly cold. We are about in the middle here between the winterless equatorial climates and the summerless polar climes. As our map shows, almost all the areas of C climate lie just beyond the Tropics of Cancer and Capricorn. They never experience the vertical ray of the sun, but in summer, they come close.

Different types of mid-latitude climate are recognized on several bases. Some have cooler winters than others. One really significant variation in the temperate regime is the so-called *Mediterranean* climate, marked as Cs on the map and found in five major areas on the globe. Here is an anomaly. Although rainfall is usually associated with the high-sun season, the *s* means that here, the summer happens to be the dry season. We get a clue as to why this might be so from the westerly position of the five Cs areas in Spain and the Mediterranean, California, Chile, South Africa's Cape Province, and Australia. We may be seeing here the effect of the northward and southward shift of the belt of westerly winds, which bring cyclonic and rainy weather to mid-latitude areas. With the arrival of winter, this belt moves somewhat toward the equator, bringing frontal precipitation to the Mediterranean-type areas, but when summer returns, the westerlies shift poleward, leaving the Mediterranean areas dominated by the high-pressure areas associated with the subtropics.

Elsewhere, the mid-latitude climates are characterized by equable conditions, with temperatures in the warmest month averaging typically in the high 60s or lower 70s and in the coolest month in the 30s or 40s. Again, there are variations: Florida's winter is warmer; central China's is colder. In terms of precipitation, Fig. 2.11 shows that the warm temperate climes generate 20 to 40 inches, in places more through orographic effects (notably in Chile and the United States-Canadian Pacific margins). The Mediterranean regime, with its low-sun rainfall, also has annual total rainfall averaging less than normal for mid-latitude climates.

Snow Climates (D)

Sometimes the D climates are identified as *continental* climates, for they seem to develop where there is a lot of landmass, as in Eurasia and northern North America. Note that no D climate exists in the Southern Hemisphere at all! The snow climates' regime is marked by just that, cold, snowy winters and comparatively short, cool summers. The annual temperature range is often very large. In the Dfa type, for example, the warmest month (July) may average 73°F, but the coldest month (January) only 11°F. Most D climates have cooler summers than that, though, and even colder winter months.

The precipitation in the snow climates is relatively low (Fig. 2.11), especially in northwestern Canada and in Siberia. On the other hand, there is much less loss through evapotranspiration, as there is enough moisture to sustain dense forests (Fig. 1.13). A closer look at Fig. 2.11 shows that there is a considerable range in the amount of precipitation occurring in the continental climatic areas. Northeast Canada, the western U.S.S.R., and smaller sections of inner Eurasia receive over 30 inches annually. Elsewhere, fewer than 12 inches are recorded, and some stations report fewer than 8.

Ice Climates (E)

Northern Canada's islands, the north coast of Alaska, Greenland, and northern marginal Siberia are dominated by a climatic regime marked by the absence of a summer and by cold temperatures. Only four months of the year record mean temperatures above freezing, and in many locales only two or three months reach these, and even then the mean is always below 50°F. Precipitation is extremely low, comparable to desert conditions. Only tundra-type vegetation can sustain itself here, and the "milder" variety of ice climate is often called the *tundra* climate; where no vegetation and permanent ice prevail, the climate is aptly called *ice cap.*

One other climatic area appears on Fig. 2.12. It is marked by an H, for high elevation. In some parts of the world, for example, in the Himalayas, Rocky Mountains, Andes, Ethiopian Highlands and certain other places, elevations are so high that it is inappropriate to classify the prevailing climatic conditions by general standards. Here the overriding element is the coldness of the air. Large parts of the H zone lie under permanent snow. But even where there is no permanent snow, the day air is cool and night air often bitterly cold. At 10,000 to 15,000 feet and even higher, the air is also thin; precipitation is not great. With their rugged terrain and thin soils, these areas are not among the world's most productive or most densely populated.

MAN AND THE ATMOSPHERE

In the preceding pages we have outlined, in the most general of terms, the life support system with which man found himself endowed on this planet. It is an incredibly complex system, with the finest balance among innumerable elements. Three and a half billion people consume its oxygen continuously, as do factories, furnaces, fires; yet there has been no measured change in the atmosphere's oxygen content. Cities, farms, and manufacturing plants use immense quantities of water, but the system keeps replenishing the supply. Automobiles, smokestacks, coal burners pour tons of dirt into the atmosphere, but even over the dirtiest city a brisk breeze still manages to produce a beautiful, blue sky. It appears as though nothing can permanently injure the troposphere.

So it seems. But it may be that while we are busy seeking solutions to the energy crisis foreseen by some experts, while we drill bore holes into the continental shelf, while we search for new alloys, we are on the verge of damaging the most vital part of our legacy of abundance. We can survive without food for several weeks, without water for several days. But we can survive without air for only a matter of minutes. This earth may still conceal many resources, but the thin ocean of air at the bottom of which we live is all the air there is, and all there is ever likely to be.

Pollution of the Atmosphere

By our growing numbers and our drive for progress, we are making giant sewers out of the air and the oceans. Somehow the atmosphere seems to be able always to cleanse itself, even when we blacken the skies with soot. But in the end, that may prove to be an illusion. True, the air disperses even the densest smoke and the most acrid chemical fumes. Some of the waste pouring into the atmosphere, however, is producing irreversible changes in the troposphere, changes that may ultimately affect the radiation balance and thus the whole life support system on this earth.

TYPES OF POLLUTION Man is not the only source of air pollution. Volcanic eruptions, forest and grass fires, and dust storms also dirty the atmosphere. This natural pollution, however, does not cause permanent damage to the troposphere.

We experience pollution of the air and its consequences in our personal lives (for example, when someone smokes in our presence), at work (in an office or a factory), and in our community in general. In our North American society, automobiles in the early 1970s were contributing about 60 percent of all the pollutants sent into the atmosphere, and factories about 20 percent. The automobiles contribute a great deal of carbon monoxide as well as hydrocarbons and nitrogen dioxide; these last two are the main elements in a now-familiar urban phenomenon, smog. Industries such as steel mills, refineries, and chemical plants produce another gas, carbon dioxide, among many pollutants. As we will note later, carbon dioxide in the atmosphere plays a major role in increasing the greenhouse effect in heavily industrialized urban areas. Thus the outpouring of carbon dioxide may have an impact on global climate also.

As we noted in the first chapter, much of the electricity produced today is still derived from burning coal and oil; the day of the nuclear plant is still in the future. While this is so, power generators spew huge quantities of sulfur dioxide into the atmosphere, nearly 2 million tons annually in the New York area, for one example. Sulfur dioxide has played a prominent role in several pollution disasters of the past decades, and undoubtedly will do so again.

There are other types and sources of air pollution. Who has not seen a municipal refuse dump sending a column of smoke into the air? Or has not seen city buses emitting ugly, black exhaust fumes? Even the winter fire in the fireplace

That is not a cloud layer hovering over the city of Chicago, but man-made smog so intense that breathing it causes you to inhale as many pollutants as you would if you smoked four packs of cigarettes a day. Until the atmospheric pressure changes and the smog is wafted away, the city suffers this discomfort—and it grows worse daily as automobiles, factories, and other sources continue to pour pollutants into the air. Smog alerts are now a part of city life, along with power blackouts, water shortages, and other discomforts.

or the summer barbecue—multiply it by millions and the cumulative pollution figures are staggering.

METEOROLOGY AND POLLUTION When we discussed the processes that cause air to rise (convectional, orographic, and frontal), we thereby identified the ways in which pollution in the air is dispersed. When the air is forced to rise, the pollutants that have accumulated near the bottom of the troposphere are scattered through the whole atmosphere. No matter how dirty the air over a city, the passage of a cold front can clear the skies.

There are times, however, when this clean-up mechanism takes a long time to arrive, and atmospheric conditions actually favor a buildup of pollutants. It is not difficult to recognize such conditions. We discussed the nature of high-pressure cells, in which the air has a tendency to subside, to press down on the surface. When this occurs, there is obviously no rising of air and no quick removal of pollution. We have to wait until the anticyclone has been pushed eastward and a new system, we hope a frontal cyclone, has taken its place. Summer high-pressure systems, then, can be a mixed blessing. They bring rainless, sunny days and clear nights, but those days and nights may turn smog-ridden and hazy.

There is a meteorological condition that is especially hazardous when it develops and lingers over a big, industrial city. This is called an *inversion.* The term is not difficult to interpret. We know that under normal circumstances temperature in still air decreases with altitude about 3.5°F for each 1000 feet. When the temperature fails to decrease this way but actually *increases* with height, the situation is upside down, and there is an inversion. How can such a condition develop? There are several ways. First, we have seen that under clear skies the surface loses heat rapidly at night through ground radiation. The layer of air adjacent to the ground is the first to be cooled, too, through contact with the cold surface and through its own loss of radiation upward. Thus a cool layer of air perhaps several hundred feet thick lies on the ground, but the air farther up has not yet been cooled. It has not lost as much radiation outward. Hence the lowest layer may be cooler than the layer above it, which means that the temperature increases with height. The result often is that a dense fog develops, a dirty fog full of pollutants. Most commonly this eye-searing fog is broken up by the sun at mid-morning, but there have been cases where the sun failed to penetrate the fog, could not heat the lowest air layer, and could not insolate the ground. The inversion intensified, and more pollutants continued to be poured into it. In London in 1952 hundreds of people died as a direct consequence of a situation such as this.

What we just described is a *ground inversion,* but it is also possible for an inversion to develop at higher altitudes. Imagine a situation, for instance, where the lapse rate is not the usual, steady 3.5°F per 1000 feet, but as follows: surface 80°F; 1000 feet, 73°F; 2000 feet, 68°F; 3000 feet, 65°F, 4000 feet, 64°F. Such a deviation from the normal, the *environmental lapse rate,* can occur in a given location through the mixing of air at upper levels or through the transmission of ground radiation under certain circumstances. Let us now follow the fate of an air parcel that begins to rise and cool according to the dry adiabatic lapse rate. At the surface it is 80°F. By the time it is at 1000 feet, its temperature is $80 - 5.5 = 74.5$°F. Since the stationary air there is 73°F, the rising air is warmer and continues to rise. At 2000 feet, the quiet air is at 68°F, and the rising air still is warmer, $74.5 - 5.5 = 69$°F. But at 3000 feet, the stationary air is 65°F, while the rising air is $69 - 5.5 = 63.5$°F. Now the rising air has become cooler; if it reached 4000 feet, it would be fully 6°F colder than the still air. It therefore begins to subside, flowing back toward the surface. This means that there is a "roof" on the convectional process, and pollutants will not be carried beyond 3000 feet above the surface. Again, with smokestacks and automobiles pouring their dirt into this thin layer, dangerous pollution levels will develop. This, incidentally, is the kind of condition that often evolves while an anticyclone, a high, prevails over an area. The wait for a clearing front will be an anxious one.

DISTRIBUTION OF POLLUTION Specialists who study and advise on our pollution problems recognize (1) point or single sources, such as a single factory or refinery, (2) area or multiple sources, such as clusters of industries, suburbs, and downtown buildings, and (3) linear sources, such as four-lane, heavily traveled highways. The degree to which a single source can pollute a wide area is soon proved through the study of smoke plumes emanating from factory stacks. The cumulative effect of such plumes, apart from their impact on the air itself, may be a lowering of the value of houses on the prevailing downwind side and the general deterioration of neighborhoods as a result. Area sources, for example, a suburb where thousands of homes have wood-burning fireplaces, have the disadvantage that the pollutants are put in the air very near the ground, the worst place for this to happen because it is breathing level. As for line sources, it was reported in March, 1972, that automobiles on the entry ramps of Los Angeles freeways could not achieve full power to enter the flow of traffic because there was not enough oxygen to render the engines fully effective. What of the drivers with open windows?

Clearly it is the urban areas of the "developed" world that face the greatest pollution today. In a book entitled *Atmosphere, Weather, and Climate* (1970), authors R. G. Barry and R. J. Chorley report that the concentration of pollutant nuclei in the British countryside averages about 9500 per cubic centimeter, but in London it is normally 150,000 per cubic centimeter; in some cities it can reach 4 million. (A cubic centimeter is a little more than one-sixth of a cubic inch!) Moreover, it is in the temperate climatic regions that atmospheric conditions favor persistent pollution: here develop the high-pressure cells with their subsidence, and here the sun can be too weak to break through inversions. Thus cities like Tokyo, Johannesburg, Los Angeles, and London are frequently afflicted by combinations of high-pollutant emission and low atmospheric-removal capacity.

But pollution is not the monopoly of Western, industrial cities. It would be difficult to find a more pungent example of the point source than the house of an African family where a wood fire burns right in the middle of the living place and the smoke collects under the roof, seeping into the air but bringing streams of tears to uninitiated eyes before doing so. The initiated eyes? Glaucoma is one

Factory smokestacks belching pollutants into the air in Kawasaki, part of the Tokyo-Yokohama industrial region, Japan. Note the "pollution cone" extending from the factories. Prevailing winds play their part in maintaining a pollution cone such as this. The land values of the residential areas under pollution cones almost daily may be lowered compared with nearby urban areas that are spared because the smoke plumes miss them most of the time.

of Africa's serious diseases. In 1960, authors writing about political tensions in a part of what is today Tanzania titled their book with what they saw as most pervasive about the scene, *Smoke in the Hills.*

This leads us to the question geographers like to ask: what is the regional distribution of air pollution today, and can the future spatial pattern of incidence be predicted? The first part of the question has been partially answered. Air pollution is not confined to the cities identified previously, but also affects whole areas of urban-industrial concentration including the Ruhr (Germany), Rotterdam and the lower Maas (Netherlands), the Eastern Seaboard (United States), the Witwatersrand (South Africa), and so forth. The second part of the question is more difficult because it involves not only a calculation of the tonnage of pollutants poured into the atmosphere but also a prediction of the number and position of stagnant anticyclones. Some attempts to forecast pollution incidence have been made, and understandably they invariably point to Appalachia and southern California as the most pollution-prone areas of the future. In the East it is Appalachia rather than, say, the New York area, because a stationary anticyclone centered over West Virginia-Kentucky can blanket an entire region from Alabama to Massachusetts with pollutants, assimilating dirt from Birmingham to Boston.

If pollution was a local problem in the days when an English king forbade the burning of coal in London on penalty of death, it is now a global phenomenon. The forerunner of worldwide pollution crises appears to have occurred in July, 1970, when cities as far apart as Rome, Tokyo, Sydney, and New York reported that the air was bad and that people with breathing problems were suffering severely. In June, 1972, the skies of southern Florida, not one of the industrial heartlands of the world, were colored by a yellow haze that blocked out much of the sun's summer rays. Meteorologists found that a high-pressure cell with a rather easterly location was wafting to Florida the pollutants generated by industries of the northeast, and it was on a path that would take it all the way across the western Atlantic. We now produce enough pollution to span the world.

EFFECTS OF AIR POLLUTION There are various ways of appraising the impact of air pollution on man and his environment. We will dispense here with its effects on our health; it is hardly necessary to emphasize that air pollution is bad for people. Carbon monoxide is a lethal gas. Sulfur dioxide in excess makes people ill. Tobacco smoke has been linked to lung cancer. Our present interest, however, is in the way air pollution affects the climate, our life support system, locally and globally.

It is noteworthy that some textbooks of climatology have recently begun to include chapters on "urban climate." Indeed, by their considerable production of heat, by their emission of smoke, gases, and dust into the atmosphere, and by their relief and its effect on air movement, urban areas modify their own weather and climate. It has been estimated that some European cities lose between one-third and one-half of their incoming winter radiation because the sun's rays cannot penetrate the smog dome over them. Some of the loss, however, is made up by the smog particles' interception of outbound ground radiation at night, which makes night temperatures usually somewhat higher than in the far suburbs and beyond. In Chicago, for example, the night temperature in winter at outlying O'Hare International Airport is sometimes as much as 10°F below that of the Loop. Overall, the annual mean temperature in a large city can be expected to be between 1 and 2°F higher than in the surrounding countryside. In terms of fog, this has become a prominent urban climatic phenomenon. With so many dust particles in the air around which water droplets can form, cities experience from 60 to 70 percent more foggy days than the countryside, where the air is less polluted. And it has been proved that cities, contributing as they do to the uplift of air both by their physical mass (tall buildings) and by the heat they generate, produce excess rainfall as well. Estimates range from 5 to 12 percent; not surprisingly, cities also have more cloud cover than the nearby countryside.

All these are local urban climatic changes brought on by the cities and recorded in the cities themselves. But is it possible that the cumulative effects of the air pollution of many cities might be changing the climate of the earth as a

---

AIR POLLUTION: CASES

MEUSE VALLEY, BELGIUM, 1930 The valley of the Meuse River in December, 1930, became a death trap for dozens of people when an inversion confined a sulfur dioxide fog produced by steel mills, factories, and chemical plants. Hundreds more fell seriously ill.

DONORA, PENNSYLVANIA, 1948 Donora is in the valley of the Monongahela River. Its air is frequently polluted by numerous factories and plants. In October, 1948, an inversion interrupted the valley's natural ventilation system and pollutants, chiefly sulfur dioxide, made the air acrid and almost unbreathable. Half of Donora's 12,000 people fell ill, and many probably lost years of their lives. Some were even less fortunate: twenty people died from the smog that began on a Tuesday and held on until Sunday night.

LONDON, ENGLAND, 1952 Cool, stagnant air under high pressure, a temperature inversion at low altitude, hundreds of thousands of coal fires in private homes, and industrial emissions combined to produce a smog consisting of smoke, sulfur dioxide, and other pollutants that lowered visibility to a few feet. The fog remained for four days, and death rates from respiratory diseases rose dramatically. At least 4000 people died as a direct result, and an estimated 15,000 died from complications during the next three months.

ELSEWHERE Death-dealing pollution episodes have been reported by New York (1953, 1963, 1966), London (1956, 1957, 1962), Tokyo (1970, 1971), and other places.

---

whole? Although there is no agreement among scientists, there are those who argue that the answer is yes and that far-reaching changes will come sooner than most of us expect. One problem is that world temperatures fluctuate in a cyclic manner in any case, and it is difficult to determine whether recorded changes are owing to this natural oscillation or are pollution-induced. We know, for example, that the mean world temperature from 1880 to 1940 rose by about 0.7°F and that after 1940 it dropped about 0.4°F. This decline after 1940 is especially interesting since sunspot activity, meaning an increase in the intensity of solar radiation, has continued to grow. Is the pollution in the air responsible for this apparent anomaly?

These few facts underscore the problems facing scientists trying to predict the effect our pollution is having on the world climate. The obvious conclusion is that we are indeed causing a cooling trend, in part through the increased outward reflection of incoming solar radiation by all those pollutants floating in the atmosphere. A decrease in temperature of only ½°F or so in 30 years does not seem much until you realize that a full-scale ice age could be brought on by a drop of only 4 to 5°F. Some scientists warn that we may be bringing back the ice of the Pleistocene in just a few hundred years, perhaps less.

But there are observers who interpret the facts differently. True, they say, the temperature is presently dropping just under ½°F per 30 years (since 1940), but it would have been more had it not been for all that atmospheric pollution. These scientists worry that we are upsetting the balance inherent in the cyclic changes of temperature that have been recorded before, and that when a warming trend begins again, it will warm us up so much that the ice caps will melt and the oceans will drown large areas of all the continents. Their arguments are based on calculations involving chiefly the carbon dioxide we are pouring into the air. Carbon dioxide, a by-product of combustion processes, has little effect on incoming solar radiation, because it is nearly transparent to those rays. But it does absorb ground radiation, and radiates energy back to the surface. Calculations have shown that increased carbon dioxide content in the atmosphere results in a warming of the lower troposphere. The amount of warming depends on the quantity of water vapor also present in the air, but it is estimated that a 10 percent increase in carbon dioxide concentration results in a warming of 0.36°F. At present, we are causing a carbon dioxide increase in the atmosphere of about 2 percent per decade, but the rate of increase is speeding up. Again, it does not take much extrapolation to arrive at the conclusion that we may be warming up the atmosphere enough to melt the polar ice caps.

Are these estimates matters of distant concern, fit for the geologic time scale but not for the calendar? Many experts do not think that it will be long before we come to face the consequences of our unconcern. Some have even suggested that the inundation of low-lying continental areas will start in 80 years or so and

FIG. 2.13 Potential world coastal inundation. [Modified from R. J. Kopec, "Global Climatic Change and the Impact of a Maximum Sea Level on Coastal Settlements," *Journal of Geography, 70,* 9 (December, 1971), pp. 546–547]

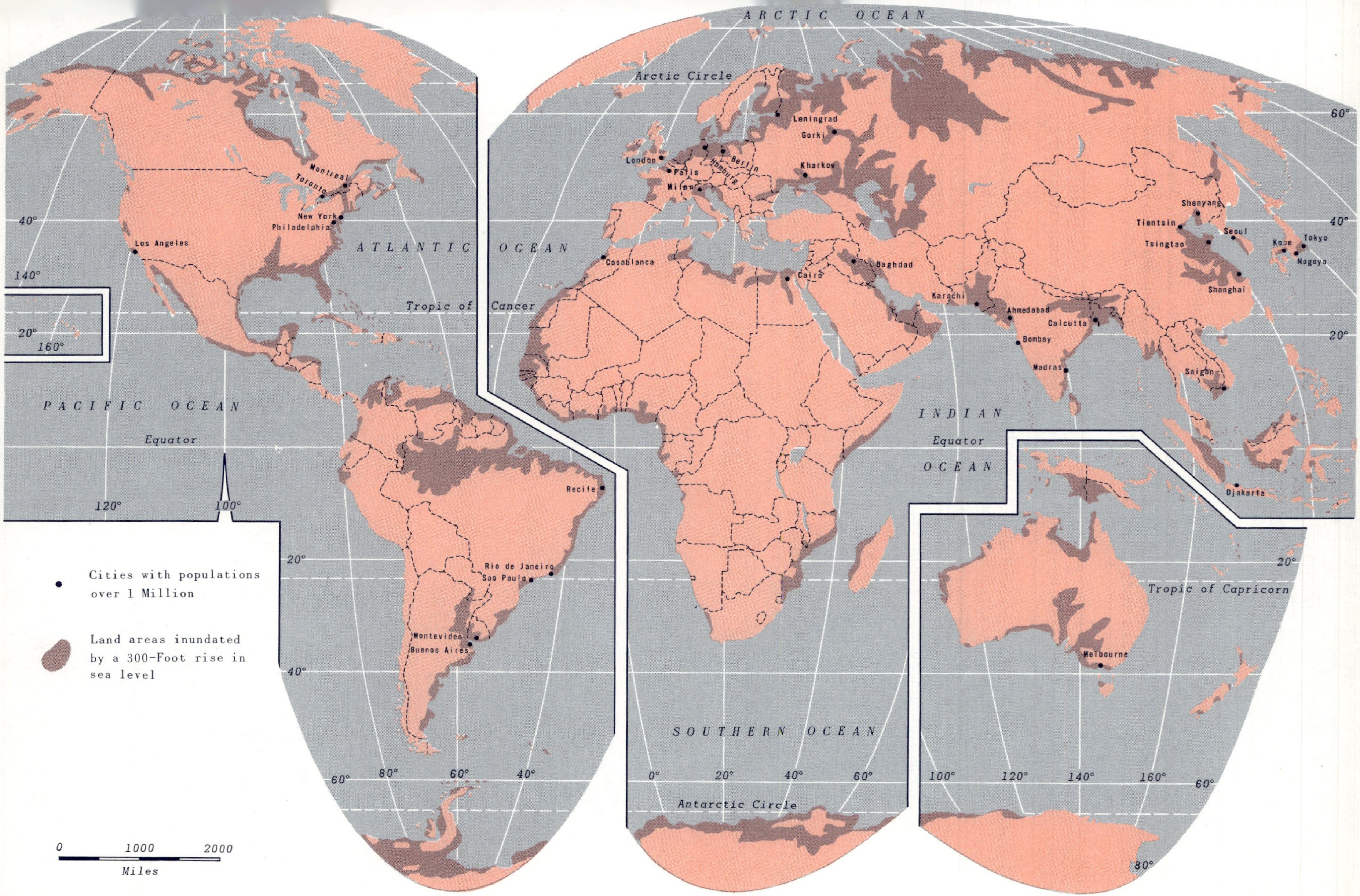
ARCTIC OCEAN
Arctic Circle
Leningrad
Gorki
London
Berlin
Hamburg
Paris
Milan
Kharkov
Montreal
Toronto
New York
Philadelphia
Los Angeles
ATLANTIC OCEAN
Casablanca
Cairo
Baghdad
Karachi
Ahmedabad
Calcutta
Bombay
Madras
Shenyang
Tientsin
Tsingtao
Seoul
Kobe
Tokyo
Nagoya
Shanghai
Saigon
Djakarta
Tropic of Cancer
PACIFIC OCEAN
Equator
INDIAN OCEAN
Recife
Rio de Janeiro
Sao Paulo
Montevideo
Buenos Aires
Melbourne
Tropic of Capricorn
SOUTHERN OCEAN
Antarctic Circle
Cities with populations over 1 Million
Land areas inundated by a 300-Foot rise in sea level
0 1000 2000
Miles

that we had better get familiar with a new map of the outlines of the continents, a map showing 300 feet of land submerged by risen waters (Fig. 2.13). Many of the world's great cities, including New York, London, Paris, Bombay, Calcutta, and Buenos Aires would be marked only by the tops of tall buildings sticking up above the waves.

All this may or may not come to pass, but one point is clear: there are great potential dangers in the carelessness with which man has been treating the atmosphere. We do not know what effect massive pollution of the air may have. Cooling or heating of the atmosphere are only two possibilities. Already the effect on individual persons of smoke inhalation has been proved destructive. We have records of thousands of excess deaths in cities struck by pollution disasters. There was a time when smoking was viewed as safe and relaxing. There was a time when it was unthinkable that urban populations could be trapped in a killing smog emanating from their own cars and industries. And now some other unthinkable prospects are coming into view.

Of all the natural wealth with which man was endowed on this planet, the atmosphere and the life support system of which it is a part has probably been treated with less caution than any resource. While we have been concerned over the future of energy sources, of alloys, of metals, we have allowed our vast reservoir of fresh air to deteriorate. To millions of people on this globe who enjoy the benefits of material abundance, there is one thing scarce: fresh, clean air.

CHAPTER THREE

Infinite Humanity: The End of Evolution?

THREE

Late in 1972, the population of the world was 3.8 billion. Sometime near New Year's Day, 1975, it will pass the 4 billion mark. And if it continues to grow at the present rate, about 2 percent per year, the earth will have to support twice as many people, 8 billion, in the first decade of the twenty-first century.

Right now about 130 million babies are born every year, and about 50 million people die. This means that we are adding 80 million inhabitants every 12 months, over 200,000 a day. By the time you have finished reading this chapter, there will be 10,000 more people on this earth than there are at the moment you are reading this sentence.

Can this go on indefinitely? Obviously not. Will current rates of population growth level off automatically? In some parts of the world, yes, but there are few signs of it in those countries where the numbers are greatest and the annual increment is most substantial. In general, the richer nations of the world have growth rates under the 2 percent average, but the poorer ones grow at even faster rates. And there are several times as many poor people in this world as rich. Are we, then, headed for a population crisis? Ask the families of those in India who died of hunger last year or the survivors of the half million who perished in the 1971 Bangladesh typhoon. For them and for hundreds of millions of others, the population crisis is already here.

EXPONENTIAL GROWTH

When we think of growth, we tend to view it as a *linear* process, that is, a process whereby something increases by a uniform amount during a series of equal time periods. If you have $100 and add $10 to it the first year and each successive year, your $100 will become $200 after 10 years, $300 after 20 years, and so on. This is linear growth. But if you manage to invest your initial $100 in such a way that it returns 10 percent interest (also $10 the first year), it will be worth $259 after 10 years and a staggering $673 after 20 years. *That* is what exponential growth does.

Another way of looking at exponential growth is by considering the *doubling time* involved. For example, our $100 took just over 7 years to become $200, then 7-plus more years to become $400, and after another 7 years it amounted to $800. When the growth rate is 10 percent, therefore, the doubling time is about 7 years. The doubling time of our world population growth rate, 2 percent, is 35 years. Every 35 years the population doubles. If in 1975 it is 4 billion, there will be 8 billion people in 2010, and 16 billion in 2045, and 32 billion by 2080, a mind-boggling prospect indeed. Something will have to change.

Population Explosion

Looking in the other direction, to the past, we can see why the term *population explosion* has come into use in recent decades. It is estimated that the entire world's population at the time of the birth of Christ was about 250 million and that it took nearly 17 centuries for that population to double to the 500 million ($\frac{1}{2}$ billion) it is thought to have been about 1650. But it took only 170 years for that 500 million to grow to 1 billion, which was reached about 1820. Just over a century later, in 1930, the world population was 2 billion, and the doubling time was down to 110 years. Now we have a doubling time of only 45 years, for it is certain that the 4 billion mark will be reached in 1975, and our 2 percent growth rate indicates that the next doubling time will be 35 years, still less. Those estimates in the previous paragraph, then, may have been optimistic. Unless the growth rate declines, the doubling time will fall below 35 years. The history of man has not only been one of increasing numbers but also one of increasing *rates* of increase.

It is difficult these days to pick up a newspaper that does not contain a story relating to population pressure, family planning, contraception, abortion, and similar topics. Although there has long been concern over the population question, it has never had the exposure and deliberation it gets today. Governments in many parts of the world are encouraging family planning and subsidizing birth control clinics. Is this so because we are at a stage in the curve of exponential population growth where the next prospect—4 billion *more* only 40 years from now—is utterly unthinkable? Or is it because we are becoming conditioned to crisis reporting as the only way we will pay attention to a serious issue?

Malthus

The world at large is finally ready to take the population question seriously enough to search for ways to take collective, massive action, but there have long been prophets of doom. As long ago as 1798, a British economist named Thomas R. Malthus published a long paper entitled *An Essay on the Principle of Population as It Affects the Future Improvement of Society* in which he sounded the alarm: population was increasing faster than the means of subsistence. Malthus recognized the nature of exponential growth and pointed out that population increases at what he called a geometric rate. The means of subsistence, he argued, grow only at an arithmetic (linear) rate. Malthus issued revised editions of his *Essays* from 1803 to 1826, and he responded vigorously to a barrage of criticism (for he condemned the social order in England). He suggested that the population growth in Britain might be checked by hunger within 50 years from the first appearance of his warning.

Malthus could not have foreseen the multiple impacts of colonization and migration, and undoubtedly he would not have believed the United Kingdom could sustain between 50 and 60 million people as it does today. Nor was he correct about the linear increase of food production. It, too, has grown exponentially as acreages under crops have expanded, improved strains of seed have been cultivated, and fertilizers have been applied. But the rate of increase of food production has not kept up with the rate of population growth, and although Malthus was wrong in terms of timing and detail, the essence of his argument appears still to have merit.

But Malthus had his critics, too. Karl H. Marx, the nineteenth-century socialist, suggested that the poverty of the masses foreseen by Malthus had less to do with their numbers than with the capitalist system governing them. Far from proposing a reduction in the rate of population growth, Marx wanted to alter the control and use of natural resources by taking them from the hands of the few and distributing them among the many. Others have criticized the Malthusian doctrine on the grounds that the earth has proved capable of ever-expanding food production. There are still untapped resources both on land (the Amazon and Congo-Zaïre basins could be made productive, for example) and in the sea (the oceans contain a storehouse of food). Only technological progress needs to be made to record these advances. There is the old refrain: if we can place men on the moon,

---

MALTHUS AND DARWIN

Thomas Robert Malthus was born in 1766 on the Surrey (England) estate owned by his father. He went to private schools and entered Cambridge University's Jesus College. Theology was his initial aim, and his first job was as clergyman at the parish of Albury in Surrey. But even while his career was going in that direction, he was at work on his famous *Essay*, whose first edition appeared just a year after he took the Albury position. Writing and lecturing began to take up most of Malthus's time, and he left the clergy. In 1805 he took a job more appropriate to his interests, professor of modern history and political economy at the Haileybury College operated by the East Indian Company.

Karl Marx was not the only author to respond to Malthus's writings. Charles Darwin read the *Essay* and related Malthus's concept of the struggle for existence to the principle of the survival of the fittest—the process of natural selection was now understood. Darwin acknowledged that Malthus's work had a role in the development of his theory of evolution.

Malthus's contributions to economic theory are often overlooked, but they are substantial. Malthus was the first economist to outline the law of diminishing returns as it relates to farming, and he wrote about various aspects of theories of value. His writing was not quite as incisive in these areas as it was in his *Essay*, but it was highly original just the same.

Malthus died at Haileybury in 1834.

---

then we can cultivate stubborn tropical soils and fish the deepest depths. Incidentally, these pronouncements are usually made by people who have good breakfasts before they sit down at their typewriters and who have good dinners at night.

Those who feel the weight of Malthus's concerns (even if not every detail of his argumentation) are sometimes called the neo-Malthusians, and their number is growing. Their anguish can be read from the titles of their writings: Osborn's *Our Plundered Planet,* Vogt's *Road to Survival,* Borgstrom's *The Hungry Planet,* and the Club of Rome's *Limits to Growth,* among many others. Let us examine some of the bases for their fears and the remedies they propose.

WORLD POPULATION DISTRIBUTION

Over considerable areas of this earth, the population figures we have are informed estimates at best. But we do have a rather good, if general, picture of the distribution of people on this globe (Fig. 3.1). Our map shows that there are three major clusters of large population and several areas where the population is very sparse. The three large agglomerations all lie in Eurasia: in descending order of magnitude, they are in eastern Asia (centered on China), southern Asia (mainly India), and Europe. Smaller but still substantial population clusters lie in east-central North America and in Southeast Asia. The sparsely peopled areas include the Sahara-Arabian Desert, northern Canada and the Arctic, sections of interior Eurasia, much of Australia, and Antarctica.

Whenever geographers see an uneven distribution like this, they look for explanations. And with what we know about the physical world, it would seem easy to conclude that population has grown rapidly in the more fertile, better-watered, lower-elevation regions, and that desert drought, arctic coldness, and mountain barrenness have repelled human occupation. Compare Fig. 3.1 with Figs. 1.5, 1.11, and 2.12 (pages 12, 39, and 85); it appears that these correlations do indeed prevail. But when we look at the map somewhat closer, we observe that there are some interesting exceptions. Mountainous areas supposedly are sparsely occupied because of their rarified air, poor, thin soils, and cold climate, as well as their inaccessibility and difficult terrain. But note that Ethiopia (where the Blue Nile begins) has a significant population cluster where Fig. 1.5 tells us there are high, precipitous mountains. Note also that South America displays population agglomerations in the Andes with more people per square mile than in adjacent, lower areas. And follow the 60-degree parallel from empty northern Canada across the Atlantic into Eurasia, and we find that this same environment (according to the maps we use for comparative purposes) supports many more people there than in North America. Now we must begin to wonder: can a difference in environments *really* explain such huge contrasts in population numbers as between India and tropical South America?

Certainly the environment is part of it. But individual societies and cultures have their own capacities, their own ways of dealing with their natural environments. The Ethiopians, the Andean Indians, and the Chinese have developed

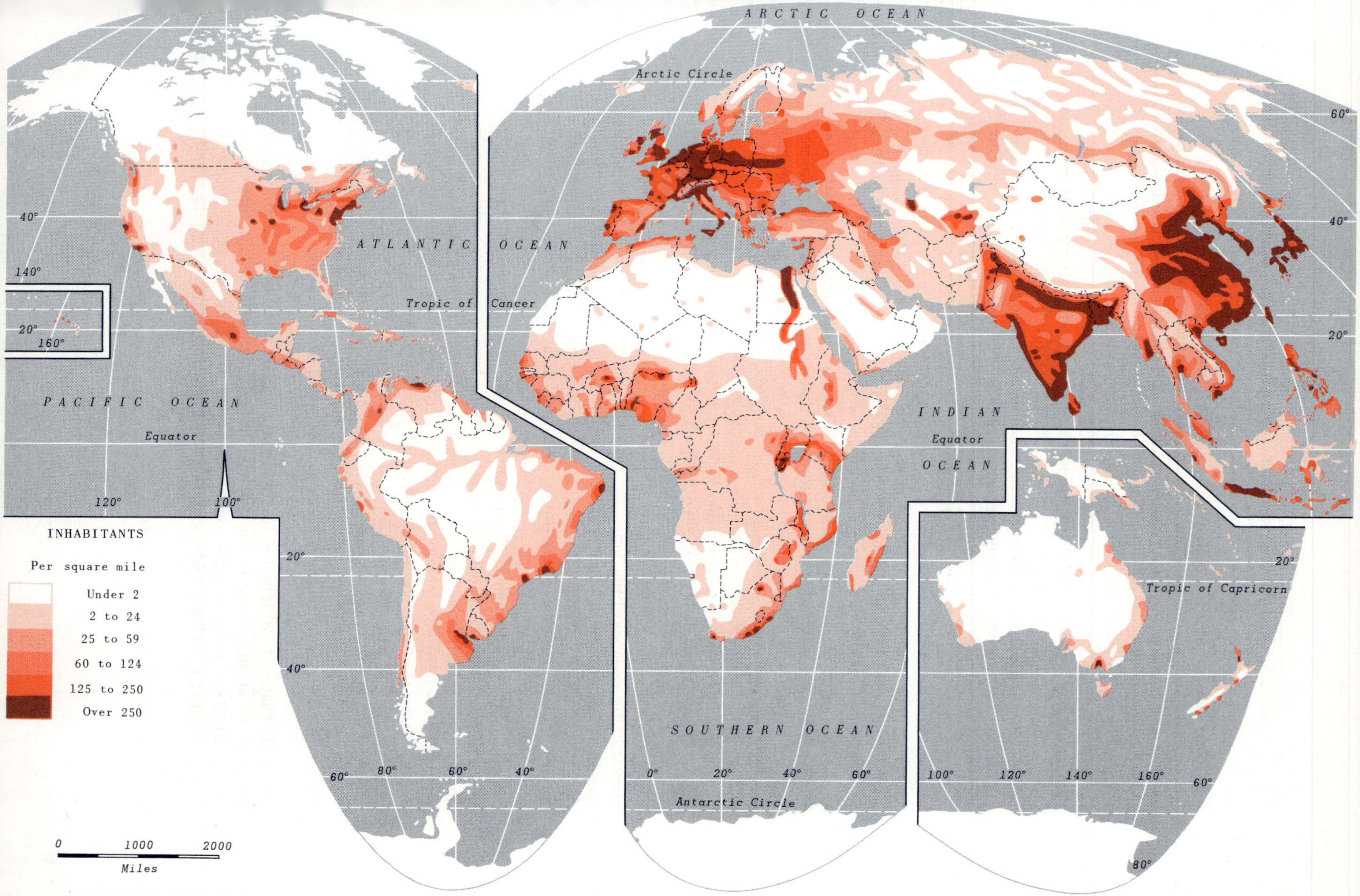

FIG. 3.1 World population distribution.

different levels and kinds of technology and different economic systems, and they have different demands and different capacities to meet them. Historical geography, too, has much to do with present population distributions. The Ethiopians fled into their mountains when they were threatened by enemies, and once there they successfully defended themselves and thrived. The Andean Indians are remnant clusters of once-powerful empires that would have stood out on a fifteenth-century population map of South America as strongly as the lower Nile valley does in Africa today. The Chinese spread from a northern core area, unhindered for many centuries. The Europeans had an ocean to cross. Before we generalize about world population distribution, therefore, we ought to view the nature of culture and the technological achievements of human societies. We shall do so in the next chapter.

Our North American vantage point tends sometimes to obscure the realities, so it is useful to place the major and minor population clusters we have identified in the context of world and continental totals. Before China was admitted to the United Nations, it was often said that the UN was hardly representative of mankind since it excluded a quarter of the world's population. And indeed, if the highest of many wide-ranging estimates of China's population are nearest the truth, China's 900 million people constitute just that. Its growth rate is also a matter for speculation, and it is variously reported to be between 1.2 and 1.8 percent annually. If the truth is somewhere in the middle, China is adding more than 1 million persons *a month* to its already huge total.

The eastern Asian population cluster includes, in addition to China, the densely peopled islands of Japan (106 million), populous Korea (50 million), and the core of North Viet Nam (22 million). The world map (Fig. 3.1) shows quite clearly the immense concentration of people in the coastal and river plains of China, especially the lowlands of the Yangtze and the Hwang. Northward a wedge of dense population extends into Manchuria; to the south the near-coastal zone from Shanghai to the Hanoi area constitutes a ribbon of population agglomeration. Both in terms of numbers and in contiguously occupied area, this is the world's largest cluster of population.

Although this eastern Asian region contains several of the world's largest cities, the vast majority of its inhabitants live on the land as farmers. Only in Japan and Taiwan are there more people in cities and towns than in rural areas, and those two countries have only about 120 million of the region's 1.1 billion residents. Again, figures for China are not yet available, but next-door North Viet Nam in 1966 was only 10 percent urbanized, and in South Korea in the same year fewer than one out of three people lived in towns. In China the figure is almost certainly similar. Thirty percent may be a high estimate, despite Shanghai (10 million), Peking (7 million), and perhaps 20 cities with more than 1 million residents.

What this means, of course, is that the rural areas of eastern Asia are very crowded. Shanghai, which was just a fishing village until the mid-nineteenth century, lies at one corner of the delta of the Yangtze River, a delta that covers about 20,000 square miles and accommodates some 50 million people. To the north lies the North China Plain, the lowland of the Hwang River, also one of the world's most overpopulated agricultural areas with over 1200 people per square mile of *cultivated* land. Wherever the terrain is not too steep and the land is irrigable, there are people. The influence of relief is well shown in southern China, where the Southeast Uplands appear as a crescent of comparatively sparse population.

Between China and southern Asia, Southeast Asia's population distribution is marked by comparatively small, high-density clusters separated by expanses of sparse population. As Fig. 3.1 shows, these clusters are positioned on the coasts in the river deltas of South Viet Nam (the Mekong), Thailand (Chao Praya), and Burma (Irawaddy). But in southern Asia the map displays a population agglomeration comparable in scale to eastern Asia's. Here the pivotal country is India, which has fewer people than China (nearly 600 million) but a higher rate of growth (2.6 percent annually). Adjacent to India lie three populous states: Bangladesh (70 million), Pakistan (60 million), and Sri Lanka, or Ceylon (14.5 million). At present rates of increase, southern Asia will have 1 billion inhabitants at the end of the 1970s.

In southern Asia as in eastern and Southeastern Asia, the population is concentrated in the lowlands of the major rivers, in this case the Ganges-Brahmaputra and Indus plains, and along the coast. Dense population prevails along practically the entire rim of the Indian peninsula. Elsewhere, too, the limits of the region are clearly marked on the map as the population density falls off dramatically. To the north, the Himalayan ranges rise above the Ganges lowland; just a few hundred miles from crowded northern India lies virtually empty country in the Asian interior. To the west, the desert takes over not far from the west bank of the Indus, and population densities drop from 125 to 250 per square mile in the valley to under 25 in the arid areas of western Pakistan, mountainous Afghanistan, and Iran. And to the east, mountainous areas separate Bangladesh from Burma, and there, too, there is less population than in the Ganges delta.

Southern Asia shares with eastern Asia its primarily rural character. There are large cities with several million inhabitants, including Calcutta, Bombay, Delhi, and Karachi, but fewer than one in five of the inhabitants of this region lives in cities. Most live in small villages closely spaced over the countryside. In Bangladesh, a country of 55,000 square miles (about the size of Iowa), there are 70 million people

India, more than perhaps any other country, symbolizes the crowding of our planet. Its cities are overfilled and teeming, and its rural areas are in places the most densely peopled in the world. The photograph shows a small segment of a crowd of several hundred thousand people on the banks of one of the country's major rivers during a religious festival.

and only one major city, which has well under a million residents! The farmlands are incredibly crowded; even if all the people of Bangladesh were absolutely evenly distributed over its area, there would be 1220 people for *every* square mile. Of course, even Bangladesh has some less heavily populated areas, but that means that there are other areas where the average is far exceeded.

Despite this enormous crush of people, there are still areas in southern and eastern Asia where the population is quite sparse and where it is possible to travel for miles without seeing people. A comparison between Figs. 3.1 and 2.12 suggests why this should be so in southern Asia. Interior India, especially in the west, is dry enough to be mapped as steppe country. It is, in addition, plateau territory, as Fig. 1.5 reminds us. At present it does not seem likely that these less heavily peopled areas will soon play a role in relieving population pressures in India. The Indian farmer cultivates as he has done for centuries, and nature's barriers remain as effective as ever.

The third large population agglomeration on the Eurasian continent differs in many ways from the two just discussed. From the British Isles into Soviet Russia and from Denmark to Italy, Europe is densely populated, more so, even, than India. In the Netherlands, which has just under 13,000 square miles, there are nearly 13 million inhabitants. That produces an average density per square mile (1000) more than twice that of India (490). But a look at European population and area figures soon proves that high densities are nothing unusual. In the United Kingdom it approaches 600; in West Germany, 650; in Italy, 450; in Belgium, 820. And counting "European" Russia, that is, the Soviet Union west of the Urals, the European population cluster has a dimension very nearly that of southern Asia with some 670 million residents.

Some of the differences between the Asian population clusters and that of Europe emerge when we study the spatial features they display. In the first place, when we compare the population and physiographic maps, it becomes clear that there is far less correspondence between plainlands and population concentrations than there is in Asia. True, on Fig. 3.1 we can discern the Alps and Pyrenees as rather sparsely peopled zones, but, on the other hand, there are some fairly rugged areas that are enormously crowded, for example, along the Polish-Czechoslovakian boundary. From this we may conclude that the terrain means less relative to population distribution than it does in Asia. A second feature of the European population region is its distinct internal clustering, evident even on a small-scale map such as Fig. 3.1. There is not the contiguity of China and India; in Europe there are several cells of extremely heavily concentrated population separated by rural areas that are distinctly less crowded. What we are seeing here, of course, is the effect of Europe's high degree of urbanization. In Asia the large cities, populous as they are, lie in rural areas that themselves have densities approaching the level of suburbanization. In Europe, on the other hand, a much larger number of people are concentrated in the cities and the farming areas are much less densely peopled. Note the contrast with Asia when we identify the percentage of people living in cities in the United Kingdom (80), West Germany (78), Belgium (69), France (70), and the Soviet Union (56).

It is no coincidence that the east-west axis of dense population that extends from the United Kingdom across the middle of Europe to beyond Kiev in the U.S.S.R. (Fig. 3.1) corresponds closely to the European coal belt shown on Fig. 1.16. Population distribution in the European region relates in large measure to the location of the power resources that fired the industrial revolution, from Britain's Midlands and Germany's Ruhr to the Donetz Basin north of the Black Sea. Large cities, the major centers of industry, hold the majority of the population in this zone. The rise of industrial Europe confirmed other urban places in their historic primacy, notably those that formed large markets (such as London and Paris) and those that benefited from the intensified circulation of raw materials and products. Thus Rotterdam emerged as the world's leading port at the mouth of one of Europe's great water-transport routes, the Ruhr-serving Rhine.

But it would be wrong to suggest that environment has nothing at all to do with Europe's population distribution. Long before the industrial revolution, Europe already was a populous region by world standards, and in preindustrial Europe most of the people were farmers. Evidence of those times is written in the modern map, not only by the density of the population on the fertile lowland soils of the Netherlands and Denmark, Germany and France, but also by the sparseness of population in the cold northlands of Scandinavia and Finland and the comparative emptiness of drought-ridden interior Spain. Ireland, northern Scotland, and much of eastern Europe all bear witness to the same reality. Over most of the continent, however, the industrial-technological revolution that transformed European culture also complicated man-land relationships to such an extent that the Asian generalizations simply do not apply here. We shall consider this point in greater depth in the next chapter.

The three population regions we have considered so far contain nearly 2.6 of the world's 3.8 billion people. The next ranking cluster lies in North America, and it incorporates the east-central United States and southeastern Canada. The contrast between urban-centered densities and rural openness is stronger even than in Europe. Figure 3.1 suggests that the North American population region is dominated by one major, two lesser, and several minor urban foci. The major focus is the urban complex along the Eastern Seaboard, from Boston to Washington, which includes New York, Philadelphia, and Baltimore. This is the region called *megalopolis* by urban geographers who note that it is but a matter of time before the whole area coalesces into an enormous megacity. The two lesser foci lie in the Midwest. One is focused on Chicago and includes Milwaukee, and the other includes Detroit and Cleveland. Among the other concentrations that emerge on the map are those centered on Pittsburgh, St. Louis, and Minneapolis-St. Paul.

In terms of areal extent, the North American population region, though not as large as eastern or southern Asia, or Europe, is still of substantial dimensions. But the total population is of the order of 170 million, about a quarter of the smallest of the Eurasian clusters. When we realize that about three-quarters of these people live in urban centers, we gain an impression of the contrast between the occupance of the farmlands of the Hwang and Ganges rivers and, say, the Mississippi. This does not mean that farming areas in North America are not intensively utilized; it is another manifestation of differing cultures and technologies. It is true, however, that comparatively more of North America is very sparsely populated than Eurasia. Contrast the high-latitude zone in European Russia with that of Canada!

In terms of population numbers, the Southeast Asian region (including Indonesia and the Philippines) is larger than that of North America, but it is so fragmented that it is hardly a cluster. Rather, it consists of a set of nuclei, among which the Indonesian island of Djawa (Java) with over 70 million people is the largest. Figure 3.1 shows clearly that even the land-based nuclei are almost insular in character, comprising the countries' capital cities and their farmland surroundings. If this suggests that Southeast Asia is less crowded than either China's habitable areas or India's, that impression is correct: the region has long been able to export rice to its neighbors.

All the population agglomerations we have so far discussed (with the single exception of Djawa) lie in the Northern Hemisphere. South America, Africa, and Australia have nothing comparable, and our map reflects that situation quite clearly. South America's 1972 population of 155 million equals that of Japan and Korea; Africa's 360 million, just 60 percent of India's alone, occupy a continent that constitutes one-fifth of all the world's land. And Australia, an epitome of emptiness, has 13 million inhabitants, many fewer than China's gain in a single year. In South America and Australia, the distribution of population is markedly peripheral, that is, the majority of the people live on or near the coasts of those continents. South America's largest cities, Sao Paulo, Rio de Janeiro, and Buenos Aires, lie on the eastern margins of the landmass. Santiago (Chile) and Lima (Peru) lie near the west coast. Caracas (Venezuela) is not far from the north coast, and only Bogota (Colombia) is positioned somewhat farther into the interior. These cities are the foci of South America's relatively small population; similarly, 85 percent of Australian people live in such cities as Sydney, Melbourne, Brisbane, Adelaide, and Perth, all of them coastal centers. Later, we shall consider the factors involved in these patterns.

Only Africa has anything comparable to the crowded riverine plains of Asia: the Nile valley and delta and its 35 million residents. The pattern is similar, but not the scale of things. The Nile's gift is but a miniature compared to the Ganges and the Yangtze. But etched into the dryness of the Sahara Desert are a ribbon and triangle of rich, watered soil clothed in a blanket of green—and being strained to support a mass of farmers who cultivate every foot of it. In the remainder of Africa, above-average population densities exist in West Africa (where Nigeria, with 60 million people, is Africa's most populous country) and in a belt in East Africa from Ethiopia to the far south. The region of Lake Victoria is one of the nuclei of population, as Fig. 3.1 indicates. However, all this barely begins to resemble the enormous Eurasian agglomerations of people.

In many parts of the world, life in the rural areas is not much more comfortable than it is in city slums. This family of thirteen (three are away, working in the fields) occupies the one-room dwelling shown, located in rural Mexico. The house is made of corrugated cardboard coated with tar; the roof is made of sheets of corrugated iron. There are no sanitary facilities, and a nearby stream serves for washing and drinking. Is it surprising that the city beckons?

CONCEPTS OF DENSITY

Repeatedly in our discussion of the world's population distribution, we have referred in passing to "sparse" and "dense" populations and have spoken of total populations in terms of the total areas of the countries in which they live. It is important to understand that all references to population density, verbal or cartographic, involve some generalization and, therefore, some distortion of the truth. Atlases and other statistical sources, for example, usually give figures for countries' total area, total population, and *population density,* the population averaged out per square mile. The United States, with 3,675,633 square miles of territory, presently has a population approaching 210 million and an average density of 57 per square mile. This is *arithmetic* density, and we arrived at that average by simple calculation. We took nothing into account, not the clustering of people in the east-central region and in big cities, not the emptiness of much of Alaska, not the sparseness of population in much of the west. Arithmetic density, then, is the coarsest of all such estimates. Without other data at hand, it really means very little. Take the case of Egypt, with its 35 million inhabitants. The statistics say that Egypt has a total area of 386,000 square miles, so the average density works out to 90. But what meaning does that average really have when we know that Egypt's millions are crowded onto the irrigable lands of the Nile River, leaving the rest of the country virtually empty? Arithmetic densities can mislead us considerably.

Of course, there are ways to refine arithmetic density figures. In the United States, we could map densities by state, which would begin to give some idea of the eastern concentration of people. But even such a map would distort reality. Better still would be a map by county, which would begin to show the urban concentrations of population. Thus the smaller the unit of measurement, the more useful the concept of arithmetic density can become. Unfortunately, such detailed information as we would require is available for comparatively few countries.

How do we obtain a more valid measure of population density than the national arithmetic average? One way is to relate the total population, not to *all* the territory of the country, but to the available *cultivable* land there is. This has disadvantages too, for some cultivable land is much more productive than other farming areas, and no account is taken of forests, natural grazing lands, and other productive locales. Even so, this measure has great advantages. Consider the case of Egypt again. Now the ratio is between the population and the Nile zone on which most of the people directly depend. This is the *physiological* density, and for Egypt the figure is 3458. This means that in Egypt, there are nearly 3500 people for every square mile of arable land. That is about twelve times what it is for the United States, a much more accurate measure than a comparison of the two countries' arithmetic densities would provide. Probably the highest physiological density in the world prevails in Japan, where the arithmetic density (715) appears high but is still less than that of England and Wales, Bangladesh, the Netherlands, and several other countries. But Japan's physiological density proves to be 4300 per square mile, far higher than Egypt and also much more than Europe's leading country, the Netherlands (3040).

Still another measure of population density is the *agricultural* density. This measure differs from the physiological density in that the calculations exclude the urban residents, so making possible a more nearly accurate estimate of the pressure of people on the rural areas. This measure is especially useful where farmers make up the great majority of a country's population. Also used is the concept of *settlement* density, in which the figure represents the number of square miles in a country for each city with 100,000 people or more. In the United Kingdom there is such a city per 2000 square miles, in the United States, per 27,000 square miles, in China, per 53,000 square miles. When the figures are available, it is also instructive to derive a country's *urban density* (that is, the number of persons per unit of urban area) and its *rural density* (the number of people per unit of land under cultivation). But such refinements require information beyond the data-gathering capacities of many countries.

More on Urban Density

The rise of the megacity is a pervasive phenomenon attending the explosive growth of world population during the twentieth century. It is a condition of industrialization: four of five Englishmen (and women!) and Germans live in urban centers, three

This is what those rural residents pictured in the previous photograph might find if they did move to the city. All over the world, people are migrating into urban areas that cannot accommodate them. Often the circumstances of life are even worse than those shown here, in a section of one of Peru's major cities. Note the drainage-sewage ditch in the right foreground.

of four Japanese, seven of ten Americans. Tokyo, New York City, London—their populations, in the 10-million range, are larger than those of many countries. Consider it: there are more people in New York than in Sweden. There are more people in Tokyo than in Portugal. There are more people in London than in Austria.

But huge cities have emerged in the less-industrialized world, too. Calcutta and Cairo, Karachi and Canton, each contain millions of residents, their numbers growing with each passing month. In several parts of the world, notably in eastern North America but also in western Europe, southern Africa, India, and Japan, there are areas where sprawling cities and their suburbs are coalescing and whole urban regions are forming.

In large cities, people are often crowded closely together, in apartment buildings, tenements, and in teeming slums. People arrive from outside the city to add to the natural increase of the already resident population; no housing expansion could keep up with the flow, and acres of shacks are thrown up overnight, without even the barest of facilities. From Lima to Lagos, from Mombasa to Madras, hundreds of thousands of persons live in substandard housing. And this situation is not confined to cities of the developing world. Inadequate housing exists in every major city of the United States, too.

Over the past decade, a number of social scientists, including psychologists, sociologists, and geographers, have begun to prove that there is more to high urban density than the physical limitations imposed by housing and facilities. When circumstances force people to live closely packed together in an urban environment, they are placed under an emotional stress many of them cannot bear. Noisy streets cause loss of sleep. Congested stores irritate. Traffic jams have tempers flaring. It is a vicious circle: as the city grows, so do the stress-producing conditions.

In the Middle Ages, a city could be a dangerous place to live because famine and disease epidemics took their heaviest tolls there. Today a new sort of epidemic may be threatening urban dwellers, a double-barreled one. The stresses of urban life can generate hypertension, ulcers, sexual failure, and glandular malfunctions, and they appear to be related to the incidence of suicide attempts and withdrawal symptoms. In aggregate these are related to a breakdown of the social order, a high incidence of aggressive behavior, and violence. The urban dweller who manages to adjust to the stresses of city life may be victimized by the behavior of others less able to absorb them.

It has proved difficult to measure the stressful effects of urban life on city residents. Psychologists have experimented with animal populations of rats, cats, and monkeys, placing them under the sort of crowded circumstances humans in cities sometimes face: too many individuals in a living space, no escape from confined surroundings, no place of privacy for mothers and their newborn. Dramatic results were recorded. Aggressiveness became a dominant trait in some individuals; others, unable to find a sanctuary from the constant threat imposed by their violent neighbors, died, not from the attacks they suffered, but from physiological trauma arising from the ever-present threats, harassment, and defeat. Mothers failed in their maternal functions, and the young died in large numbers. Males either withdrew from sexual activity or developed deviant behavior. In one experiment, a researcher actually built a simulated tenement building, complete with rooms, hallways, and stairs. He then populated the structure with a comparatively small number of mice. In the beginning, the individuals distributed themselves through the building with ample space for all. But then the colony began to multiply. For a while the structure could accommodate the increase, but eventually it began to become overcrowded. As it did, troubles arose in the community. Fights broke out. Older individuals even cannibalized the young. The number of stillbirths increased. Mothers abandoned babies. All the orderliness that had marked the community prior to this phase disappeared.

We cannot conclude, of course, that these experiments produced results that can be applied to human populations. After all, we could argue, humans have the capacity of choice and they could leave circumstances such as those described above. But is this always so? What are the options for people driven or born into poverty, trapped in a slum, deprived of education, without marketable skills? The word *trapped* we just used is a cultural reality in many parts of the world; the trap is as effective as the cages in those experiments.

And there is some evidence that the consequences of such entrapment are not altogether unlike what we might expect after study of the psychologists' experi-

Slum and ghetto conditions are not unique to cities in the non-Western world. Dozens of families live behind the grimy walls and fire escapes shown in this photograph of a section of a street in East Harlem, New York.

ments. There are correlations between overcrowding and mental health problems; for example, urban congestion is related to a comparatively high incidence of attempted suicide. The cramped living and depersonalization of life in large cities have much to do with alcoholism, drug addiction, and delinquency. Crimes of violence, steadily rising in practically all large urban areas, multiply in part because increasing numbers of individuals feel that they have no stake in the social order, no accountability, no commitment. Alienation and despair can lead individuals to acts of transgression; whole communities can explode into destruction and revolt. Then it does not matter that many and perhaps most of the victims are their own, that the dwellings that are burned are in the slums themselves, not in the comfortable, roomier areas of the city. High urban densities coupled with deficient living conditions may not have the same effects on humans that crowding has on test animals, but the effects are deeply destructive, beyond doubt.

A Note on Density Maps

The representation of population distribution and density presents problems. Several methods can be used to depict population densities on maps; Fig. 3.1 is one of these. On that map, six categories are defined in the legend, categories that are arbitrarily established and within which there can be much variation. We know, for example, that there are areas where population densities are over 1000 per square mile. That is four times the highest density on the map, and yet these areas are not identified. The density range from 60 to 124 per square mile, quite large, leaves us in the dark whether areas so shaded are on the high or the low side of that part of the scale. Still, Fig. 3.1 has advantages over maps based on political units. At least it shows (as in Egypt) where the major clusters of population are within individual countries.

Another way to represent population distribution and density is by the *dot map.* Here again a decision has to be made. How many people shall be represented by one dot? The fewer people, obviously, the more accurate the map. But scale has to be taken into consideration. Figure 3.2 shows the entire world population by the dot method, and obviously one dot must represent many thousands of people, or there would not be space for them all. Dot maps do not show specific regions where certain per-square-mile densities occur, but they do give an accurate impression of the clustering of population.

DEMOGRAPHIC CYCLES

We come now to the subject that is most crucial of all: population growth and the factors influencing it. The natural growth of population is the difference between the number of births and the number of deaths. These two measures, the birthrate and death rate, are commonly expressed as number of individuals per thousand. In statistical tables, they are called the *crude birthrate* (the number of live births per year per thousand) and the *crude death rate* (the number of deaths per year per thousand). For example, in the United States in the year 1972, the crude birthrate was 16.6 and the death rate 9.6. If there had been absolutely no immigration during that year, the rate of increase would have been 7.0 per 1000, or 0.7 percent. Actually, taking all factors of population change into consideration, the rate of growth in the United States at the beginning of this decade was just under 1 percent. Compare this with an African country, Zambia, where the 1970 birthrate was 51 and the death rate 20, or a natural increase of 31 per 1000, one of the highest in the world. Another region of high increase is South America: in Colombia in 1970 the birthrate was 44 per 1000 and the death rate just 11.

These days we speak of population *change* almost synonymously with population *growth,* so familiar has the condition of increase become. But population change has not always been positive. In world history there have been times when populations declined greatly, and the number of deaths exceeded the number of births. Earlier in this chapter we described the slow increase of world population until the early nineteenth century; what we did not report was what held the population down. There were times when people succumbed by the hundreds of thousands to epidemics and plagues. Europe between 1348 and 1350 was ravaged by the bubonic plague, and fully a quarter of the entire population died; in the epidemic and its aftermath, many cities and towns were left with less than half

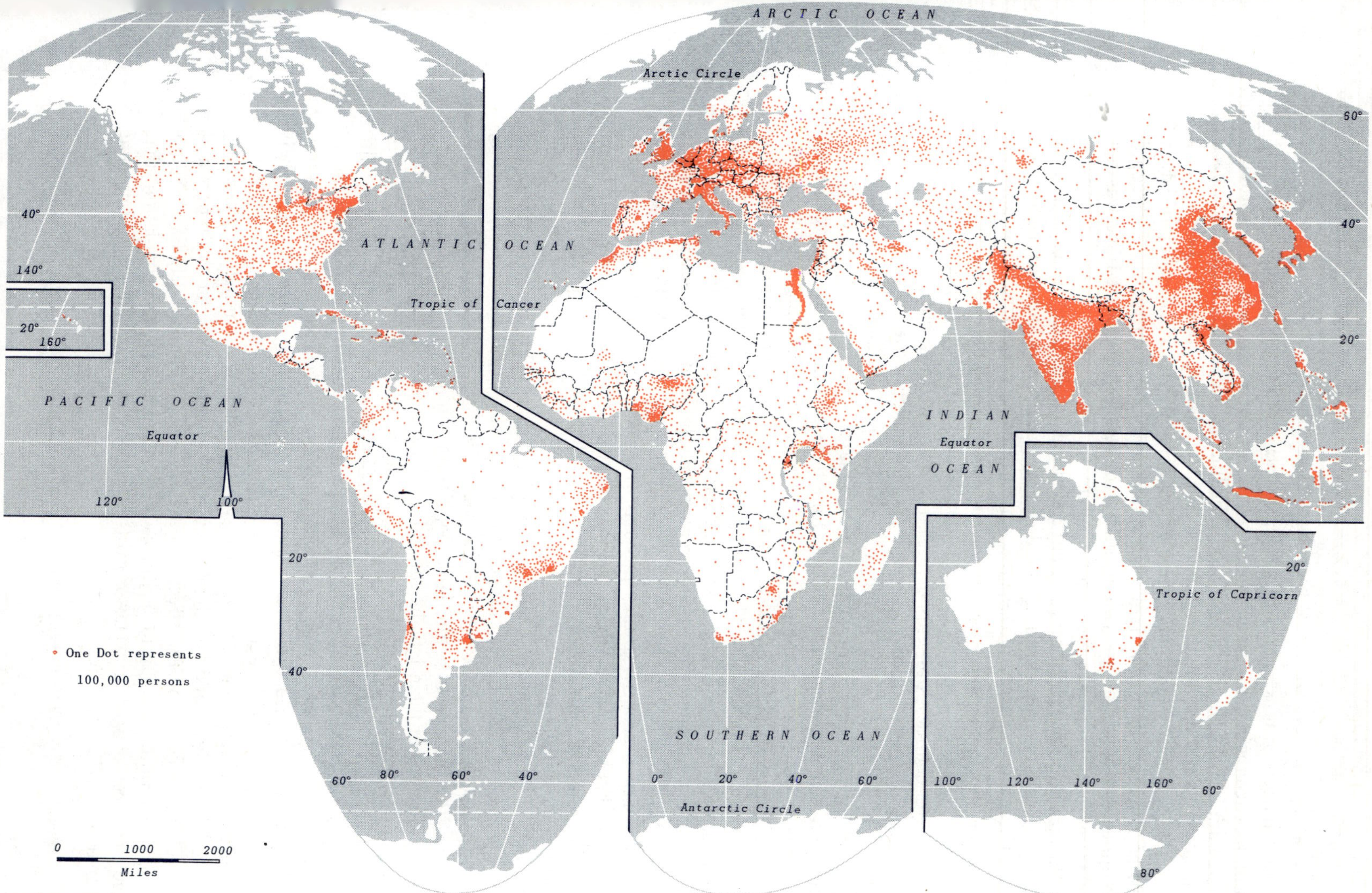

FIG. 3.2 World population: a different view.

their inhabitants. It is estimated that the population of England, which was nearly 4 million when the plague began, was just over 2 million when it was over. A student of population change in the second half of the fourteenth century would have been more likely to talk of annual population *decrease* than of growth!

If it was not an epidemic, then famine could reverse a period of population growth. There are records of famines in India and China during the eighteenth and nineteenth centuries in which millions of people perished; Europe was not safe from such disasters either. From Britain to Russia the ravages of unusual weather periodically caused crops to fail and people to die from hunger. At other times gains were largely wiped out by destructive wars, of which Europe and the world has seen many. Thus the slow and steady increase of world population sometimes shown on graphs depicting its growth from 250 million at A.D. 0 to 500 million in 1650 and 1 billion in 1820 is something of a misrepresentation. It shows an average of countless ups and downs. Birthrates were high, but death rates were high, too; and there were times when there were many more deaths than live births.

But then things began to change. In Europe (we know much less about Asia, Africa, and the Americas) there was a marked increase in the growth rate during the eighteenth century, and this time there was no major setback to erase the gains made. This was the time of the agricultural revolution, when farming methods improved, yields increased, storage capacities were expanded, and distribution systems were bettered. The industrial revolution began to wipe out the old Europe as well: sanitation facilities made the towns and cities safer from epidemics, and modern medical practices began to have widespread effect. Disease prevention through vaccination introduced a new era in public health. Death rates declined markedly. Before about 1750 they probably averaged 35 per 1000 (births averaged under 40), but by 1850 the death rate was about 16 per 1000. Consider what this means in terms of growth rate: if in 1750 the birthrate was 39 and the death rate 35, then the natural increase was 4 per 1000 or 0.4 percent. In 1850 births were still high, perhaps 36 per 1000, but the death rate was 16. Thus the natural increase was 2.0 percent. The change is especially spectacular in the context of doubling time. In 1750 it was of the order of 150 years; in 1850 only 35 years.

What happened to the large population numbers Europe was now generating? Millions of them left the squalid, crowded industrial cities and the farms as well to emigrate to other parts of the world, to North and South America, to Australia, to South Africa, and elsewhere. They were not the first to make the journey. Adventurers, explorers, merchants, and colonists had gone before them. Those early emigrants in many areas had decimated the indigenous population, by conquest, by forced exodus as slaves, and through the introduction of their diseases, against which the local peoples had no defense by natural immunity. But when the European colonization started in earnest during the nineteenth century, the Europeans brought with them their newfound methods of sanitation and medical techniques, and the effect was the opposite. In Africa, in India, in South America, death rates began to decline as they had in Europe, and populations that had long been caught in cycles of gains and losses commenced new, permanent growth—at increasing rates.

We can speak in only the most general terms about the populations of indigenous America, Africa, Asia, and Australia before this phase of Europeanization. About the time of the first European contact, it is speculated that there were probably fewer than 25 million people in all of North and South America; in Africa south of the Sahara there may have been 70 million. In China in the mid-seventeenth century the population may have been less than 200 million; India probably had less than 100 million inhabitants. But there is no doubt about the consequences of European involvement. It reduced the impact of periodic natural checks on the growth of numbers and swept these areas up in the explosive increase prevailing today.

Stages in the Cycle

Demographers who have studied population growth in various parts of the world have noted that the high rates of increase now occurring in much of the "developing" world are not necessarily permanent. Today in Europe, for example, the situation is very different from what it was a century ago. In the United Kingdom in 1970, for example, the birthrate was 17.1 and the death rate 11.9, producing a rate of increase of just over 0.5 percent. That last figure is reminiscent of the preindustrial period, when there also was a small difference between birth and death rates, but both were high. Now both are low. It is in the intervening stage, when birthrates remained high but death rates were lowered rapidly, that Britain's population explosion took place. It is not difficult, then, to discern three prominent stages in the United Kingdom's demographic cycle: (1) marked by high birthrates and high death rates and a small rate of growth; (2) marked by continuing high birthrates, but declining and low death rates, and a high rate of growth; (3) marked by continued low death rates, but declining and low birthrates, and a low rate of growth. As Fig. 3.3 shows, the second stage can be divided into two phases, one of increasing growth rate (A) and one of decreasing growth rate (B).

FIG. 3.3 Stages in the demographic cycle.

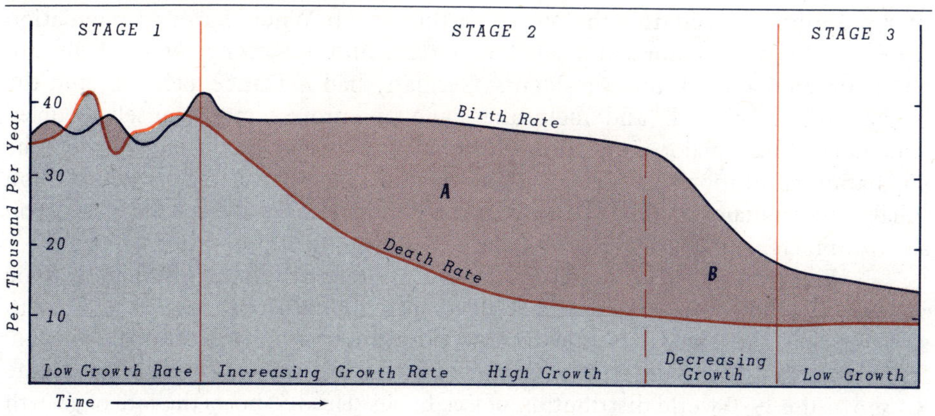

This, then, is what happened in the United Kingdom and in much of Europe. Europe as a continent has a population growth rate of under 1.0 percent, after generating its own population explosion during which the United Kingdom's population grew from about 6½ million in 1750 to 35 million just after 1900. If this is the rule, then we have reason to be optimistic. It would appear that the population "bomb" will eventually fizzle out, and that in due course the European model, low birth and low death rates, will prevail everywhere.

Or will it? In Europe, the birthrate came down in large part because of the effects of industrialization, urbanization, and modernization. In much of the developing world, the majority of the people have not been seriously affected by such changes, and if the world is to wait for them to occur, the population crisis will be there. The example of what happened in Ceylon underscores the situation. There, the population before 1945 increased irregularly, but probably at less than 1 percent per year. Birthrates were high (32–35), but death rates were also high, mainly because Ceylon was susceptible to virulent epidemics of malaria. In 1934 and 1935 the death rate was as high as 33 per 1000 as tens of thousands of people died from the fever. But then, in 1945, a massive attack was launched on the malaria mosquito with the aid of a pesticide called DDT. The plan was executed in 1946, and the results were dramatic: the mosquito was practically wiped out, and the incidence of malaria dropped precipitously. So did the death rate. In 1945 it had stood at 22, by 1947 it was 15, and by 1954 it was an unprecedented 10. In 1970 Ceylon reported a death rate of 8 per 1000. But what about the birthrate? In 1970, it still remained at 32, and Ceylon's population was increasing at no less than 2.4 percent. Ceylon is not an urban society and not an industrial one. It may never be, but in Ceylon there will not be time to find out. In Europe, it took about 150 years for the demographic cycle to pass through its two-phase second stage, the stage of explosive growth. During that time, the population of another island, Britain, increased from 6½ million to 35 million, not counting the effects of emigration. In Ceylon, the second stage began in 1946. If it lasts 150 years, the population on *that* island by 2096 will be 248 million. That is close to 10,000 people per square mile.

This raises another difference between the European model and the situation prevailing in certain other parts of the world. When Europe's population revolution began (for such it was), the starting numbers were not great. Britain had between 6 and 7 million residents. Germany had 7, France, Belgium, and the Netherlands combined (and including some of Europe's most densely peopled lands), 18. Asia's major population clusters were already much larger when the population revolution commenced there: China may have had over 200 million inhabitants, India over 100. Superimpose the growth rates of the second stage upon such numbers, and the increments take on astronomical proportions.

It appears unwise, therefore, to assume that all countries' demographic cycles will follow the sequence that developed in industrializing Europe *or* to suppose that the explosive growth now prevailing in Ceylon, India, Pakistan, Zambia, Ecuador, and other developing countries will go away with time. Figure 3.4 shows the 1970 world distribution of live births; Fig. 3.5 shows the rate of growth

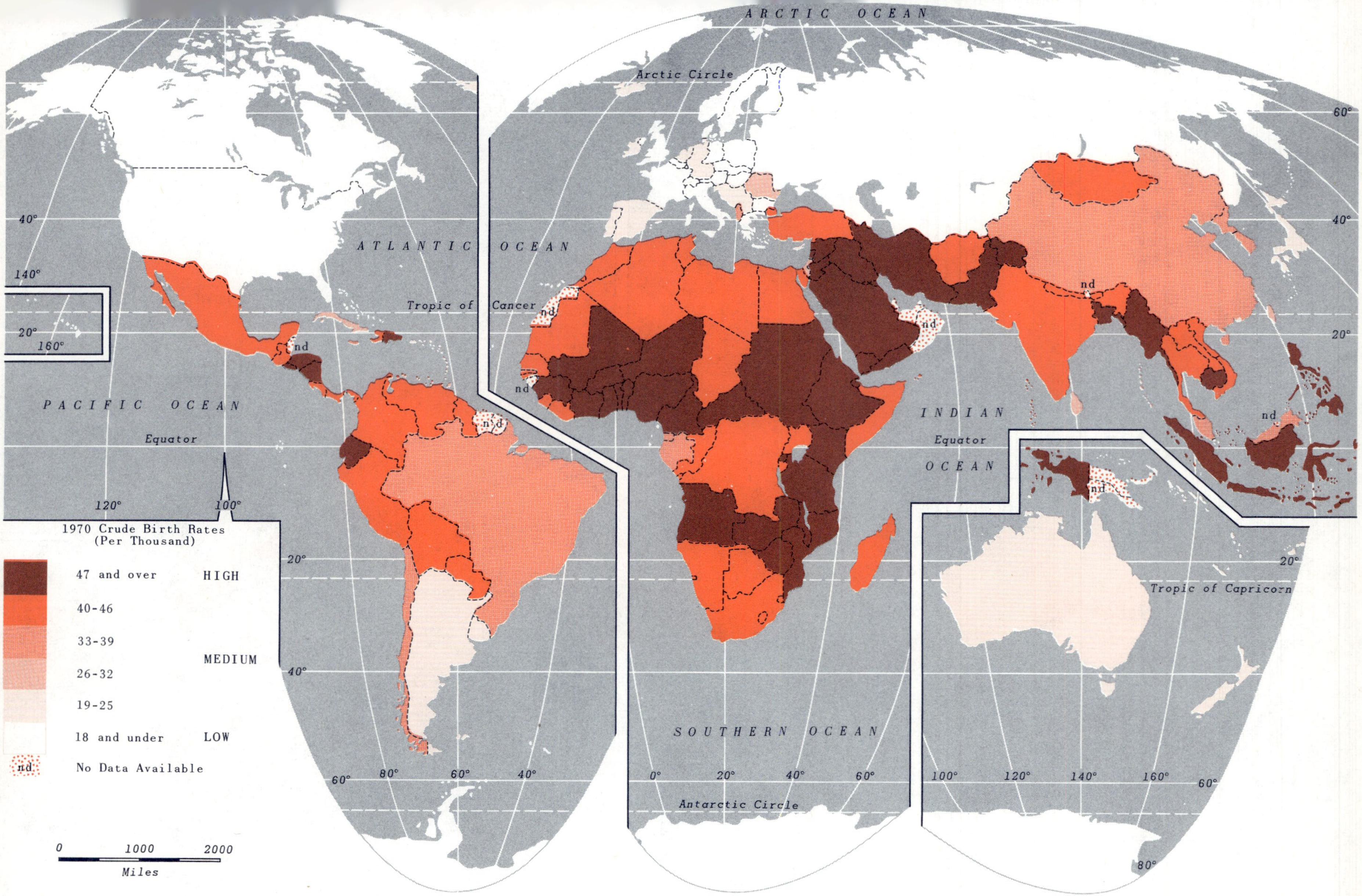

FIG. 3.4 World crude birth rates.

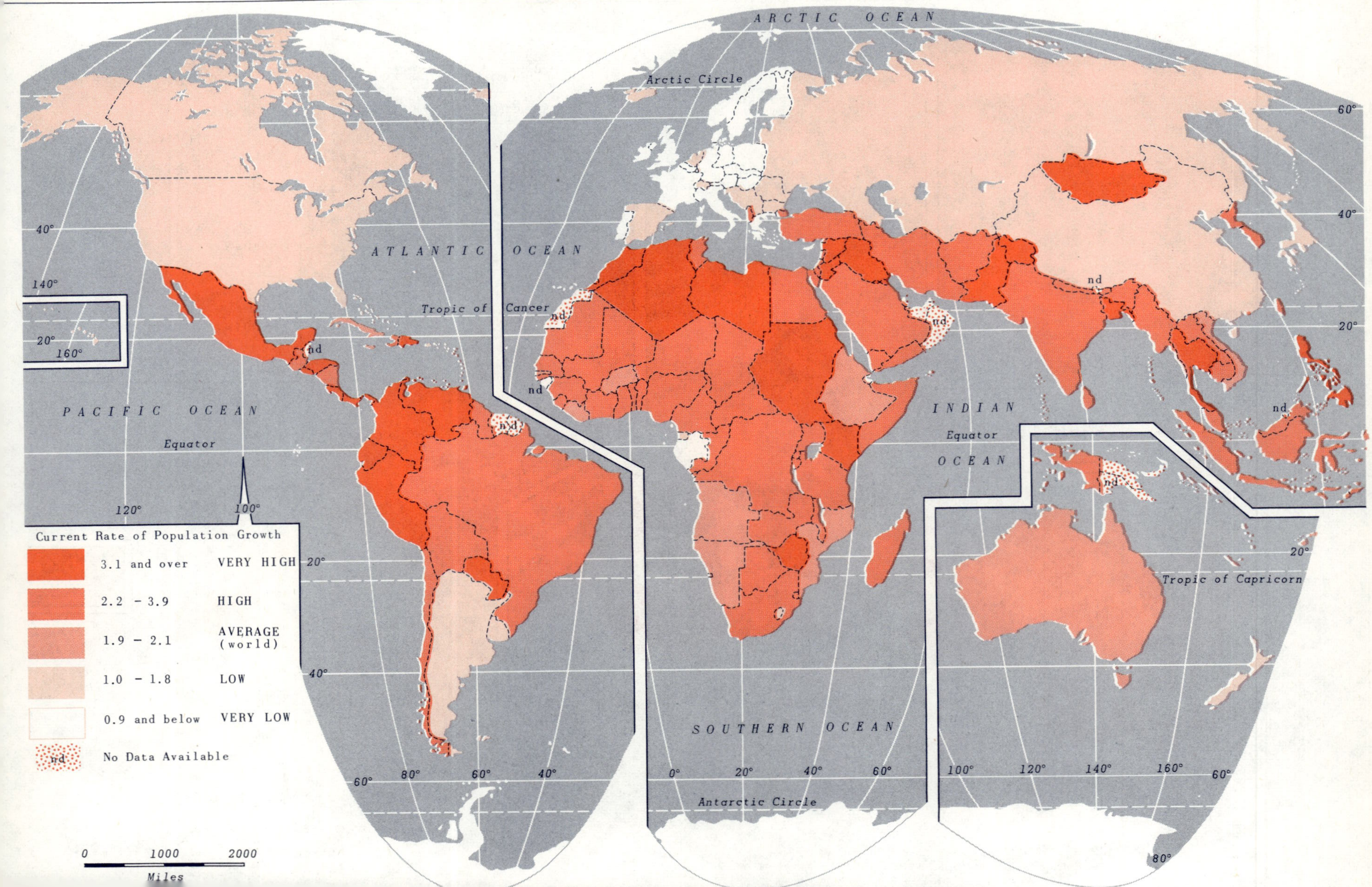
ARCTIC OCEAN
Arctic Circle
ATLANTIC OCEAN
Tropic of Cancer
PACIFIC OCEAN
Equator
INDIAN OCEAN
SOUTHERN OCEAN
Tropic of Capricorn
Antarctic Circle
nd
Current Rate of Population Growth
3.1 and over VERY HIGH
2.2 - 3.9 HIGH
1.9 - 2.1 AVERAGE (world)
1.0 - 1.8 LOW
0.9 and below VERY LOW
No Data Available
0 1000 2000
Miles

of populations in the world's countries. There is not much doubt about their similarity; in only very few countries are both birth and death rates still high. (Gabon, Africa, is such a case). Note the correspondence between tropical South America, tropical Africa, and southern and eastern Asia; lower rates of growth prevail in North America, Europe and the Soviet Union, Australia, and southern South America. But compare this map to Fig. 3.1. Where the lower rates prevail, there are also, generally, fewer people. Where growth rates are high are some of the world's major clusters of population.

Population Structure

Just what was it that brought birthrates down in industrializing Europe? It was not just a matter that life in the cities made children a more expensive and difficult proposition; there were other factors. The age at which people married rose. Abor-

---

THE CENSUS

The modern census is a complex process whereby governments collect, organize, and publish data concerning the peoples they govern. Population counts were taken even in ancient times. For example, the Romans conducted counts for purposes of taxation and military conscription. European countries during the fifteenth and sixteenth centuries began taking periodic population counts, but the beginnings of modern census taking did not occur until the eighteenth century, when Sweden and several other European states instituted more systematic and wider-ranging data collection procedures. The goal became not merely a head count, but information on births, deaths, age, sex, migration, place of residence, and more.

The United States carried out its first census in 1790 and takes a census every 10 years (there was one in 1970; the next complete census will be in 1980). During the nineteenth century most European countries also developed systems of periodic census taking. But even today there are countries that have yet to take a census or have just taken an initial census: Afghanistan, Ethiopia, Laos, the Somali Republic, and South Viet Nam are among the latecomers. This is so despite strong United Nations support for the worldwide application of a standard census: it is vital to have basic data on total population, sex, age, marital status, place of birth, education, language spoken, income, place of residence, family structure, and fertility.

The U.S. Bureau of the Census is an agency of the Department of Commerce. The Bureau conducts not just one, but several different kinds of census, including those of agriculture (every five years), housing (along with population, every decade), government (every five years), business, industry, and transportation. From the periodic data record, it is possible to determine changes and trends, population shifts, regional growth and decline, and government can appropriately allocate aid and assistance.

---

tions became possible for more women. Birth control methods spread. The confinement of urban living quarters persuaded families to remain childless or to have only one or two children.

What happens when a population generates fewer offspring? The *structure* of that population changes. This means that the percentage of people in different age groups is altered from what it was before. The age structure of a population is represented by the age-sex pyramid (Fig. 3.6). The horizontal axis shows the percentages of people in various age groups and the vertical axis represents the succession of age groups or classes, usually in five-year units. The vertical line separates males (left) from females (right). In a population having a rather rapid rate of growth, the age structure is rather like a pyramid. Figure 3.6 depicts the situation in Mexico, which in 1970 still had a birthrate of 44 and a death rate as low as 10. Fully 44 percent of Mexico's population, then, is under fifteen years old! Mexicans tend to marry young, and they have large families, averaging nearly five children per couple. Mexico's present-day population pyramid looks quite a bit like that of England during the height of its population explosion, say, around 1850.

But as we noted, when the number of births declines, the pyramid reflects it, and it may cease to look like a pyramid. Consider the pyramid of France in the early 1960s (Fig. 3.7). It is beehive-shaped, and far fewer people are in the 0–14 age groups than was the case in Mexico, which signifies low birthrates. The death rate is also low, as reflected by the very slight indentations for the progressively older age groups. This beehive-shaped structure is seen in other European countries as well, including the United Kingdom and Sweden. In the case of France, there is a strong indentation in the 45–49 age group. This is the age group (the pyramid dates from the early 1960s) that was in its twenties during the Second World War,

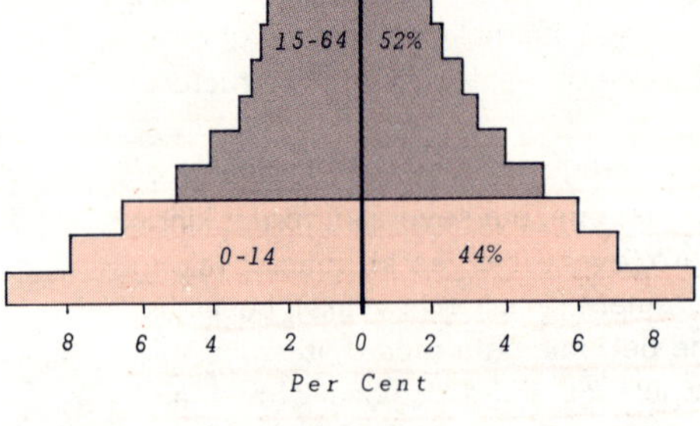

FIG. 3.6 Age-sex pyramid: Mexico.

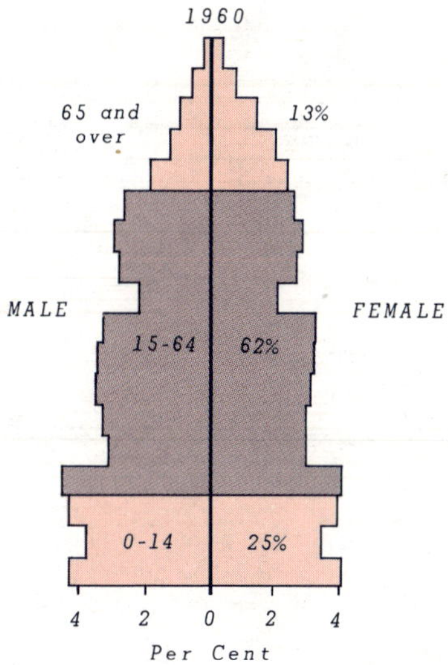

FIG. 3.7 Age-sex pyramid: France.

and the losses in population through military deaths and civilian casualties were great enough to be etched in the pyramid.

The population pyramid of Japan (Fig. 3.8) shows the effects of a rapid drop in the number of births. During its era of modernization, expansion, and military successes Japan's leadership encouraged families to have several children, and limiting families was actually opposed. But modernizing Japan also had growing urban centers, which tended to reduce the birthrate somewhat. The combination of circumstances tended to stabilize the rate of growth. With the end of the Second World War, Japan found itself with many hundreds of thousands of refugee nationals who had left the colonies; soldiers came home and rejoined their families, and the American occupation was attended by the introduction of improved medical services and public health. The cumulative effect was an unprecedented increase in the birthrate and a simultaneous drop in death rates; and Japan's rate of growth, which over the decades had averaged about 1.3, suddenly rose to 2.0 percent per year. That represents a doubling time of 35 years, and with Japan's population already about 70 million at that time, it was a crisis situation.

In 1948, the Japanese government took action, and the results show clearly in the population pyramid. The so-called Eugenic Protection Act legalized abortions for "social, medical, and economic reasons." Contraceptives were made available, and family planning clinics were set up throughout the country. Although contraception and female sterilization (also made possible) helped reduce the birthrate, it was the enormous number of abortions that really brought it down. In fact, so many abortions were performed, perhaps 7 or 8 million in a decade, that the Japanese authorities began to worry about the effect on the well-being of the nation and encouraged contraception by means of propaganda and educational programs.

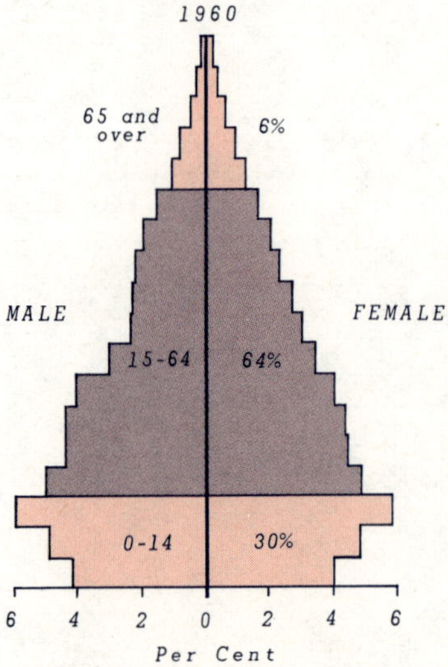

FIG. 3.8 Age-sex pyramid: Japan.

In any case, the birthrate, which in 1947 had been over 34 per 1000, was down to 18.0 just one decade after the implementation of the Eugenic Protection Act. It has been between 17.0 and 18.0 since; the death rate has declined from 14.2 in 1948 to 7.5 in 1958 and 7.0 in 1970. Thus Japan's population, growing at about 1 percent per year, increases by about 1 million annually.

Japan's population pyramid (1960) shows the effect of the rapid decline in growth just described. Above the 10–14 age class it is pyramidal and similar to developing countries, but the lowest age groups show a pattern similar to that of France. If present circumstances continue, Japan's pyramid will slowly change to correspond to the beehive model.

The population pyramids provide efficient statistical summaries of the condition of a given population, and specialists can quickly detect what kinds of changes the population structure is undergoing, the weight of youth or the aged, and any imbalance between the sexes at crucial ages. In the examples we have just discussed, the populations are all national populations. But population pyramids can also be used to interpret relationships among different groups of people within one country. It may be, for example, that a country's total population includes one or more minorities. While the majority population shows the beehive structure, the minority may very well be represented by the regularly shaped, triangular pyramid reflective of higher birth and death rates or the pyramid with bowed sides (like that of Mexico), which depicts high birthrates and declining death rates. This tells us that the minority population is growing more rapidly; its economic situation is likely to be quite different from that of the majority. Other kinds of internal differences emerge when the population structures of urban and rural residents are compared. Still other cases have special qualities: a population pyramid of Hiroshima, for example, could be interpreted only in light of what happened there at the close of the Second World War.

Migration

Before leaving the topic of population growth and change, we should examine one other major factor: people's mobility. When we spoke of the period of explosive population growth in nineteenth-century Europe, we did not fully account for a process that kept the total increase far below what it might have been, the departure of millions of Europeans to other parts of the world. This emigration began slowly, and over several centuries prior to the 1830s perhaps $2\frac{1}{4}$ million Europeans left to settle overseas in newly acquired colonies. But then the rate of emigration increased sharply, and between 1835 and 1935 perhaps as many as 75 million departed. Britishers went to North America, Australia, New Zealand, and South Africa; from Spain and Portugal many hundreds of thousands emigrated to Middle and South America. Early colonial settlements grew to substantial size even in such places as Angola, Kenya, and Java. True, some millions of Europeans eventually returned, but the net outflow from Europe was enormous.

We cannot calculate exactly what the effect would have been on Europe's population growth if all those people had remained in their home countries, but clearly this opportunity to leave unproductive farm and crowded city represented a great release for Europe's burgeoning population. Here again is a factor that ought to caution us against comparing second-stage Europe to second-stage India or Ceylon. There are Indian population clusters in several parts of the world (East and South Africa and the Caribbean), but their numbers are infinitesimal compared to the

---

PUSH-PULL CONCEPT

Cultural geographers have long been interested in the process of human migration. Migration is so essentially geographic a phenomenon that it has occupied the attention of many. One concept that has been used often relates to the forces that make people migrate. Sometimes people leave their abodes because something pushes them away. It may be an oppressive government, unacceptable social conditions, discrimination against a minority by a majority, and so forth. At the same time there may be attractions in the distance: the appeal of the glamorous, skyscrapered, job-generating city or some colonial territory where land may be had cheaply and labor to work it even more so. The problem, of course, is not only to recognize these push-pull factors for what they are, but also to measure their strength in some way. That, as the literature will prove to you, has been a tall order.

South Asian population cluster. Europe's emigrants, on the other hand, represent a substantial fraction of the total.

This European emigration has had no counterpart in modern world history in terms of scale and numbers, but it is not the only major migration flow to have occurred over the past several centuries. The Americas were the destination of another substantial movement of people: Africans, transported in bondage, but migrants nevertheless. This migration began during the sixteenth century, when Africans were brought to the Caribbean. In the first decades of the seventeenth century they arrived in small numbers on the plantations that were developing in coastal eastern North America, and so they were among the very first settlers in this country.

It will never be known how many Africans were forced to emigrate to the Americas; estimates range from 12 to 30 million. As the plantation economy evolved, labor was needed. Before the 1830s, the European population in the Americas remained quite small, and so the flow of African slave labor grew. There was a time, before the flow of European immigration intensified, when there were five to six times more blacks in the Americas than whites. The heaviest concentration of Africans in bondage was in the Caribbean and northern and eastern South America, and of course there are several Caribbean countries today where the descendants of these people constitute the overwhelming majority of the total population. Actually, despite the substantial black population of the United States now, the number of enslaved Africans who were taken to this country was quite small compared with those who came as forced immigrants to the Caribbean and South America. By 1800 the black population of the United States was just 1 million.

It is appropriate to reflect on the consequences the quest for slaves had in Africa itself (Fig. 3.9). Europe's emigrants mostly were able to leave by their own choice (some did go as exiled prisoners or as indentured laborers), and by their departure they relieved to some extent the very pressures that forced them to leave. In Africa, on the other hand, there was no such freedom, and far from improving the demographic situation, the forced emigration greatly damaged the societies involved. In Europe, families left as a whole and joined communities in America that could to some extent cushion the effects of migration-shock; in Africa, families were destroyed, children were orphaned, and in America the African slaves faced not only terror but also loneliness. In Africa the communities subjected to slave raiding, if they survived at all, found themselves without most of the younger men; in America, there long was a shortage of women even when freedom came. Imagine the age-sex pyramid of coastal West African societies afflicted for years by the slavers. It took generations to repair the damage.

What has just been said about two great migration flows, Europeans to overseas settlements and Africans to the Americas, suggests that there are different *types* of migrations. The most obvious classification is into *voluntary* and *forced* migrations. Examples of both types can be observed in the 1970s. The emigration of Cubans to the United States, which has continued over a number of years, in a sense is a voluntary migration. The decision in 1972 by the government of Uganda to oust some 50,000 residents of Asian ancestry led to a forced migration, mainly

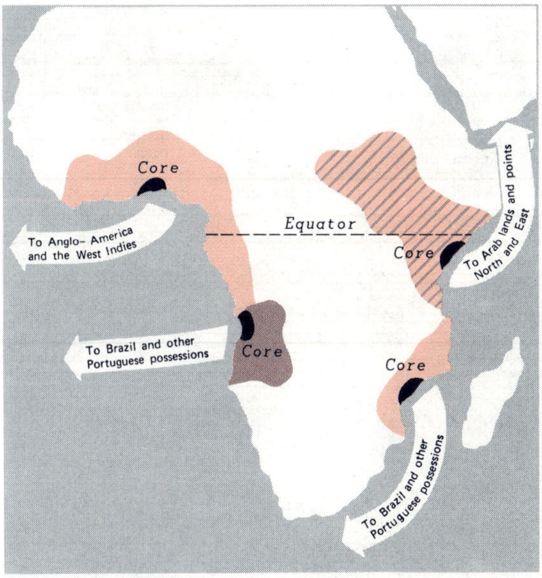

FIG. 3.9 African source areas for slavery. (Modified after D. K. Fellows, *A Mosaic of America's Ethnic Minorities*, Wiley, New York, 1972, p. 18.)

to the United Kingdom. From numerous parts of the world, Jews are migrating to Israel, a voluntary movement. Any analysis of the population structure of Israel would have no meaning without reference to the country's immigration flow.

When we consider population statistics, then, and the "rate of growth" column does not correspond to the difference between the annual births and deaths per thousand, we should look for a migration process among the factors possibly responsible. People everywhere are on the move almost all the time, and, it seems, more so every year. It was recently calculated that the average person in the United States moves 14 times during a lifetime. That may be more times than in any other country, but people everywhere have mobility. They move to different farmland, from farm to city, from inner city to suburb, from city to city. The economic cooperation among European countries that has been one of the significant modern developments in that region has also stimulated movement: Italian waiters serve in Dutch restaurants, Spanish factory workers seek work in Belgium and France. Thus there is *internal* as well as *international* migration. And those migrations have caused reactions among the local residents in the "host" country. The United Kingdom received tens of thousands of nonwhite immigrants from the Caribbean, Africa, and Asia, people to whom the rights of a British passport had been extended during the colonial period. Now exercising their rights, Jamaicans, Guyanans, Indians, Pakistani, and others arrived in England to challenge locals for jobs and residences. Friction was not long in coming; calls for a quota system were not, either.

Later, in another context, we shall consider some geographical implications of the migration process again. Meanwhile, this element in the complex of population change should be kept in mind in our evaluation of population structure.

One kind of voluntary migration is that in which people move from farms and fields to towns and cities. Frequently the new urban immigrants' first stop is in a slum area. Such squatter settlements are often populated by other people who have migrated from the same area, and the resulting sense of community can soften the stresses of urbanization for the new immigrants. Provision of even the most meager facilities, a water faucet, some building materials, can make life bearable.

HUNGER—THE ULTIMATE POVERTY

We live in a society in which advertising space for weight reduction clinics and various remedies for obesity exceeds that given to reports about the prevailing condition of this world's inhabitants, malnutrition. It is a sad reality that there are many more ill-fed than well-fed people in today's world and that the lives of hundreds of millions of people are afflicted from childhood until death by gnawing hunger, incomplete mental and physical development, lack of energy, and susceptibility to diseases. Even in the richest country in the world there are such instances.

Adequate nutrition is not only a matter of eating enough food. The diet must also be balanced, and in large parts of the world that balance is lacking. People need carbohydrates (derived from staples such as rice, corn, wheat, and potatoes), proteins (from meat, poultry, fish, eggs, dairy products), vitamins (from fruits and vegetables, as well as other sources), fats, and minerals. The amount of food intake is measured in terms of calories, which are units of energy production in the body. It is impossible to generalize about the number of calories we need without adding many qualifications: people in desk jobs need fewer calories than people engaged in manual labor, and young adults need more calories than old persons or children. People of large stature require more calories than smaller people. Men need more calories, on an average, than women. But even with these considerations in mind, it is clear that there are whole countries whose populations are badly underfed. The per person daily calorie intake in the United States, Canada, Australia, and New Zealand exceeds 3200; for such countries as Ceylon, India, the Philippines, and Ecuador it is 2000 or fewer. Generally, 2300 calories is taken as the minimum daily consumption level. Even 2500 is still low, but, if we can rely on published figures, about half the people of the world survive on fewer calories than this.

The situation would not be so bad if the people who must survive on too little food could at least have a balanced diet. But they cannot. With few exceptions, the countries where there is a low calorie intake are also the countries where protein is in short supply. This is not surprising when we remember that it takes food to raise the animals that could offset protein deficiencies, food that cannot be given up to animals when people are going hungry. There are places in the world where certain taboos limit people's access to the meat products that might offset their dietary imbalance, but most of all it is shortage of surplus feed that creates the barrier. The result of this is that even persons whose calorie intake is marginally adequate are actually malnourished. There is "hidden hunger" even in the areas identified as "moderate" in terms of calorie supply on Fig. 3.10. A caution about this map: even published statistical information about calorie intake, especially for the developing countries, can be unreliable. The map should be taken for a general impression only. Even so, it is worth comparing Figs. 3.10 and 3.1. Two of the three major Eurasian population clusters lie in the "low" category, the two largest.

It is especially tragic that undernutrition and malnutrition prevail in countries where so large a percentage of the population is young, in the 0–14 age group. A child's brain grows to about 80 percent of its adult size in the first three years of life, and an adequate supply of protein is crucial in this development. If there is not enough protein, brain growth is inhibited and mental capacities are permanently impaired. There can be no subsequent recovery of the loss. Both mental capacity and physical growth are adversely affected by inadequate nutrition, and so the unfortunate youngster born into an environment of deprivation faces a lifelong handicap. In large regions of several countries, the slums of Middle and South American cities, the desert margins of Africa, extensive rural parts of India, whole populations are so afflicted, their hopes for progress dashed not long after birth.

FIG. 3.10 Calory consumption in the world.

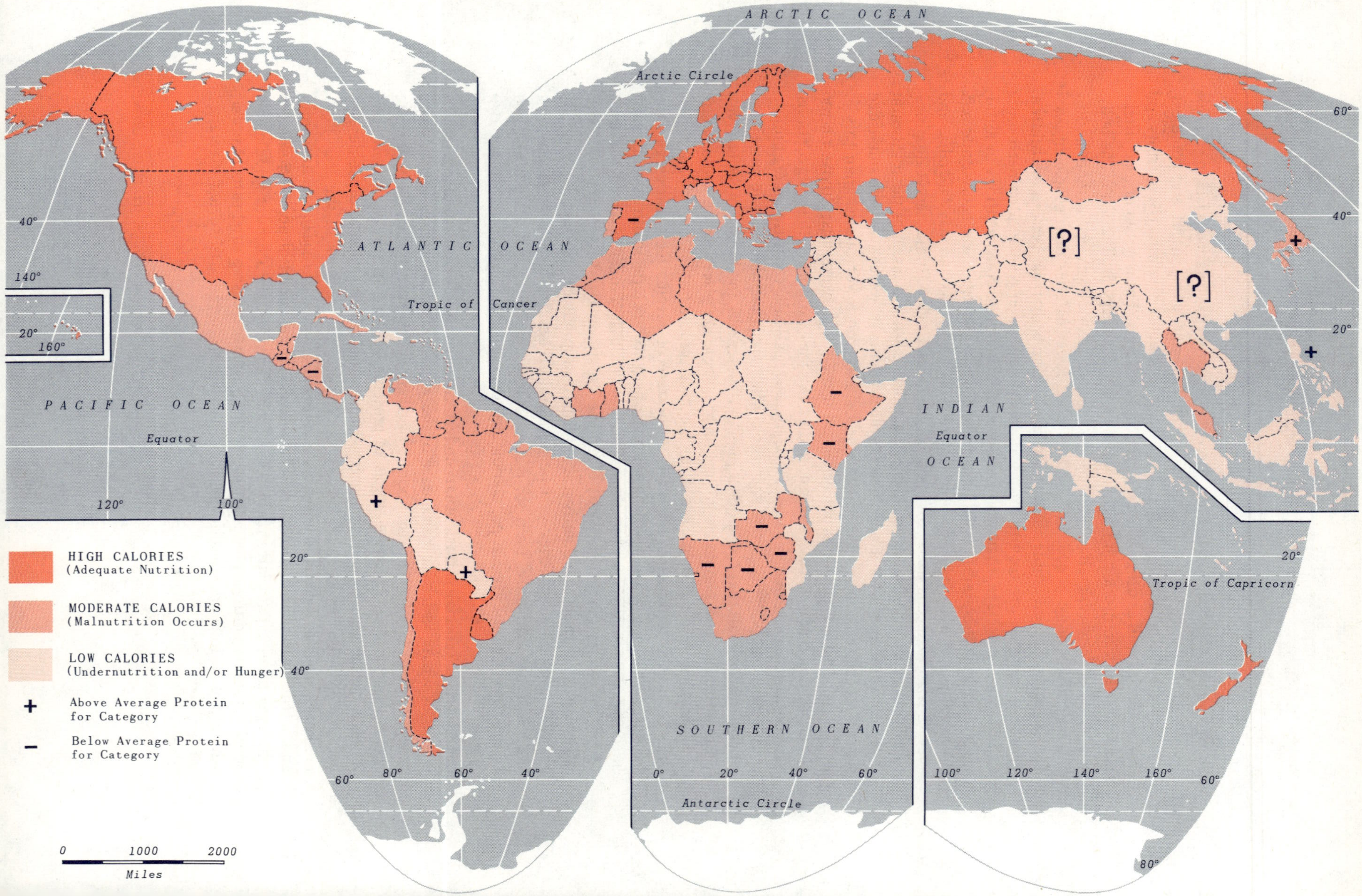

Hunger and Sickness

People who are inadequately fed are susceptible to numerous debilitating diseases. Again the impact on children is devastating. Kwashiorkor, a condition rampant in protein-poor tropical and subtropical countries, causes a child to develop a distended belly, a grotesque irony, while the skin discolors and hair loosens and develops a reddish tinge. Eventually liquids begin to collect in swelling limbs, the digestive system fails, appetite is lost. Apathy takes over, and the child dies. These countries' high infant-mortality rates are directly related to the incidence of kwashiorkor, the greatest killer of small children there. Significantly, a child can develop kwashiorkor even when he or she receives enough total calories. Kwashiorkor is a result of malnutrition, not necessarily undernutrition. It tends to develop when a mother stops breast feeding her child (perhaps because a new baby has been born), and the child is put on a starchy diet.

When there are not enough calories *and* a gross lack of protein, a child is likely to develop marasmus. The body is thin and bony, the skin shrivels, and the eyes appear huge in the tiny, drawn face.

Kwashiorkor and marasmus are among a host of other threats facing the undernourished. Besides calorie inadequacy and protein deficiency, hungry people may have vitamin deficiencies. Low vitamin A intake is related to diseases affecting the eyes, and these, too, take a heavy toll of children. Beriberi, which affects the nerves, digestive system, and the heart, is related to vitamin B inadequacy. It is frequent in eastern and southern Asia. Insufficiency in vitamin C can produce scurvy; and vitamin D is needed to ward off rickets.

This depressing (and far from complete) list of diseases directly related to hunger does not include the many other, infectious, diseases that prey upon the ill-protected, malnourished body. Cholera, yellow fever, hookworm, malaria, and numerous other maladies ravage people already weakened by their imbalanced diets, and they take a heavy toll if not of life, then of energy and longevity. And so well-fed tourists and travelers from the comfortable world of plenty see people sleeping on sidewalks, resting on their jobs, working at a snail's pace, and they conclude that those people are not better off because they are lazy, there in Guatemala, in Peru, in Gabon, in Niger, in India, in the Philippines.

Food for Billions

The fact is that the world today comes close to producing a quantity of food which, if equally distributed among all the earth's inhabitants, would come very close to satisfying the 2300-calorie minimum established earlier. Obviously something happens to prevent this from taking place. In the first place, a shockingly large amount of the food grown and raised each year is lost through decay in storage or transportation, through destruction by mice, rats, and other rodents and pests, and through misfortunes such as the British dock strike of 1972, which left hundreds of thousands of tons of fruits, vegetables, and other foodstuffs rotting on docks and in warehouses. It is difficult to estimate the amount of this loss, but it has been calculated as between 10 and 20 percent of total world production, and in some areas the loss may be as high as 50 percent. Then there is the matter of distribution. The world's food production is simply not distributed to ward off hunger everywhere. Things have improved a great deal, and grain shipments to famine-threatened areas can now stave off disaster when millions would have died in another period of history. But areas still remain where the supply of food is chronically inadequate or where *seasonal* hunger occurs over and over. What prevails, though, is the fact that even if 2300 calories, minus the losses incurred through spoilage and destruction, were available to every person on this globe, it would be inadequate quantitatively and qualitatively. There can be no doubt about it: if all the world's food were allotted evenly to all people, everyone in this world would be malnourished.

The obvious answer to this aspect of the population crisis, then, seems to be the production of more food. But will more food expand the limits to population growth, thereby contributing to an eventual, still greater shortage of food? We are back to Malthus's original warning, but in recent years another dimension has been added. We are destroying streams to create irrigation ditches. We are cutting down forests to clear still more land. We are causing soils to erode and grasslands to be overgrazed. We are exhausting the fishing grounds of the oceans. We must ask

This is how millions of people must receive their food each day—through relief efforts and emergency distribution systems. Droughts, floods, wars, and other disruptive events cause countless thousands of refugees to leave their homes to seek help elsewhere, as these people from Bangladesh did. More often than not, the area where they seek assistance is already at the subsistence level, and international organizations must provide support. The early 1970s were marked by severe and widespread droughts in India, the Soviet Union, Africa, and elsewhere.

ourselves not only whether we can produce enough food to supply added billions of people, but also what we will do, in the process, to our earthly environment, our life support system.

The "Green Revolution"

In Chapters One and Two we considered some of the environmental characteristics of our earth, and we found that there are limits to the agriculturally productive areas. Forests are still being cut over to make way for new farmlands, and some marginal areas have not yet been brought into production, but in the next several decades we will not discover any new Ganges Plains or Nile Deltas. In the 1960s the optimists among demographers, who would argue that technological progress will overcome population-food crises, scored a point when there was a real breakthrough in the search for new, more productive strains of cereals. "Miracle rice," as one such new variety aptly came to be called, permitted the Philippines to end more than 50 years of dependence on rice imports; Ceylon's rice production rose by 32 percent in two years. Wheat, too, shared in this Green Revolution: Pakistan and India in three years increased their wheat harvests by 50 percent. Over a longer period, Mexico, where the wheat experiment succeeded, tripled its production of this grain. The miracle rice, developed at a research station in the Philippines, has a short maturing time, a high-yield capacity, and a high response to fertilizers.

But even the Green Revolution has not been able to eradicate hunger in our world. It was a reprieve, but it was able to close the gap between population and food needs only partially, despite its dramatic effects on yields. Pessimists pointed out that much of the spectacular increase in yields had to do with exceptionally good weather during the first years of introduction. Others feared the impact of the use of fertilizers on the environment. Economists pointed out that many farmers will not have the capital or credit to participate in the technology of the Green Revolution. And the inexorable march of population growth, they argued, would soon eat up the gains made and produce essentially the same situation as prevailed before, only this time involving many more people.

In Chapter Seven, we shall report more fully upon the state of the world's agriculture, per-acre and per-worker productivity, and potential yield increases. It is only natural, given the unabated growth of population, for the search for increased food production to continue. But it is equally obvious that the cycle has to be broken. The cost of the search is rising too. Authors Lester C. Brown and Gail W. Finsterbusch, writing in a book entitled *Man and His Environment: Food* (1972), comment:

> *Man's quest for food is converting literally hundreds, perhaps thousands, of freshwater lakes and streams into algae-laden swamps throughout North America, Europe, and now increasingly in those poor countries where fertilizer usage is beginning to climb. The costs of rescuing or reclaiming dying or eutrophic lakes is enormous. Lake Erie, the best known victim, has received nutrients from both agricultural and industrial sources. An estimated $50 billion, the equivalent of*

Experimentation with new varieties of rice (as well as sorghum, wheat, sweet potatoes, and other crops) has produced new strains that are much more productive than older ones. The search is always on for better fertilizers and in particular areas for especially fruitful crop-soil-fertilizer combinations. One such area is northeastern Thailand, where great potential for increased farm productivity exists. Shown here are rice fields of the Kalasin Experimental Farm and workers transplanting rice.

> *the Department of the Interior budget for the next twenty-two years at the current level, would be required to restore Lake Erie to its original state as a freshwater body . . .*

What is happening, they add, is that the Green Revolution's fertilizers are threatening eutrophication in numerous freshwater lakes and streams in the developing world, depriving villages of fish, a critical element in protein-deficient diets.

In Chapter Two we discussed the modifications we may be bringing to our climate through our technological progress. We are doing the same thing in our quest for more food. Marginal farmlands are being forced into production, and thousands of acres turn into dust bowls, contributing to the amount of particulate matter in the air and hence to possible changes in the solar-energy balance. Plans to cut severely into the Amazon Basin's forests have raised fears that the oxygen supply they transmit into the atmosphere will be reduced. Rainmaking attempts and the diversion of rivers for irrigation purposes are all experiments that threaten change in the climates of the world, change that cannot be predicted and, once set in motion, perhaps not reversed.

There is another element in the search for more food, one that we will touch upon again in Chapter Ten: the intensified use of the oceans and seas as sources of proteins. There have always been demographers who have argued that when the real food crisis comes, the matchless riches of the seas will save the situation. And indeed the seas have yielded enormous amounts of valuable protein-rich nutrients for man, amounts that have risen through the decade of the 1960s and early 1970s. Once again man's infinite capacity for technological innovation is both his ally and his enemy: veritable floating factories can catch in a few days what whole fleets used to catch in a month. But in the process the oceans and seas are being exhausted, not harvested. There is talk about ecology, restrictions on catches, and the survival of species, but the scramble for the oceans' wealth continues all the same.

There are some spatial aspects to the fishing industry that render the picture rather less bright than some experts would have us believe. The oceans and seas may cover 70 percent of our earth, but not everywhere do they contain teeming schools of fish. Like people on the land, the fishes of the oceans tend to agglomerate and cluster in certain regions. Much of the open ocean is as empty as the desert. Shallow areas, continental shelves, estuaries, submerged banks, and zones where water wells up with strength are among the zones where the fish fauna concentrate. For centuries man has pursued them there, in the North Sea, in the North Atlantic, off Japan, off Peru. Until quite recently only a few species were endangered, as whales are today, by overexploitation. But now radar-equipped, machine-driven

A natural landscape transformed: paddy fields on the Kowloon Peninsula, near Hong Kong, China. Every foot of available land in this alluvial river valley has been meticulously prepared for the production of food crops. It has been so for hundreds of years: before this area was leased to the British by China in 1898, part of the crop was sent to the Imperial Household in Peking.

Harvest from the sea: commercial fishing boats in the North Atlantic off the United States coast. Experts are concerned that the search for the world's fish will lead quite rapidly to the perhaps permanent depletion of this resource.

vessels can secure such enormous hauls that there are already areas where such common fish as sardines, herring, salmon, tuna, and cod are being depleted beyond the level of potential recovery. Larger fish eat smaller fish. When a whole region is cleared of small fish, larger fishes can be expected to die out. Increasingly, young species, not yet full grown, are caught, and the schools of fish that once awaited the seasonally operating fishing fleet are no longer there.

ZPG

The letters *ZPG* stand for "Zero Population Growth," the name (and objective) of an activist organization formed in 1969 by a group of concerned citizens including, among others, the biologist Paul Ehrlich, author of *The Population Bomb* (1968). ZPGs rapid growth (it counted 10,000 members after just one year of existence) reflects the growing awareness of the negative consequences of crowding on our planet. Hunger, environmental deterioration, unwanted children, unchanging living standards, and associated prospects led ZPG to adopt programs calling for legalized abortion, limitations on family size (no more than two children per family), tax incentives for small families, and government support for birth control. Most campuses have ZPG representation.

Numerous observers predict that rational harvesting of the seas, if it ever comes about, will be achieved only after permanent damage has been done. The seas, as we shall see in Chapter Ten, are still mostly open to all comers. And who are the comers? The countries who need protein desperately? Or the technologically advanced countries that have floating factories and huge fishing fleets plying all the oceans?

If the ocean fisheries form a dark picture, there is hope in another direction. Aquaculture, the raising of fish in net enclosures, is in its infancy, but it may be a form of protein production that will attain importance in the future. It has gone quite far in Japan, where some hundreds of thousands of tons of fish are produced this way. This is a drop in the bucket when the annual world catch is between 50 and 60 million metric tons, as it has been for some years. But fish contribute importantly to the protein supply in Japan's imbalanced diets, and as fishing grounds fail, the aquaculture industry is likely to grow.

POPULATION POLICIES AND PLANNING

The realization is spreading that the world's explosive population growth, with its attendant rise in demand for more foodstuffs, keeps this globe in a state of perpetual crisis, political, environmental, and otherwise. Even in our society, where progress

Aquaculture, the "farming" of fish in artificial ponds, helps in the battle to eliminate protein deficiencies in poorly provided areas and could reduce the pressure on marine resources. One of the world's most protein-deficient areas is in the state of Orissa, India. To compensate for this, the Orissa government has embarked on a major program of aquaculture. Now village fish ponds are a major protein source.

is equated strongly with economic growth, increasing size, and "development," and where the crush of population is not nearly what it is in may other areas, the goal of a stabilized world population has emerged as an objective of great priority. In 1970, the Family Planning Service and Population Research Act, a manifestation of this phenomenon, was passed by Congress. This legislation facilitates the diffusion of family planning information and the distribution of means of birth control. Public attitudes toward abortion have changed quite drastically in this country over the past decade, and these changes are also reflected by changes in the laws of several states. Locally, our newfound concern over numbers is revealed by public opposition to further suburbanization, high-rise construction, and even to in-migration: the *Miami Herald,* for example, has carried letters from readers objecting to any further influx of northerners in search of retirement condominiums. As many demographers have pointed out, public attitudes must change before population planning can be effective.

In other parts of the world, population policies have also changed, but their effect has so far been limited. The prime example of population control is still Japan, which went through a stage of precipitous decline in the birthrate, mainly through abortion. The task over the past decade in Japan has been to convert from the abortion form of control to other means, and to family planning. China has reportedly engaged for some years in a policy of restriction by supporting late marriage, two-child families, sterilization, and abortion. Population growth in China may now be as low as 1 percent annually, although no official figures are available. (And even with this, China would add a population the size of Belgium to its total each year.) India, too, has begun to support birth control and family planning, as has Pakistan. Here the problems are several: neither India nor Pakistan is capable of investing as much money into birth control clinics and educational efforts as is needed. India began a modest program in 1952, based on education about the rhythm method. After 20 years, the impact is slight.

In Africa and South America, as well as Middle America, there has been comparatively little progress in the adoption of population control. In Africa this relates to nationalist ideologies that in the colonial aftermath equate strength with numbers and to prohibitions of former (French) and remaining (Portuguese) colonial powers. In Middle and South America the Roman Catholic Church continues to militate against birth control, but here, too, the concept of strength-in-numbers still prevails. In European countries dominated by the Roman Catholic Church, where birthrates have declined, it has happened through late marriage, illegal abortions, and permitted (rhythm) means of control as well as through forbidden means. And in the communist countries of Eastern Europe and the U.S.S.R., the social order functions to deter large families: women tend to work away from the home more frequently, late marriage is common, and family planning information is widely available.

All this suggests that population control is least effective where the crisis is felt most strongly and where what little progress made annually is consumed by the additional mouths that must be fed. But it is difficult to disseminate the required information and devices to illiterate populations that live in places where there has been little or no change in ways and means of living. Where death rates have not begun to decline noticeably, people are not likely to be much interested in methods to curb the birthrate. Where farmers have for centuries struggled in the same way to wrest a meager harvest from the soil, where new equipment is unknown and fertilizers cannot be afforded, the measure of greater productivity appears to lie in more manpower—larger families. It is a vicious cycle, and until it changes, population control will be deterred.

We ought to keep in mind another aspect to the population question. To many in the underprivileged world, external aid and encouragement toward birth control appears to be just another case of neoimperialism. It is all aimed, it seems, at reducing the competition by people in the underdeveloped countries for the resources the developed countries consume at such enormous rates. Thus population control may be opposed on ideological grounds. In Africa, it is noted with interest that the South African government encourages white families to have more children at the same time that it propagandizes for birth control among the black South Africans. If the "wisdom" of the developed countries in the area of population control is to prevail, then the example will have to be set at home before the concept can credibly be exported.

POPULATION AND RESOURCES

In these first three chapters, we have discussed the contents of our world in physical and human terms. In 1959, a geographer, Edward A. Ackerman, constructed a map based on the size and technological capacity of populations, on the one hand, and the quantity and quality of resources, on the other, what he called the population/resource ratio. On these bases he established five population/resource regions, whose character will become clear by reference to Fig. 3.11. The Ackerman scheme was modified by W. Zelinsky who, in his book *A Prologue to Population Geography* (1966), outlines the five regions as follows:

1. *United States Type.* Large areas, major, varied resources, small or moderate populations, advanced technology: the United States, Canada, Australia, New Zealand, much of the Soviet Union, parts of Argentina and South Africa. Low population/resource ratio. About one-sixth of the world's population.
2. *European Type.* Smaller territories, larger population clusters, fewer and smaller resources than type 1, but highly advanced technology, skillful use of available resources, high rate of acquisition of external resources: most of Europe, Japan, western U.S.S.R. Higher population/resource ratio. About one-sixth of the world's population.
3. *Egyptian Type.* Extensive areas, large population clusters, shortage of resources, deficient technology: China, India, Pakistan, Bangladesh, Egypt, Mexico. High population/resource ratio. About half of the world's population.
4. *Brazilian Type.* Substantial areas, small to moderate population; resources would be adequate for sustaining population if developed: Brazil, Southeast Asia,

FIG. 3.11 Population/resource regions of the world. (From W. Zelinsky, *A Prologue to Population Geography*, Prentice-Hall, Englewood Cliffs, N.J., 1966, pp. 108–109.)

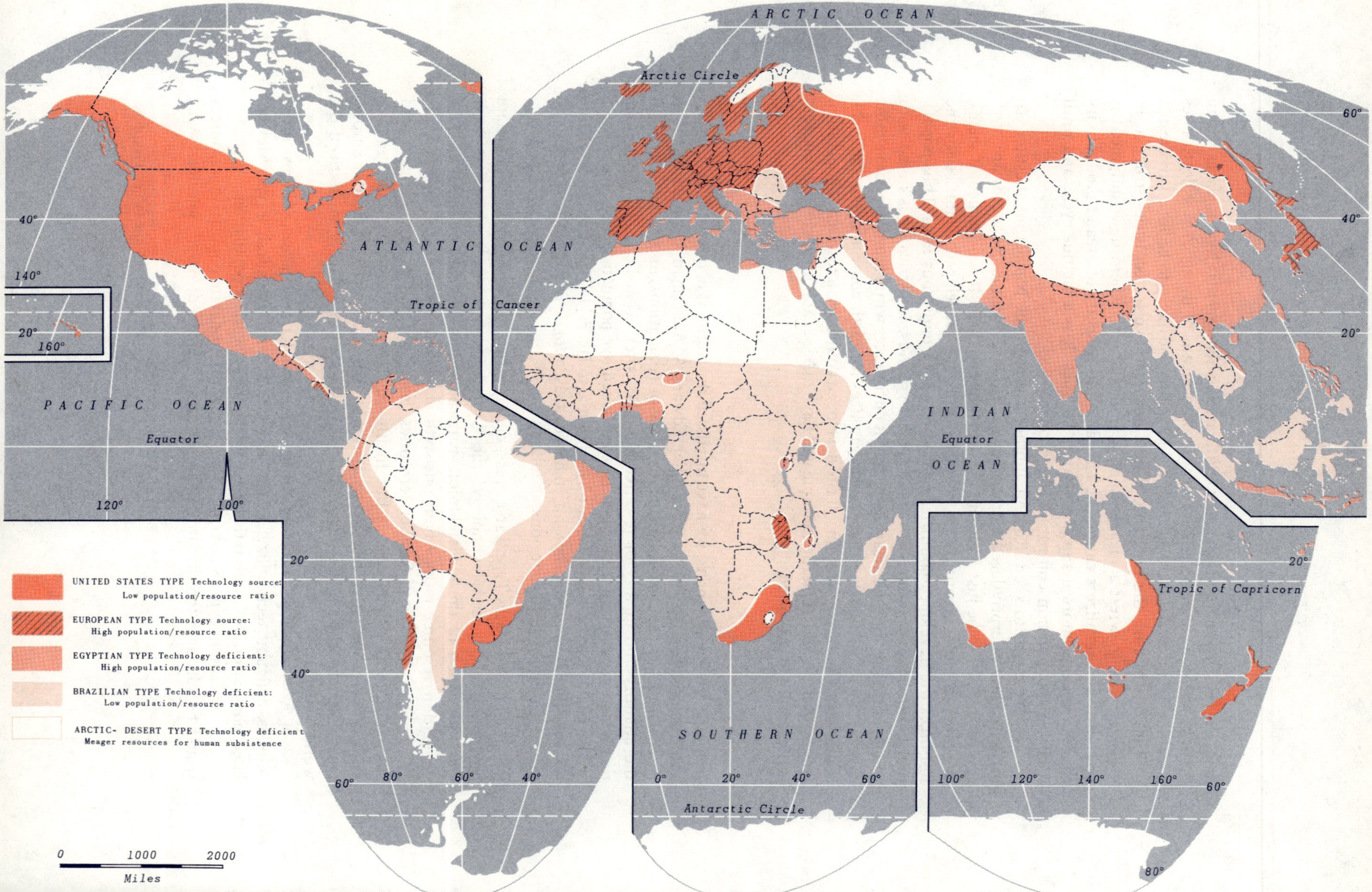

parts of Africa. Low population/resource ratio. About one-sixth of the world's population.

5. *Arctic-Desert Type.* Extensive areas, very small population, technology deficient, meager resources for human subsistence: Sahara-Arabian Desert, Amazon Basin, interior Asia, northern Eurasia and North America, Antarctic. Less than 1 percent of world's population.

This framework is not absolute, of course. Figure 3.11 is a very generalized representation of a complex world situation with many local and regional variations. But the map forms a most useful summary for our discussion of terrestrial resources as outlined in the first chapter, environmental and climatic conditions as described in the second chapter, and the present consideration of world population.

CHAPTER FOUR

Man and Culture

FOUR

There are 4 billion people on this earth, and they live their lives in thousands of different ways. Their languages differ. Their racial sources vary. Their religious beliefs are not the same. In their architecture, their systems of education, their technologies, and in countless other ways the societies of the world reveal contrasts and differences. All of the world's societies, large and small, occupy a part of the earth's surface, and thus all have regional qualities. And all of the world's societies have in some special way exploited their particular physical environments and resources. Small wonder that cultural geography as a field of investigation has attracted many practitioners.

REGIONS AND CULTURES

Repeatedly in this book, we have made use in passing of the term *region*, and now is the time to zero in on some definitions. After all, everyone has some idea of what the word *region* means. When we talk about potential conflict in the Middle East or farm problems in the Corn Belt, we have used the regional concept in its broadest sense as a frame of reference (Fig. 4.1). But as geographers, we may have to be more specific than that. Just what is the Middle East? Does it consist of all the countries of the Arab world or only certain ones, and if only certain ones, exactly which ones? Where are the precise limits of this Corn Belt? How large a percentage of farmland must be under corn cultivation in order for it to qualify as part of this region? These are not frivolous questions. If the regional concept is to be useful, it must have firm foundations.

The first thing we observe about regions is that they have *location.* By our reference to the Corn Belt (or to Indochina or the Amazon Basin or the Chicago area, all regions of a kind) we have in mind a region positioned in a particular location somewhere on this globe. We could actually identify that location by using the grid system referred to in Chapter Two to describe it by degrees latitude and longitude. That would be the region's *absolute* location, but such a numerical index would not really have much practical value. Location attains relevance only when it relates to other locations. Thus if we are to recognize the Corn Belt as a region, then we would identify it with reference to the climatic region in which it lies (Fig. 2.12) and the soil regions over which it partly extends (Fig. 1.11). Now we are saying something about the Corn Belt's *relative* location, a much more meaningful criterion.

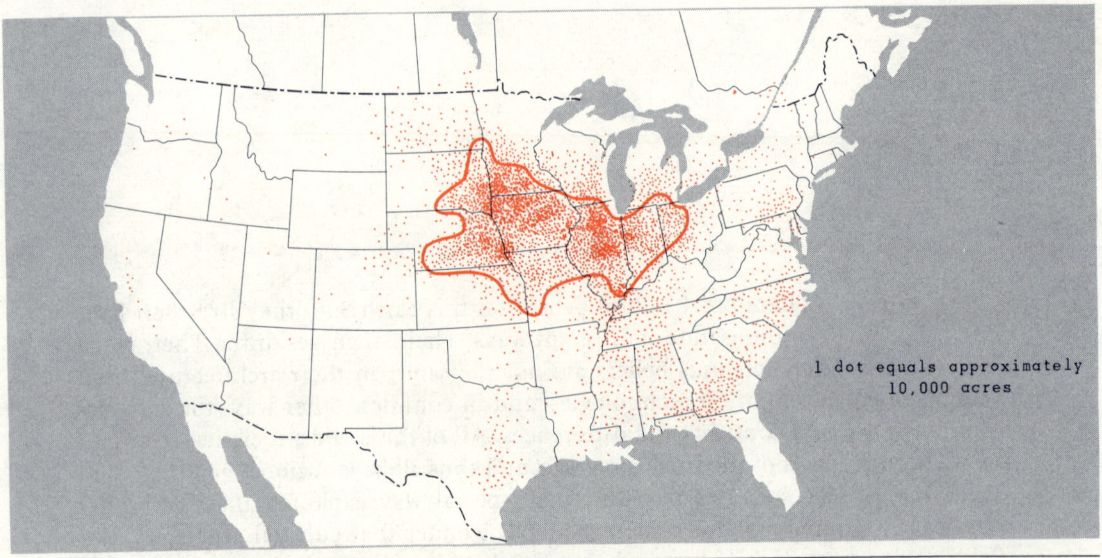

FIG. 4.1 Single-category formal regions: the United States Corn Belt.

A second quality of regions is that they have *area.* This seems too obvious to even mention, but herein lie some difficult problems. If a region is to have areal extent, then its limits must be definable. We would probably all agree that the Rocky Mountains form a region in the United States. But taking a map and drawing a line exactly where the Rocky Mountains end and other regions (such as the adjacent Great Plains) begin is not as simple as it might seem. Mountainous regions often peter out in a marginal belt of lower hills before they give way to true plains. Thus a decision has to be made whether those hills lie beyond the limits of the mountain region or are part of it.

A third characteristic of regions is their internal sameness, or *homogeneity.* Each time we mentioned examples of regions, there was something to unify: a high intensity of corn farming in the Corn Belt; forests, heat, humidity, flatness in the Amazon Basin (Fig. 4.2); high relief, slopes, rugged topography in the Rocky Mountains. Professor R. Harshorne in a book entitled *Perspective on the Nature of Geography* (1959) defines a region as "an area of specific location which is in some way distinctive from other areas and which extends as far as that distinction extends." This distinctiveness may be physiographic, as in several of our examples, but it can also be cultural. A certain type of architecture may prevail over an area, and we recognize a region. In eastern Canada, the French language is spoken by a majority of people in a region centered on Montreal. In Chapter Two we discussed several climatic regions. In Chapter Three, when we spoke of important population clusters in the world, we thereby regionalized population on the basis of high density.

Thus a region is a part of the earth's surface, in some specific locale, in which certain predetermined criteria are met. The boundaries of the region are determined by the criteria; they are not something we would go and search for in the field. Take our Corn Belt example again. We would have to establish some criterion (50 percent of farmland under corn, as an instance) and then draw our region's boundary on that basis. Someone else might very well argue that 60 percent is a more appropriate criterion to use. That will change the region's boundaries. Thus the Corn Belt is not an absolute region; it is an artificial construct, a device to use for further discussion. That is true of all regions: they are intellectual devices designed to function as organizing concepts in geography.

The kind of region we have so far discussed, in which sameness, or homogeneity, is the rule, is called a *formal* region. But regions may be conceptualized on other bases. For example, a large city has a substantial surrounding area for which it supplies goods and services, from which it buys farm products, and with which it interacts in numerous other ways. Its newspapers sell in the smaller towns in this area, the city's manufacturers distribute their products wholesale to subsidiaries there. Any map of road traffic, or of telephone calls, or of radio-station listening will confirm the domination of the large urban center in this, its tributary area, or hinterland. Here again we have a region, but the region is not characterized by homogeneity, but rather by the city-centered system of interaction that creates it. Appropriately, such a region is referred to as a *functional* region (Fig. 4.3). And

FIG. 4.2 Formal region: the Amazon Basin.

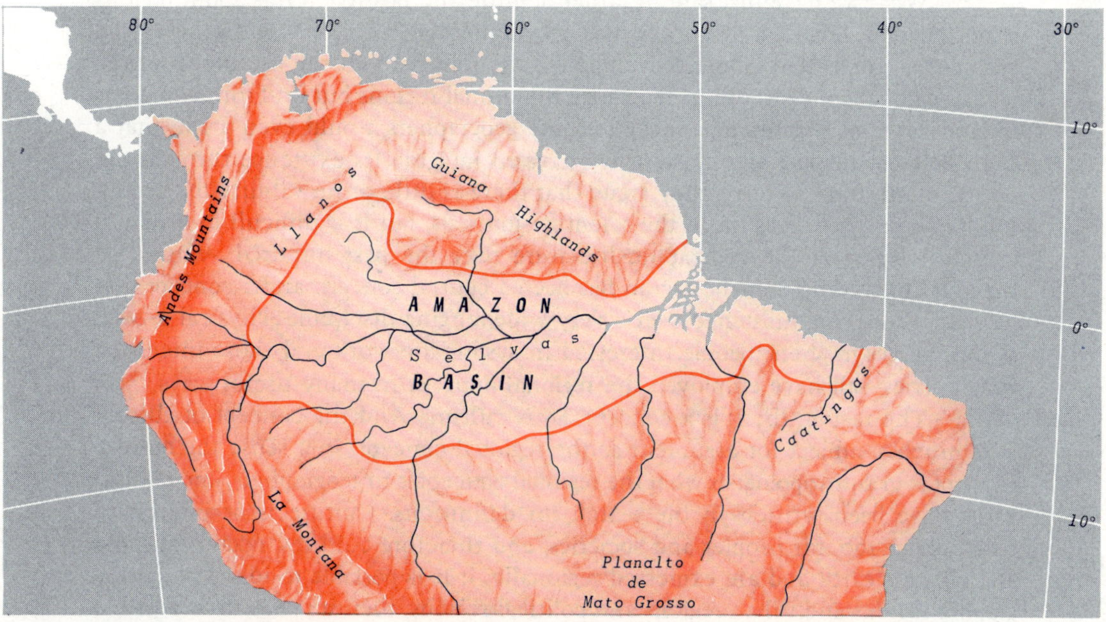

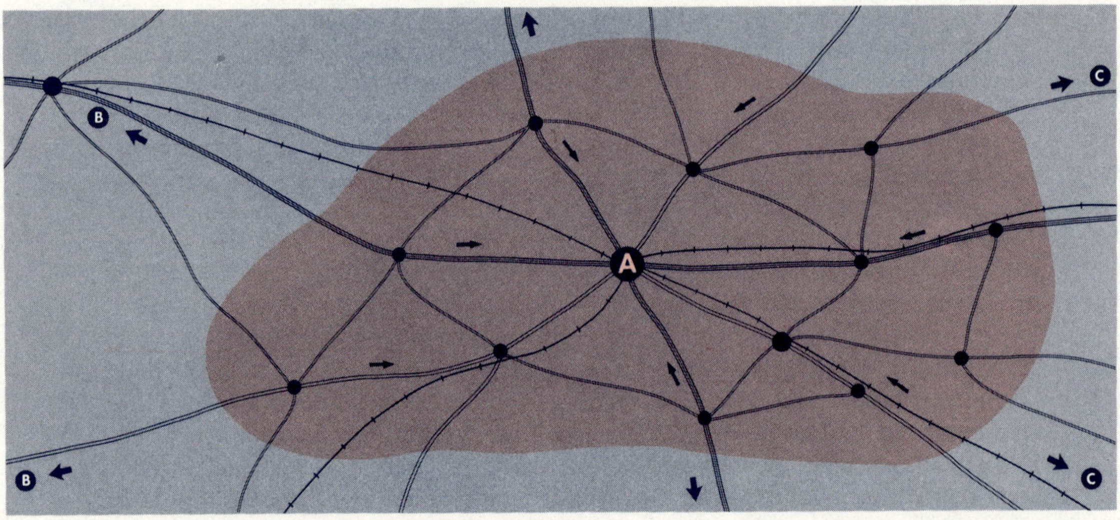

FIG. 4.3 Functional region: the hinterland of a city.

obviously it can only be recognized through the analysis of the many ways the city dominates the activities within it. Where the city's influence wanes lie the boundaries; another city's strength may be taking over. Once again the functional region is an artificial, a mental, construct, and "finding" it depends on the criteria we set ourselves before we begin our search for it.

What we have just said in rather definite and positive terms about regions is not beyond debate, and no one is a geographer for very long before he or she becomes aware of that. Some geographers have doubts about the usefulness of the whole regional concept; others recognize functional regions but find formal regions unproductive as organizing concepts. Some geographers would even argue that a Corn Belt region does not exist: if we have to imagine its existence through the setting of arbitrary criteria, they argue that there is little or no validity to the end result. The dispute goes on in the scientific journals, but it is difficult to read any substantial part of the geographic literature over the past hundred years or so without concluding that the region has been a central concern and a concept that has had a unifying influence over a very wide-ranging discipline. Physical as well as human geographers, people interested in systematic work and those interested in countries—all have employed the regional concept in many useful ways.

The region, then, is a concept. So is *culture.* And the concept of culture is as much a subject for argument as the regional concept. In a sense the culture concept is to anthropologists what the regional concept is to geographers, and endless pages have been written in the search for definitions. Culture is of interest to geographers, too, for peoples and societies, and therefore cultures, occupy specific areas of this earth and are expressed regionally. If it is possible to identify physiographic regions and urban regions and political regions, then it ought to be possible to establish culture regions. But first we should get to know something more about culture itself.

A culture is the way of life of a people. But right away even as simple a definition as this causes trouble: is it their *actual* way of life ("the way the game is played") or the standards whereby they give evidence of *wanting* to live, through their statements of beliefs and values ("the rules of the game")? Anthropologists differ on this, and in the process their own definitions become quite involved. E. A. Hoebel in his book *Anthropology: the Study of Man* (1966) says that culture is "the integrated system of learned behavior patterns which are characteristic of the members of a society and which are not the result of biological inheritance. . . . culture is not genetically predetermined; it is noninstinctive. . . . [culture] is wholly the result of social invention and is transmitted solely through communication and learning." Importantly, this definition tells us that our patterns of behavior are not generated by some primitive, basic instinct but are acquired through learning. This would seem to contradict some currently popular notions about one such supposed instinct, territoriality. A more complex and frequently quoted definition of culture was written in 1952 by anthropologists A. L. Kroeber and C. Kluckhohn in an article about the problems of defining the concept: "Culture consists of patterns, explicit and implicit, of and for behavior and transmitted by symbols, constituting the distinctive achievements of human groups, including their embodiments in artifacts. . . . the essential core of culture consists of traditional (that is, historically derived and selected) ideas and especially their attached values; culture systems may, on the one hand, be considered as products of action, and on the other as conditioning elements of further action." Such definitions do require a bit of dissection, but the overriding point they make is that culture consists of prevailing ways of life, ideas, and values, all with a common heritage.

Culture, then, consists of people's beliefs (religious, political, and so forth), institutions (legal, educational, governmental), and technology (equipment, skills). It is expressed in the way people communicate, in the way they perceive and exploit their resources, in their architecture and art, and in the way they organize that part of the earth that is theirs.

Cultural Landscape

A culture gives character to an area. Aesthetics play an important role in all cultures, and often a single scene, a photograph or a picture, can tell us in general terms in what part of the world the picture was made. The architecture, the mode of dress of the people, the means of transportation, perhaps the goods being carried reveal enough to allow us a guess.

The people of any given culture, when they occupy a part of the earth's surface, transform their land by building structures upon it, creating lines of contact and communication, tilling the fields. There are few exceptions: nomadic peoples may leave a minimum of permanent evidence, and some peoples living in desert margins (such as the Bushmen) and in tropical forests alter their natural environ-

A section of the urban landscape (the cityscape) of Lisbon, Portugal. The tiled roofs, columned public buildings, central square, and inlaid sidewalks give an unmistakable impression of the cultural landscape of Mediterranean Europe.

ment little. But most of the time, there is change: asphalt roadways, irrigation canals, terraced hillslopes. This composite of man-made features is conceptualized as the *cultural landscape,* a term that came into general use in geography in the 1920s. The geographer whose name still is most closely identified with this concept is Carl O. Sauer. In 1927, he wrote an article entitled "Recent Developments in Cultural Geography," and there he produced a deceptively simple definition of it: the cultural landscape, he said, constitutes "the forms superimposed on the physical landscape by the activities of man." But when man's activities cause change in the physical or natural landscape itself, does the physical landscape then become a cultural landscape? For example, when a dam is built in the upper course of a river, that dam can affect the whole character of that river downstream, even hundreds of miles away. It can alter the strength of the river's flow and the rate of deposition of sediments in a delta. Does that mean that the river is no longer part of the natural landscape and therefore has become a cultural landscape feature? Similar issues are raised by man-induced erosion of untilled soil and by regenerated, formerly cutover forests. Anyone who is interested can trace the debate in the geographical literature. For our purposes, perhaps the best definition is the broadest, namely that the cultural landscape includes all identifiably man-induced changes in the natural landscape, involving the surface as well as the biosphere.

The concept of cultural landscape has been an organizing construct for the cultural-historical school of geography. Geographers working in this area are interested in the origins of cultures, their former distributions, their dispersal and spread, and their present character. The reconstruction of past cultural landscapes in an area, at several stages in its history, can answer many questions about the influences that such an area was receiving and transmitting, the prevailing technology, directions of change, and sources of innovations.

A cultural landscape consists of buildings and roads and fields and more. But it also has an intangible quality, an atmosphere that is often so easy to perceive and yet so difficult to define. The smells and sights and sounds of an African marketplace are unmistakable, but try to record those qualities for purposes of comparative study! Geographers have long grappled with this problem of the less tangible characteristics of the cultural landscape, which are often so significant in producing a regional personality, a gestalt, as a school of German psychologists in the early part of this century came to call such a mentally perceived configuration involving many different impressions. J. Gottman, a European geographer, put it as follows in *A Geography of Europe* (1969):

> *To be distinct from its surroundings, a region needs much more than a mountain or a valley, a given language or certain skills; it needs essentially a strong belief based on some religious creed, some social viewpoint, or some pattern of political memories, and often a combination of all three. Thus regionalism has what might be called an* iconography *as its foundation: each community has found for itself or was given an icon, a symbol slightly different from those cherished by its neighbors. For centuries the icon was cared for, adorned with whatever riches and jewels the community could supply. In many cases such an amount of labor and capital was invested that what started as a belief, or as the cult of or even the memory of a military feat, grew into a considerable economic reinvestment around which the interests of an economic region united . . .*

Obviously there is hardly a place in the world where these comments are as appropriate as in the Europe Gottmann was writing about. There, many centuries have gone into the evolution of distinct and rich cultural landscapes and into the molding of those systems of beliefs and values that contribute so strongly to the personality of the culture.

Culture Hearths

As long as man has existed on this earth, there have been places where he has done well, where he has succeeded, where invention and effort were rewarded by an increase in numbers, growing strength, comparative stability, and general progress. And there have been areas where he has not done well at all. Those areas of success and progress were the places where the first larger clusters of population developed, both because of natural increase and because people from elsewhere were attracted there. The increase in numbers brought about new ways to exploit the resources locally available; it also generated power over resources farther away.

Progress was made in farming techniques and consequently in yields. Settlements could expand. Society grew more complex, and there were people who could afford to spend time not in subsisting, but in politics and the arts. The circulation of goods and ideas intensified. Traditions developed and ways of life that became the example for other areas far and near. These areas were man's early *culture hearths,* the sources of civilization, and from here radiated outward the ideas, innovations, and ideologies that would change the world beyond.

Among the early culture hearths to develop were the Tigris-Euphrates area (Mesopotamia), the Nile valley (Egypt), the Indus basin (in what is now Pakistan), the Wei-Hwang Rivers lowland (China) and, in America, the Mexican plateau and the central Andes mountains. In Africa, a culture hearth emerged in the western savanna (Fig 4.4). We shall refer to these areas again when considering the regionalization of culture in Chapter Five, but it is appropriate to identify them here. Mesopotamia, "between the rivers," lay juxtaposed to the Fertile Crescent, one of the places where man first learned to domesticate plants. It was one of the first regions to benefit from the new techniques emanating from the Crescent; the opportunities for irrigation afforded by the river regimes added new dimensions in planning, cultivation, and harvesting. Substantial settlements developed, and Mesopotamia found itself at the focus of a whole network of routes of trade and movements across southwest Asia. Stimuli came from the Mediterranean, from Persia, and from the north, and innovations were diffused as far away as Egypt and the Indus, and farther still. Not all those inventions were quickly adopted wherever they could have been. The wheel and the chariot, in use in Mesopotamia not long after 4000 B.C., found adoption in Egypt more than 2000 years later.

Ancient Egypt, too, was one of the world's earliest culture hearths. The scene of this culture hearth lay above the delta and below the first cataract, where the Nile valley lies surrounded by inhospitable country. The region was open to the Mediterranean, but otherwise it was rather inaccessible to overland contact. In contrast to Mesopotamia, which was something of a marchland, the Nile valley was a natural fortress of sorts. There, the ancient Egyptians converted the security of their isolation into progress. The Nile waterway was the area's highway of trade and association, its lifeline. It also sustained its irrigated agriculture and was a great deal more predictable in its cyclic regime than were the Tigris and Euphrates. By the time ancient Egypt finally began to fall victim to outside invaders, from about 1700 B.C. onward, a full-scale civilization had emerged, whose permanence and continuity are reflected by the massive stone monuments its artist-engineers designed and created.

Much of the archeological record of the Indus civilization has been buried beneath the present water table in the Indus River valley, but there is no doubt that here, too, was a quite sophisticated urban culture, with large and well-organized cities in which houses were built of fired brick and consisted of two and sometimes even more levels. There were drainage systems, public baths, brick-lined wells. Protective fortifications were laid out around the major cities, of which Mohenjo-Daro and Harappa are best known. As in Mesopotamia and the Nile valley, consid-

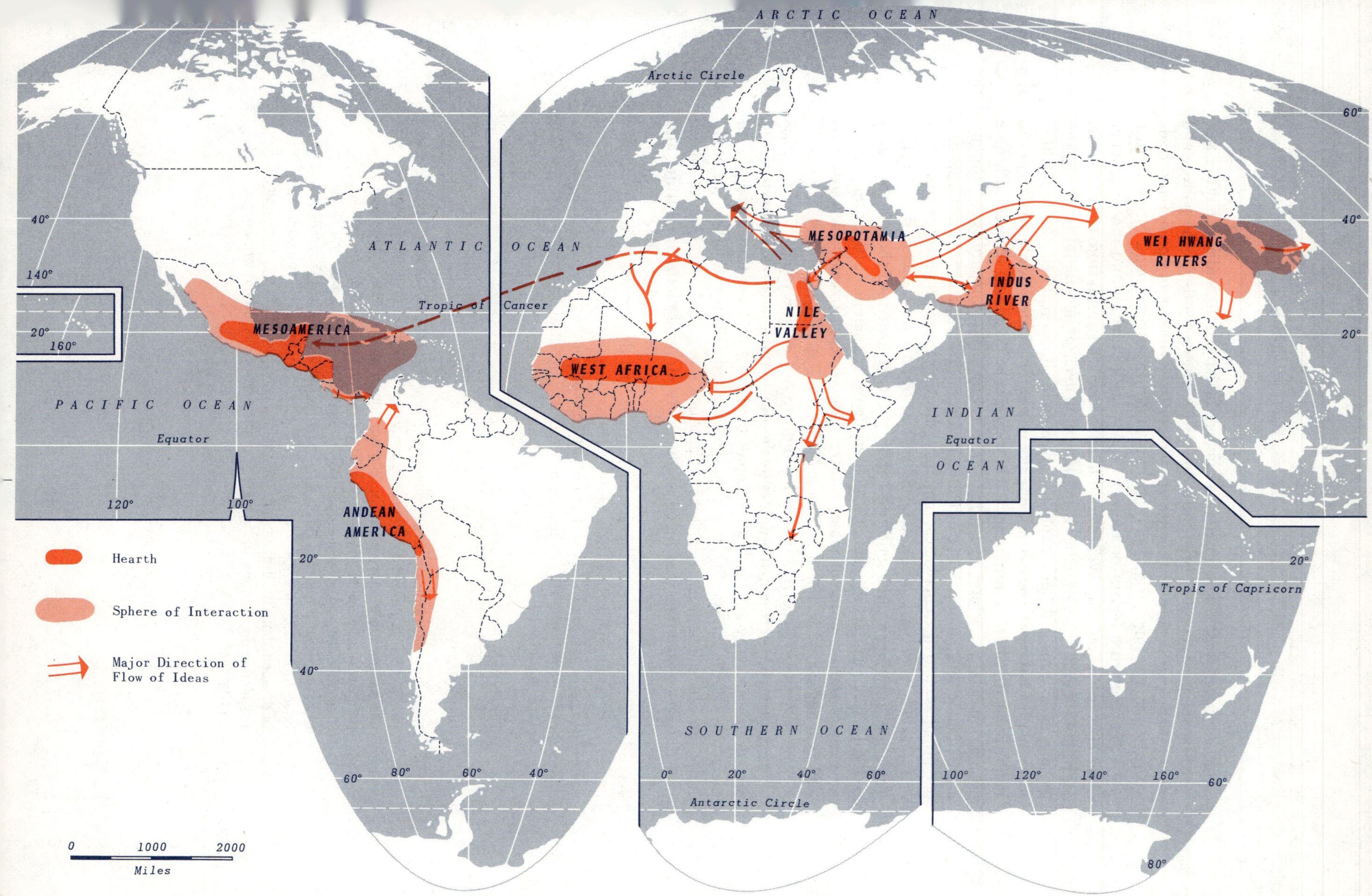

FIG. 4.4 Early culture hearths.

erable advances were made in the technology of irrigation, and the culture was based on the productivity of the Indus lowland's irrigated soils. In the pottery and other artistic expressions of this literate society lies evidence of contact and trade with the other major lowland civilizations of Southwest Asia and North Africa.

The culture hearth of ancient China was located around the confluence of the Hwang (Yellow) and Wei rivers. There, it received stimuli from the southwest Asian cultures in areas of agriculture, irrigation techniques, metal working, and possibly even writing. Whether and how these innovations reached northern China still remain matters of conjecture, but the character of the Hwang-Wei region as a culture hearth of prime importance is beyond doubt. There is a record of Chinese civilization 5000 years long, and from 2000 B.C. onward that record is continuous, quite reliable, and proof positive of China's cultural individuality. In fact, this part of China has yielded archeological remains indicating a human presence many tens of thousands of years ago. Chinese cultural attributes were disseminated across Korea and into Japan, and southward from the core area into the region that is modern China. China still remains focused upon its ancient northern heartland, the only modern country of major stature whose lineage traces back spatially as well as temporally directly to its oldest origins.

Savanna West Africa received stimuli from the Nile valley. The savannas of West Africa lie between the desert to the north and the tropical rain forest to the south. For many centuries the peoples of the desert have traded goods with the peoples of the forests. The forest dwellers needed salt; the desert people could easily provide it. The forest people could produce ivory, gold, spices, hides, wood; from the desert they received metal goods, many kinds of cloth, dried fruits, and much else. The peoples of the West African savanna found themselves positioned very advantageously between the trading partners, and their towns grew into market centers. From the Nile valley these peoples received some political ideas: about the divine kingship, about the power inherent in a monopoly over the use of iron, and about the use of military forces to exact tribute and maintain control. But

---

HIERARCHY

Hierarchy was originally a religious term (*hierarch*, "religious leader") that denoted the organization of a ruling body of clergy into ranks. Today we use the term to identify an order or gradation of phenomena, each grade being subordinate to the one above it and superior to the one below; thus a hierarchy is like a stepladder. It can be established on the basis of many criteria and can be applied to many phenomena. Geographers have tried to define a hierarchy of urban settlements (as an example), both according to the amount of their commercial activity and their respective populations, using such terms as *metropolis*, *city*, *town*, *village*, and *hamlet*.

West Africa was itself also a place of invention and innovation. Some anthropologists say that West Africa, along with Southwest Asia, Southeast Asia, and Middle America, was one of the four major agricultural complexes of the ancient world, where crops were domesticated and farming techniques invented and refined. And so, in West Africa, a series of states of impressive strength and amazing durability arose, which welded diverse peoples into unified entities. Their influence radiated far into the forests and into the desert, and southeastward into black Africa.

Tropical Middle America also was the scene of impressive cultural achievements. From central Mexico to Costa Rica lay a region of highly developed indigenous culture with agricultural specialization, urbanization, permanent transport networks, writing, various art forms, religion, and other features. Here lay the great Mayan and Aztec civilizations, in the culture hearth known as *Mesoamerica*. The Mayan state was theocratic, with a complex religious hierarchy; like the ancient Egyptians, the Maya built magnificent palaces, huge temples, and large pyramids (it has been suggested that Egyptian mariners somehow crossed the Atlantic and contributed their knowledge to Mesoamerica). When the Spaniards arrived, much of Maya civilization was already in ruins, the cities overgrown with tropical vegetation. But on the Mexican plateau the Aztec civilization thrived. The Aztecs' leading city, Tenochtitlan, was a metropolis for its time, counting perhaps 100,000 inhabitants and serving a vast hinterland. The Maya cities had been primarily ceremonial centers, but the Aztec cities and towns were true central places (see Chapter Nine), interacting with major spheres of influence. In these urban places Aztec culture recorded remarkable accomplishments, in architecture, the arts, and other spheres. But it was in the rural areas that the culture had its most lasting impact. Dozens of crops were domesticated, some of them staples in the diets of many millions of people to this day. And the *chinampa* method of cultivation added uniquely to the productivity of the Aztecs' lands. By this method, practiced along the shallow margins of the area's lakes, layers of aquatic vegetation and lake mud were piled

---

*MESO*AMERICA

The word *meso* actually means "middle" or "in between," and thus Mesoamerica is another way of saying Middle America. But it would be confusing to identify both the old culture hearth and the geographic region as Middle America; so Mesoamerica is used to denote an historically significant part of Middle America.

THEOCRATIC

A *theocratic* state is one whose government is deemed guided by divine action. The rulers of such a state are priests, and since their actions represent divine will, they are final and beyond question.

---

upon each other to create thick, layered mats of alluvial and vegetative material so fertile that several crops could be grown in them each year. The method is still in use today.

In South America, the mountainous country of the high Andes witnessed the rise of an important culture hearth. The roots of this culture appear to lie in coastal Peru, from where it may have been transmitted into the *altiplanos,* the valleys between the ranges of the Andean chain. These individual intermontane basins came to sustain discrete regional cultures. One of these was centered on Cuzco, destined to become the focus of Inca civilization. Gradually other *altiplanos* were brought under Inca domination, and eventually the Incas built an empire that extended over much of the Andes region (Fig. 4.4). The Incas were superb engineers and extremely effective colonizers. They managed to create a network of communications over their difficult territory by building roads and bridges, they terraced the steepest of slopes and farmed them, and they brought the unifying element of Inca culture wherever they subjugated neighboring peoples. Although they did not match the Aztecs in calendrics, mathematics, or the arts, the Incas did achieve much in fields of architecture (Cuzco had impressive fortresses and public buildings), economics, administration, and agriculture. Cuzco, their ancient headquarters, was a city of perhaps as many as 250,000 people, 200 miles inland from the ocean and 11,000 feet above sea level! Even today, with these dimensions, it would be among Andean South America's most substantial urban centers. Certainly the Inca civilization ranks among the great ancient culture hearths.

This brief discussion of the world's older culture hearths will be useful to us again in Chapters Five (when we consider the modern regionalization of world cultures) and Seven (where, in our analysis of world agriculture, we will have reason again to view agricultural origins and dispersals). But obviously, except in China where the old focus still prevails, the culture hearths of today are no longer those of the past. Elsewhere, we now look to Western Europe, to the eastern United States, to the western U.S.S.R. for the innovations and ideas that are shaping the modern world. Western Europe became a culture hearth in recent history, disseminating its culture to virtually all parts of the world and superimposing the products of its revolutions in technology, agriculture, and ideology upon the indigenous peoples of the Americas, Africa, and Asia.

Acculturation

When cultures of different strengths make contact, the stronger culture prevails. The culture of the weaker society may be somewhat changed, considerably modified, or even completely transformed, but in every case the stronger culture will contribute certain of its qualities to it. This process by which a culture is changed substantially through interaction with another culture is *acculturation.*

Actually, the process of acculturation is not the one-way street our definition might suggest. Stronger cultures do impose many of their attributes upon weaker ones, but they themselves may well adopt some of the weaker culture's properties. To take a case relating to the previous discussion of ancient culture hearths: when the Spanish invaders reached the Middle American mainland, they soon were able to overthrow Aztec society. They were helped by enemies of the Aztecs who had long faced the threat of being carried off to be sacrificed to the sun-god. The Spaniards' culture began to prevail. Towns were transformed, a new religious order was introduced, and economic realities changed drastically. Acculturation was in full swing. But Spanish culture underwent some changes too, subtle as they were. In their architecture, the Spaniards reflected some of the motifs of Indian buildings. Indian crops were taken back to Iberia. Domesticated animals were acquired. The Spaniards began to wear clothing that revealed the influence of Indian color and cut. A few Indian words got into Spanish use. Indian culture made its contributions, too.

Some societies, such as the southern African Bushmen, the Amazonian Indians, and other peoples isolated by desert, forest, mountains, or distance have experienced little acculturation. At the other end of the scale there are whole societies changed enormously by the process. Japan is the outstanding example. In the case of Japan, a national leadership arose that was bent on achieving power and equality for the nation by purposefully adopting European technological innovations. In the process, the society was transformed, and a new Japan arose to emulate and surpass its Western competitors. This is not to suggest that nothing of the old Japan remains. On the contrary, the fabric of original Japanese culture has proved very durable, and it survives alongside of and interwoven with the new order.

Some societies have been fortunate enough to be able to select the elements of an alien culture they would adopt, but others have experienced acculturation through superimposition. To a greater or lesser degree the recent history of much of America, Africa, Asia, and Australia has been a history of Europeanization, as the beliefs, values, and practices of European cultures were imposed on indigenous societies. European-style cities arose, European languages became the official languages of distant countries, European religions were promoted, and European standards of law and justice were invoked. In spheres of life ranging from mode of dress to education, the image of Europe became a veneer over local cultures. African schoolchildren learned more about European history in their classrooms than about Africa's past. Asian families began to send their children to British and French schools and universities, for progress was now dictated by European rules. American Indians found their lands carved into huge estates, and if they were not lucky enough to be admitted as squatters, there were fences to keep them out. Changes in practices governing the possession and use of land were perhaps the most difficult of all elements in the acculturation process for the recipient society to accommodate. We shall return to this theme when we examine the regionalization of the world's cultures in Chapter Five. The case of black Africa is especially relevant in this connection.

The Europeanization of the colonial world involved the coming of Christian religions to Africa, the Americas, and Asia. Angola and Moçambique, Portuguese colonies in southern Africa, have become overwhelmingly Catholic (among non-African religions, Catholicism has a virtual monopoly), just as is the case in Portugal itself. The Dutch Reformed church gained a foothold in South Africa, the Anglican church in parts of the former British Empire. The photograph shows the Catholic cathedral in Lourenço Marques, Moçambique's capital city.

Culture and Environment

The distribution of old culture hearths inevitably leads to questions about the environment under which they thrived. Some ancient cultures lie revealed today by ruins in the desert and the tropical rain forest. Could the cities and other settlements have prospered under such conditions, or have climates changed so much that where there is desert today there may have been ample water supply some thousands of years ago? And in any case, what are the optimum environmental conditions for the evolution and growth of civilizations? Present-day world culture hearths seem to lie in temperate climatic regions, not in the tropics, or in the desert, or in near-polar zones. Does this mean that we can generalize about "optimum" climatic conditions for the human success story?

Long ago, Aristotle thought so: he argued that the peoples of cold, distant Europe were "full of spirit, but wanting in intelligence and skill; and therefore they retain comparative freedom, but have no political organization, and are incapable of ruling over others. Whereas the natives of Asia are intelligent and inventive, but they are wanting in spirit, and therefore they are always in a state of subjection and slavery. But the Hellenic race, which is situated between them, is likewise intermediate in character, being high-spirited and intelligent . . ." (as he is quoted from a 1921 translation, *The Works of Aristotle*). And many geographers after

Aristotle have held similar views. How easy it is to view people living in cold climates as "hardy, but not very intelligent," those of the warm tropics as "lazy and passive," and those in the intermediate zones as productive and progressive. Here is how Ellsworth Huntington, a twentieth-century geographer, stated it in *Principles of Human Geography,* published in 1940: "The well-known contrast between the energetic people of the most progressive parts of the temperate zone and the inert inhabitants of the tropics and even of intermediate regions, such as Persia, is largely due to climate. . . . the people of the cyclonic regions rank so far above those of other parts of the world that they are the natural leaders."

The doctrine expressed by these statements is *environmentalism* or, more specifically, *environmental determinism.* It holds that man's behavior, individually and collectively, is strongly affected by and even controlled or determined by the environment under which he lives. It suggests that the climate is the critical factor in this determination: for progress and productiveness in culture, politics, and technology the "ideal" climate would be, say, that of Western Europe or the northeastern United States. The people of hot, tropical areas or cold, near-polar zones might as well forget it. Their habitat decreed that they would never achieve anything even approaching the accomplishments of their mid-latitude contemporaries.

And so, for a time, geographers set about their attempts to explain the distribution of present and past centers of culture in terms of the "dictating environment." In the early part of this century that was a major theme in the field, the manner in which the natural environment conditions human activities. Quite soon, though, there were geographers who doubted whether those sweeping generalizations about climate and character, environment and productiveness were really valid. They recognized exceptions to the environmentalists' postulations (for example, the Maya empire in Mesoamerica may have arisen in tropical rain forest) and argued that man was capable of much more than merely adapting to his natural environment. As for the supposed "efficiency" of the climate of Western Europe, this was an interesting idea but not scientifically proved. And surely it was best not to base laws of determinism on inadequate data in the face of apparently contradictory evidence.

These arguments helped guide the search for answers to questions about the relationships between man and his environment in different directions, but the unqualified environmentalist position was still held by some in the past several decades. For a more modern approach, it is interesting to read S. F. Markham's *Climate and the Energy of Nations* (1947), still an environmentalist statement, but with the benefit of Huntington's massive writings and the works of several others behind it. Markham thought that he could detect in the migration of the center of power in the Mediterranean (from Egypt to Greece to Rome and onward) the changing climates of that part of Europe during the most recent several thousand years of glacial retreat. He argued that Egypt was much cooler in its prime than it is now; when the Greeks rose to prominence, Greece's climate, too, was more conducive to the generation of energetic people. Rome was so cool that public hot baths were a part of daily life. In all this, Markham saw the northward movement of isotherms. Like his predecessors, he tried to determine the "optimum" temperature for physical and for mental activity, and he was sure that these temperatures prevailed in turn in Egypt, Greece, and Rome (and later in Western Europe).

But geographers grew increasingly cautious about such speculative writings, and they began to ask their questions about man and his environment in new ways. There still is the old interest in how man is affected by his natural environment, but if there are generalizations to be made, they ought to come from detailed, carefully designed research. Everyone agrees that man's activities are in certain ways influenced by his environment, but man appears to be more the decision maker in this context than the subordinate the environmentalists believe him to be.

Environmental Perception

Out of these deliberations has come a whole new field of geography, in which interest focuses *not* upon the manner in which environment conditions human behavior, but on the way man perceives his environment and acts on the basis of his perceptions. We all have an image of the world around us, and we could argue that our decisions and actions are based on this image rather than on reality. If this is the case, then our concern when we analyze behavior ought not to be with the objective realities of our environment, but with these images we have in our head. What goes into the formation of these images? Undoubtedly an Indian's view of the American Southwest would be quite different from most white Americans' image of that region. A black person living in an urban ghetto perceives it quite differently from a white person who does not live there. Note that we have begun to involve in our discussion not only the natural environment, but also the social environment. Environmental determinists of the past mostly failed to recognize that the environment that interacts with man consists not only of landscapes and climates but also of attributes created by man himself, all elements of a cultural environment.

Now, when we want to study human activities in the context of environment, the magnitude of the problems involved begins to emerge. Experience, the accessibility of information, psychological state, physiological condition, cultural traditions, social constraints—all these factors come into play. And this is only a short list. The longer it becomes, the less likely we are to engage in the business of sweeping generalizations about climate and behavior!

Out of this research, we may hope, it will be possible for us some day to know not only how we Westerners perceive a non-Western culture, but also how the people of that culture themselves view it. If Europeans had known more about the way Africans view their lands and herds, then the shock of contact between the cultures might have been less. That land could lie unfenced, open to be used and exploited by all, was inconceivable to the white man. That land should be taken as personal property and parceled into private holdings was equally inconceivable to Africans. To the white man who urged the African cattleman to weed out the old and sick animals so that the quality of the herd would improve and overgrazing would cease, an African replied, "It is like suggesting that you take your billfold and throw out the dirty, worn money; your wallet will be much cleaner and less bulky." He could conceive of our money economy no more than Europeans could grasp the idea that in some parts of Africa, the number of cattle is all that matters to the owners, not how old, sick, or unproductive they are. It is all a question of how resources are perceived.

RACE AND CULTURE

When you, as a traveler, enter a country for a visit you will be asked to fill out a questionnaire to be presented, with your passport, to the immigration officer. As often as not, one of the questions will be, "What is your race?" And the appropriate answer, of course, is "human." All the people of this world belong to the same subspecies. We all have far more features in common than there are differences between individuals or groups.

But the answer "human" on the questionnaire is not the one that is wanted. In fact, it may get you into trouble. The term *race* has come to refer to an undeniable reality of our human existence. People differ physically from each other. And what is more, those differences have regional expression. Even after centuries of movement and migration, mixing and intermarriage, and even in our mobile world, peoples with distinct physical traits remain clustered in particular areas of the world. Thus we use the term *race* in quite a different way: we speak of the European, African, and Polynesian *races* of man. It may not be quite correct to do so (indeed, some anthropologists have proposed that the whole concept of race be abandoned), but it is an inescapable fact of life.

Race, then, is first and foremost a biological concept; it refers to people's physical features. A racial group such as those just mentioned (European, African) is recognized because it has a distinctive combination of such physical traits, the product of a particular genetic inheritance. This inheritance has been determined by many centuries of isolation and inbreeding, during which a certain dominant set of genes, a gene pool, evolved for each racial group. Man presumably evolved from a common stock, but after man radiated outward from the source area (perhaps Southwest Asia, northern, and eastern Africa) spatial and social isolation began to play its role in generating discrete gene pools and racial groups in Asia, Australia, Africa, Europe, and the Americas. Humanity differentiated into what used to be called the white, black, red, yellow, and brown races. As the concept of race was refined, other terms emerged: Caucasoid, Negroid, Amerindian, Mongoloid, Australian. Classifications are still changing (Fig. 4.5). Anthropologists have recently been using a nine-unit racial classification of man: (1) *European,* (2) *Indian,* peoples of the Indian subcontinent, (3) *Asian,* peoples of China, Japan, inner Asia, Southeast Asia, Indonesia, (4) *African,* peoples of Africa south of the Sahara, (5) *American,* the indigenous, Indian populations of the Americas, (6) *Australian,* the original peoples of Australia, (7) *Micronesian,* (8) *Melanesian,* and (9) *Polynesian.* The last three groups are peoples of the Pacific Ocean's islands. The Euro-

FIG. 4.5 Generalized map of the racial groups of man.

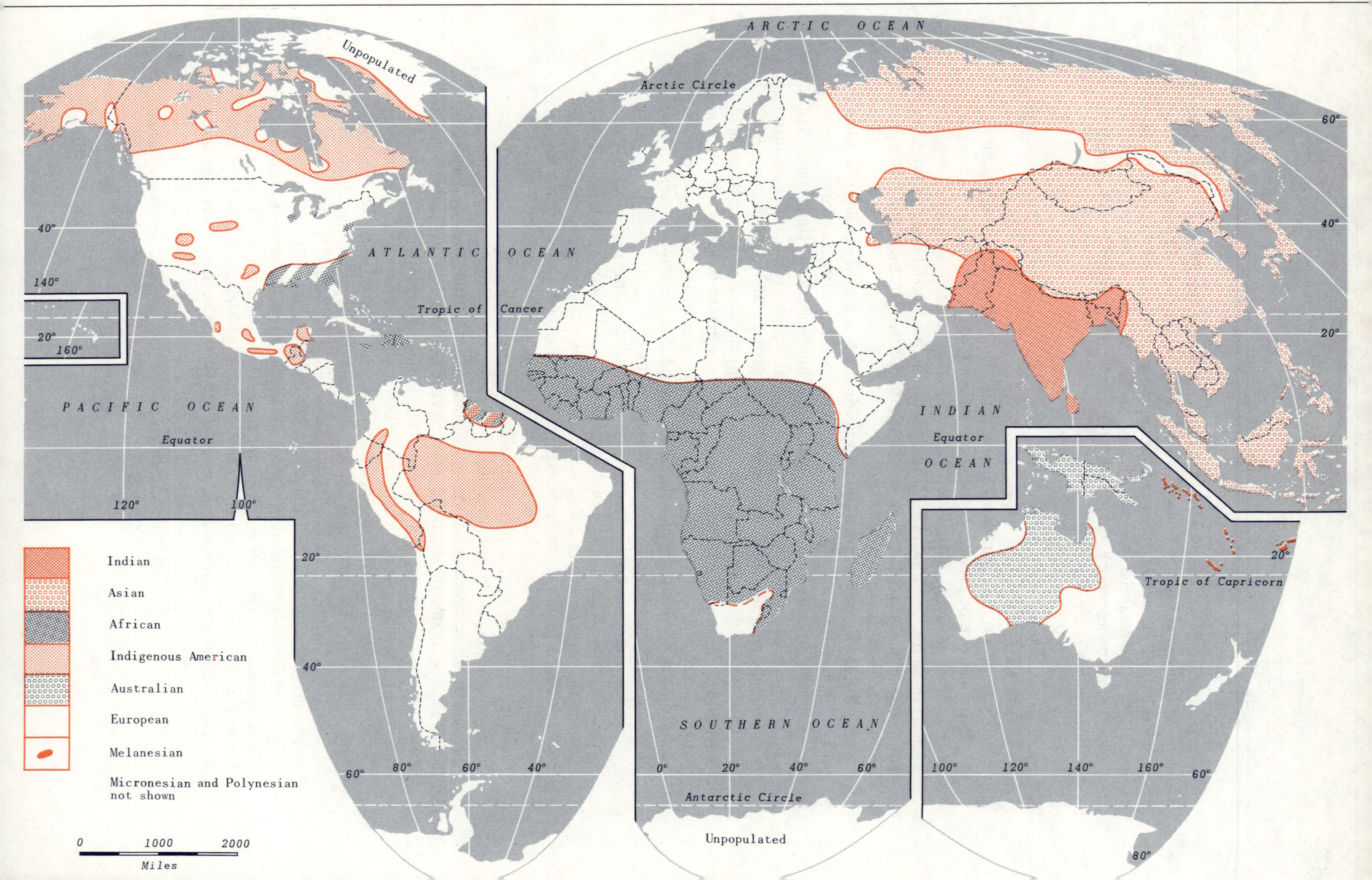

pean race includes not only Scandinavians, Russians, Germans, and Italians, but also the peoples of Southwest Asia (such as the Iranians, Syrians, Saudi Arabians) and North Africa (Egyptians, Algerians, Moroccans).

Physical Features

What are these physical traits that form the bases for racial classifications of this kind? They are quite familiar to us, especially one, the cloak of color we all wear. Skin color is the most pervasive of physical traits, and our differentiation by color has bedeviled human relationships for uncounted centuries. Skin color is a matter of pigmentation, a protective element against strong radiation. The more pigment, the darker the skin. But there are internal variations within groups. There are dark-skinned northern Europeans and light-skinned Asians. The first image in the one race's or culture's perception of another, however, is still likely to be skin color. This is despite the fact that the color of human skins varies from white to nearly black, through all the intermediates.

Another somewhat obvious physical feature is hair, straight or woolly, fine-textured or thick. Asians tend to have straight hair, Europeans, often curly hair, and Africans, woolly hair. Again, there are differences in people's stature or height. Although this factor is obviously related to adequacy of nutrition, it is clear that there are genetic forces at work as well. Under similar living conditions, peoples have developed quite different average heights. In the eastern Zaïre (Congo) and neighboring areas, three peoples of different stature live together or nearly so: the tall Watusi, the small Pygmy, and the medium-height Bahutu. Still another, less obvious physical trait is the shape of the head. Anthropologists for a long time thought that the *cephalic* index (which is the ratio of the breadth and length of the skull) might yield substantial insights into the evolution of races. But although it was possible to record a clustering of prevailing head shapes (Japanese have round heads; Western Europeans have long heads), it did not prove possible to use the data for much more than speculation.

Other physical traits, including the shape of the eye, of the nose, of the lips, and the degree of protrusion of the jaw, provide data to compare, but little in the way of evidence for the racial differentiation of man. Anthropologists could do little more than classify peoples according to individual traits and map their distribution; meaningful covariances were difficult to find. But more recently, another physical feature has been studied with more success. This is one we normally cannot see at all, the blood and blood types.

We all know that various blood types exist. The classification is based on the presence or absence of agglutinative agents, which cause red blood cells to clot when they are mixed with other blood that does not contain them. There are four blood types, A, B, AB, and O. The letters A and B denote blood carrying those types of agents, AB carries both, and type O does not have either. When the existence of blood types was discovered, data soon began to accumulate from all over the world on this newly recognized trait. And since a person has only one blood type, the resulting statistics were much less hazy than those based on skin color or height. A population might vary in height from 5 feet 1 inch to 6 feet 4 inches, with 5 feet 9 inches the average; the skin color might average brown, but there might be variation from very light to very dark. But the blood type could be reported in hard figures: 70 percent O, 20 percent B, 10 percent A.

The world distribution of blood types holds much interest, although it remains subject to different interpretations. The O type is strong, for example, among America's Indians (over 70, often 90 percent); in Northeast Asia it is still over 70 percent, and among Mongoloid peoples in East Asia it is 30 to 40 percent. Does this mean that the affinity between American Indians and Asians is confirmed? Some researchers think so. They feel that the O type was diluted in Asia by contact with other genes, but that the American Indians, by migrating into the Americas, retained their O type almost unchanged. This view holds that the "original" blood type was type O and that A, then B and AB types evolved and began to supplant the O type. And indeed, the O type remains strong in comparatively isolated areas of the world, such as central Africa and northern Australia. It might also explain the prevalence of the A type in Western Europe and the B type farther to the east, a later intrusion. But the same theory raises questions. Although the O type dominates in large areas of Africa, the B type occurs there as well. Where did that B gene come from? How did it also get to islands of the Pacific Ocean and even into the Americas? Many scholars now believe that man ranged far more widely over this earth in ancient times than was long believed and that very early migrations are the answer. Few believe that parallel evolution in far-flung parts of the globe produced the situation. Whatever the eventual answers, the analysis of blood types has added enormously to the arsenal of evidence in the search for the origins and early dispersals of man.

It would be pointless to try to draw a map of the world showing the sources of man and arrows to suggest routes of migration. Scholars have changed their views several times and will again. One map that is frequently seen has three major areas of dispersal: in East Asia (the Mongoloids), in West Asia (the Caucasoids), and in Africa (the Negroids). The Negroid source area is usually placed in the vicinity of Nigeria/Cameroun, although the evidence is growing that African sources lie elsewhere. Africa, apparently, was *not* populated from this source, first westward into West Africa and later, by Bantu, southward into central and southern Africa. It is a convenient explanation, but the African source may well lie far to the west, in what is today Mauritania. And there may have been more than one source area. For example, the Bantu may have emerged from quite another region of the continent. As long as the uncertainties are of this scale, maps are premature.

Race and Culture

Culture, we stated earlier in this chapter, is acquired, learned, not inherited. Yet it is so easy to generalize about correlations between race and culture that this truth is all too easily forgotten. And from these correlations spring doctrines of racial superiority and inferiority, the roots of racism. When you postulate that Western society has advanced further than other peoples because the white race is superior to other races, you assume that the cultural capacity of white people is greater than that of others, a proposition that has not been proved.

Certainly there have been scholars who have sought to prove that certain races are inherently superior. Some still are working on the proposition. But there is no basis for concluding that the races of man differ in their innate capacity for cultural development. True, peoples differ in their physical size and height, their skin color, hair type, and in other ways. But these are not grounds for assuming that one racial group is better adapted for functioning under modern conditions than others. And, as anthropologists have proved, the "closer to the ape" analysis of human physiology does not sustain the thesis that white peoples are more developed than some of their contemporaries. All races have traits that resemble their distant ancestors, Europeans more than some others. Once again skin color pervades the argument more than it ought to. When it comes to such traits as the head hair and body hair, Europeans are far more apelike than Africans or Asians.

These are sensitive topics, but this has not deterred scholars from searching in other directions for evidences of racial superiority, for example, by calculating brain sizes and correlating these data with intelligence. There has been some evidence that larger brain size may be related to greater intellectual capacity, but larger brain size in turn is connected with better nutrition and general physical well-being. The average brain size of several non-Western peoples (for instance, in Africa) is greater than that of Europeans. And there is evidence that Neanderthal man had a larger brain than modern man, or one at least as large. This may mean that it is not brain size, but the functional differentiation of the brain that has to do with intelligence.

And then, of course, there is the "evidence" from intelligence tests. How many times has it been reported that a group of white schoolchildren in Massachusetts or some other northeastern state did better in a test than a group of minority children elsewhere? And how often have the wrong conclusions been drawn and popularized by uncritical reporting? Until racial inequities in plural (multiracial) societies such as that of the United States are removed, intelligence tests will continue to reveal different levels of intellectual achievement between racially discrete groups. Those differences do *not* have to do with the inherent ability of the children being examined; instead they have to do with the circumstances of their upbringing, their education, access to stimulation, and cultural environment. And there is something else about these tests. They are practically always prepared by people of one cultural experience, but applied to people with another cultural environment. Anthropologists have long warned against generalizations drawn from such cross-cultural testing when it is carried out, say, in Africa or Australia; but even within our own society tests that have cultural bias are applied every day.

If it is impossible to prove the racial superiority of one race over another, it is also impossible to disprove the existence of differences between the races in terms of mental functioning and psychological capacities. Psychologists have reported that there are differences in the learning abilities of different breeds within the same species of animal; the same could be true for man. Still, there is no conclusive proof for such differences (and difference does not mean superiority). To prove them, it would be necessary to eliminate the cultural experience variable from the test. No one has succeeded in doing that.

All this does not help us much in answering some of the questions posed by the map of cultures of the world. For example, some of man's races are culturally far more advanced than others. Why? And why have some of the cultures that once led the world fallen into disrepair? And again, why were some societies static for a long period of time, suddenly to blossom into major cultures? Here we are, still asking the questions the environmental determinists also sought to answer. The answers may have eluded us so far, but at least we have an idea of where *not* to search for solutions. Apparently solutions and answers do not lie either in predestination by environmental conditions or in race.

We simply do not know what stimulates man to achieve and accomplish in certain areas and at given times in his history. We, with our contemporary view of the world, find it all too natural that Western society is complex and advanced and that indigenous Australians and South American Indians have less advanced cultures. But in the context of human history it is only yesterday that Europeans were excelled by other cultures elsewhere in the world, in Asia, in Africa, and in the Americas. In another age, we would have had quite a different impression and quite different explanations. Who, at the height of the power and splendor of Mesopotamia, would have predicted that barely a shadow of all that achievement would remain a few centuries later? Who, in the days of the Roman Empire, would have imagined that Rome's distant, backward colony Britain would before long generate an empire to span the world? And who, a hundred years ago, would have predicted for Japan, an isolated, poor, island country off Asia, a meteoric rise and an explosion of cultural development? There are lots of explanations now, with the benefit of hindsight. But all those explanations do not seem to give us the confidence to predict where it will happen next. In Ethiopia? In Brazil?

When we speak of race and culture, we must do so with great care, for our knowledge is still very limited and incomplete. The racial composition of Japan did not change when the country experienced its enormous burst of cultural development (the climate did not either). When it comes to racial identifications, there may be greater differences *within* a racial group in some categories than *between* racial groups. In this respect race shares with the concept of the region its primary quality as an intellectual construct, not reality.

LANGUAGE AND CULTURE

Language is a vital element of culture. It is our means of communicating and in literate societies the use of written words increases greatly the efficiency with which culture is transmitted from generation to generation. This is not to suggest that peoples who do not write (preliterate societies) cannot transfer their culture from one generation to another; but writing is an enormous asset and, indeed, a necessity in complex, advanced cultures.

What is a language? The term is defined in many different ways. Look in any dictionary. For our purposes, the question of definition is not terribly important as long as we realize that a language is a language no matter how large or small the vocabulary or how large or small the number of people using the language. As we discuss language in the spatial context of culture, we should also remind ourselves now and then that languages are always changing. That reminder should hardly be necessary, considering how our own use has changed over the past few years. Again, a language exists in several ways. In advanced cultures there may be an official, standard language, but many speakers will use a variation of it containing some incorrect words ("He ain't seen nothin' yet"). This may be *slang,* with roots in the social structure of society; but there are also special uses and accents with regional expression. These are *dialects.* A trained ear can tell much about where a speaker lives, say, in the United States; it often does not even take a trained ear to recognize a Southerner or a Bostonian.

Language Distribution

In the context of culture, we are obviously interested in the way languages are distributed in the world, what the forces creating this distribution are, and how the pattern is changing. But first, let us consider briefly the problem of classification. Depending on the way language is defined, between 2500 and 3500 languages are in use in the world today, over 1000 of them in Africa alone. Clearly we cannot begin to study the relative location of around 3000 items in the world without bringing some order to the jigsaw. Languages are classified with terms also used in biology, and for the same reasons: certain languages are related to each other; some are not. Languages grouped together in a *family* are thought to have a shared origin; in the *subfamily* their commonality is more definite. These are in turn divided into language *groups,* which consist of sets of individual languages. As Fig. 4.6 shows, English as a language is part of the Indo-European family, the largest of the families in terms of the number of speakers. Indo-European languages are spoken in Europe, Russia, India, North and South America, and several other areas, by about half the world's people all told. English is one of the Germanic (sometimes called Teutonic) group of languages within this family, along with German, Dutch, and the Scandinavian languages. French, also an Indo-European language, belongs in the Romance group, with Italian, Spanish, and Portuguese. Linguists postulate that a lost language they call Proto-Indo-European existed somewhere in east-central Europe and that the present languages of the Indo-European family evolved from this common heritage. In the process, as vocabularies grew and changed and peoples dispersed and migrated, differentiation took place. Latin arose out of this early period, to die out and be supplanted by Italian, French, and the other Romance languages.

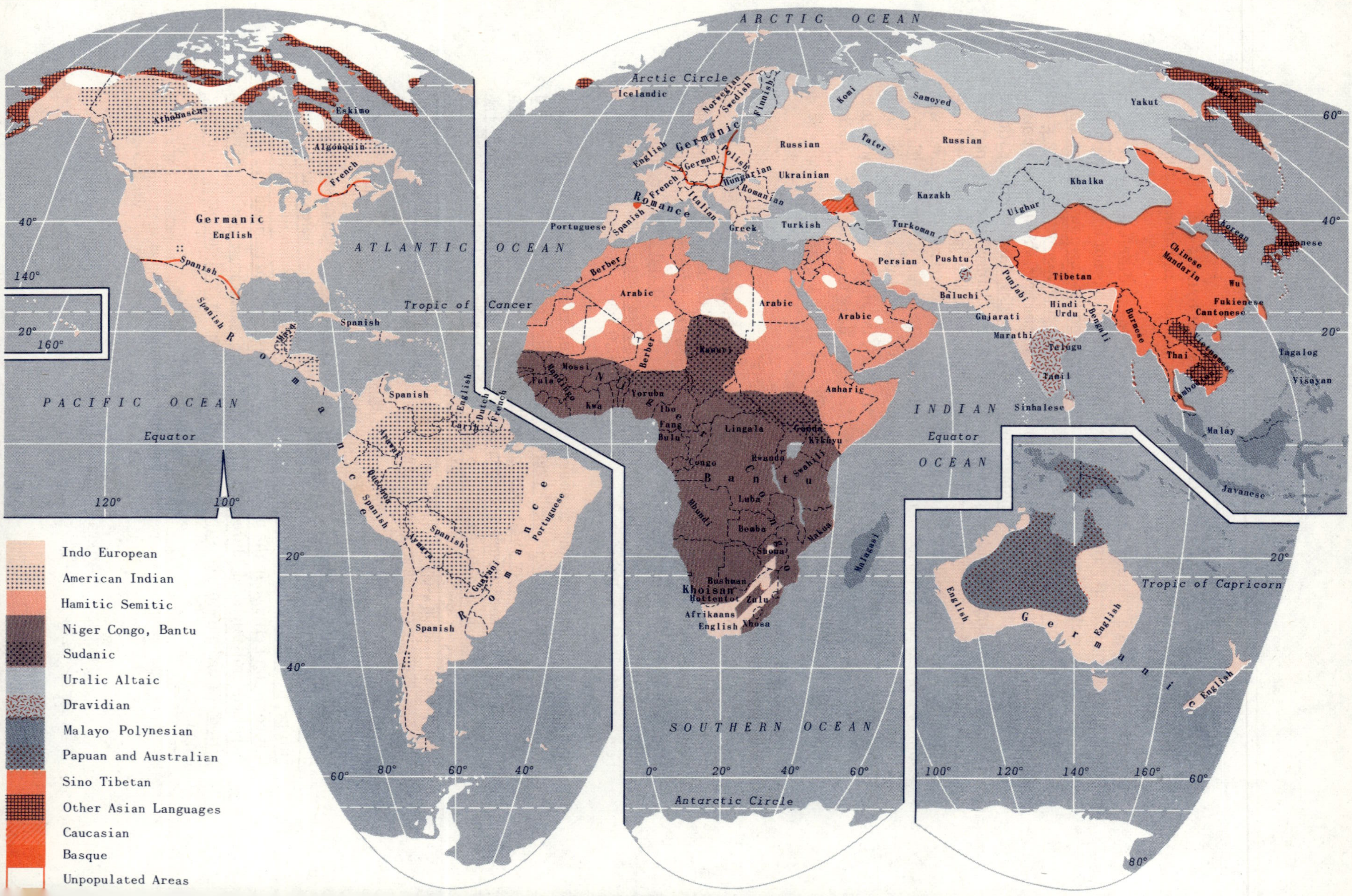

ARCTIC OCEAN
ATLANTIC OCEAN
PACIFIC OCEAN
INDIAN OCEAN
SOUTHERN OCEAN
Arctic Circle
Tropic of Cancer
Equator
Tropic of Capricorn
Antarctic Circle
Germanic
English
Spanish
Romance
Portuguese
French
Eskimo
Athabascan
Algonquin
Maya
Carib
Dutch
Arawak
Quechua
Aymara
Guarani
Icelandic
Norwegian
Swedish
Finnish
German
Polish
Hungarian
Romanian
Italian
Greek
Russian
Ukrainian
Komi
Samoyed
Tater
Kazakh
Turkish
Turkoman
Khalka
Uighur
Yakut
Persian
Pushtu
Baluchi
Punjabi
Hindi
Urdu
Bengali
Gujarati
Marathi
Telugu
Tamil
Sinhalese
Tibetan
Chinese Mandarin
Wu
Fukienese
Cantonese
Burmese
Thai
Vietnamese
Cambodian
Korean
Japanese
Tagalog
Visayan
Malay
Javanese
Malagasi
Berber
Arabic
Amharic
Mossi
Mandingo
Fula
Kwa
Yoruba
Ibo
Fang
Bulu
Kanuri
Niger Congo
Lingala
Ganda
Kikuyu
Rwanda
Swahili
Congo
Bantu
Luba
Mbundu
Bemba
Makua
Shona
Bushman
Khoisan
Hottentot
Zulu
Afrikaans
Xhosa
Indo European
American Indian
Hamitic Semitic
Niger Congo, Bantu
Sudanic
Uralic Altaic
Dravidian
Malayo Polynesian
Papuan and Australian
Sino Tibetan
Other Asian Languages
Caucasian
Basque
Unpopulated Areas

Figure 4.6 tells us not only the modern story of human migration, which brought Indo-European languages to the Americas, Australia, South Africa, but also something of the past in Eurasia itself. Note that Indo-European languages are spoken as far east as India; Hindi is one of the languages in this family. Does this mean that India may have been peopled substantially from the west, through Southwest Asia and Iran? It does. The oldest inhabitants of the Indian subcontinent were driven into the tip of the peninsula by newcomers from the northwest, and the linguistic contrasts are part of the proof.

Languages in Africa

The promise of the spatial study of languages and dialects in historical and cultural geography is especially evident in Africa. Perhaps as many as a thousand languages are spoken in black Africa, and linguists have been working for decades to record many of these; most of them were unwritten. From the resulting data, it has been possible to gain some significant insights into Africa's past. The languages of black Africa, as Fig. 4.6 shows, are mostly of the Niger-Congo family, which extends from far West Africa all the way to the south. This Niger-Congo family can be subdivided into seven subfamilies, one of which is the large Bantu subfamily whose languages are spoken by most of the people living near and south of the equator. The oldest languages of black Africa are the *Khoisan* languages, which share a click sound. These include the Bushman languages, and they are spoken still by some tens of thousands of people in southwestern Africa. Perhaps they once were the main languages over much of Africa, but they are comparatively insignificant now.

How can languages help us in reconstructing the cultural evolution of Africa? Consider what has happened in Europe where the Romance language family has differentiated into, among others, Italian, French, Spanish, and Portuguese. Even within these languages we can recognize differentiation, such as between Castilian and Catalan Spanish, and between northern and southern French and Walloon. Such differentiation develops over time, and it is reasonable to assume that the more time that elapses, the greater the differences between languages. If, therefore, the peoples of a large region speak languages that are somewhat different, but still closely related, it is reasonable to argue that they have migrated into that region or emerged there relatively recently. On the other hand, languages that are of recognizably common roots yet strongly different must have undergone modification over a long period of time.

Among the languages of the Niger-Congo family, those of the Bantu subfamily are much more closely related to each other than those of other subfamilies. It may be deduced, then, that the Bantu peoples of central and southern Africa are of much more recent origins than West Africans. Even a little general knowledge

FIG. 4.6 Language areas of the world. (After a map in *Man's Domain*, McGraw-Hill, New York, 1968, p. 11.)

of African names can suggest these affinities: The word *Bantu* should actually be spelled BaNtu ("people"), with the *Ba* a prefix. Sometimes the prefix is retained in common usage, sometimes not. The *Wa*tusi, for example, are sometimes called the Tusi or Tutsi. The people of southeastern Uganda are the *Ba*Ganda or Ganda. The famous Zulu are actually the *Ama*Zulu. In stories about Rhodesia you often read about the *Ma*Shona or Shona. Remember Bechuanaland, now called Botswana? That should have been *Be*Chuanaland, or Chuanaland, or, as it is written in some places, Tswanaland. The point is becoming clear: *Ba, Ma, Wa,* and *Ama* are not very far from each other linguistically. But the BaGanda live at one end of black Africa, and the AmaZulu practically at the other. You will not find the same thing in West Africa.

There are other pieces of evidence to support the conclusion that the peoples of Bantu Africa have a shorter history than those of West Africa, but the linguistic evidence and orally transmitted traditions provide the basis. Within Bantu Africa itself, too, linguistic data have been interpreted to mean that the Bantu did not simply spread from West Africa into the central and southern areas, but emerged in the region perhaps several millenia ago. Some of their migrations have been reconstructed from linguistic and oral data.

LINGUA FRANCA A *lingua franca* is a language used in trade, commerce, and general communication among peoples who speak other languages of their own. In East Africa, Swahili is such a language, and it is used in Kenya, Tanzania, Uganda, and even the eastern Zaïre (Congo) and northern Moçambique. Swahili is essentially a Bantu language, much modified by the tongue brought to East Africa by Arabic traders. Various forms of Swahili are spoken in different parts of the region, but it serves effectively as a means of communication in a multilingual area. In West Africa, the Hausa language also has the characteristics of a *lingua franca.* It is used in commerce and other contact across much of the West African Sudan, from eastern Senegal to northern Nigeria. Along the coast of western Africa there developed a pidgin English called Wes Kos, also a kind of *lingua franca* in the ports and coastal towns of that area.

LINGUA FRANCA

The term *lingua franca* derives from "Frankish language" and applies to a trade tongue spoken in Mediterranean ports that consists of a mixture of Italian, French, Greek, Spanish, and even some Arabic. But its meaning in the chapter is "common language," a language that can be spoken and understood by many peoples although they speak another language at home. Thus in Africa, when the colonizers came, English became the *lingua franca* over large areas, even though the African peoples speak hundreds of African languages. French became the *lingua franca* in other areas.

In the Caribbean, too, and in the islands of the western Pacific, pidgin based on English has come to serve as a *lingua franca;* its use extends even to non-English-speaking territories such as, for example, Dutch-influenced Surinam. Elsewhere, *linguae francae* facilitate communication among hundreds of millions of people: Hindi in India, Chinese in much of eastern Asia, Malay from Southeast Asia to New Guinea. And in much of the formerly colonized world the long-prevailing European powers introduced languages that have become limited *linguae francae:* French across West Africa and Southeast Asia, and English in parts of West, East, and southern Africa as well as India and Pakistan.

MULTILINGUALISM Some countries have no effective *lingua franca;* instead they have a linguistic fragmentation that may reflect a strong cultural pluralism and the existence of divisive forces. This is true in formerly colonial states where peoples of diverse tongues were thrown together by the force of foreign interests; there are numerous examples in Africa and Asia. But it also exists elsewhere. As Fig. 4.7 shows, Indian languages are spoken by more than half of the people in substantial areas of Guatemala and Mexico, although those countries tend to be viewed as Spanish speaking. A substantial section of Canada is French speaking, and Canada's multilingualism mirrors a cultural division of considerable intensity. In Belgium, a fairly sharp line divides the northern, Flemish-speaking half of the country from the Walloon-speaking south. In the examples just given, the linguistic fragmentation is regional in character, but this is not always the case in multilingual countries. In South Africa the white population of just under 4 million speaks Afrikaans and English, but both languages are spoken throughout the white areas (although interior

FIG. 4.7 Indian languages in use in Middle America.

Multilingualism prevails in many countries. Canada, for example, is a *bilingual* country. Although a majority of Canadians speak English, a substantial segment of the population is French-speaking. As a minority in Canada, the French Canadians have demanded recognition for their language and culture. The photograph shows one result of this insistence: public signs are printed in both English and French.

rural areas of the Orange Free State and the Transvaal are dominantly Afrikaans speaking). If there is regionalism to the linguistic pattern of South Africa, it lies in the Bantu languages spoken there. As elsewhere in Africa, the black peoples of South Africa speak numerous, discrete languages (about two dozen in that country). Thus Zulu is the major African language in Natal, Xhosa is important in the Transkei, and so forth.

South Africa, India, and Switzerland (where German, French, and Italian are used) are multilingual countries in which several languages are spoken. Belgium, Canada, and Guatemala, where two languages prevail, are *bilingual* (two-language) countries. Actually, many more countries are multilingual and bilingual than one might believe, and this has to do with the relative position of the speakers of the languages involved. When asked, most people will say that Guatemala is a Spanish-speaking country, even though half the people speak Indian languages. In Guatemala, the Indians, despite their numbers, do not yet have much say in the affairs of their country. That Canada is perceived as an English-speaking country (and is often categorized as part of "Anglo" America) reflects the subordinate position of the French-speaking community there. That South Africa should be known as a bilingual country reflects the dominant position of the white population of that strongly plural society; it negates millions of speakers of dozens of other tongues. Close scrutiny of a large-scale map produces some real surprises. France, surely, is a one-language country, when French even spills over into Switzerland and Belgium? Not quite. France's northeast corner still speaks German, its island of Corsica speaks Italian, Basque and Spanish are spoken in southern areas, and the peninsula of Brittany still speaks Breton, an ancient language that has resisted the advance of modern French.

Why do political boundaries of the world not conform to linguistic boundaries? We have already seen part of the answer: this is a mobile world, in which people move and migrate and carry their cultural attributes into alien lands. In addition, conflicts between adjacent countries have caused boundary shifts that still separate people who speak essentially the same tongue. The border between Germany and France is a case in point. It has been relocated several times following the wars in which those two countries found themselves on opposite sides. Still another factor involves the strength of an emerging language and the retreat of a declining one. Breton, the language still spoken on France's western peninsula, is related to Irish and Scots Gaelic as well as Welsh. These Celtic languages once extended over much more of Europe, but modern English and French replaced them. This fate has overtaken many languages, and it is still happening today. Many of Africa's several hundred languages will disappear, while others will grow stronger. The Khoisan (click) languages that now survive only in the southwest of that continent were supplanted by Bantu languages in the same manner that the Celtic languages were pushed back.

There is a counter to these factors that favor replacement, namely the determination of a minority group to retain its identity and to sustain its culture and especially its language. The world is seeing more of this during the second half of the twentieth century as minority groups become conscious of their strength. Thus languages that might otherwise be doomed are retained and fostered, and the process of attrition, even if it is not halted, is at least delayed.

Obviously the whole area of language in cultural geography affords innumerable opportunities for research and study. Language is a vital element in the

---

FRANCOPHONE

Francophone is used to identify countries or areas where French is the *lingua franca* or the official language or perhaps the language of the elite, the people who make the important decisions. For some time the term *Francophone Africa*, for example, has been used to identify those African states formerly under French colonial control whose educated people communicate in French. They also know their African languages, of course. It is no longer appropriate to say *French Africa*, and *French-speaking Africa* seems to imply that everybody there speaks French, which is not true. So "French sound is heard." Quebec, of course, is Canada's Francophone region.

---

reconstruction of past cultures; the transmittal of oral literature and its interpretation opens up whole new possibilities. The professional storyteller in an African village is not just a picturesque figure of incidental interest. He holds in his tales the history and psyche of the people of whom he speaks. The study of dialects and the spatial character of modifications in words can tell us much about peoples' movements, their external contacts or isolation, former distribution, and more. And then there is the relevant issue brought up by so many in arguments about the value in learning a foreign language. Language can reveal much about the way peoples view reality, their own culture, as well as other cultures. In their structure and vocabulary and in their capacity (or inability) to express certain concepts and ideas, languages reflect something of the way people think and perceive their world. There are African languages that have no word, no term, for the concept of a god. Others (in Asia) really have no system for the reporting of chronological events, no time scale. It would be difficult for us to understand those peoples' perception of the world about them, preoccupied as our culture is with the supernatural and dating and timing. Learning the language would be a first, timid step. But the rewards could be enormous.

RELIGION AND CULTURE

In many parts of the world, especially in non-Western areas, religion is so vital a part of culture that it practically *constitutes* culture. Thus it is not surprising that we would have difficulty in defining exactly what a religion is. The phenomenon manifests itself in so many different ways, in the worship of the souls of ancestors living in natural objects such as mountains, animals, or trees, in the belief that a certain living person possesses particular capacities granted him by a supernatural power, and in the belief in a deity or deities, as in the great religions. In some societies, notably those of the Western, industrialized, urbanized, commercialized world, religion has become a rather subordinate, ephemeral matter in the life of many people. But in Asia and parts of Africa religious doctrine may exert tight control over behavior, during the daytime through ritual and practice, and even at night in prescribing the orientation of the sleeping body.

If we cannot define religion satisfactorily, we can at least note several of its properties. Apart from the doctrines and beliefs relating to the god or gods, there is usually a set of more or less complex rituals through which these beliefs are expressed. These rituals involve the use of the religion's literature, such as the Bible or the Koran. Religions, especially the major ones, have produced complex and vast organizational structures with a hierarchy of officers. These people police an approved set of standards of behavior, a code of ethics peculiar to each religion.

From this brief statement of properties it is evident how greatly a culture may be dominated by the prevailing religious faith. Modes of dress, the kinds of food people eat, commercial practices, and even the location and structure of houses may be determined by the rules of religion. And it is hardly necessary to remind anyone of the daily influences of a religious heritage when our calendar, our holidays, our architectural styles, many place names, and even the slogan "In God

We Trust" on the money we spend come directly from that source. Even in societies that are seeking to divest themselves of religion and religious influences (the Soviet Union, for example), those very influences continue to mark living routines and the cultural landscape as a whole. Imagine, then, the role of religion in an almost totally Roman Catholic society such as Spain. In Islamic society, too, it is religion that sustains and perpetuates the culture. Religious doctrine has determined that much of India is vegetarian (and overrun by holy animals), that the eating of pork is taboo in Moslem countries (eliminating the profitable pig in domestic economies), that eating meat on Fridays was inappropriate for Roman Catholics.

It is natural that our first questions should relate to the locational characteristics of the major religions. Where are the sources and what are present distributions? Let us begin by identifying the world's major religions in general terms. We recognize seven, in two clusters, one west of India and the other from India eastward (Fig. 4.8). To the west are Judaism, Islam, and Christianity. To the east, the map shows, are Hinduism, Buddhism, the Chinese religions, and Shinto. The Chinese religions, which extend from China into Korea and Viet Nam, are mainly constituted by Confucianism, Taoism, and Buddhism. In terms of numbers of adherents, Christianity leads with between 900 and 1000 million. Islam is next, with perhaps 500 million faithful, and Hinduism follows with some 450 million. There are about 230 million Buddhists, and 70 million Shinto adherents. These figures are estimates, of course, for there are no reliable data. Many people profess more than one faith, and there are different kinds of participation in a faith. This problem is especially relevant to China, where Confucianism is a way of life as much as a faith. Does this mean that the religion has 400 million faithful?

The major religions, distributed today as shown on Fig. 4.8, originated in a remarkably small area of the world. Judaism and Christianity began in what is today the Israel-Jordan area. Islam arose through the teachings of Mohammed, a resident of Mecca in western Arabia. The Hindu religion, which has no central figure and is a complex, multifaceted faith, originated in the Indus region long before Christianity or Islam. Buddhism emerged from the teachings of Prince Siddharta, a man who renounced his claims to power and wealth in his kingdom in northeastern India to seek salvation and enlightenment in religious meditation. The Buddha's teachings constituted to a great extent a reaction to Hinduism, which by the sixth century B.C. had overspread much of southern and southeastern Asia.

These source areas coincide quite strongly with the culture hearths shown on Fig. 4.4, and there can be no doubt that while developments in other spheres were occurring there—urbanization, irrigated agriculture, political growth, increasingly complex social orders, legal systems—religious systems also became more sophisticated.

Hinduism

Chronologically, the Hindu religion is the oldest of the major religions and one of the oldest extant religions in the world. It emerged, without a prophet, a book of scriptures, and without evolving a bureaucratic structure comparable to that,

FIG. 4.8 Religions of the world. (After a map in *Man's Domain*, McGraw-Hill, New York, 1968, p. 10.)

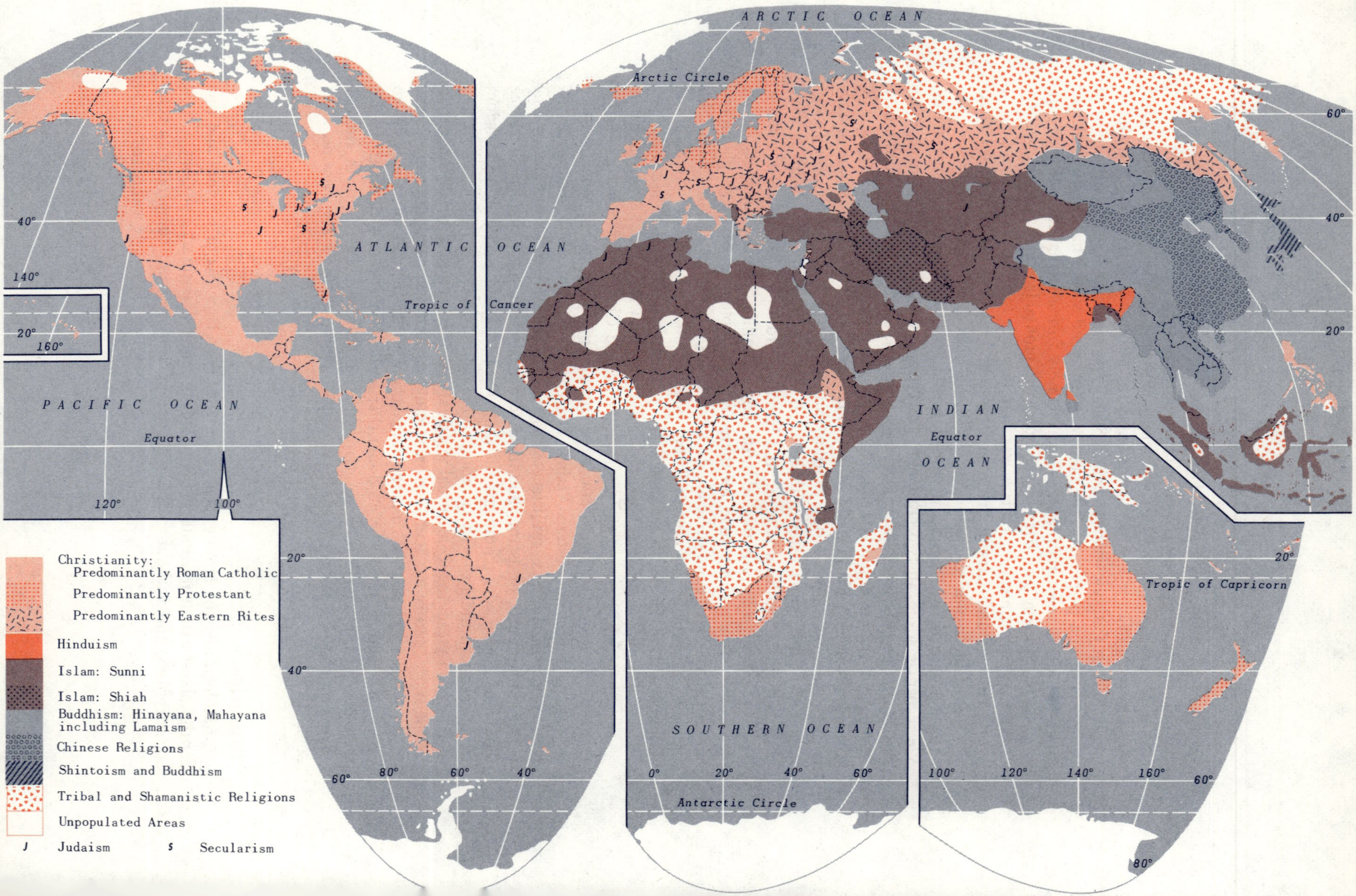

say, of the Christian religions. The scene of Hinduism's rise appears to have been the Indus valley; the time may have been more than 4000 years ago. The fundamental doctrine of Hinduism is karma, which involves the concept of transferability of the soul. All beings are in a hierarchy; all have souls. The ideal is to move upward in the hierarchy and then to escape from the eternal cycle through union with the Brahman, the top of the ladder. A soul moves upward or downward according to its behavior in the present life. Good deeds and adherence to the faith lead to a higher level in the next life. Bad behavior leads to a demotion. All souls, those of animals as well as humans, participate in this process, and the principle of reincarnation is a cornerstone of the faith: mistreat an animal in this life, and chances are that you will *be* that animal in a future life.

Hinduism's doctrines are closely related to the caste system, for castes themselves are steps of the universal ladder. But the caste system locks people in social classes, reducing mobility, an undesirable feature manifested especially in the lowest of the castes, the untouchables. The untouchables until a generation ago could not enter temples, were excluded from certain schools, and were restricted to performing the most unpleasant tasks. The coming of other religions to India, the effects of modernization under colonialism, and especially the work of Gandhi, India's famous leader, loosened the social barriers of the caste system and bettered the lot of the 70 million untouchables.

Hinduism, born on the Indian subcontinent's western flank, spread eastward across India and, even before the advent of Christianity, into southeastern Asia (Fig. 4.9). It would penetrate by first attaching itself to prevailing tribal faiths and then supplanting them. Later, when Islam and Christianity emerged and were vigorously promulgated in Hindu areas, Hindu thinkers sought to assimilate certain of the new teachings in their own religion. For example, the elements of the Sermon on the Mount now form part of Hindu preaching, and Christian beliefs led in some measure to a moderation of the caste system's extreme rigidity. In other cases the confrontation with other faiths led to the formulation of a compromise religion. The monotheism of Islam led to the rise of the *Sikhs*, who disapproved of the worship of all sorts of idols and who also disliked the caste system, but who nevertheless retained concepts of reincarnation and karma.

The Hindu faith today remains primarily India's religion, but remnants still exist in Southeast Asia (where Islam later replaced much of Hinduism). During the colonial period, many hundreds of thousands of Indians were taken to other areas of the world, such as East and South Africa, the Caribbean, and northern South America. In those locations, they continue to practice the Hindu religion.

Buddhism

Buddhism also was a reaction to the less desirable features of Hinduism, as Prince Siddharta, the Buddha, perceived them in the sixth century B.C. The prince, who was in line to the throne of his middle-Indian kingdom, renounced his worldly attributes and decided to seek salvation through meditation and teaching. The

Religious shrines dominate the cultural landscapes of societies in various parts of the world, signifying the importance of religion in culture. These two photographs, from widely different areas of the world, have this in common. In the photo at left, the cathedral of Köln, Germany, towers over the city's buildings and the Rhine River. The photo at right shows the famous Blue Mosque, the shrine of Ali, a focus of the city of Kabul, Afghanistan.

Buddha ("enlightened one") was perhaps the first prominent Indian religious leader to speak out against the caste system. Salvation, he said, could be attained by anyone, no matter what his caste. Fundamentally the Buddha saw the world as a place of ignorance and misery. Enlightenment came through knowledge, especially self-knowledge, the elimination of covetousness, craving, and desire, the principle of complete honesty, and the determination not to hurt another person or animal. The tradition of karma still prevailed.

Following the death of the Buddha in 489 B.C., the faith grew but slowly until, during the middle of the third century B.C., the Emperor Asoka became a convert. Asoka was the leader of a large and powerful state that covered India from the Punjab to Bengal and from the Himalayan foothills to Mysore. Asoka not only ruled his country in accordance with the teachings of the Buddha, but he also sent missionaries to the outside world to carry the Buddha's teachings to distant peoples. Buddhism now spread as far south as Ceylon and later, west toward the Mediterranean, north into Tibet, and east into China, Korea, Japan, Viet Nam, and Indonesia, over a span of some 10 centuries. But while Buddhism spread to distant lands, it began to decline in India where it had arisen. During Asoka's rule there may have been more Buddhist adherents in India than Hindus, but after that period the strength of Hinduism began once again to assert itself. Today, Buddhism is practically extinct in India, though it still thrives in Ceylon, in several Southeast Asian countries, and, in various forms, in Nepal and Tibet and in Korea. Along with other faiths it also exists in Japan.

Chinese Religions

While the Buddha's teachings drew converts in India and the issue of the transmigration of souls was debated there, a religious revolution of another kind was taking place in China. Chinese thought centered on two basic issues: the oneness of man and nature and the reverence for the souls of ancestors. Lao-tsu's teachings focused on the former. Man should learn to live in harmony with nature, viewing himself as but an insignificant element in the great universal order. This, the doctrine of Taoism, suffered much falsification as time went on and it became a cult of the masses. Lao-tsu himself was worshipped as a god (among many), something he would have disapproved. People, animals, even dragons became the objects of worship as well, and a sort of Taoist witchcraft emerged. Confucius emphasized the need to maintain the best possible relationships between people, such as between parents and children and between rulers and their subjects. His was a very practical and useful philosophy, and during the Han dynasty (second century B.C.) it became the state ethic. Worship of and obedience to the emperor constituted central elements of Confucianism, and as the national creed it prevailed until the twentieth century.

Buddhism was introduced into the Chinese philosophical scheme during the Han dynasty as well, and it blended easily with the domestic belief systems, with their reverence for departed souls and appreciation for nature. Thus China developed a quite discrete religious posture, taking elements from several teachings to create, if not a distinct single faith, certainly a Chinese religious way of life.

Emperor worship, a reverence for nature, and a strong feeling for land and nation are elements that forged the *Shinto* religion of Japan. Before Western powers made contact with Japan, Buddhism had made great inroads there. But the events of the nineteenth century in East Asia along with Japanese determination that their country would not be subjugated brought Shinto back as a religious-political force. Following Japan's defeat in the Second World War, Shinto's role as the official state religion was terminated and the doctrine relating to the emperor's divine descent rejected.

Judaism

The oldest major religion west of India, Judaism, grew from the belief system of the Jews, one of several Semitic, nomadic tribes that traversed Southwest Asia about 2000 years B.C. The Jews led an existence filled with upheavals. Moses led them from Egypt, where they suffered repression, to Canaan, where an internal conflict arose and the nation split into two, Israel and Judah. Israel was subsequently wiped out by enemies, but Judah survived longer, only to be conquered by the Babylonians. Regrouping to rebuild Jerusalem, the Jews fell victim to a sequence of alien powers and saw their holy city destroyed again in A.D. 70. The Jews were driven away and scattered all over the region and eventually into Europe and much of the rest of the world. For many centuries, indeed until late in the eighteenth century, they were persecuted, denied citizenship, driven into ghettos, and massacred. As the

FIG. 4.9 Diffusion of the four major religions.

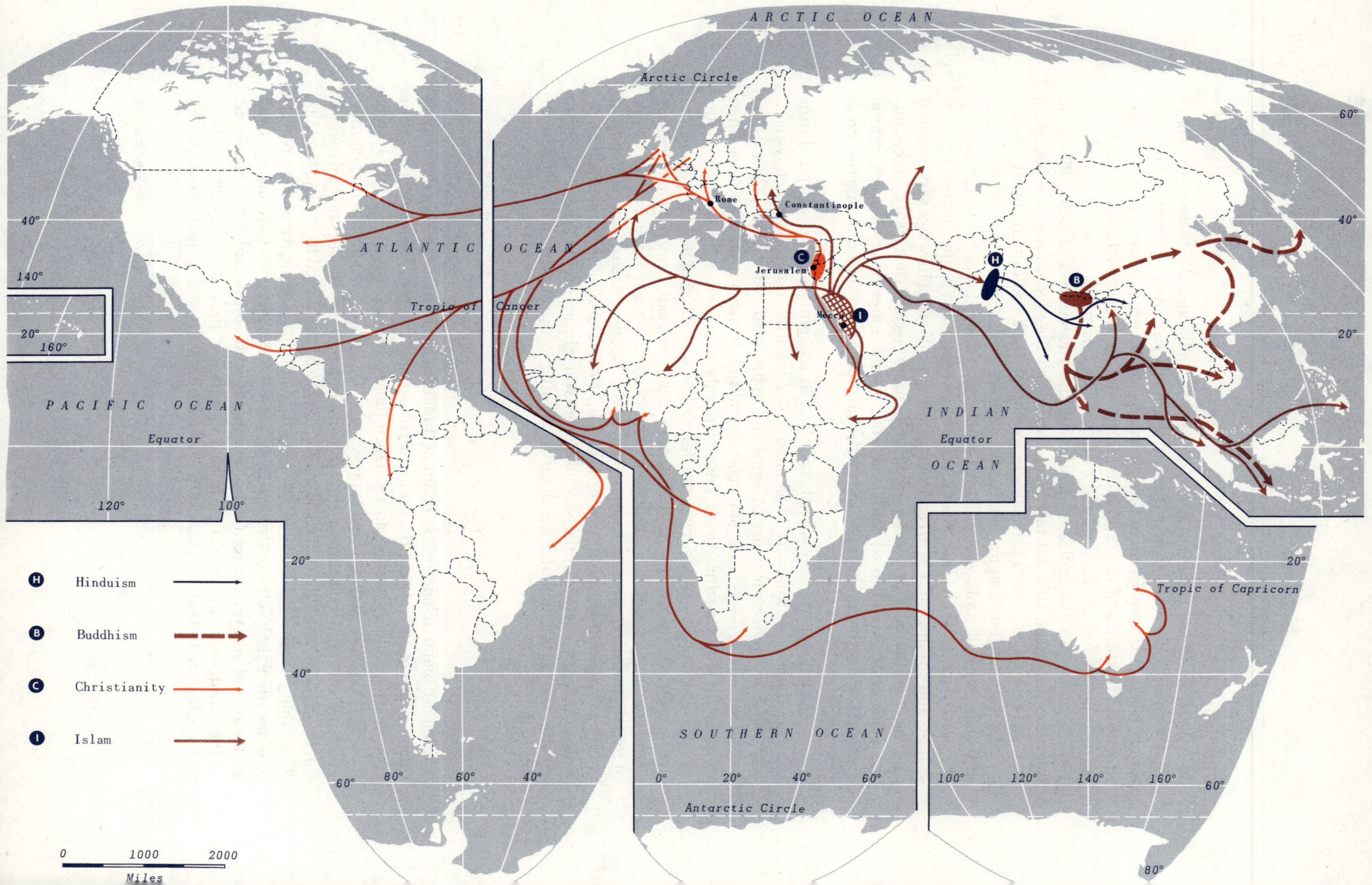

twentieth century has proved, the world has not been a safe place for Jewish people even in modern times.

In the face of such constant threats to their existence, it has been the faith that has sustained the Jews, the certainty that the Messiah would come to deliver them from their enemies. The roots of Jewish religious tradition lie in the teachings of Abraham, who united his people and during whose time the faith began to take shape. Among the religions adhered to by the ancient Semitic tribes, that of the Jews was unique in that it involved the worship of only one god who, the Jews believed, had selected them to bear witness to his existence and works. But their religion also showed the many contacts with other beliefs and other peoples. *Zoroastrianism*, which arose in Persia during the sixth century B.C., gave Judaism its concepts of heaven and hell, angels and devils.

Modern times have seen a division of Judaism into several branches. During the nineteenth century, when Jews began to enjoy greater freedom, a Reform movement developed that was aimed at adjusting the faith and its practices to current time. But there were many who feared that this would cause a loss of identity and cohesion, and the Orthodox movement sought to retain the old precepts as effectively as possible. Between these two extremes is a branch that is less strictly orthodox but not as liberal as the reformers. And the objective of a homeland for the Jewish people, which emerged during the nineteenth century also, produced the ideology of Zionism, a determination that Jews would not be absorbed and assimilated by other societies and a posture of alertness against anti-Semitism. Zionism's goal of a Jewish state became reality in 1948 when Israel was created on the shores of the Mediterranean.

Christianity

The Christian religion had its beginnings in the Jews' search for deliverance from Roman oppression and the appearance of Jesus. Many saw in Jesus a manifestation of God, but probably even more hoped that he would be a temporal as well as a spiritual leader and secure freedom as well as salvation. Among the Apostles was Paul, a Jew with a Greek education who, after the crucifixion, began offering the teachings of Jesus to non-Jews. Paul played a central role in organizing the Christian church and in disseminating Jesus' teachings to the European Mediterranean world. After Paul's death the church continued to grow, but at the cost of many lives because Roman authority resisted the intrusion. A crucial event in the development of the Christian faith was the conversion of the Emperor Constantine, and from the fourth century A.D. onward it was the state religion. The Roman Empire soon was to decline and break up, and when the sector centered on Rome fell on hard times, the eastern half, with Constantinople at its heart, became the focus. It was there, in the eastern realm, that Christianity thrived and from there it radiated into other areas including the Balkan Peninsula. Today the Eastern or Orthodox churches still form one of the three major branches of the faith, despite the blows it sustained when Constantinople fell to the Turks and Islam invaded eastern Europe (in the fifteenth century) and again when the church in Russia was threatened by the rise of communism (twentieth century).

In Rome, the papacy was established, and here came to be the center of the second branch of the faith, Roman Catholicism. In the Middle Ages the power of the church was at its zenith, with all the excesses that involved. A reaction was inevitable. It came during the fifteenth and sixteenth centuries with the teachings of Luther, Calvin, and others. Widespread friction and open warfare marked the deteriorating relationships among Christians of different views on the continent. But the Protestant movement, the third major branch, could not be denied.

The worldwide dispersal of Christianity was accomplished by the era of colonial acquisition upon which Europe embarked at about this time. Spain invaded Middle and South America, bringing the Catholic faith to those areas. Protestant refugees, tired of conflict and oppression and in search of new hope and freedom, came in large numbers to North America. A patchwork of missionary efforts produced mixed conversions in much of black Africa. Catholicism made inroads in Zaïre (Congo), Angola, and Moçambique. A very small percentage of the people in formerly British India were converted to Christianity; Catholicism scored heavily in the Philippines while the islands were under Spanish control.

Islam

The faith of the Moslems is the youngest of the major religions, having been born from the teachings of Mohammed in the seventh century A.D. According to Moslem belief, Mohammed received the truth directly from Allah in a series of revelations, during which he involuntarily spoke the words of the Koran, the Moslems' holy book. Mohammed, a student of religion even before this event, admired the monotheism of Christianity and Judaism; he believed that Allah had already manifested himself through other prophets (including Jesus). But he, Mohammed, was the real and ultimate of prophets.

Islam brought to the Arab world not only a unifying religious faith, but also a whole new set of values, a new way of life. Along with observance of the "five pillars" of Islam (repeated expressions of the basic creed, frequent prayer, a month of daytime fasting, almsgiving, and at least one pilgrimage to Mecca), the faith prescribed and proscribed in other spheres. Alcohol, smoking, and gambling were forbidden. Polygamy was tolerated. Pork should be avoided. Mosques made their appearance in Arab settlements. They not only served as places for the Friday prayer but also as social gathering places to bring communities closer together. Mecca became the spiritual center for a divided, far-flung people for whom a joint focus was something new. The stimulus given by Mohammed, spiritual as well as political, was such that the Arab world was mobilized almost overnight. The prophet died in 632, but his faith and fame continued to spread like wildfire. Arab armies formed, they invaded and conquered, and Islam was carried throughout North Africa. The Moors joined the cause and a successful invasion of Iberia was organized. Eastward the faith was carried into what is now Pakistan, and northward into present-day Soviet Asia. By the middle of the eighth century the Arab empire extended from western India to Morocco and from Turkey to eastern Ethiopia. The original capital was at Medina in Arabia, but it was moved, first to Damascus and then to Bagdad, in response to the strategic successes. In the fields of architecture, mathematics, and science the Arabs far overshadowed their European contemporaries, and they established institutions of higher learning in many cities, including Bagdad, Cairo, and Toledo (Spain). The faith had spawned a culture.

OTHER EXPRESSIONS

Architecture

In educational institutions and systems, particular technological achievements, and a host of other ways culture is expressed. The Moslems brought to European Iberia (and later to the Balkan Peninsula) not only their faith, but also something more tangible and visible in the cultural landscape, their architecture. We know it today as "Spanish" or "Mediterranean" architecture, but its arches and pillars and elaborate decorative features are rooted in the Moorish period.

It is difficult to specify exactly the characteristics of architecture in a culture, for there are different expressions or styles within cultures and different purposes for the structures built. In black Africa, people in certain areas build their houses from branches and mud, with thatch roofs; elsewhere it is grass. In the Transkei, houses are built with round walls and a conical roof; in Zululand they are beehive-shaped. In the Transvaal the Ndebele tend to build square or rectangular houses. In Kenya, the Kikuyu build a round house, but the Gyriama construct a strongly rectangular longhouse. Travel south along the Nile River, and the square dwellings of the Arab world give way to the round houses of black Africa.

We can, therefore, study the building materials and the floor plans of the homes of a society and, using the data derived, possibly contribute to the discovery of past dispersals and migrations. In the United States, for example, log houses were first built by German settlers in Pennsylvania, from whom the English immigrants appear to have taken over the technique. The English, in turn, carried the type toward the leading edge of white settlement and as far south as Georgia. In Tanzania, East Africa, the mixture of floor plans and building styles evinces a rather intensive intermingling of tribal peoples.

The way people use available materials and seek to achieve optimal living space and shelter at the same time is one facet of architecture. There is another. Every culture seeks also to construct buildings (religious, ceremonial, public) that epitomize their capacities in this sphere just as their highest standards of language, ethics, and so on, do in others. These buildings have artistic qualities, they require special skills of their builders, and they involve the use of particular materials, perhaps materials not normally used in the building of ordinary houses. Whether churches, temples, or shrines, or city halls, libraries, palaces, or museums, such buildings may reflect characteristic styles (such as that alluded to as Mediterranean) and often dominate the urban landscape of cities and towns. From the Parthenon of ancient Athens to the Victorian elaborateness of imperial Britain, architecture mirrors the aspirations and confidence of a society, and even some of its ideals.

Music

The culture of a people is in part a description of their way of life. Music is an ingredient of this description. In preliterate societies, the role of music is vital because songs and plays describe and transmit the history of the people. Elsewhere, too, music tells of the way people worship their god or gods, and it expresses cheerfulness and sorrow, hope and dismay. It does not take very much musical knowledge to recognize a Viennese waltz, a Polish polka, or Spanish and Latin American rhythms. Note that we immediately begin to attach regional identifications to the musical terms. Indeed, there is regionalism to music. French music such as that of Debussy or Ravel has unmistakable qualities that are an essential element in the whole of French culture. And the music of India, of Japan, of the Arab world conjures up the sights and smells of those places unfailingly.

Music as a cultural phenomenon should be of great interest to cultural geographers. Numerous opportunities present themselves for research and study into diffusion, for example. Musical instruments are invented, then often borrowed by other people when contact between peoples occurs. Musical ideas, melodic, structural, rhythmic, stylistic, are generated and dispersed. Music also changes as culture changes. In Europe, the baroque, classical, romantic, and modern periods parallel phases of cultural expression in other fields. In recent decades we have even seen attempts to gear musical expression, somehow, to the "folk culture" of a whole nation. In the Soviet Union, the famous composer Shostakovitch temporarily fell from official grace because his music was allegedly beyond the grasp of the masses and somehow "not representative" of the nation's life and aspirations. In countries where ideological fervor is great, the importance of the role of music is reflected by attempts to suppress certain kinds of music and to support and foster other music. Americans saw on their television sets the strident music and ballet that have become part of the new culture of China when the 1972 presidential visit to Peking took place.

The United States itself affords some of the most interesting examples of the relationships between music and cultural geography. There has been much interaction between the European schools of music and American composers such as, for example, Aaron Copland, whose musical descriptions of American life and countryside are quite comparable to, say, those of Britten and Delius in England. But there is another major tradition in this country, and it stems from quite another source, Afro-America. Black people during the period of slavery needed their music to survive their ordeal, and their traditions in this context were African. But they also made contact with European folk music, church hymns, popular songs, military marches, and so on. The African traditions of strong rhythm and powerful singing

Architecture and music making are among expressions of cultures. These two photographs remind us of the different approaches to music that can be taken. The upper photo shows the informality and intensity of a group of jazz musicians at work; Afro-American culture is being stated here. The photo below shows the stiffness, formality, and discipline of a Western symphony orchestra, in this case the London Philharmonic.

were welded to the melodic qualities of European music, and a distinct Afro-American musical tradition began to emerge, including blues and dance music. It was from this dance music that jazz, the most distinctive type, evolved.

Improvisation, of course, is the hallmark of jazz. At first such improvisation was essentially embellishment, the addition of brief new phrases to a melodic line. But it developed into a unique and complex form whose practitioners, such as Ella Fitzgerald, spin variations as intricate as the rhythmic underpinnings of West African music. And these rhythmic peculiarities, too, were transferred to the Afro-American scene. In general, European ears were used to rather complex harmonics but fairly simple beats. African and Afro-American rhythms are more complicated and, when subjected to syncopation, give jazz another distinctive quality.

Jazz, and the music of black America in general, are more than just one facet of Afro-American cultural expression. In a sense the music of black America *is* black America, with its roots in Africa, its qualities of release born of the desperation of the era of slavery, its chromaticism a virtue of black musicians wherever in the world they make music, and its performers' virtuosity a universal standard.

How is all this related to cultural geography? Consider the birth of a popular musical phenomenon that occurred in the United States in the 1950s. It evolved through a union of (white) country and western music and (black) rhythm and blues. Through a series of modifications in time and by place, it made its way from the shacks of the rural South to the recording studios of New York. There could hardly be a better example of the process of spatial diffusion of a cultural phenomenon, and the sequence is described by Professor Larry Ford in "Geographic Factors in the Origin, Evolution, and Diffusion of Rock and Roll Music," published in the *Journal of Geography* of November, 1971. For a practical application of some of the ideas touched upon in this chapter, this article is worth a trip to the library.

CHAPTER FIVE

The Regionalization of Culture

FIVE

Five thousand years is but a fraction of the total time man has existed on this earth. The most recent five millenia may constitute less than 1 percent of man's time span, but consider what has happened during this period. The ancient empires, which were the roots of civilization, came and went. The Roman Empire rose and fell. The great religions emerged and diffused all over the world. Languages grew and differentiated. The human population mushroomed from less than 100 million to nearly 4 billion. Urbanization was embryonic in 3000 B.C.; today huge cities exist from China to Egypt, India to Argentina. The wheel had just been invented; today we are beyond the threshold of the era of space travel. On the surface, man's mobility is beyond any comparison even with what it was a single century ago. Students from Africa attend United States and Soviet universities. Soldiers from Korea and Australia fight in a war in Viet Nam. Artists from Spain perform in four South American countries within a single week. An invention from Japan is on the British market in a matter of days. Races, cultures, languages, and technologies are mixing and interacting as never before.

Does this mean that it is no longer possible to identify regionally discrete cultures or culture realms on this globe? Hardly. In the Old World the legacies of the ancient civilizations can still be read on the cultural map; in the New World, where they were swamped by the white man's invasion, regional cultures of a new order now prevail. Anyone who is at all observant about the world at large can name some of these large realms: Europe is one of them, and so are Southeast Asia and black Africa. Such conceptualizations as the "Chinese sphere" and the "Middle East" frequently come up in conversation about world issues and problems. Admittedly they are broad, sweeping generalizations, and there will be disagreement over the exact limits of those areas. Are these perceived "regions of intuitive consensus" useless because they are subjectively and imprecisely defined? Of course not. When we speak of them, we are using categories of phenomena that seem to be of overriding importance in our impression of the regionalization of the world. We are responding to images of difference in ways of life, traditions, ideas, and values between various parts of the world. We have discussed several of these in the preceding chapters. Language, religion, racial grouping, relative location, physical environment—all these play a role. So do some factors we have not yet gone into: political organization, power, economic geography. As we shall see, these last phenomena are more easily discussed when we have a world regional framework firmly in mind. And this is our present objective.

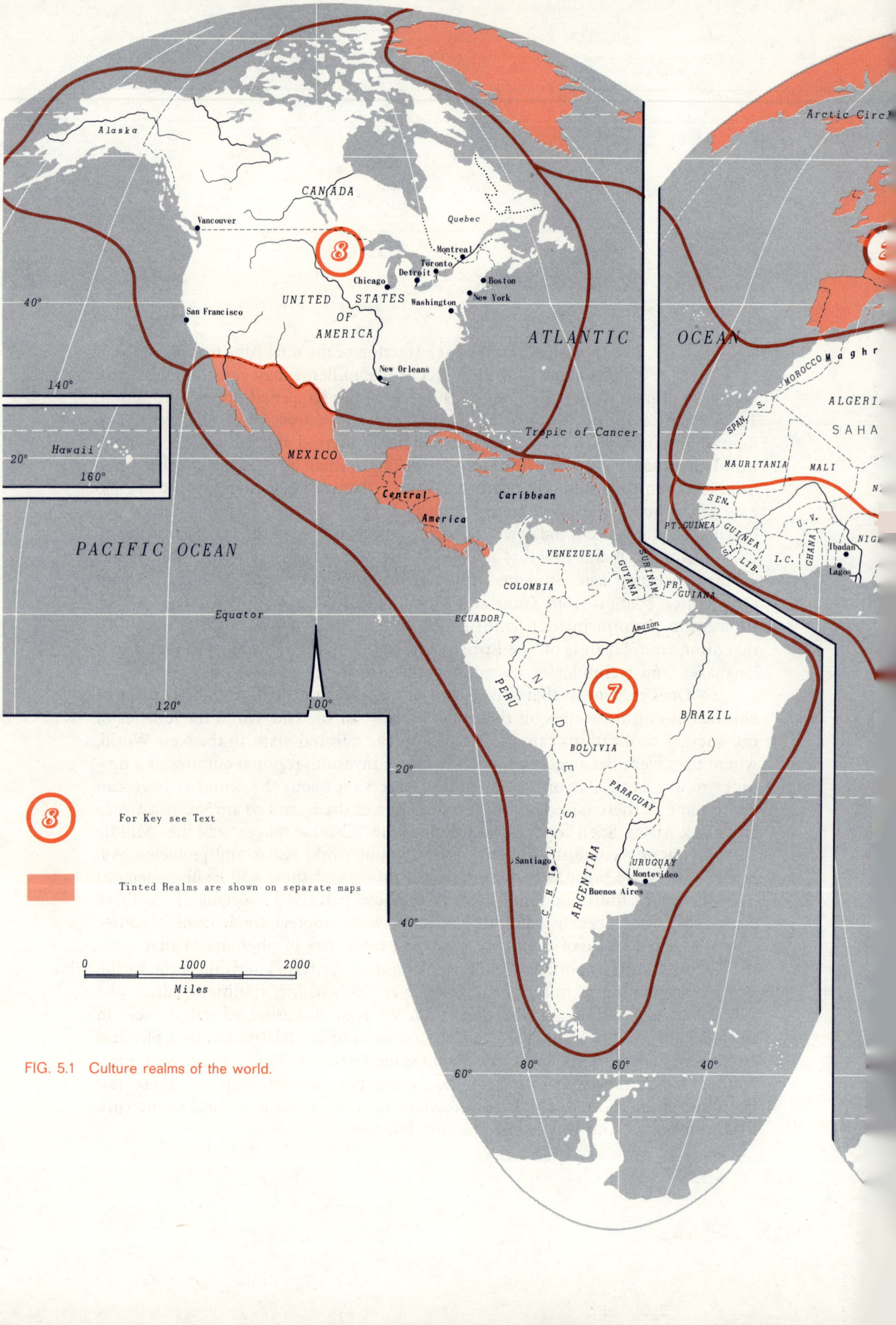

FIG. 5.1 Culture realms of the world.

ARCTIC OCEAN
Leningrad
Moscow
Kiev
U.S.S.R.
Kazakh S.S.R.
Tashkent
Kirghiz S.S.R.
Samarkand
Ashkhabad
Turkmen S.S.R.
Caspian S.
Black S.
Georgia
Arm.
TURKEY
IRAN
AFGHANISTAN
MONGOLIA
CHINA
INDIA
Euphrates
Middle East
SEA
EGYPT
ARABIA
ARABIAN DESERT
SUDAN
Nile
ETHIOPIA
Horn
SOMALI R.
INDIAN OCEAN
Equator
KENYA
UGANDA
Nairobi
MT KILIMANJARO
Mombasa
TANZANIA
Zanzibar
ZAIRE
ZAMBIA
MALAWI
MOCAMBIQUE
Salisbury
RHODESIA
BOTSWANA
Johannesburg
Kimberley
SOUTH AFRICA
MALAGASY REP.
NEW GUINEA
BISMARCK ARCH.
SOLOMON IS.
FIJI IS.
NEW HEBRIDES
NEW CALEDONIA
Tropic of Capricorn
Darwin
AUSTRALIA
Brisbane
New South Wales
Sydney
Canberra
Victoria
Melbourne
Adelaide
Perth
Auckland
NEW ZEALAND
Wellington
OUTHERN OCEAN
arctic Circle
1
3
4
5
6
9
10
11
12
60°
40°
20°
20°
40°
60°
100°
120°
140°
160°

WORLD REALMS

Geographers, anthropologists, historians, and others have proposed systems of regional division of the human world and have written chapters and articles backing up their arguments with lists of criteria and justifications. A study of all these schemes indicates a rather surprising degree of consensus about the culture realms of the world; disagreements are usually a matter of detail. Culture realms do not have sharply defined limits; rather, they merge in transition zones that may be quite broad. But we must mark the realms on maps, and therefore lines of separation are necessary. About the exact position of these lines there is always argument (as there is about the limits of many regions of different kinds). Often the issue is minor. Should the boundary between the Arab-dominated, Islamic realm and black Africa be several dozen miles north of the position shown on a given map where it crosses Nigeria? Sometimes a more difficult question arises. For example, consider the island of Madagascar, off the southeast coast of Africa. The people of this island, which constitutes the Malagasy Republic, number approximately 8 million. A very substantial percentage of this population (no one is sure precisely how large) has a Southeast Asian rather than African ancestry. The influx from Indonesia and the Malay Peninsula is believed to have begun more than 20 centuries ago and continued intermittently until after the arrival of the Europeans. The Southeast Asians came to dominate the life of the island, even though (Bantu) Africans also arrived in considerable numbers. Add to this the absence of the wildlife of eastern Africa (just 250 miles away) and the presence of some animals with Asian affinities, and a case begins to develop for excluding the island from the African realm and including it in a region of insular units stretching across the Indian and Pacific Oceans, a Malaysian-Melanesian-Micronesian-Polynesian realm. As we shall see later, the case is probably not strong enough, and the Malagasy Republic ought to be included in the African realm, but it is a debatable issue.

All this is to emphasize that our scheme, presented in map form by Fig. 5.1, is also subject to argument and modification; it is no synthesis of other regionalizations. It has the advantage that it is comparatively simple and consists of only a dozen world realms. But simplicity means generalization, and about some of the generalizations involved here we could argue productively for many hours.

Referring to Fig. 5.1, a problem that always arises is the naming of the realms. This would seem at first glance to be a minor issue, but the fact is that names can be misleading and even irritating to many people. Take a case close at hand, the realm numbered (8) on the map. This is the region often called "Anglo America." The word *Anglo* refers to the dominantly English heritage of this culture realm, but there are millions of people in this region who do not share this particular background. In the United States, neither the black population nor the Indian Americans have such affinities. In Canada, the great majority of the people in the province of Quebec have a non-Anglo cultural inheritance, tracing their lines of descent to France, not England. Numerous other examples exist on a smaller scale. And although it is true that English has become the prevailing language in this realm and many English traditions were transferred here, millions of people who have been left out of the mainstream of development have begun to express their sensitivities to just such terminology as Anglo America, saying that it reinforces their sense of having been left out. The preferable alternative, then, is to identify realm (8) as North America, which says something about location, not cultural content. We will run into this problem again. How do you suppose the descendants of the Incas, the Asians in Surinam, the Afro-Americans of Guyana and Jamaica, and the French-speaking black population of Haiti feel about being part of *Latin* America? With this issue in mind, let us examine the 12 realms. In the discussion that follows, the numbers in the text correspond to those on Fig. 5.1.

NORTH AFRICA AND SOUTHWEST ASIA (1)

From Morocco on the shores of the Atlantic Ocean to Afghanistan in Asia, and from Turkey on the periphery of Europe to the Somali Republic in the Horn of Africa lies a vast realm of enormous historical and cultural complexity, central to the map and at the heart of the world. It lies juxtaposed between Europe, Asia, and Africa, and it is a part of all three. Throughout history its influences have radiated to these continents and to practically every other part of the world as well. This is one of man's source areas. On the banks of the Euphrates and the Nile arose early, complex civilizations; in its soils plants were domesticated that are now grown from the Americas to Australia; its paths were walked by prophets whose teachings are followed by hundreds of millions of people to this day.

This realm is often characterized in a few words, stressing one or more of its dominant features. For example, it is often called the "dry world," since it contains the Saharan and Arabian deserts. Compare region (1) on Fig. 5.1 with the same general area on Fig. 2.12, and the suggestion that this is the dry world seems reasonable. But culture realms are about people, not physiography, and a further comparative look, this time at Fig. 3.1, suggests that the population clusters here live in moist zones, the Nile Delta, the shores of the Mediterranean Sea, in the Tigris-Euphrates Basin, in oases. Although the Nile Delta does lie in a region that is climatically a desert, the 30 million people who are concentrated there live in a patchwork of cropland and waterways, and the adage "dry world" would hardly apply. True, this is a region where water is almost always at a premium, where peasants often struggle to make soil and moisture yield a small harvest, where nomadic peoples follow their animals across dust-blown plains, where oases are islands of sedentary farming in a sea of aridity. But it is also the land of the Nile, the great Gezira irrigation scheme of the Sudan, and the crop-covered slopes near the Mediterranean. This is a dry world, but not everywhere.

North Africa and Southwest Asia also are often referred to as the "Arab world." Again, this implies a uniformity that does not actually exist. In the first place, the name *Arab* is give loosely to the peoples of this realm who speak the Arabic language (and some related languages as well), but ethnologists normally restrict it to certain occupants of the Arabian Peninsula, the Arab "source." Second, although it is true that Arabic is widely spoken over a region that extends from

Mauritania in the west to southwestern Iran in the east, there are many areas where it is not used by most of the people. In Turkey, for example, Turkic is the major language, and it does not have Semitic or Hamitic roots. In Iran, the Iranian language belongs to the Indo-European family (Fig. 4.6). In Ethiopia, Amharic is spoken by the ruling plateau people; in Israel it is Hebrew. Other "Arab-world" languages that have their own identity are spoken by the Tuareg of the Sahara Desert, the Berbers of Northwest Africa, and the peoples who live in the transition between North Africa and black Africa. Thus there are, in this "Arab world," numerous elements that are distinctly non-Arab.

The realm we are discussing here is also often called the "world of Islam." Indeed, the Islamic religion today extends well beyond the limits of this realm: it is the dominant faith in Bangladesh and Pakistan, it has strength in Indonesia, it survives even in eastern Europe in places. But although Islam is certainly a common denominator in North Africa and Southwest Asia, the "world of Islam" is not entirely Islamic either. In Israel, Judaism is the prevailing faith. In Lebanon, perhaps as much as half the population adheres to an old form of Christianity. In Ethiopia, the Amharic-speaking ruling class, having managed centuries ago to stave off the Islamic onslaught, also practices an ancient Christian religion. Coptic Christian churches still exist in Egypt. Hence the connotation "world of Islam" is not altogether satisfactory: the faith extends beyond the realm's borders, and the region itself counts a number of exceptions.

Perhaps the most frequently heard name of this realm is the "Middle East." This must sound quite odd to someone, say, in India, who might think of it as a Middle West rather than a Middle East! The name, of course, reflects its source, the Western world, which saw a "Middle" East in Egypt, Arabia, and Iran, and a "Far" East in China and Japan. Still, the term has become popular and is frequently used by officials. In view of the complexity of the realm, its transitional margins, and its far-flung areal components, at least the name "Middle East" has the merit of being imprecise.

Thus we remain with the most appropriate (and most cumbersome) name for this realm, North Africa and Southwest Asia. As Fig. 5.1 shows, the region has land boundaries only in Africa and in Asia; elsewhere, Mediterranean, Black, and Caspian seacoasts and Atlantic and Indian ocean shores form the margins. But the concept of the transitional nature of the boundaries of culture realms is underscored both in Asia and in Africa. In Asia, Pakistan presents a problem, and many scholars would include Pakistan in the present region on several grounds, principally the country's overwhelming adherence to Islam and its dry-world physiography. On the other hand, Pakistani are not Arabs, and despite its present-day difficulties with India, Pakistan has tended from the days following the Indus civilizations to the present to look eastward rather than west. Many Pakistani came to the country from India; the Punjab in Pakistan is an extremity of a much larger core region that extends into India. For many centuries, and especially during the British colonial period, the area that is now Pakistan has been a South Asian, not a Southwest Asian entity.

Transitions

In Africa, too, problems of this sort arise. There are places where the transition from North Africa to black Africa is quite sharp, for example, in the Sudan. There, change comes quite strikingly: the physical appearance of the people, their faith, their language, the way they build houses. The central Sudan is North Africa; the southern Sudan is black Africa. Westward, however, the sharpness of the contrast fades. When Islam reached the old savanna states of West Africa, the leaders were converted and the religion gained a large number of black African adherents. Between the margins of the Sahara Desert and the coastal forests of West Africa lies a wide belt of transition and change, where dry world conditions yield, where black African languages prevail, but where Islam dominates, as it does, for example, in northern Nigeria. It is best to view the line westward from the Sudan as representing a broad zone of culture change and interdigitation.

In one way or another we all have an image of the "Middle East," an image perhaps fostered by the impressions gained during the Arab-Israeli conflicts, by the famous photograph, taken from space, of the Nile Delta and Sinai Peninsula, perhaps even by films such as *Lawrence of Arabia.* If that image includes a sense of contrast between wealthy, city-based landlords and merchants and poverty-stricken, debt-afflicted peasants, an impression of simultaneous groping for the strength of nationalism and submission to forces of division and internal conflict, it would hold much truth. Apart from a substantial share of the oil reserves of the world, the region is not well endowed with natural resources. Except for the quarrel with Israel, there is little to unite the causes of the far-flung peoples of North Africa and Southwest Asia. (The Israeli problem does not especially arouse Morocco, Ethiopia, or Turkey.) On the other hand, there are serious divisions: between Egypt and Jordan over ideology, between Ethiopia and Somalia over territory, between the Kurds and Iraq over autonomy, among Moroccans over the monarchy.

Regions

Repeatedly we have observed how far-flung the realm is, extending some 5000 miles across two continents. Not surprisingly, it has strong regional contrasts. The heart of the Arab realm is also the stronghold of the Arabic peoples, including Egypt, Syria, Iraq, Jordan, and Arabia. This cluster of countries at the eastern end of the Mediterranean is the region most often referred to as the "Middle East." Here the criteria we enumerated earlier for this realm apply most consistently. But move northward from this core area, and we reach Turkey, the modernizing, Europe-oriented, populous but peripheral Islamic state. Look eastward, and there is Iran, dry-world, Islamic, but non-Arab. South lie the almost empty deserts of the Sahara and Arabia, and beyond, the Sudan and the Horn of Africa, involved as much with black Africa or more than they are with the "Arab world." And to the west lies the Maghreb, an area with distinct qualities as well. Despite the strength of the Berber element in its population (about half of the 40 million in the four countries), the region's lengthy contacts with Europe, the incursion of black African strains, and the physical separation from the "Middle Eastern" core, the Maghreb perhaps most nearly represents the norms of the Arab heartland. It was one of the leading avenues of Arab penetration and Islamic conversion. Later, in response to the colonial period, it was the scene of revolution and nationalist fervor. In the still-surviving monarchy of Morocco, the socialism of Algeria, the progress of Tunisia, and the radicalism of Libya the whole range of the realm's social, political, and cultural realities is mirrored.

EUROPE (2)

Europe merits identification as a culture realm despite the fact that it occupies a mere fraction of the total area of the Eurasian continent, a fraction that, moreover, is largely made up of the peninsular western extremities of that vast landmass. Certainly Europe's size is no measure of its present greatness, and it would be difficult to argue that Europe does not deserve consideration as a discrete unit of world significance. Probably no other part of the world is or ever has been so packed full of the products of human achievement. Innovations and revolutions that transformed the world originated in Europe. Over centuries of modern times the evolution of world interaction focused on European capitals and European states. Time and again, through internal wars, through the loss of colonial empires, through the threat of superior external competition, Europe proved to contain the resources needed for rebounding and renewed progress.

Among Europe's greatest assets is its internal diversity. From the warm shores of the Mediterranean to the frigid Scandinavian Arctic and from the flat coastlands of the North Sea to the grandeur of the Alps, Europe presents an almost infinite range of natural environments. An insular and peninsular west is contrasted against a more continental east. A resource-laden backbone extends across Europe from England eastward. Excellent soils produce harvests of enormous quantity and variety. The North Sea is proving today to be still another source of natural wealth in an area where Europe has been deficient—gas, oil, energy. But Europe has its poor areas too. Even here countries exist with "underdeveloped" economies.

Europe's diversity is not confined to the continent's physical makeup. The European realm includes peoples of many different stocks, peoples grouped under familiar names such as Latin, Slavic, and Germanic. Within these groups, however, there is great variety. And Europe has its minorities as well. The Hungarians and the Finns are examples. In recent decades, Europe has received immigrants from many parts of the world, all of whom contribute to the human mosaic. Indonesians come to the Netherlands, Asians and West Indians to the United Kingdom, Algerians to France. In terms of languages, religions, architectures, music, and everything else, Europe has an almost infinite range of expressions.

This diversity has been an advantage to Europe in uncountable ways. Very early in the region's history there were indications of what was to come when areas with different products began to trade and exchange their goods, and *functional specialization* began. Using the numerous riverine waterways, bays, gulfs, and

EXHIBITION
DB
432-1932

comparatively narrow seas, the Europeans developed a network of commercial interaction that has been the realm's hallmark ever since, and whose tentacles eventually extended around the world. Today Europe is a realm dominated, especially in the west, by great cities, intensive transport networks and mobility, enormous productivity, dynamic growth, a large and in many areas very dense population, and an extremely complex technology.

It is not difficult to recognize Europe as one of the world's major culture realms, but its boundaries are a matter of debate. Some geographers argue that the western U.S.S.R. is so much a part of Europe, culturally, historically, and otherwise, that the realm terminates at the Ural Mountains east of Moscow. Others have suggested that Turkey has become so Europe-oriented and modernized that it has become a part of the European realm. On the other hand, the view has been expressed that the Soviet Union has so stamped Eastern Europe with its ideology that the boundary of Europe now excludes that area (and East Germany as well). But in considering what constitutes the European realm, we should try to balance our impression of the permanent, overriding, formative qualities of the region against probably temporary conditions. Eastern Europe has been a changeable area. It has seen many overlords come and go. Still it is Europe.

As Fig. 5.1 indicates, the boundary of the European realm begins, in the north, along the border between Finland and the Soviet Union. It follows this line into the Gulf of Finland, and traverses the Baltic Sea to the northern end of the Poland-U.S.S.R. boundary. From this point it coincides with the western boundary of Russia, entering the Black Sea at the mouth of the Danube River. It then turns southwestward into the Aegean Sea, separating Bulgaria and Greece from Turkey. The European realm thus defined is detailed in Fig. 5.2.

Divisions

It is a matter of common knowledge that the European realm subdivides into several major regions, one of which, Eastern Europe, we just mentioned. The proximity of that region's countries, their historical association, cultural similarities, common habitats, and similar economic pursuits lead us to recognize Eastern Europe as a region of Europe. On the same grounds, we can discern a region in northern Europe (Norway, Sweden, Finland, Denmark), in the south in Mediterranean Europe (Portugal, Spain, Italy, Greece), in the west in Western Europe (France, the Netherlands, Belgium, Luxemburg, West Germany, Switzerland), and offshore in the British Isles (Fig. 5.2).

Western Europe constitutes the heart of the realm. Here Europe's industry, productivity, and dynamism are epitomized. Here lie Europe's greatest cities. Here, too, Europe is at its most mobile and active. This is Europe's melting pot. Almost any map tells part of the story: more railroad and road mileage, more industries (such as the German Ruhr complex), more urbanization, more farm production, more people than any other European region. The shock waves sent out by the

FIG. 5.2 The European realm.

industrial revolution had their strongest impact here, for the resource backbone we discussed in Chapter One traverses Western Europe from one end to the other.

Eastward, a different Europe exists. Eastern Europe, more remote from the sources of change, less endowed with natural resources, caught between powerful forces east (Russia) and west (Germany), severely divided internally by language, religion, and in a host of other ways, has lagged behind the West in almost every context. The seven states of Eastern Europe are Poland, the largest of all, Czechoslovakia and Hungary, both landlocked, Romania and Bulgaria, which face the Black Sea, and Yugoslavia and Albania, political mavericks that have Adriatic coastlines. Poland and Czechoslovakia probably have the lion's share of the region's resources (a significant industrial area is developing along their joint border), but even there the contrasts with the West are evident. Eastern Europe is Europe at its most remote, continental, agrarian.

Northern Europe—Scandinavia and Finland—is the realm's least populous region. With well under 25 million people, just slightly more than Yugoslavia has, northern Europe reflects its high-latitude position, its comparative isolation (all of the region but Denmark is separated from the rest of Europe by water), its often rugged, difficult terrain, and its limited opportunities. Northern Europe's relative location, which has entailed not only isolation and remoteness but economic disadvantages of other kinds as well, did have some positive consequences. The common cultural heritage of the countries in northern Europe is strong. They may have been at odds with each other through their histories, but they were never overrun by an outside power. Except for Finnish, the major languages of the region are mutually intelligible. There is overwhelming adherence to the same (Lutheran) church; religious conflicts comparable to those of other parts of Europe do not exist. And in the political evolution and the socioeconomic development of the countries of northern Europe, too, there is much similarity. Democratic, representative parliaments emerged early (Iceland's is reputed to have been the first); individual rights have long been carefully protected. A socialist philosophy dominates economics and planning.

Southern, Mediterranean Europe presents numerous contrasts to the comparatively frigid north. But, as Fig. 5.2 shows, there are also similarities. Southern Europe, like Scandinavia, lies mainly on peninsulas and islands. And, like Scandinavia, Mediterranean Europe was rather bypassed in the apparent allotment of natural resources needed in modern times. In a region heavily dependent upon farming, mountains and upland country prevail, slopes are frequently steep, the soils thin, rocky, and often poor. For its agricultural productivity southern Europe depends mainly upon it river basins and valleys, and upon coastal lowlands. Those areas sustain most of the people, and so Mediterranean Europe's population is much more clustered than Western Europe's.

Finally, we recognize the British Isles as a discrete region of Europe. We do so despite the United Kingdom's points of similarity to Western Europe and its recent inclusion in the mainland-based European Economic Community (the

Common Market). Earlier in this chapter we determined that we would use long-term criteria in our scheme rather than current ones. For centuries insular Britain has stood aloof from Europe proper, exporting ideas, people, technology, and head-quartering a world empire. Britain was a source area and an island in more ways than one.

To ensure correct usage of the term, the *British Isles* includes the large island of Britain, smaller Ireland, and a host of still smaller islands all around. The term *British Isles* goes back to the time when the British ruled not only over their own island, but over Ireland as well. In 1921, after a bitter struggle, the British yielded and the Republic of Ireland was born. But in the northeast corner, where English and Scottish Protestants had settled, London kept control. That area, Northern Ireland, has for years now been afflicted by strife between the Protestant majority and the minority Catholics. Ireland, overwhelmingly Catholic, is in a difficult position. The other island, Britain, also has political subdivisions: England, Wales, and Scotland. These three, along with Northern Ireland, constitute the United Kingdom.

The United Kingdom is one of the most highly urbanized countries in the world. Over 90 percent of the people live in cities and towns. As the source of the industrial revolution, the United Kingdom is also a country of industrial specialization, where cities are known for particular products (just as Detroit is known for the automobile industry). Only 1 out of 25 gainfully employed persons makes a living out of farming. Most Britishers work in commerce, manufacturing, transportation, or other urban-oriented activities. With 60 million people, the United Kingdom needs huge quantities of food imports and, as its own natural resources become worked out, an ever-growing amount of external raw materials.

Europe's Core

In additon to Europe's five major regions, Fig. 5.2 displays still another regionalization of the realm, a core area that includes southern Britain, France, Benelux (Belgium, the Netherlands, and Luxemburg), West Germany, northern Italy, and the Alpine countries (Switzerland and Austria). It is not difficult to guess the criteria that permit us to identify such a European heartland. This is Europe at its most densely populated, highly urbanized, heavily industrialized, and strongly specialized. Served by Europe's most intensive transport and communications networks (road, rail, water, air), the exchange of goods and materials in the core area is at peak efficiency. Productivity is enormous. Great cities with massive hinterlands are coalescing into veritable urban regions, megalopolises. Here in the core area Europe produces the capacities to withstand the modern competition of the United States, the Soviet Union, and Japan. Countries here have lost world empires, and yet they disproved all predictions of doom. Here, too, the tedious struggle toward political integration has the highest stakes. A United States of Europe, focused upon this core area, would be a power of world stature.

INDIA AND THE INDIAN PERIMETER (3)

The south of Asia consists of three protruding landmasses. In the southwest lies the huge rectangle of Arabia, and in the southeast lie the slim peninsulas and elongated islands of Malaysia and Indonesia. Between these there is the familiar triangle of India, a subcontinent in itself. Bounded by the immense Himalaya ranges to the north, by the mountains of eastern Assam to the east, and by the rugged, arid topography of Iran and Afghanistan to the west, the Indian subcontinent is a clearly discernible physical region (Fig. 5.3). Its population of nearly three-quarters of a billion constitutes one of the great human concentrations on earth. The scene of one of the oldest civilizations, it became the cornerstone of the vast British

FIG. 5.3 The Indian realm.

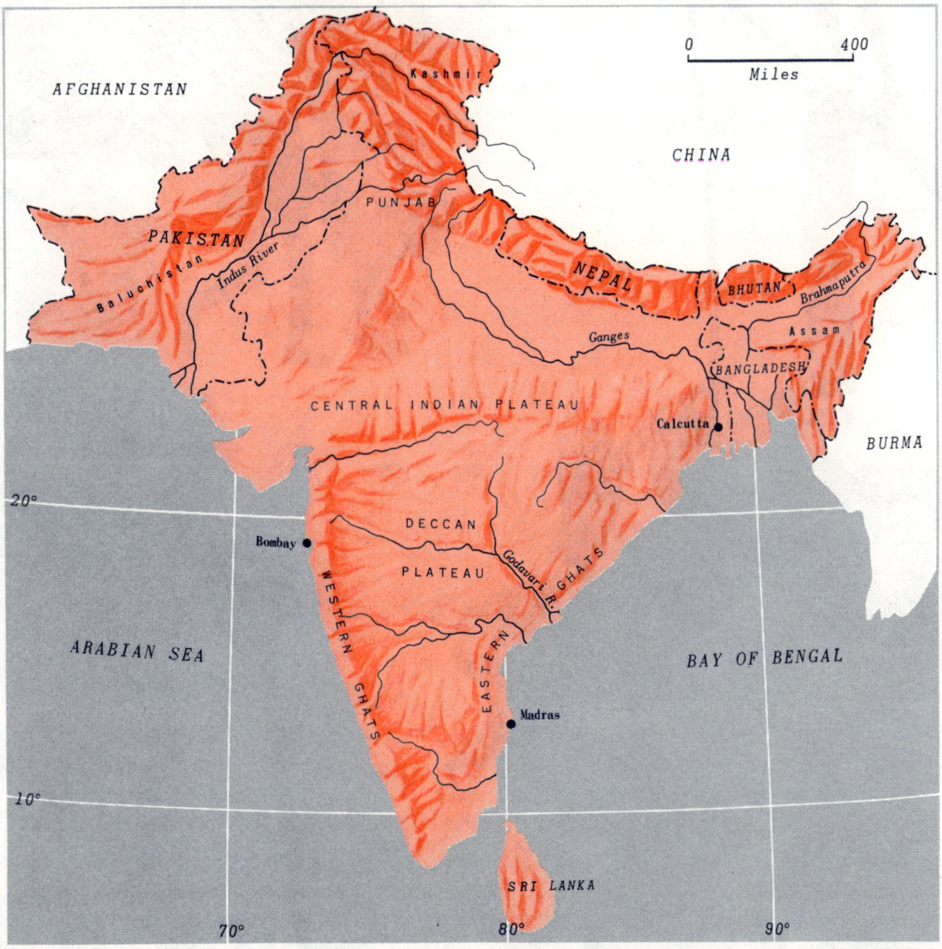

colonial empire. Out of the colonial period and its aftermath emerged four states: India (550 million), Bangladesh (75 million), Pakistan (65 million), and Sri Lanka, formerly known as Ceylon (15 million).

Europe was a recipient of Middle Eastern achievements in many spheres; so was India. From Arabia and the Persian Gulf came traders across the sea. The wave of Islam came over land, across the Indus, through the Punjab, and into the Ganges valley. Indeed, India throughout its history sustained impacts from the west and northwest. It was from those directions that the realm's modern inhabitants came, driving the ancient residents, the Dravidians, southward into the tip of the peninsula. A reference to Fig. 4.6 shows a line across India dividing the Indo-European languages from the Dravidian languages. That line fairly reflects the leading edge of the Aryan invasion of the subcontinent.

The introduction of Islam complicated the religious picture greatly. Two major faiths, Hinduism and Buddhism, had already arisen there. And with Islam came political subjugation as well, as the Mogul Empire thrust Moslem rulers into power. Into this complex situation Europe inserted itself, timidly before 1500, tentatively during much of the 1600s, and forcefully during the 1700s and after. British and French chartered companies fought over primacy in India, and eventually Britain declared a colony in India in the mid-1800s. The administration of such a huge region was no simple matter. To make things worse, European officials prior to the Crown's takeover had made numerous deals with Indian kings, princes, regional governors, and feudal rulers, guaranteeing them various degrees of autonomy. London saw no alternative but to honor these treaties, and so India became an almost chaotic jigsaw of colonial dependencies and traditional, feudal units. It is something of a miracle that, when the realm became independent in 1947, it fragmented into (initially) only three countries.

The foundations of modern India, Pakistan, Bangladesh, and Sri Lanka were forged during the colonial period. On the negative side, the British wiped out much of the indigenous industrial base: India had been southern Asia's chief manufacturing area, especially in metal goods and textiles. The British also intercepted the active trade carried on by Indian merchants to Southwest and Southeast Asia, and the whole pattern of Indian commerce changed. Soon India was exporting raw materials and importing manufactured goods. India's handcraft industries declined; and after the first stimulus the export trade in agricultural raw materials also suffered as other parts of the world were colonized and tied in trade to Europe. At the same time, the population growth rate soared, but without remedies for the problems this entailed. Of course, famine and hunger were recurrent problems in the Indian realm before the Europeans arrived; droughts, floods, and consequent epidemics have always afflicted the region. The colonizers built roads and railroads so that a particular area threatened by famine could be supplied from elsewhere. That was an innovation, a mobility precolonial India had not known. But the unprecedented increases in population drove the demand for food ever further beyond what India and Pakistan could produce, so still there were famines. In 1943–1944 more than $1\frac{1}{2}$ million people died as a result of the failure of the 1942 winter rice crop; famine and epidemics were rampant, especially in Bengal.

On the positive side, British engineers laid out irrigation canals (in addition to roads and railroads) that brought millions of additional acres of land into cultivation. Settlements were founded that grew into major cities and bustling ports, including Calcutta (6 million), Bombay (6 million), and Madras (3 million), three of India's largest urban centers. Modern industrialization was brought to India on a limited scale. In education an effort was made to combine English and Indian traditions; the westernization of India's elite was supported through the education of numerous Indians in the United Kingdom. Modern practices of medicine also were introduced. In addition, the British administration tried to eliminate features of Indian culture that were undesirable by any standards, such as the burning of wives on the funeral pyres of their husbands, female infanticide, child marriage, and the caste system. Obviously the task was far too great to be successfully achieved in less than four generations of rule, but India's government has continued the efforts initiated by the British in these spheres.

But the Indian realm alone contains one-fifth of the earth's people, and the majority of this population felt the colonial impact very little, if at all, except perhaps through the tax collectors and through new laws governing land ownership. India may be known for its large and crowded cities, but in the early 1970s there were only about 2000 cities and towns over 5000 people, and less than 20 percent of the population was urbanized. Thus the overwhelming majority of Indians continued to be village dwellers as they had been for countless centuries. And in their villages, more often than not, they were remote from the progressive influences the colonial period brought, little affected by programs of social improvement, mired in a debt that had always been a way of life, confronted by the constant threat of starvation, and weakened by the ever-present condition of malnutrition.

If there is any scene that comes to the mind's eye in the context of India, it is the huge number of cattle that wander about, perhaps as many as 250 million of them, mostly emaciated-looking, undernourished, weak animals. In the drier parts of the country, goats and sheep, perhaps 125 million, seem to be everywhere. All these animals may be a liability rather than an asset. True, they produce milk and manure, but in their huge numbers they compete for the available food with a human population that can ill afford such competition. And anyway, the milk production of such poor-quality animals is low. Here is a burden imposed by an element of Hindu culture that, of course, forbids the keeping of animals for slaughter. The Indian government has begun programs of sterilization, selective breeding, and the expansion of pastureland. But the roots of this problem lie deep in the region's dominant belief system. The colonial period produced only a veneer of modernization, hardly touching such basic matters as these.

THE CHINESE WORLD (4)

China is a nation-state as well as a culture realm. Here, in the largest country in the world, lies one of the oldest, perhaps the oldest, of continuous civilizations. It was born in the upper basin of the Hwang Ho (Yellow River) and now covers an area in excess of $3\frac{1}{2}$ million square miles and has some 900 million residents. Alone

among the ancient cultures we identified in Chapter Four (see Fig. 4.4), China's culture hearth spawned a major, modern state of world stature, while the strands of the source remained intact. Chinese people still call themselves the "People of Han," referring to the great dynasty that prevailed from 202 B.C. to A.D. 220, which was a formative one in the evolution of the country. But the cultural individuality and continuity of China were already established 2000 years before that time, and probably longer.

In the lengthy process of its evolution as a regionally discrete culture and a great nation-state, China has had an ally in its isolation. To the north lie the great mountain barriers of eastern Siberia and the barren country of Mongolia. To the northwest, beyond Sinkiang, the mountains open, but only into the vast Kirgiz Steppe. To the west and southwest lie more great mountains: the Tien Shan, the Pamirs, the forbidding Plateau of Tibet, and the Himalayan wall. And to the south there are the mountain slopes and tropical forests of Southeast Asia, the same that mark India's eastern flank with Burma (Fig. 5.4). But more telling even than these physical barriers to contact and interaction is the factor of distance. China has always been far from the source areas of innovation and technical change. True, it was itself such an area, but compared with Southwest Asia's culture hearths, its external contacts were limited. China is still a distant land. Overland from the heartland of its Eurasian neighbor, the Soviet Union, is a long and tedious journey by rail. Direct surface connections with India are virtually nonexistent. Apart from a nearly finished railroad to Burma, the only long-existing rail connection to any part of Southeast Asia is the line to Hanoi. And China's lengthy Pacific coastline looks into the world's largest body of water. Japan is China's only major industrialized neighbor.

From its northern source area the Chinese domain grew and expanded, southward across the Yangtze Kiang to the Hsi Kiang (West River) and beyond. Only the Mongols, who ruled in China from 1280 until 1368, broke the sequence of indigenous Chinese ruling dynasties. Under the last of these, the Manchus (1644–1911), China's territory reached its maximum extent, incorporating most of mainland Southeast Asia except Thailand and Malaya, Tibet, East Kazakhstan (now part of the U.S.S.R.), Mongolia, Korea, and southeastern Siberia between Manchuria and the ocean. But the last century of the Manchu dynasty also saw China succumb to European pressure in economic and political fields. Territory was lost, annexed by foreign powers, or "leased" away. After centuries of cultural integrity, economic security, and political continuity, the Chinese world at last lay open to the aggressions of foreigners whose innovative capacities China had denied to the end. Ships flying European flags lay in the ports of China's coasts and rivers; the smokestacks of foreign factories rose above the urban scene of its great cities. The Japanese were in Korea; the Russians entered Kazakhstan and Manchuria. Opium addiction was widespread.

All this led to ferment in Chinese society. The authority of the Manchu rulers was so weakened that feudalism emerged. The Chinese were divided among themselves over whether to modernize and compete as Japan was doing or to

FIG. 5.4 The Chinese realm.

preserve the old order as much as possible. Civil strife broke out, and in 1911 the emperor's garrisons were attacked and massacred all over China. The old order was gone.

A half century was to pass before China could turn inward again, more than 40 years of continued civil war between the Nationalists, led by Sun Yat-Sen and later by Chiang Kai-shek, and the Communists, who were on the rise under Mao Tse-tung. There was more than internal conflict: the Japanese armies penetrated during the Second World War and made the best of China's fragmentation. For a time there was a three-cornered war, with both Chinese factions battling each other as well as the Japanese. Following the defeat of Japan, the civil conflict was resumed and the Communists prevailed, chasing the Nationalists to a last outpost on Taiwan.

Thus, at mid-century, the new China began to take shape. From the foreign invaders the Chinese had learned that power lies in industrial productivity. From the Nationalists the Communists had learned the inherent dangers in ineffective national control. Thus the Peking government of the 1950s set about reorganizing China as it had never been organized before. It was time for the Great Leap Forward.

For a quarter of a century, China reformed, but it afforded the outside world only the slightest glimpse of what was occurring. In the late 1960s it was estimated that there were perhaps 500 Westerners resident in China, a country of 900 million people. Foreign journalists could enter only sporadically. Statistics were simply not available. But in 1972, there were signs of change. United States government leaders visited Peking, accompanied by a large number of reporters; other groups, including professors, soon followed. The bamboo curtain appears to be lifting, slower perhaps than suggested by the drama of a United States presidential call in Peking, but lifting nevertheless.

Has one generation of communist administration brought permanent change to China? It seems so. Reform in agriculture initially eliminated the system of tenancy and parceled out the holdings of landlords among landless village families. When the results (in terms of output) were deemed inadequate, a collectivization program was begun, and villagers were pressed into cooperatives. These cooperatives in fact were socialist collectives; the Chinese peasant, who had briefly held some land under the Agrarian Reform Law of 1950, had lost it again to the state. Then in 1958 the collectivization program was carried one step further. Over 120 million peasant households, most of them already organized in collectives, were reorganized into about 26,500 People's Communes numbering about 20,000 persons each. This was not a paper reorganization; it was a massive modification of China's whole socioeconomic structure on an unprecedented scale. This was the Great Leap Forward, China's transformation from a socialist to a truly communist society.

We know now that the attempt at communization succeeded only partially, in part because of error, in part because of resistance by the farmers, and in part because its implementation was attended by several years of drought and bad crop yields. Still, the effect of the system on the face of China was great. Organized into production teams with military designations (companies, battalions, brigades), workers by the thousands tackled projects such as irrigation dams, roads, dikes, and so forth; fences and hedges between the lands of former collectives were torn down and the acreages consolidated. Villages were leveled, others enlarged. New roads were laid out to serve the new system better.

While there was reorganization in the agricultural sector, the Chinese Communist government invested heavily in industry. The first Five-Year Plan (1952–1957) poured almost half of the state's investment into industry; only 3 percent went to agriculture. Later, those priorities were changed to give agriculture a greater share, but the plan gave a good indication of the nature of China's new aspirations. And certainly China has the resources to make progress in industry. There is ample coal, self-sufficiency has been achieved in petroleum, steel production from domestic sources has increased at least tenfold in the last 20 years, and alloys are plentiful.

But for all its new industries and its great cities (such as Shanghai, 10 million, and Peking, 8 million), China still remains a realm of crowded farmlands, carefully diked floodplains, and intricately terraced hillslopes where crops of rice and wheat are meticulously cultivated and harvested and where the vast majority of the people bend daily to the soil. As Fig. 3.1 shows, most of China's near-billion people are concentrated in the country's east, and in the lowlands of its major river systems. Westward into Asia lies an unfamiliar China, with deserts and rugged, snow-covered mountains. Still partially unexplored, there may lie the resources to sustain China's forward thrust.

SOUTHEAST ASIA (5)

Asia's southeastern corner of peninsulas and islands is bounded by India to the northwest and by China to the northeast. This is the Southeast Asian realm, its western coasts washed by the Indian Ocean and its eastern islands stretching far into the Pacific. From all these directions, Southeast Asia has been penetrated by outside forces. From India came traders and from China came settlers. Across the Indian Ocean arrived the Arabs to engage in commerce and the Europeans in pursuit of empires. And from across the Pacific came the Americans. Southeast Asia has been the scene of countless contests for power and primacy; the competitors have come from near and far.

Southeast Asia is nowhere nearly as well-defined a culture realm as India or China. It does not have a single, dominant core of indigenous development. Culturally Southeast Asia is diverse in every respect: there are numerous ethnic and linguistic groups, various major religions, and different economies. Even spatially the realm's discontinuity is obvious: it consists of a peninsular mainland section and hundreds of islands, which form the archipelagoes of Indonesia and the Philippines. The mainland area, smaller than India alone, is divided into no less than eight political entities (Fig. 5.1), and the islands, in addition to Indonesia and the Philippines, contain two of the world's remaining dependencies, namely Portuguese Timor and British-protected Brunei, a sultanate in Malaysia. This political fragmentation underscores not only the cultural complexity of the realm, but also its history of European colonial intervention. Except Thailand, none of the states on the map of Southeast Asia today was independent when the Second World War came to an end.

Southeast Asia is a realm of many countries, but it is far less populous than either India or China. This has to do with the realm's overland approaches, which have major obstacles (the Chinese came mainly by sea), and with the limited

agricultural opportunities the land provides. Most of the people are directly dependent upon the soil. Dense tropical rain forest (Fig. 1.13) and considerable relief (Fig. 1.5), combined with seasonal drought problems and poor soils, make Southeast Asia anything but a farmer's paradise. The people are concentrated in the valleys and deltas of the major rivers, in the areas of fertile volcanic soils on the islands, and in the areas where the Europeans established thriving plantations. Predictably, nearly all the region's leading cities lie in the lower courses of the rivers, in or near the deltas. Hanoi, Bangkok, and Rangoon are so positioned; Saigon lies near the Mekong Delta (Fig. 5.5).

Indochina

For the time that France maintained a colonial empire in what today is a set of four states, North and South Viet Nam, Cambodia, and Laos, the French referred to that area as "Indochina." That is a good name for the whole mainland realm of Southeast Asia, for it identifies two of the major sources of influence that have affected the realm. The great majority of Southeast Asia's inhabitants have ethnic affinities to the people of China, but it was from India that the realm received its first strong, widely disseminated cultural imprint. The Buddhist and Hindu faiths were introduced, and Indian architecture, art (especially sculpture), writing and literature, and social structure as well. The building styles of the ancient Indian temples can still be detected in the architecture of the region today. Also by way of India came the Moslem faith, which gained a major stronghold in Indonesia and numerous footholds elsewhere. Among the latter is the Moslem minority in the Philippines, clustered on the island of Mindanao and frequently at odds with the Catholic-dominated state.

Those are some of the "Indo" elements of Indochina; in many respects the "china" influence has been even greater. The majority of the modern inhabitants of Southeast Asia's mainland have ethnic ties with the Chinese; Mongoloid strains are strongest in the northern part of the region while Malay features predominate in Indonesia's distant east. Chinese modes of dress, plastic arts, boat types, and numerous other cultural attributes were adopted in Southeast Asia. And during the past century a renewed Chinese immigration brought Chinese skills and energies, especially to the cities, where substantial Chinese sectors developed. The Chinese attained positions of considerable wealth and power in the region, strengthening further the impact of their culture on this realm.

If there is anything wrong with the term *Indochina,* it is that it does not take into account the third major influence that has penetrated and modified the culture of this realm, the influence of Europe. The Spanish took control of the Philippines, the Hollanders gained a vast empire in Indonesia, the British occupied the Malay Peninsula and a part of Borneo (largest of the Indonesian islands) as well as Burma, and the French controlled all the territory east of Thailand and south of China. Only Thailand itself escaped colonial subjugation.

Initially the Europeans sought to develop and exploit the resources of

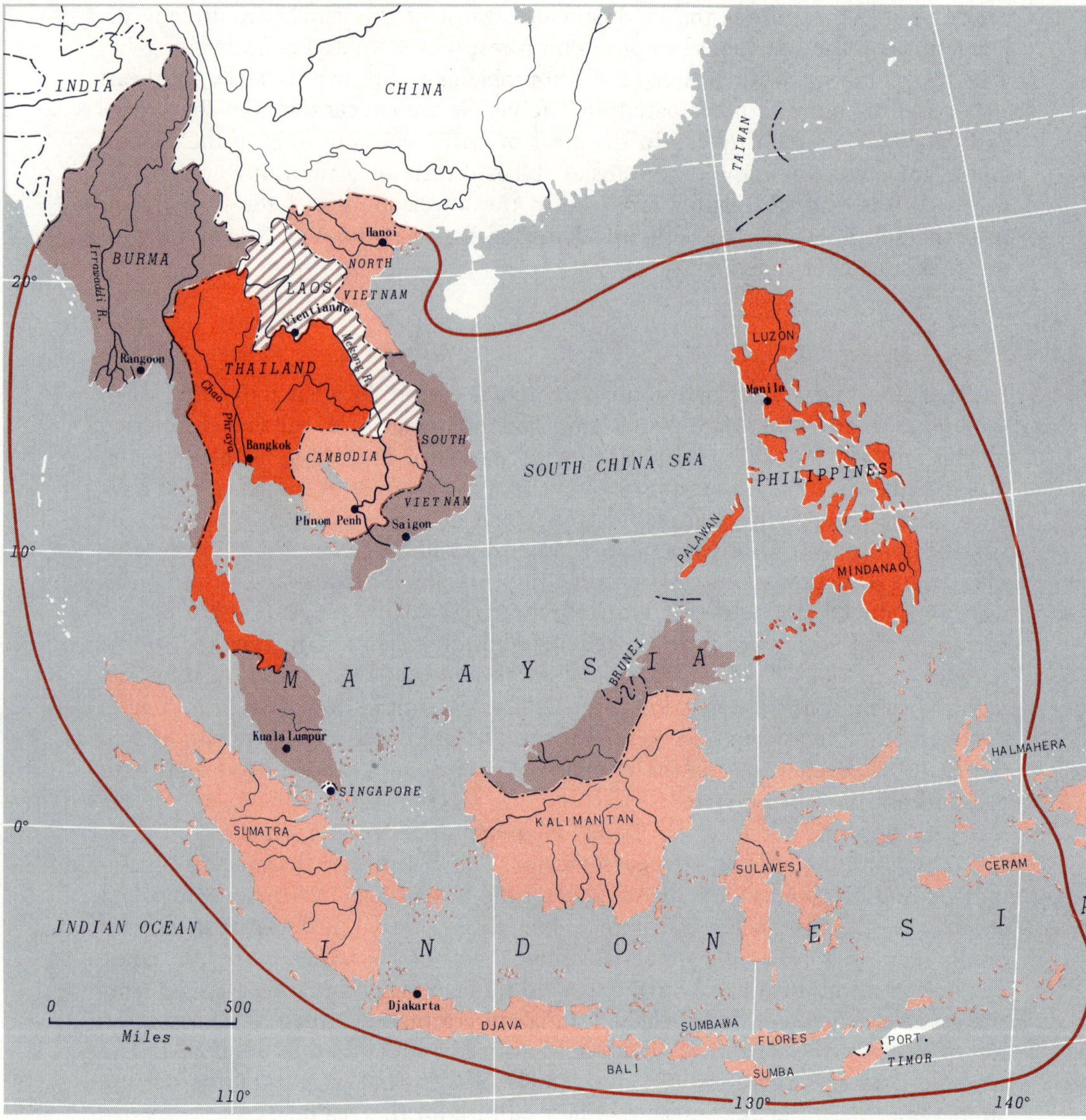

FIG. 5.5 The Southeast Asian realm.

Southeast Asia; the acquisition of territory was a secondary matter. But European spheres of influence expanded over time, and during the nineteenth century there was a rush to consolidate empires and establish firm control. This process was accompanied by accelerated economic development in several spheres. The introduction of plantations completely revolutionized the economies of such areas as

Malaya, Java (or Djawa, the most populous island of Indonesia), part of Sumatra, and elsewhere. Forced-crop and forced-labor systems required the cultivation of stipulated crops on predetermined acreages. The practice turned Java into one great farm estate. And here as in India, the European presence was attended by an increase in the rate of population growth. Until the Europeans came, the rice produced in Southeast Asia was almost all consumed by the families of the people on whose land it was grown. The Europeans built and expanded cities and towns, and they laid out transport routes to link these cities to the rice-growing areas so that the people in the cities would be supplied with their needs. Rice cultivation became a major cash proposition in those areas. The Europeans also found and brought into production new areas for additional rice yields.

The new areas have become major centers of population concentration in Southeast Asia. European planning and draining techniques, British and French capital, Chinese and Indian labor combined to prepare the lands. In the process, the realm was transformed. In the delta of the Irrawaddy River in the hinterland of Rangoon, Burma, in the delta of the Chao Praya, below Bangkok, Thailand, and in the delta of the Mekong River, south of Saigon in Viet Nam, Southeast Asia's new breadbaskets were created. There was more rice even than the cities needed, and it could be sent to the labor forces on the plantations and even exported to the food-deficient countries outside of Southeast Asia.

Southeast Asia, then, is a realm of many influences and exposures, where high population densities in comparatively isolated clusters are contrasted against nearly empty, sparsely peopled terrain, where conflict still prevails between the ancient peoples of the hills and mountains and the modern inhabitants who displaced them, and where external competitors still vie and fight for authority. In the cities, still bearing the strong evidence of their European stimulation, substantial Chinese minorities bear witness to the impact of the giant to the north.

BLACK AFRICA (6)

Across the Sahara Desert, through the Sudan and Ethiopia, and along the border between Kenya and the Somali Republic, lies a transition zone between culture realms, between the Arab world in Africa and black Africa itself. It is indeed a transition zone that separates these realms and not a distinct boundary (as Fig. 5.1 might suggest), for the Sahara Desert notwithstanding, there has been much contact between Arab and black Africa over the centuries of history. The ancient African states of the West traded with the North by caravan across the desert. Islam arrived and hundreds of thousands of black Africans became Moslems. Northern Nigeria today is black Africa, but it is also a part of the world of Islam with some 30 million adherents. Arabs and Persians reached the coast of East Africa and founded settlements, some of which (such as Mombasa and Zanzibar) have become prominent centers.

Black Africa today is emerging from an era of contact of another kind, colonial control. The white man came, first to build revictualing stations for the

trade routes to the Indies, then to establish beachheads for the slave trade, and still later, during the nineteenth century, to carve out territorial holdings. And so the world came to know the Gold Coast and the Belgian Congo, German East Africa and Portuguese Guinea. Among the attributes Africa got from that period is the now-familiar framework of boundaries that fragment the realm into several dozen countries. Among the things Africa lost (apart from independence) were people by the millions to the slavers, raw materials by the tons to the European profiteers, and land by the thousands of acres to white farmers and ranchers.

The colonial era is now ending. Only Portugal holds a substantial empire near the major bastion of white control, South Africa. In the capital cities the Europeans founded and built, African governments now rule. Plantations and farms are run as cooperatives following the withdrawal or the ouster of their alien owners. Mines and mineral resources have been nationalized. Names have been changed: the Gold Coast is Ghana, the Belgian Congo has become Zaïre, and German East Africa now is called Tanzania. If there are common goals in Africa, they relate to the elimination of the last vestiges of colonialism, the termination of South African control in Namibia (South West Africa), and ultimately the ending of white minority rule in South Africa itself.

Despite the European scramble for Africa in the second half of the nineteenth century, black Africa is no realm of great riches. True, there are the gold of the Johannesburg area and the Orange Free State and the unparalleled diamond mines of Kimberley. In Northern Rhodesia (now Zambia) and the Congo (Zaïre) the Europeans found copper deposits of enormous value. In Kenya and Rhodesia there were fertile farmlands to be cultivated. Elsewhere, too, profits could be made from the mines and the soil and the low-cost labor so readily available. But these islands of opportunity lay, and still lie, in an ocean of territory where a comparatively sparse population can wrest a living from the soil only with the greatest of efforts. Here in Africa, situated astride the equator, with great escarpments and mountains and endless plains, where the overwhelming majority of the people depend directly on their land and their livestock for survival, the great problem is not excess moisture, as in so many tropical zones, but drought. For months the savanna is dry and parched, the soil hardens into a cementlike cake, the grasses turn gray and burn, crops of corn wither, goats and cattle die. Then, when the life-giving rains come, they beat down on the hardened surface and the water runs off into gulleys and streams. It takes weeks for the soil to get its moisture, and by then the gulleys have widened into crevasses and tons of tons of soil have been carried away.

With some noteworthy exceptions, black Africa is a realm of farmers. A few people still depend upon hunting and gathering alone, but even the Pygmies trade for vegetables with their neighbors. Along the coast, especially the west coast, and the major rivers, some communities depend primarily on fishing. But otherwise the principal mode of life is farming. And this subsistence way of life was little changed by the colonialists. Tens of thousands of villages all across Africa never were brought into the economic orbit of the European invaders, and life in these settlements went on more or less unchanged. Where Africa is moister and warmest (corresponding approximately with the rain-forest region, Fig. 2.12), root crops form the staples: yams, cassava, sweet potatoes. In some areas of Zaïre, bananas form the staple food. In the savanna and plateau regions, grains predominate, grown in time-honored ways with the hand hoe and the digging stick. It is a back-breaking, low-yield proposition, but the African farmer does not have much choice. Where he does, in the Kenya Highlands, on the fertile slopes of Mount Kilimanjaro, on the cotton fields of Uganda and the cocoa farms of West Africa, he proves himself to be a successful cultivator.

Africa's herders more often than not mix farming with their pastoral pursuits, and very few of them are actually "pure" herdsmen. In Africa south of the Sahara there are two zones where pastoralism prevails, one extending along the West African savanna-steppe and connecting with the East African area (where the Masai drive their herds) and the other centering on the plateau of South Africa. Especially in East and South Africa, cattle are less important as a source of food than they are a measure of the wealth and prestige of their owners in the community. In West Africa the pastoralists are willing to sell their cattle for meat, but they face the disadvantage of being far from the major, coastal markets. But cattle are not the only livestock in Africa. There are millions of goats. They are everywhere, in the savanna, in the forest, in the steppe, even in the desert. They always seem to survive, and no African village would be complete without its share. They are an asset as a source of food, but they denude the countryside.

Against this image of rural Africa we should contrast the bustling indigenous cities such as Kumasi (Ghana) and Kano (Nigeria), places of substantial size and importance long before the Europeans came to reorient Africa's social, political, and economic directions, and the headquarters the Europeans built, Nairobi, Salisbury, Kinshasa (formerly Leopoldville), and Lagos. Black Africa has no cities to rival Calcutta or Shanghai. The realm's largest urban centers (Ibadan, Johannesburg) are in the 1-to-2 million class. In West Africa's savanna zone, those indigenous cities rose to importance as centers of trade; the European-built cities, serving as points of transfer for products and as foci for developing areas, reflect by their limited dimensions the restricted qualities of their hinterlands.

In the historical geography of Africa, the west and north of the realm have been focal areas. It is thought that the Bantu invasion of recent millenia spread southward from a northern or north-central source; West Africa was a place of political progress and agricultural innovation. The second half of the twentieth century witnessed West Africa again in the forefront of the drive against colonial subjugation. Ghana was the first truly black African state to regain its independence. At the same time it was southern Africa that held the greatest wealth and attracted the largest number of Europeans, there in the south where black Africa could offer least resistance. From the vineyards of the Cape Province in the hinterland of Cape Town, through the Orange Free State with its gold and coal, Johannesburg and the Witwatersrand, Rhodesia with its minerals and good farmlands, into Zambia and

Zaïre with their copper fields, lies the greatest reservoir of natural resources in Africa. In 1973, all of it except Zaïre's and Zambia's copper was still in white-controlled areas.

MIDDLE AND SOUTH AMERICA (7)

Middle and South America is the realm often referred to as "Latin" America because of the Iberian imprint placed upon it during the European expansion. Portugal in Brazil, and Spain in much of the remainder of this realm, founded vast colonial empires. Later the American dependencies freed themselves from Iberian control, but the cultural impress was there to stay. Spanish is spoken from Mexico to Argentina, and Brazil's nearly 100 million people speak Portuguese. The Catholic church prevails throughout Ibero-America. Systems of land ownership, tenancy, taxation, and tribute were transferred from the Old World to the New.

Actually, there are two regions in this realm, one somewhat less "Latin" than the other. South America has only three comparatively small territories that were not endowed with Iberian cultural qualities: Guyana, Surinam, and French Guiana, on the north coast between Venezuela and Brazil. Guyana was a British dependency, Surinam in 1973 still was nominally attached to the Netherlands, and French Guiana remains among the least-developed parts of the shrinking empire. Elsewhere on the continent, the Andean Indian populations from Colombia to Bolivia and the numerous black (but more strongly acculturated) people of Brazil and the Caribbean coast are recipients rather than generators of the Latin culture; much of interior, Andean Peru is hardly "Latin" America. In Middle America the exceptions are far more numerous, especially if the region is defined as on our map (Fig. 5.1) to include not only the countries from Mexico to Panama but also the islands of the Caribbean. On the mainland, Belize (British Honduras) has non-Latin attributes; in the Caribbean, Jamaica, Trinidad, and Tobago, and several smaller islands are British influenced; Haiti has a black population of French cultural background; and the Netherlands retains several islands as well. But even in this Middle American region Latin influences are widespread and pervasive: on the mainland from Mexico to Panama, in the islands in Cuba, and in the Dominican Republic and Puerto Rico.

South America

South America might more appropriately be named *Southeast* America because most of it lies not only south but also east of its northern counterpart. Practically all of the continent lies east of the longitude of Miami, Florida! Lying so far eastward, South America on its western flank faces a much wider Pacific Ocean than does North America. From its west coast to Australia is nearly twice as far as from San Francisco to Japan. South America has virtually no interaction with the Pacific world or Australasia, not just because of the vast distances involved,

but also because of its position opposite the insular and less populous south, in contrast to North America, which faces Japan and the crowded East Asian mainland. The additional factor of culture comes into play because Australia and New Zealand, South America's nearest Pacific neighbors, share with North America the imprint of English and northwest European culture.

As if to confirm South America's eastward and northward orientation, the western margins of the continent are rimmed by one of the world's great mountain ranges, the Andes, a giant wall that extends from the very southern tip of the South American triangle to Colombia in the north. The continent lies open to the north and east, and its proximities have brought colonists from Europe, labor from Africa, and economic interests from the United States. As Fig. 3.1 reminds us, the major population clusters in South America are positioned along the east and north coasts, overshadowing those of the west.

Latin-influenced South America is fragmented into 10 countries, of which Brazil is the largest. Indeed, Brazil by itself constitutes a region in South America, not only in terms of sheer size but also in terms of population numbers (nearly half of the continent's total) and its Portuguese cultural heritage. Brazil is South America's eminent non-Andean country. It incorporates the drainage basin of the continent's greatest river, the Amazon, as well as the whole eastern upland of the continent. Closer than any other South American country to Europe and Africa, Brazil has received more immigrants from both. Never densely populated by Indian peoples, it now has fewer Indian inhabitants than most of its neighbors. No South American country exemplifies to a greater degree the peripheral location of population that is so characteristic of the continent as a whole.

Along the northern coast, Colombia, Venezuela, Guyana, Surinam, and French Guiana can on several bases be grouped together regionally. This is "Caribbean" South America, where the Europeans came to establish plantations, where black labor was imported, and where thousands of Asians (mainly Indians) were brought as indentured laborers. The economic geography of Venezuela and Colombia shows diversification, but the plantation economy still prevails in the three other territories.

The countries that have substantial Indian (American Indian) minorities can also be grouped regionally. "Indian" South America consists of Peru, Ecuador, Bolivia, and Paraguay. About half the population in these countries is Indian, though it is difficult to be precise since it is often impossible to distinguish between Indian and "mixed" people of strong Indian character. Here incomes are low, as is productivity; here prevails the grinding poverty of the landless peonage, a problem that looms large in the future of the realm. These, too, are South America's least urbanized countries. And here live millions of people to whom the Spanish language is still alien, to whom the white man's religion is another element of deculturation, and to whom decorous Spanish styles of architecture are pretty meaningless when a decent roof and a solid floor are still unattainable luxuries. After the three Guianas, this is South America's least "Latin" region.

Several reasons exist for grouping South America's three southern countries, Argentina with its neighbors Chile and Uruguay, in one region. This is the continent's most strongly European region, with few Indian or other nonwhite elements in the population. Economically these countries are quite advanced. The Argentine Pampa is one of the world's most productive agricultural zones; the core area of Argentina is served by one of the continent's two relatively intensive transport networks (the other is in eastern Brazil). Middle Chile, centered on Santiago, north-central Argentina, focused on Buenos Aires, and Uruguay, whose capital is Montevideo, lie in a belt outside the tropics, in temperate South America, another element they share.

Middle America

Middle America includes three sectors: Mexico, larger in territory and population than all the rest together, the Central American Republics (Guatemala, El Salvador, Honduras, Nicaragua, Costa Rica, Panama, Belize), and the Caribbean islands

FIG. 5.6 Middle America.

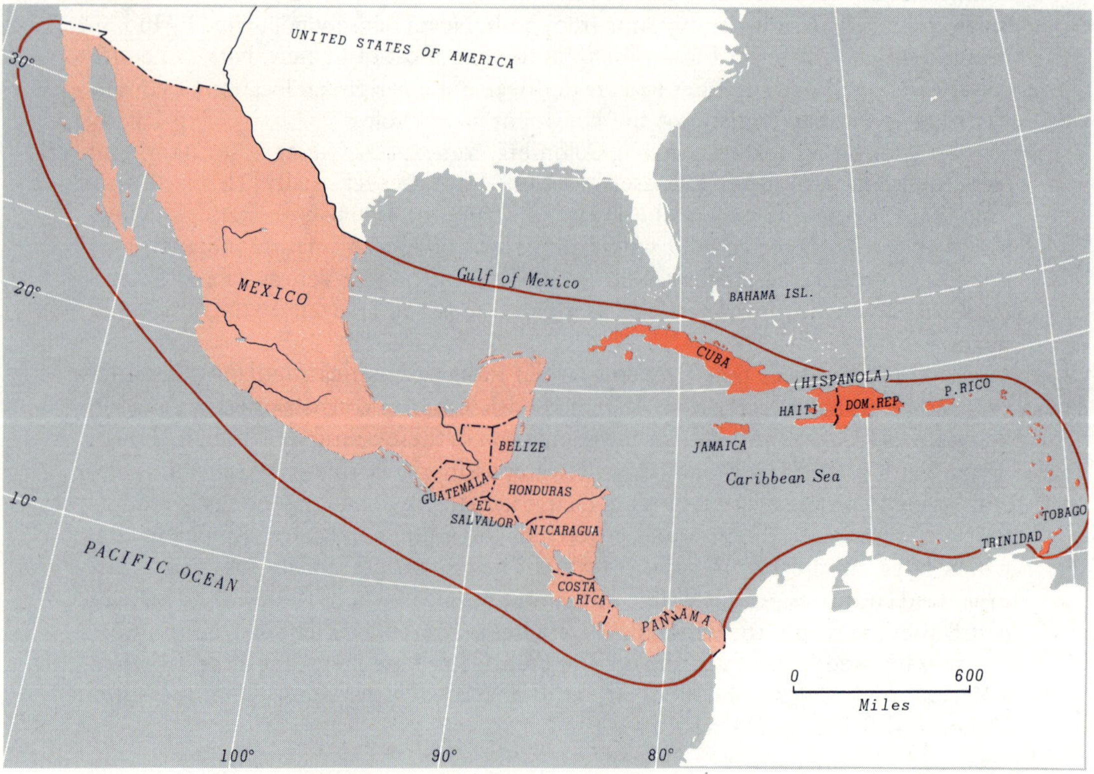

(Cuba, Hispaniola, Puerto Rico, Jamaica, and the smaller isles) (Fig. 5.6). Mexico is Spanish-influenced America's largest country, but its population of 40 million includes some 12 million Indians. Mexico was the scene of the great Aztec empire, and the impact of Indian culture on Mexican life is strong. Among the republics of Central America, Guatemala has a population that is about 55 percent Indian, a population that has been far less involved in the culture of that country and that remains quite pure Maya and Quiché. Elsewhere the majority of the people are *mestizo* (mixed European and Indian), with "pure" white and "pure" Indian minorities. But quite a different situation prevails in the Caribbean, where the original Indian population was wiped out and replaced by Africans, Europeans, and Asians. Cuba and Puerto Rico share with the mainland their Spanish heritage, but Jamaica is British influenced (with an overwhelmingly black population), Haiti has a French imprint and is also dominantly peopled by Afro-Americans, and Trinidad-Tobago has a significant Asian population sector. Nowhere is the concept of the plural society more powerfully exemplified.

NORTH AMERICA (8)

The North American realm also is known by another name, and one not much more appropriate than "Latin" America. In this case the name is "Anglo" America, referring to the English heritage that is strong in the two countries that constitute the region, Canada and the United States. But, as we noted earlier, both countries have plural societies. At first this might not seem to be the case. Each country has received immigrants in large numbers from many cultures for more than a century (although immigration to the United States declined sharply after restrictive legislation was passed following the First World War). These immigrants have to a varying degree been absorbed into an American culture that was different from any of the parts, earning the United States the label of "melting pot." Three or four generations were usually sufficient for United States or Canadian cultural characteristics to prevail over those of the grandparents' home country. Absorption was occasionally retarded by choice, as in the case of concentrations of farmers descending from German settlers in Pennsylvania, Michigan, and Wisconsin or the population enclave of those with Polish background in Hamtramack, Michigan, a city within the city of Detroit. It may be severely retarded by the non-European background of the immigrants, leading to a Chinatown in San Francisco and in Chicago, Puerto Rican ghettos in New York, and Mexican-American ghettos in the Southwest. In spite of the fact that these individual groups are still recognizable after several generations, members of these clusters (and others like them) also gradually develop cultural characteristics that identify them as distinctly American or Canadian to an observer from another culture region.

Even so, Canada and the United States are plural societies in that they contain sizable minorities that differ significantly from the dominant culture even after many generations of residence on the continent. In Canada, the great majority of the inhabitants of the Province of Quebec are of French ancestry and still

CINE ROYALE
NATIONAL
SHIRT SHOPS
Coast to Coast
CANAL JEWELRY CO.
NATIONAL
SHIRT SHOPS
SEYMOURS
CREDIT CLOTHIERS
FLAGG
BROS
DIAMONDS
U.S. MAIL

maintain many cultural similarities with the French. When the Canadian Constitution was written, French rights and privileges were guaranteed. But history and geography conspired to keep Canada's French sector concentrated, aloof, unassimilated. Compare New Orleans, where a French heritage was submerged in the melting pot, with Montreal, where it was not.

In the United States, the population sector with dominantly African ancestry found it necessary to develop a culture of its own within the larger cultural sphere of the dominantly white population. Black America has a culture quite different from the one that prevails in this realm, but black American culture is quite different from African cultures too. For many decades, even centuries, the black population of America was denied access to the white culture, and the resulting presence of a dual culture now constitutes a divisive factor in this country as a whole.

The North American realm consists of two of the most strongly urbanized and industrialized countries in the world. In the United States there are more than 60 cities with populations of more than a half million, within which live fully half the people of this country. Canada, with a population about one-tenth as large as the United States, still has three urban agglomerations in excess of 1 million: Montreal, Toronto, and Vancouver. And in the United States, the vast majority of the remainder live in towns larger than 10,000 people. Among the population movements that have prevailed in North America over the past century, the rural-to-urban migration has gathered momentum enormously in recent decades. In the United States people have also been migrating from the South to the North and from the East to the West. In the process, with the majority of the realm's residents still concentrated in the eastern areas of the region, the cities of the Atlantic Seaboard are coalescing into a vast megalopolis that dwarfs similar megacities elsewhere. From Boston to beyond Washington, and including Baltimore, Philadelphia, New York, and intervening areas, suburbs are meeting, surface and air transport networks are intensifying, skylines are changing, and rural remnants are disappearing. Skyscrapers, traffic jams, waiting lines, noise, pollution of air and water—these are some of the attributes of our American technocracy.

The flight from the farms to the cities has drastically changed the balance of labor involved in agriculture and industry. Late in the nineteenth century as much as three-quarters of the population were engaged in some rural activity, as farmers, tenant farmers, or in the raising of livestock. Now, less than 100 years later, only about 3 million farm families remain to produce food for the United States' 210 million people. Despite this, actual farm output has continued to rise, and herein lies one of the qualities of the North American realm: the efficiency of its agricultural production, its mechanization, and its high per-person yield.

Farmers benefited from the technological progress that is a hallmark of North American culture, but North America is also well endowed with the natural resources to make great productivity possible. As we noted in Chapter One, good, humus-rich soils extend in a wide zone through the heart of the continent, and there grow the two most important crops in America, corn and wheat. The Wheat

Belt extends from the central United States into the Canadian Great Plains. The land is flat enough to permit the use of mechanized equipment; this is true also for several of the specialty crops grown in the region, such as cotton and tobacco. But America's productivity is hardly confined to these products of the soil. From the vineyards of California to the citrus groves of Florida it yields bounty upon bounty.

The industrial resources North America provides are also rich and varied. Extensive coalfields, one of the world's major clusters of oil fields, and large iron ores exist in this region; and exploration in some areas is still yielding new discoveries (for example, in Alaska and on the continental shelf). Few minerals required by modern industry are not found in this realm. Bauxite (for aluminum), tin, and manganese ores are the only major resources not found in quantity. The industrial rush forward has severely strained the supplies, and North America imports many commodities from faraway places to supplement them. But few areas of the world supplied as diverse a resource complex as did North America.

How do the peoples of other cultures perceive North America, and especially the United States? There is growing awareness not only of America's wealth and power, but also of its enormous consumption. Here in America, 6 percent of the people of this earth are consuming nearly 50 percent of all the commodities consumed in the entire world each year. Can the earth afford such a corner on luxury in one realm? Many think not. But that realization and the worldwide presence of United States interests—economic, political, military—combine to affect the images of America in the rest of the world.

AUSTRALIA (9)

Beyond the islands and archipelagoes of Southeast Asia in the vast expanse of water of the Pacific and Indian oceans lies the island continent of Australia. Smaller even than Europe, Australia might never have come to be recognized as an individual continent and, with New Zealand, as a distinct culture realm. If Australia were populated by peoples of Malayan stock with ways and standards of living resembling those of Indonesia and mainland Southeast Asia and had a comparable history, then it is quite possible that it would be viewed today only as an exceptionally large island sector of the Asian continent. But just as Europe merits recognition as a continental realm despite the fact that it is merely a peninsula of Eurasia, so Australia has achieved identity as the island realm of Australasia. Australia and New Zealand are European, white man's outposts in an Asian-Pacific world, as unlike Indonesia as Britain and America are unlike India.

Although Australia was spawned by Europe and its people and economy are Western in every way, Australia as a continental realm is a far cry from the crowded, productive, populous European world. Apart from the fact that its area is under 3 million square miles, which is only three-quarters that of Europe and smaller than the Soviet Union, Canada, China, the United States, or Brazil, its

SA
117
A
12
70
3

population is really diminutive. It is a mere 13 million. This gives a population density of about 4.5 and suggests that Australia is a virtual population vacuum on the very edge of teeming Asia. But the physiological density is actually quite high. So much of Australia is arid or semiarid (see Fig. 2.12) that only about 8 percent of the continent is agriculturally productive, and much of this moister part of the continent (mainly the east and north) is too rugged for farming. By some calculations, only 1 percent of Australia's total area is prime land for intensive agriculture; even if this figure is 50 percent too low, the physiological density would exceed 210 per square mile. That may not be awfully high either, but it is far more representative of the real situation than the 4.5 per square mile for Australia as a whole. And it does suggest that Australia, although capable of absorbing many more people, is no feasible outlet for Asia's huge overpopulation. Australia's *total* population is less than the annual *increase* of China's alone!

Apparently Australia never was a very heavily populated realm. The indigenous Australians who were there when the first Europeans arrived numbered well under a half million; after crossing into Australia from New Guinea they migrated southward along the coasts. The white man first saw Australia in the 1640s but did not begin permanent settlement until the 1770s. At that time New Zealand also entered the European orbit as missionaries, traders, whalers, and sealers founded bases there. Australia's remoteness was a big factor in the course of European settlement. It was selected as a site for the creation of penal colonies for deported convicts. Thus Sydney, today a city of nearly 3 million, was founded just 200 years ago. Eventually, however, free settlements began to grow larger than the penal colonies, and the immigration of free settlers was encouraged. During the nineteenth century two phases of development catapulted Australia to prominence: the growth of the pastoral industry on the seemingly endless grasslands stretching beyond the hills overlooking the east-coast settlements and the mineral age that began with the discovery of gold in New South Wales and Victoria.

Australia during the nineteenth century consisted of seven competing colonies, seven coastal towns with their hinterlands. The administration centered in each of those ocean ports was devoted to particular, local interests that were often in conflict. For example, the territories of Victoria and New South Wales, centered on Melbourne and Sydney respectively, were locked in a battle over the issue of protective tariffs (wanted by Victoria's manufacturers) versus free trade, preferred by New South Wales. Notwithstanding all the problems, however, a federal state was founded in 1901, with a new capital at Canberra.

Politically, then, Australia is a young state. It remains a frontier country in other ways as well. Australia's "empty heart" still exists, although transport systems finger timidly inward from coastal termini to inland cattle stations and farming settlements. Only in the south is there a cross-continental railroad, connecting Sydney on the east coast to Perth on the western shore, but a north-south connector from Darwin to Adelaide is still in the planning stage. On Australia's pastures move nearly 200 million sheep and 20 million head of cattle, and wool and beef figure importantly in the external trade. Vast fields of wheat, benefiting from irrigation and heavy mechanization, render Australia one of the world's chief exporters of that commodity. And it is unlikely that Australia's age of mineral discovery is over. Australia is a leading producer of lead and zinc, coal is plentiful, substantial iron ores and other metals exist, the promise of petroleum and natural gas is bright, and there is still more. Australia has everything it takes to attain the status of industrial power except population numbers.

Not surprisingly, those same towns that were the capitals of the early colonies have become the country's major industrial centers. Rural Australia, despite its great productivity, is not densely settled; Australia is a highly urbanized country, with well over 80 percent of the people living in cities and towns. The old capitals, at the foci of the railroad networks of the individual territories, also were the main ports for overseas as well as coastwide shipping. They had the best amenities for industry (a concentrated labor force, adequate power supply, water, access to government), and they were also the major domestic markets. Here, too, comes the flow of immigrants Australia encourages (though selectively; there was for a long time a white-Australia policy). New immigrants numbering 200,000 arrive each year.

We tend to have an image of Australia as a realm of impressive, large, attractive cities, wide expanses, huge herds (perhaps also the pests, rabbits and kangaroos), and beautiful coasts. There is much truth in that picture: Australia has large cities for a country with only 13 million people. Sydney leads with 3 million, Melbourne has 2½ million, Brisbane and Adelaide, 1 million each. In the seven largest cities are concentrated 9 of Australia's 13 million people! In this as well as many other respects, then, Australia is more like the United States and the core area of Europe than any closer realms. With its British heritage, its homogeneous population, single language (excepting only the indigenous, small minority), and type of economy, Australia's identification as a realm rests on its remoteness and spatial isolation.

New Zealand shares numerous characteristics with Australia, including a British heritage, an indigenous (Maori) minority, a pastoral economy, a small local market, and great, expensive distances to the world markets where its products must compete. New Zealand also shares the spatial phenomenon of peripheral development. In contrast to Australia, whose peripheral clustering is related to its desert interior, the interior of New Zealand is mountainous and rugged. But on those slopes graze 60 million sheep and 8 million cattle, and wool, meat, and dairy products secure three-quarters of the country's export revenues. New Zealand, too, is an urban country. More than two-thirds of the population is concentrated in such cities as Auckland and Wellington, where the treatment and packaging of the agricultural products constitute the major industries. In the Pacific world of which New Zealand is a part, the islands' distinctions are so sharp, and the similarities with Australia so great, that the extension of the culture realm to include them is beyond debate.

СОРЕВНУЙТЕСЬ

THE SOVIET UNION (10)

The Soviet Union dominates the map of the world because of its enormous size. From the Baltic Sea to the Pacific Ocean and from the Arctic to the borders of Iran, the U.S.S.R. (Union of Soviet Socialist Republics) occupies the heart of Eurasia, and almost half the area of that vast landmass. It is surrounded by a crowd of countries from Finland to Turkey to Afghanistan to Japan, and some of these countries have lost parts of their territory to what is truly a Russian empire. With more than 8.5 million square miles, the Soviet Union covers one-sixth of the earth's land surface. It is not only a state of continental proportions but also a true culture realm.

The Soviet Union is a realm not just because of its great size. To its south lie the Japanese, Chinese, and Southwest Asian realms, all clearly differentiated culturally from the state that Russia spawned. We discussed earlier the boundary between the Soviet and European realms; there are scholars who believe that the western part of the U.S.S.R. should be viewed as a part of the latter. And certainly there are factors to support such a position. The ethnic makeup of most of the Soviet Union is much like that of Eastern Europe. Russian cities are much like those of Europe; the Russian leader, Peter the Great, actually brought European architects to Russia to build his city, Leningrad (formerly St. Petersburg), in a European style and image. Early in the country's history, Scandinavian rulers organized the people politically; later, Russia was to take temporary control of Finland and Poland. The old city of Kiev, in its medieval days of splendor, rivaled Paris as a center of European culture.

But ultimately Russia looked eastward for an empire, and following an era of expansionism quite similar to that of the "real" European colonial powers, the Russians held control over a vast region extending all the way across Siberia to the Pacific. Indeed, the Russians were in Alaska, and but for a familiar twist of historic fate, there would be a large Soviet territory in North America today. The Russian empire, built over 5 centuries of nearly uninterrupted expansion and consolidation, came to include Armenians near the border with Turkey (between the Black and Caspian seas), Uzbeks, Kazakhs, Turkmens, Tatars, and millions of other Moslems. Such Moslem centers as Ashkhabad, Tashkent, and Samarkand became Soviet cities, the headquarters of S.S.R.'s, Soviet Socialist Republics. Farther east, Mongols and Chinese people fell into the Russian orbit.

In the late nineteenth century, the Russian steamroller finally was halted as the Russians came face to face with another imperial power, Japan. The Russians had made use of Japanese-Chinese strife to annex Chinese territory, and now they were encroaching on Japan itself. War was not long in coming, and in 1905 the Russians were forced to concede defeat. It was a preamble to what was to come in the heartland of Russia itself, where a revolution against the old order was brewing. When that revolution came, in 1917, the prize was control over a vast, contiguous empire that included not only the core of old Russia itself, but also the vast expanses of eastern Asia and the colonies (for that is what they were, contiguous colonies) between the Black Sea and Mongolia.

The overriding justification for the recognition of the Soviet Union as a discrete realm, then, lies in the events of the twentieth century. The revolution, as we know too well, produced a whole new order in the U.S.S.R., sweeping away so much of what existed previously that the entire, complex nation was transformed. The capital was moved from old St. Petersburg to Moscow, and St. Petersburg was renamed Leningrad. A new political system was introduced whereby the various parts of the empire attained status as Soviet Socialist Republics, theoretically on a par with the largest of those republics, the Russian one. Thus, among others, an Armenian S.S.R., a Kazakh S.S.R., a Kirghiz S.S.R., and a Turkman S.S.R. were created in the Moslem area. In the west, where Estonia, Lithuania, and Latvia had been absorbed, similar S.S.R.'s were proclaimed, for a total of 15. Theoretically, there is equality among the republics, but in reality Russia leads, and the others look upward. Undeniably, the new system brought the far-flung empire under effective national control for the first time. For the first time, too, the Russians, recipients of European innovations, began to transmit these innovations to non-Russian and non-Slavic parts of the realm. There were elements of compromise in the relationships between the Russians, other Slavic peoples, and the non-Slavic population of the country. Certainly conflicts remain, and individual groups have expressed separatist tendencies, but in general there seems to be an awareness of the positive aspects of Soviet administration. After the terror of the Russian conquest itself the Soviet contribution in such areas as Armenia and Georgia is proof that membership in the Soviet Union does have advantages. Standards of living have been raised, educational opportunities have improved, housing projects have replaced slums, and the economy has accelerated—all this without any enforced abandonment of local culture, including language. The Soviets, of course, like to claim that their half century of control has brought unanimous approval of their system throughout the country. Although this is not so, their treatment of minority nationalities does appear to have achieved greater domestic approval than that of some colonial powers, and, indeed, of some governments of sovereign states.

Planning

In line with communist dogma, the hallmark of Soviet life in almost all spheres has been planning. What does not fit the greater scheme of things (religion, for example) is discouraged with various degrees of fervor. Soviet domestic policy makers have reached into the very heart of Russian culture to silence authors, direct musicians and other artists to write "for the people," and dictate expression in general. The Soviets anticipate that time and their continued program of Russification will eventually submerge the cultures of the various minority nationalities and produce a unified and quite homogenous state-culture.

The new order brought the most profound changes in the economic sphere. At first there was a headlong rush toward state ownership of industry and the confiscation of land from the nobility, but the results were not what was hoped for. Following a period of relaxation during the 1920s, the State Planning Commission announced the first of a series of Five-Year Plans. Planned were the collectivization of agriculture and the rapid expansion of industry. The second goal was tied to the first, for the increased efficiency of collectivized agriculture would free many workers for the factories. The collectivization of agriculture was achieved over the next 10 years, though not without revolt and resistance; and in industry enormous progress was made. Major investments were made in transportation, electrification, and other heavy-industry-oriented facilities, and the great wealth of natural resources in the huge U.S.S.R. (see Chapter One) began to pay off. Even through the Great Depression, the productivity of U.S.S.R. industry continued to climb steeply.

This is not to suggest that the planned economy of the U.S.S.R. has been an unqualified success and that there have been no setbacks. Some scholars believe that the country would have progressed even more rapidly under a capitalist system. But despite great difficulties (such as the widespread death and destruction that came with the Second World War), the Soviet Union in 50 years has risen from a backward, divided, near-feudal country to a world power with a strong, individualistic culture, a highly advanced technology, and a set of economic and social policies that have attracted world attention and, in some instances, emulation.

JAPAN (11)

In the non-Western, non-European world there is no country quite like Japan (Fig. 5.7). Suddenly the familiar phrases, "underdevelopment," "stagnation," and "subsistence," no longer apply. Japan is an industrial giant, an urban society, a political power, a vigorous nation. No city in the world is without Japanese cars in its streets; few photography stores are without Japanese cameras; many laboratories use mainly Japanese optical equipment. From tape recorders to television sets, from ocean-going ships to children's toys, Japanese manufacturers flood the world's markets. The Japanese seem to combine the precision skills of the Swiss with the massive industrial power of prewar Germany, the forward-looking designs of the Swedes with the innovativeness of the Americans. How have the Japanese done it? Does Japan have the kind of resources that helped boost Britain into its early, revolutionary industrial prominence? Are there locational advantages in Japan's position off the eastern coast of the Eurasian landmass, just as Britain benefited from its situation off the western coast? Could Japan's rise to power and prominence have been predicted, say, a century ago?

Little more than a century ago, Japan, having successfully resisted and withstood the European wave of colonialism, faced the steel ships and firepower of the United States in its harbors. Confronting the reality that the old, feudal order would not be enough to mobilize the island country against American (and European would-be) invaders, a group of reformers rose in revolt and overthrew the existing authority. This event, called the Meiji Restoration of 1868, constitutes the beginning of the country's new era. The reformers were of the opinion that only rapid modernization in Western traditions would give Japan a chance against its

ワシントン
Colorful Summer '72
ビフテキのスエヒロ

enemies. Japan, they argued, needed not only a modern political system but also modern armed forces built by modern industries; it also needed a colonial empire. And so Japan began its phenomenal growth, a growth that brought it just such an empire, involved it in a disastrous war, and eventually saw it achieve an industrial capacity unmatched not only in the non-Western world but also in much of the Western world.

Japan's military successes have been written on the map: even before the turn of the century it was ready to defeat encroaching Russia. Japan took Taiwan from China (1895), occupied Korea (1910), and established a sphere of influence in China's Manchuria. Various archipelagoes in the Pacific Ocean were acquired by annexation, conquest, or mandate. Then the Second World War gave the Japanese a golden opportunity. The colonial powers, weakened by war at home, were unable to prevent Japan's takeover in Indochina, Burma, Malaya, and Indonesia; the United States was not able to protect the Philippines either. Japan, we know, was ultimately defeated by superior American power (two cities were wiped out by atomic bombs), but that defeat, coupled with the loss of its prewar as well as its wartime empire, failed to destroy the country's progress. Japan has rebounded with such vigor that its overall economic growth rate currently is the highest in the world.

Japan's success might lead to the conclusion that the country's raw material base was sufficient to support the same kind of industrialization that had characterized England. But Japan is not nearly as well off. Good coking coal is scarce, although readily accessible deposits were enough to provide an initial boost to industry in the late nineteenth century. Japan's iron ores are miniscule compared with Britain's. Alloys are rare and insufficient. Only a few barrels of petroleum are produced from domestic sources. Japan's search for a colonial empire was no exercise in prestige alone. Japan needed resources that could be controlled and obtained as cheaply as possible.

What, then, *did* Japan have as a basis for its great leap forward? For one thing, Japan in the middle of the nineteenth century already had a widespread, productive light manufacturing industry. In home industries and community workshops the Japanese produced silk and cotton textiles, porcelain, wood products, and metal goods. At that time the small metalliferous deposits were enough to supply the limited local needs. Power came from human arms and legs and from wheels driven by water; the chief source of fuel was charcoal. Thus there was industry, there was a labor force, and there were known manufacturing skills. These were the bases on which the Japanese built. The workshops were integrated into larger units; thermal and hydroelectric power was developed. And cheap Japanese products were sent to foreign markets, where for the first time they began to beat Western products. The income was poured back into manufacturing. The upward spiral was under way.

For a time Japan was able to do what its European competitors did, extract the necessary raw materials from a colonial empire and bring them to the home industries. Korea, Manchuria, and Taiwan were sources in this context; the Japanese created factories in their dependencies to refine ores into concentrates. Labor was

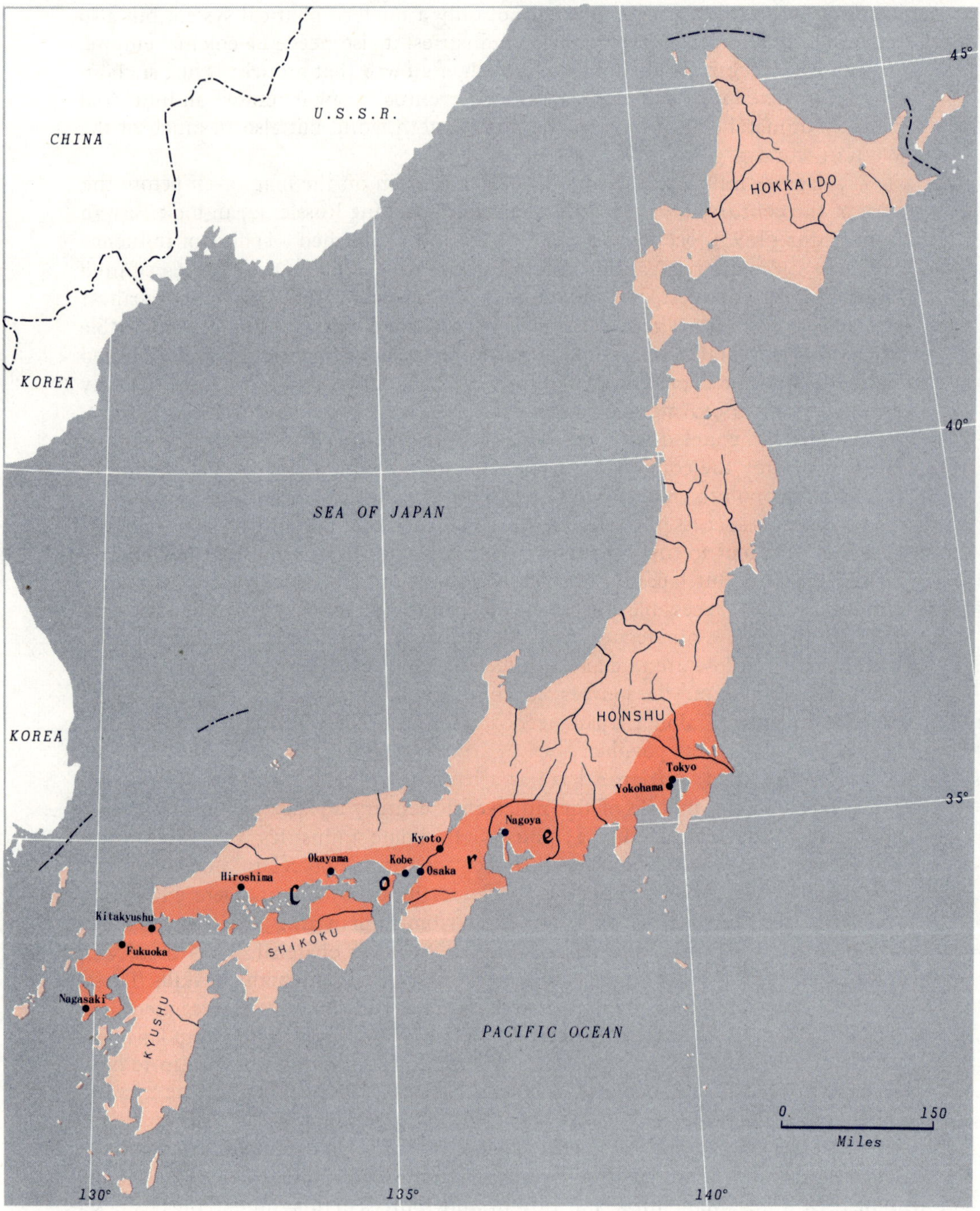

U.S.S.R.
CHINA
HOKKAIDO
KOREA
SEA OF JAPAN
KOREA
HONSHU
Tokyo
Yokohama
Nagoya
Kyoto
Okayama
Kobe
Osaka
Hiroshima
Core
Kitakyushu
Fukuoka
SHIKOKU
Nagasaki
KYUSHU
PACIFIC OCEAN
0
150
Miles
45°
40°
35°
130°
135°
140°

FIG. 5.7 Japan.

cheap. But when Japan lost its empire, the question of whether the country could fill its needs through trade rather than conquest arose. There is no doubt about the answer. Japan's iron ores come from as far away as southern Africa, Australia, and South America; its petroleum from the Persian Gulf; its coal from halfway around the world. As long as their products do as well as they do on the world's markets, the Japanese can afford to buy and even develop distant resources.

Japan constitutes a realm by virtue of its industrial strength, the transformation of its national life, and its role in the world economy today. Even with Korea, the realm so defined includes under 150 million people, one-sixth the total of adjacent China's realm. Territorially the realm is smaller even than Australia, but despite all this it would be difficult to mark Japan as merely an extension of an East Asian realm. And even the population question has been tackled with characteristic Japanese thoroughness. Faced with the threat of explosive population growth and an inability to feed so many mouths, the Japanese in the late 1940s embarked upon a campaign to limit population growth. Abortion was legalized, and birth control was advocated. In 10 years the country's birthrate dropped from 34.3 per 1000 (1947) to 18.0 (1958), and the rate of growth from 3.0 to 1.0 percent.

Even so, Japan cannot feed itself from its own land; the rugged islands of the archipelago leave too little terrain fit for farming. It is another outstanding achievement that the Japanese manage to produce more than three-quarters of their annual food needs, enough to feed over 75 million people. This is achieved by turning more than 90 percent of all the farmland to food crops (and 53 percent of this is under rice) by means of irrigation, terracing, multiple cropping, intensive fertilization, and transplanting, that is, raising the rice plants in seedbeds while another crop matures on the paddy.

These phenomenal qualities of Japan in modern times should not suggest that the country has turned its back on its old culture and that modernization is all that matters to the Japanese. Ancient customs are still adhered to and honored. Japan is still the land of *Kabuki* drama, *sumo* wrestling, Buddhist temples and Shinto shrines, tea ceremonies and fortune-tellers, arranged marriages, and traditional song and dance. There is still a reverence for older people, and despite all the factories, huge cities, fast trains, and jet airplanes, Japan is still a land of shopkeepers, handcraft industries, and home workshops. Only by holding on to the culture's old traditions could the Japanese have tolerated the onrush of their new age.

THE PACIFIC WORLD (12)

The Pacific Ocean, larger than all the landmasses of the earth combined, contains tens of thousands of islands. Some of these islands have been included in one of the eleven culture realms we have already discussed, for example, the Philippines, Indonesia, and New Zealand. We could muster grounds to support the argument that these island groups actually are part of the Pacific realm. For example, the indigenous (Maori) population of New Zealand is related to the peoples of islands in the central Pacific. Wipe away the veneer of European intrusion, and Pacific patterns are quite different from those suggested by Fig. 5.1. The problem is that the European presence or influence can hardly be said to constitute a veneer in places like New Zealand, where the Maori minority of just 200,000 is outnumbered by nearly 3 million whites.

The Pacific realm, then, is really the realm remaining after the demarcation of the Southeast Asian (5) and Australian (9) realms. But despite the extremely fragmented character of the area, it is possible to discern some regional differentiation. Thus the islands from (and including) New Guinea eastward as far as the Fiji group are called Melanesia (*mela* means "black"). The Bismarck Archipelago, Solomon Islands, New Hebrides, and New Caledonia all are part of this region. The region's identification is based upon the ethnic properties of the indigenous population. The Melanesians are black or very dark-brown people with black hair and dark eyes. North of Melanesia, that is, due east of the Philippines, lies the island region known as Micronesia (*micro* means "small"). The population of these islands shows evidence of mixture between the Melanesians and Southeast Asians. Dark brown is the prevalent skin color, and eyes are usually brown. But the islands were not so named because their inhabitants are small of stature; it is the islands themselves that are small, much smaller than the Bismarcks, Solomons, and Hebrides of Melanesia. The Marianas, Caroline Islands, Marshall, and Gilbert Islands all are part of Micronesia. And finally, there is the vast central Pacific, east of Melanesia and Micronesia and from the latitude of New Zealand to north of the Hawaiian archipelago. This is Polynesia (*poly* means "many"). Thousands of islands, many populated, others not, constitute this region. The Polynesian people, widely known for their good physique, have somewhat lighter skins than other Pacific peoples, wavier hair, and rather high noses. Their ancestry is very complex, including Indian, Melanesian, and other elements. Anthropologists recognize a second Polynesian group, the Neo-Hawaiians, a blend of Polynesian, European, and Asian ancestries.

The Pacific realm consists of various types of islands. A few, such as New Caledonia and larger islands in the Fiji group, contain rocks also found on the continents (such as granite, gneiss, and schist) and are therefore referred to as "continental" islands. Others constitute the tops of submarine volcanic mountains, and still others are coral atolls, built from the skeletons of lime-secreting organisms. The latter two types are sometimes grouped together as "oceanic" islands.

Culture in the Pacific realm reached quite a complex stage, and it is clear that the Polynesians and their contemporaries always were a people of vitality and style. In their songs and dances, their philosophies regarding the nature of the world, their religious concepts and practices, their building styles, their work in stone and cloth, and in numerous other ways the Polynesians built a culture of distinction. The omnipresence of water in their realm did not inhibit contact. The Polynesians were expert navigators and constructed maps from bamboo sticks and cowrie shells.

These, then, are 12 world realms as delimited on Fig. 5.1. Again, these realms are defined on the basis of a set of criteria that include not only cultural elements but also political and economic circumstances, relative location, and modern developments that appear to have lasting qualities. Undoubtedly there are places where our scheme is open to debate, and such a debate can itself be quite instructive. For example, it could be argued that the contrasts between Middle and South America are sufficiently great to merit the recognition of two realms rather than the one on our map. The inclusion of Korea in the Japanese realm, based on the development and historic associations of South Korea and the lesser size of North Korea (and the probability of political change) also is open to discussion. Sri Lanka (Ceylon) is viewed by some scholars as a part of the Southeast Asian realm, not the South Asian realm. All this underscores an important reality: those lines on Fig. 5.1 are not absolute or final. And in the process of making up your mind about problem areas, you learn much about diverse parts of the world.

CHAPTER SIX

The Territorial State

SIX

The living space on the earth's surface is divided today into more than 140 compartments, some extremely large, some very small. All but a few of these units are *states.* Those that are not states are colonies or dependencies of another kind, and their number has been dwindling steadily over the past several decades. At the close of the Second World War, many hundreds of millions of people still found themselves in colonies, in India, in Indonesia, in Nigeria, in Guyana. But now, in the mid-1970s, most of those dependencies have thrown off the colonial yoke and have also become states. The exceptions include the Portuguese sphere of influence in Africa and remnants of the French, British, and Dutch colonial empires in small parts of Africa, the Americas, and the Pacific.

Thus nearly all the earth's almost 4 billion people find themselves in one of those numerous states. When we look at a map of the political world, we see a jigsaw of territories delimited by boundaries. These boundaries mark the limits of states. A state is more than just a piece of the world's territory, however. A state is a most complex system of many interacting parts. The system functions to serve the people within the boundaries, the *nation.*

THE NATION-STATE

Does the incorporation of an aggregate of diverse people within a single boundary automatically create a nation? Not exactly. The term *nation* implies much more than simply existing within certain boundaries. The point can be proved rather easily through a check of some major dictionaries of the English language. The dictionary definitions seek to convey a sense of unity, commonality, homogeneity in the context of the idea of nation. Many states, however, include within their boundaries people of different racial sources and people who speak different languages and adhere to different religions. These are plural societies; some have developed far toward nationhood, others have not. There must be something those definitions fail to transmit.

The variable the definitions say little or nothing about is an intangible but vitally important one: people's attitude, their emotional posture toward the state and what it stands for. Thus we might profitably differentiate between *legal* nationhood, which is proved by birth certificate, military record, voter's registration, and other attributes of citizenship, and *emotional* nationhood, the level of commitment of the people, individually and collectively, to the state as they perceive it. No matter how homogeneous a population, serious, divisive forces will cause the national cohesion to degenerate. And a people divided by race, language, religion, and history can nevertheless be welded into a nation of strong fiber.

The *nation-state* is a term used in political geography to designate a state in which close coincidence exists between the legal perimeters of the nation and the people's demonstrated commitment to the state. No absolute measure of this is possible, of course, and many definitions of the nation-state have been written. For our purposes we shall define the nation-state as a political unit comprising a clearly defined territory and inhabited by a body of people, both of sufficient size and quality, and sufficiently well organized to possess a certain measure of power; the people shall consider themselves to be members of a nation with certain emotional and other ties that are expressed most tangibly in law and government or ideology.

This is clearly a very broad definition, but even so, several of the world's prominent, large states may not qualify as *nation*-states. And note that there is no attempt to be quantitatively precise in this model; we say "of sufficient size," and "a certain measure" of power, and "the people shall consider themselves to

ICONOGRAPHY

What makes one country appear and "feel" different from another? What goes into the creation of the "cultural atmosphere" that is so noticeable as you travel from one European country to another? The political geographer J. Gottmann tried to capture these realities in the term *iconography*, the essence of regional distinctiveness. Said Gottmann: ". . . boundaries exist because each country *feels* it is different from the other. . . . this spirit [of each nation] is always made up of many components." What are some of these components that distinguish the political region? Gottmann argues that they include "a strong belief based on some religious creed, a certain social viewpoint, a pattern of political memories. Thus regionalism has what might be called an *iconography* as its foundation: each community has found for itself or was given an icon, a symbol slightly different from that cherished by its neighbors. For centuries the icon was cared for, adorned with whatever riches and jewels the community could supply. In many cases such an amount of labor and capital was invested that what started as a belief, or as a cult or even the memory of a military feat, grew into a considerable economic investment around which the interest of an economic region united." (*A Geography of Europe*, 1969). Gottmann adds that in the Old World, stubborn regionalisms of this kind have evolved toward nationalism in modern times, causing ever more intense partitioning of the land. It is one of the tasks of political geographers to identify the strands and interwoven patterns that make up regional-political iconographies.

Contrasts in Southeast Asia are well illustrated by these two photographs. Hotel Row in Bangkok, Thailand, reflects the modernization of the major cities, most of them positioned on or near the coast. The forest village named Doi Chiang Dao, in interior northern Thailand, resembles the homes of several million Southeast Asians who have little in common with their urbanized and modernized countrymen.

be members of a nation." Obviously since there is not a state in the world where *all* the people consider themselves committed and emotionally involved in their nationality, we are saying that the vast majority of the people must feel this way. Whether it is 90 percent or 80, the definition does not say. But it does give us a sense of what the ingredients of the nation-state must be.

These ingredients include the so-called *four pillars* of the state. Our definition suggests that the state must have (1) a clearly defined territory, (2) a substantial population, (3) certain organizational qualities, and (4) a measure of power. We now examine these elements briefly.

Territory

All states occupy a part of the earth's surface; thus all states have territory. Within this territory lie the resources with which the state must work in order to survive: soils for agriculture, minerals for industry and trade. This is the space that must be organized and exploited to best advantage. When we examine the political map of the world, two aspects of territory present themselves immediately: size (that is, total area) and shape (sometimes called *morphology*).

SIZE It is tempting to assume that the more territory a state owns, the better off it is in all respects. But that is not necessarily the case. True, a very large state has a greater chance that a wide range of environments and resources exist within its borders than a small state. But much depends also on location; in Chapter One we viewed the distribution of known mineral resources and found them clustered in some areas, virtually absent in others. No matter how large a state, if it is underlain by desert sand or by a crystalline shield, the mineral-resource picture may not be all that favorable. A much smaller state that lies astride the Eurasian mineral "backbone" may be much better endowed. Then there is the question of population distribution and size as it relates to areal dimension. A very large territory may be a liability rather than an asset if the population is too small to organize and exploit it. A government must be able to exercise effective national control over the state's total area; areas that lie beyond the reach of its administration and influence can produce problems. In the first place they are not productive, and in the second, such areas may come to harbor elements hostile to the national government. Neighboring states may even seek to lay claim on such isolated reaches of territory. In recent decades the government of Brazil, for example, has been trying to "open up" the Amazonian interior and other lands west of that country's coastal heartland. Brazil is larger than the coterminous United States, but its population is less than half of the United States'. And that population is far more strongly concentrated in the east than that of the United States. The Brazilian government even went so far as to create a whole new capital 400 miles into the interior, to draw national attention westward. And now attempts are being made to build a trans-Amazonian highway. Brazil wants to integrate its whole territory into a state that for all intents and purposes, has been confined to its eastern periphery.

The range of sizes among the world's 140 states is so great that it defies generalization. At one extreme is the Soviet Union, a state larger than the whole continent of South America, whose more than 8.5 million square miles constitute one-sixth of the entire land surface of the globe. At the other extreme are states so tiny that they are referred to as *microstates.* Luxembourg, for instance, has just 998 square miles, and there are states still smaller than that. Referring to our definition of the nation-state, it is questionable whether the microstates have territory "of sufficient size and quality" to constitute the foundations of true nation-states.

Between these extremes are continent-size countries such as Australia, Canada, the United States, Brazil, and China (whose areas average over 3 million square miles), states in a class with India, Argentina, and Zaïre (around 1 million), and states of smaller dimensions such as the United Kingdom (under 100,000), the Netherlands (13,000), and Lebanon (4000). From these data it is obvious that size alone does not mean much: consider the respective roles in the modern world of the United Kingdom and Brazil, a state more than 30 times as large.

SHAPE A second quality of states' territories that is evident from the world political map is their shape, or spatial form. Some states, such as the Philippines, lie on a group of islands, their territories broken by extensive waters. Other states, for example, Chile, seem to consist of a long, narrow strip of land. When at first we look at the map, there appear to be so many different kinds of territorial forms that classification seems impossible. But when the states of the world are examined carefully, a typology does emerge. The Philippines and Chile suggest two types, and there are others.

Many states, Belgium, Uruguay, Kenya, and Cambodia, among others, have territories that are somewhere between round and rectangular. There are no islands, peninsulas, or major indentations such as bays or estuaries. These states' territories are *compact,* which means that the distance from the geometric center of the area to any point on the boundary does not vary greatly (Fig. 6.1). The compact state encloses a maximum of territory within a minimum length of boundary, an obvious asset. Quite the opposite situation prevails in the case we mentioned, the Philippines, and in Indonesia, Japan, and Malaysia. In those states the territory is *fragmented* into numerous pieces, and it is possible to go from one part to the other only by water or by air. We can recognize three different kinds of fragmented states: those whose national territory lies entirely on islands (Japan, the Philippines); those whose territory lies partly on a continental landmass and partly on islands (Malaysia, Italy); and those whose major territorial units lie on the mainland, separated by the territory of another state (the United States, whose state of Alaska cannot be reached overland from the coterminous states except through Canada).

Fragmented states have problems of internal circulation and contact, and

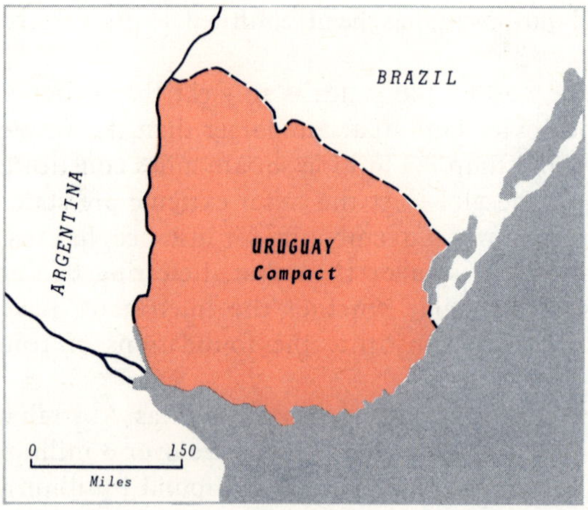

FIG. 6.1 A compact state: Uruguay.

often suffer from the friction of distance (Fig. 6.2). Far-flung Malaysia, for example, was incapable of accommodating the strong forces of secession in Singapore. After years of rising tension, East Pakistan broke away from the West. The government of Indonesia is based on the most populous island of that state, Djawa (Java), and it has had difficulties pacifying distant islands. The state of the Philippines has faced similar problems: the capital is on Luzon, the country's largest island, but

FIG. 6.2 A fragmented state: the Philippines.

there is a Moslem minority on Mindanao, another major island far to the south. The government has found it difficult to cope with the sporadic opposition mounted in Mindanao.

Quite a different spatial form is represented by Chile, Norway, Malawi (in Africa), and Viet Nam. The territory of each of these states is *elongated.* Such elongation (or *attenuation,* as it is sometimes called) also presents certain difficulties. If a state has a large territory or if it lies astride a cultural transition zone, its elongated shape may associate it with contrasting life-styles. An example is Togo in Africa, not a very large state but extending from the west coast, across the forest and savanna, to the Moslem-influenced interior. In Italy, which is an elongated as well as a fragmented state, there are north-south contrasts that are related to the different exposures of those two regions to European mainstreams of change. In Norway, the distant, frigid north, inhabited by Laplanders, is another world from the south, where the capital city and the country's core area are positioned. And Chile, of course, provides the classic example of environmental, if not cultural, contrasts (Fig. 6.3). Northern Chile is desert country, the barren Atacama prevailing there (Fig. 2.12). Central Chile has a large area of Mediterranean climatic conditions, and here the majority of the people are clustered. Southern Chile is dominated by marine west-coast conditions and by rugged topography. The effects of distance and isolation are very evident in the far, desert north and in the remote south. In many ways Chile *is* that central area between the two, the rest of the country still awaiting effective integration in the state.

A still more familiar example (to many Americans) is the case of Viet Nam.

FIG. 6.3 An elongated state: Chile.

Even separately, as North and South Viet Nam, these are elongated countries. Together, as one Viet Nam, they constitute a mere sliver of coastal territory connecting two major river basins and deltas, those of the Red River in the north and the Mekong River in the south, with Hanoi and Saigon the respective urban foci. How often, in recent years, did we hear that the "enemy poses a threat to the narrow waist of the country"? In fact, the territory's elongation is underscored by internal cultural contrasts. The French, who held control in Viet Nam for a long time, recognized three major regions: Tongkin in the north, focused on Hanoi, Annam in the center, oriented toward the ancient city of Hué, and Cochin China in the south, headquartered at Saigon. Add to this the differences between the peoples of the mountains (among them the Montagnards) and those in the lowlands, and the ribbon that is Viet Nam can be seen to have its share of diversity. Quite apart from the conflicts that have torn at the country for so many years, the territory does not make effective national control any easier.

While we are considering Southeast Asian cases, it is instructive to look at the case of Thailand. Thailand would be a compact state, except for one peculiarity. From its main body of territory there extends a long peninsula, several hundred miles southward to the boundary with Malaya (Fig. 6.4). In fact, Thailand's neighbor, Burma, has the same feature, except that its southward extension is a bit shorter. This spatial form is called *prorupt,* and anytime you see such a situation on the map a little investigation is warranted. Proruptions often have noteworthy histories. They also create special problems. In the case of Thailand, those border

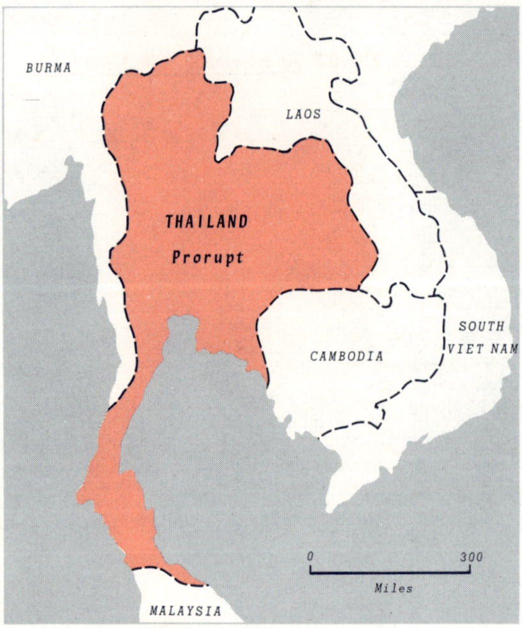

FIG. 6.4 A prorupt state: Thailand.

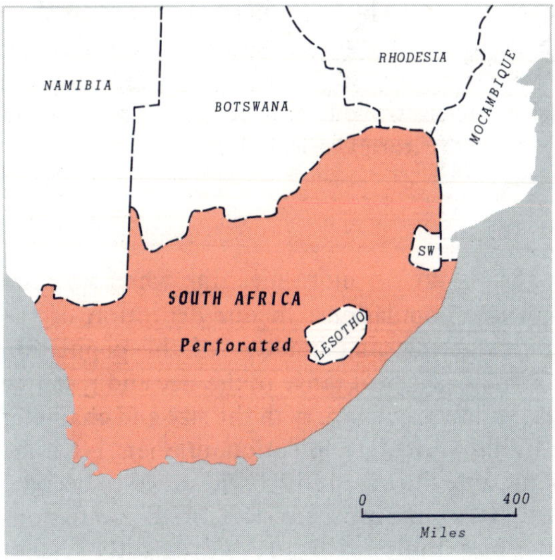

FIG. 6.5 A perforated state: South Africa.

areas near Malaya are 600 miles away from the capital, Bangkok, farther than any other area is removed from the government's headquarters. But at least there is a railroad along the Thai proruption. Not so in Burma. The railroad southward from the capital ends 300 miles short of the southern limit, and for the last 150 miles, there is not even a permanent road!

Not all proruptions lie on peninsulas. Afghanistan's proruption, the so-called Wakhan Corridor, lies squeezed between Pakistan and the Soviet Union. In Africa, the Republic of Zaïre has two large proruptions, one extending southeastward and the other connecting the main body of territory to the Atlantic Ocean. Elsewhere in Africa, the territory of South West Africa (Namibia) has a narrow but lengthy extension, the Caprivi Strip, that connects it to the Zambezi River in the heart of the continent. And in South America, a small but potentially significant proruption called the Leticia Corridor connects Colombia to the banks of the Amazon River. These proruptions have their origins in political negotiation and even territorial trade-offs. The importance of some of these extensions of territory has declined, but in other instances their significance is growing. For example, the Caprivi Strip now is under the control of the Republic of South Africa (along with the rest of Namibia). This puts South African presence right in the middle of black, independent south-central Africa.

One other spatial form should be mentioned, although its significance hardly matches that of the other four. In rare instances the territory of a state completely surrounds that of another state. That state, then, is *perforated* by the surrounded country. A look at the map of South Africa will make this clear (Fig. 6.5). The country of Lesotho (Basutoland on older maps) is entirely surrounded by South Africa. Lesotho is about the size of Belgium, so it takes a substantial piece of territory out of the heart of South Africa. Elsewhere the perforators are smaller, as in the case of San Marino, which perforates Italy. But only the South African case really has importance in political geography, for Lesotho is an independent black state and South Africa remains white-controlled. In other cases, perforation is a curiosity rather than a vital aspect of territorial morphology.

Population

The size and shape of national territories take on more meaning when we view them in association with their respective populations. In our definition of the nation-state, we referred to the "sufficient size and quality" of the population required by the system. And indeed, if there is a great range in the size and resource quality of states' territories, there is an equally great variety in the size and character of their populations. The largest state in the world, by population numbers, is China, now with 900 million persons and, sometime during the 1980s, destined to become the state of 1 billion. At the other end of the spectrum are states with populations of less than 1 million, even less than a half million. Guyana, only slightly smaller territorially than the United Kingdom, has just 800,000 inhabitants; Iceland has a mere 200,000. There are microstates with still fewer inhabitants. It would be impossible to say just what the minimum number of people is that can constitute a nation; conditions vary too greatly. For example, among Iceland's 200,000 there may be more persons with specialized education and particular skills than in a developing country with ten times the number of inhabitants. The capacities of the nation as well as its size are important.

China, then, is in a class by itself in terms of population, just as the Soviet Union is territorially. China is 40 percent larger than the next ranking state, India (over 550 million); then follow the Soviet Union and the United States, both with less than half of India's total. Around the 100 million mark are Indonesia, Japan, and Brazil. Over 130 of the world's 140 states, then, have populations under 100 million. Noteworthy among these are Bangladesh (75 million), Pakistan (65 million), Nigeria (60 million), and France, the United Kingdom, West Germany, and Italy, all between 50 and 60 million. A whole host of states have populations between 5 and 25 million, however, more than any other group.

In Chapter Three, we discussed some of the problems related to the interpretation of data on population density. Averages for individual states are not very meaningful, unless we know something about the agglomeration, the clustering of the people. Consider again the example of Brazil. With the overwhelming majority of that state's 100 million inhabitants concentrated in the eastern margins of the territory, we might easily misinterpret the statement that the average density of people per square mile there is 29. Thus we want to know not only the total population of a nation-state but also the way this population is distributed over the national territory. When we know these things, our knowledge of states' spatial forms takes on more meaning. For example, we know that Burma and Zaïre are both prorupt states. In the case of Burma, a state of 30 million people, the core area lies on the main body of territory, and the extension of area that forms the proruption is one of the least effectively integrated parts of the country. Zaïre is a much larger state territorially, but with a population of under 20 million. When, however, we consider the distribution of population in Zaïre, we realize that the two proruptions of that state are no mere extensions of area, but vitally important, productive regions. The southeastern proruption constitutes the Katanga, the famous copper-producing area centered on Lubumbashi. The western proruption connects the vast Zaïre interior to the ocean. It is the state's sole outlet to the sea. On that western proruption lie the capital city, Kinshasa, the major port, Matadi, a developing transport network, a substantial population cluster, and much more. In Burma, there may some day be problems with the isolation, the remoteness, of the proruption that lies on the Malay Peninsula. In Zaïre the problem is not so much the proruptions; it is the vastness of the intervening Congo Basin itself. The distribution of the two states' populations tells much of this story.

When, therefore, we see a compact state, our next query ought to be whether the advantages inherent in this particular spatial form have been realized, or whether the population has concentrated along one of the margins, leaving a poorly integrated, frontierlike interior. Perhaps more than one such population cluster exists, separated by zones that are barriers to contact. Nigeria is a compact state with three such clusters, two of them in the European-influenced, colonized south and one in the more remote, Moslem-traditional north. South and north long looked in different directions, but independence drew them together in a new state system. The system did not work well, and before long one of the clusters, Biafra, sought to secede from the rest. Nigeria's advantageous territorial morphology hardly counted in a situation where three population clusters lay so separated that they might as well have been islands.

Our nation-state definition refers to the "quality" of the population as well. This is a difficult matter. While it is not possible to specify requirements for a nation-state in this context, the effectiveness of a population is not expressed by numbers alone. Literacy and education, health and age structure, skills and abilities all enter the picture. All these have much to do with people's awareness, productivity, and political consciousness. The problem of achieving fuller literacy has been an especially urgent one facing newly independent states, where allegiances must be shifted from tribe and local area to the state or at least shared. The state of health, too, is a critical factor. Comparisons between countries in terms of life expectancy, daily caloric intake, dietary balance, disease incidence, and medical facilities soon indicate why available energies differ and are expended in different directions. A population that is largely engaged in day-to-day survival under a subsistence form of agriculture cannot be expected to make any great contribution to the state, economically or otherwise.

One additional aspect of population "quality" should be mentioned. Our definition suggests that plural societies can be nation-states only if national feeling, a sense of unity, overpowers the divisive forces inherent in culture contrasts. In states where this has happened, the prevailing, or leading, or numerically superior,

This photograph was taken in the Soviet Union, in Tashkent, capital of the Uzbek Soviet Socialist Republic. It shows Moslems during their Friday prayers, still held in two locations in Tashkent today. The government insists that not many Uzbek Moslems are true believers, but Islam has not been forbidden. Note, however, that there are but few young faces in this crowd; certainly the faith is not encouraged among the younger generation.

element in the population makeup has managed to commit the other sectors to national goals and to the state system in general. This is a process that takes much time, and it apparently progresses through a series of successes and reverses. Pakistan failed in Bangladesh, apparently permanently. Nigeria failed in Biafra, perhaps only temporarily. Mexico made enormous progress during the present century in extending national consciousness to its large Indian minority. On present evidence, Tanzania has enjoyed great success in welding a large number of tribal peoples into the infrastructure of a nation. The United States at mid-century is a mixture of failure and progress, and many would argue that if there is a nation-state in the United States, it still excludes the vast majority of nonwhite Americans.

Organization

The state is a system, with innumerable subsystems. From transport networks to educational systems, national health plans to tariff barriers, tax-collection methods to the armed forces, the state has a multitude of organizational arms. The taking of a census, the calling of an election, the distribution of funds to the elderly—all these evince the effectiveness of organization in the state.

In political geography, the most obvious manifestation of organization in the state is administrative. Except for some of the smallest microstates, all states are divided into subunits such as provinces, states, departments, or other designations. In the Soviet Union and China, complex hierarchies of such subdivisions were created following the appearance of the new order. In the U.S.S.R., each of the 15 Soviet Socialist Republics (see Chapter Five) broadly corresponds to one of the major nationality groups in the huge state. Most republics include smaller clusters of people with some cultural identity, and for them, autonomous Soviet Socialist Republics were created, republics *within* republics. Several lower levels of subdivision exist, even in nearly empty Siberia. It does not take long to realize that the complicated Soviet political map was created (1) to give Russia preeminence, (2) to permit culturally distinctive peoples to retain their identity while providing them with avenues for representation in Moscow, and (3) to organize population and territory in such a manner that state control over the resource base and the means of production would be effective. The Chinese system was modeled after that of the Soviet Union, also with the aim to give the minority peoples certain measures of identity and autonomy. Three levels of administrative area exist. Interestingly, municipalities (such as Peking), the 22 provinces, and 5 autonomous regions (such as Tibet, Sinkiang, and Inner Mongolia) are all at the top level. The second and third levels exist for minority peoples.

All this seems very complex compared to the system in the United States, where the states are divided into counties, and the counties in turn are subdivided into townships and parishes or incorporated municipalities. But then, the whole United States map, with its many straight-line boundaries, looks quite unlike that of the U.S.S.R. or China. We shall examine some of the reasons behind the differences later in this chapter.

UNITARY AND FEDERAL STATES When Napoleon rose to power in France, he swept the old administrative framework away and created an entirely new one, in which the country was divided into about 80 *departments* of nearly equal size. The idea was to weaken the power of large, rich provinces whose influence might dominate the state. The departments would have representation in Paris; new roads to link them to Paris were laid out. France should be focused completely upon Paris, the state's heart in a multitude of ways. What Napoleon envisaged was a *unitary* state in France, a state that would be one and undivided. Many European states are unitary states: Spain, Italy, Austria, Sweden, for instance. But in the New World and in colonial areas elsewhere, another idea took hold. There the idea was not so much unity, but voluntary association, or *federation*. The United States, Canada, Brazil, Australia are among states that were forged from diverse units that agreed to yield some sovereignty to a central government while retaining certain powers for themselves. Unitary and federal forms of politico-territorial administration reflect different systems of government and, therefore, different ways of organizing the state's living space and its people.

The economic organization of the state is a vital link in the total system. All states have resources. Some have very few; others have a wide range. Even location can be a resource. Centuries ago, the Swiss found themselves positioned near the Alpine passes through which Venice and the north of Italy had to trade with Flanders and neighboring areas. The Swiss controlled the transit of goods and founded a state in the process, a state spawned, in large measure, by the Alps' relative location. So the question is: how has the state managed to mobilize its resources, natural, spatial, and human? In Chapter Five we noted the almost total economic reorganization generated by the new order in postrevolutionary Russia and the reforms brought to China by the communist regime. Those reforms were in recognition of the failure of previous economic practices, and they were directed toward better exploitation of the available opportunities. Probably the most dramatic example of economic reorganization as a state-building process comes from Japan. The archipelago that is Japan does not have a very large or varied resource base. There were mineral resources to sustain the handcraft and home-workshop industry, but the modernization envisaged by the leaders of the restoration required the acquisition of far greater stores. The first step, the reorganization of the handcraft industries, had to be at home. Then Japanese goods would have to be sent to world markets, there to undersell competing products. With the funds gained, investments were made in the industrial sector and in another area of state organization, the military. Soon the military was capable of driving into neighboring territories (Korea, Manchuria) where resources needed in Japan were located and where goods made in Japan could be sold. The same Japan that lay isolated and remote some decades earlier became an aggressive colonial power.

Other states exist, however, whose resource base is not exploited nearly as effectively as it apparently could be. Despite the crowding of the world and the shortages faced in so many spheres, areas still lie unexplored or only partially explored. Australia has recently proved to possess some unexpected wealth in petroleum and natural gas, for instance, a commodity that the state really needs in quantity, one that has long been imported expensively. In the Amazon Basin, in Tibet, under desert sand, Arctic ice, and ocean water the earth still conceals resources. The problem for many states, of course, is that discovery does not bring unqualified benefits. Developing states that either cannot generate the capital or do not have the capacities to exploit and develop major resources may find themselves under pressure to yield to the companies of states that can and do. The developing state receives taxes and royalties, of course, but the presence of foreign agents on its territory is a reminder of inadequacies that are politically irritating.

CENTRIFUGAL AND CENTRIPETAL FORCES

Political geographers use the terms *centrifugal* and *centripetal* forces to identify forces that, *within* the state, tend respectively to pull the system apart and to bind it together. Centrifugal forces are divisive forces. They cause deteriorating internal relationships. Religious conflict, racial strife, linguistic division, and contrasting outlooks are among major centrifugal forces. During the 1960s, the Indochina war became a major centrifugal force in the United States, one whose aftermath will be felt for years to come. Newly independent countries find tribalism a leading centrifugal force, sometimes strong enough to threaten the very survival of the whole state system, as the Biafra conflict did in Nigeria. Centripetal forces are those that tend to bind the state together, to unify and strengthen it. A real or perceived external threat can be a powerful centripetal force, but more important and lasting is a sense of commitment to the system, a recognition that it constitutes the best option. This commitment is sometimes focused upon the strong charismatic qualities of one individual, a leader who personifies the state, who captures the popular imagination. (The origin of the word *charisma* lies in the Greek, where it means "divine gift.") At times such charismatic qualties can submerge nearly all else. Peron's lasting popularity in Argentina is a case in point.

The degree of strength and cohesion of the state depends upon the excess of centripetal forces over the divisive centrifugal forces. It is difficult to measure such intangible items, but some attempts have been made, for example, by determining attitudes among minorities and by evaluating the strength of regionalism as expressed in political campaigns and voter preferences. When the centrifugal forces gain the upper hand and cannot be checked (even by external imposition), the state will break up, as Pakistan did in the early 1970s.

Power

Our definition states that a nation-state must have "a certain measure of power." This does not mean power in military terms alone; rather it means overall strength, including the capacity to police and protect national interests. Power cannot be measured exactly, but obviously microstates such as Liechtenstein and Andorra do not have the kind of strength nation-states possess, even small nation-states such as Lebanon, Israel, and Belgium. The study of states' power, or *power analysis,* involves the allotment of a quantitative measure to every feature of the state deemed to have an impact on its power and strength. Some years ago (1960) Professor F. C. German wrote an article entitled "A Tentative Evaluation of World Power," published in the *Journal of Conflict Resolution.* In his calculations he included not only such items as total area, size of the population, production of steel and

The landscape of power: West Germany's massive industrial complex as it fronts the Rhine River at Duisburg. The Rhine, a veritable highway for this region, carries very dense traffic. Nearly two dozen ships and boats can be seen under way on this small stretch alone. The cluster of industries shown in the photo is only a fraction of the vast Ruhr complex that sustained Germany's war machine during the Second World War and that has again become the foundation of modern German strength.

coal and other critical commodities, but also factors for railroad density, technical efficiency of the population, "morale," food supply, and military capacity. In 1960, he concluded, the United States was the world's most powerful state, with 6459 points; next came the U.S.S.R., with 6321, then the United Kingdom, 1257, China, 999, and West Germany, 663 ½. Canada, Japan, and France followed. A recalculation for 1974 would surely show Japan near or ahead of China!

Power comes from organization. In a book entitled *The Might of Nations* (1961), Professor J. G. Stoessinger defines power as "the capacity of a nation to use its tangible and intangible resources in such a way as to affect the behavior of other nations." Again, this does not mean solely or even primarily by military presence or threat; rather, it involves possessing and producing in economic spheres. A state can win concessions or reciprocal agreements with other states through its economic strength. It can outbid other states in the competition for exploitation rights in the territories of less-developed countries by being able to offer superior conditions. Growth and development require organization; organization generates power.

THE LIMITS OF THE STATE

The politico-geographical map of the world today shows a veritable maze of boundaries, not surprising when our limited habitable space is chopped up into more than 140 states. Whenever we look at a globe, an atlas map, or a map in a newspaper or magazine to illustrate a story, we see boundaries as our frame of reference. When we refer to a country or an area where some important event is occurring, we bring to mind the layout of that country or area, defined by its boundaries. The United States is two coasts. A straight-line border separates it from Canada between the Pacific and the Great Lakes, and the Rio Grande separates it from Mexico. We could all draw it fairly accurately, we have seen it so many times.

But boundaries are more than meet the eye. For every mile of amicably negotiated boundary there is a mile of strongly contested border. As we know too well, the contests are not over yet. In Middle and South America (Guyana-Surinam, for example), in Asia (Kashmir), in the Middle East (Israel-Egypt), in Africa (Malawi-Moçambique) the disputes continue. Boundaries are under pressure in many parts of the world, and even in Europe with its old and apparently stable states numerous changes have taken place in the last 50 years alone. The boundary that now divides Germany into an East and a West appeared on the map in 1945. The city of Berlin was also divided after the Second World War, but the division was strengthened by the building of the Wall between the Eastern and Western sectors in 1961. Poland's boundaries changed after the war also, and there were adjustments along the Dutch-German and French-German borders as well.

The boundary framework with which we are familiar in the 1970s, then, is the result of innumerable modifications made over centuries of time, and although the broad outlines of this framework are likely to remain intact, every decade brings changes of detail. Still, what we see on the map today is a far cry from the situation of just one century ago. In 1875 Bolivia still possessed the northern Atacama (and thus an outlet to the sea); there was no Pakistan; the whole African jigsaw puzzle was just beginning to develop. The ill-defined "sphere of influence" was a different reality in the nineteenth century than it is today.

Frontiers and Boundaries

Sometimes the terms *boundary* and *frontier* are used interchangeably, as synonyms. But there is a difference, and an important one. The frontier throughout history served as a zone of separation, an area between states, a territorial "cushion" to keep rivals apart. Such a frontier might have been a dense, vast, almost impenetrable forest, or a swampland, or a disease-infested river basin, or a range of mountains. Sometimes a frontier was defined by distance as much as by inaccessibility. The world map of the mid-nineteenth century was full of frontiers, in the Americas, in Africa, in Asia and Australia, even in Europe. But then the effects of Europe's technological and political revolutions, the colonial scramble, explosive population growth, and the rush for resources and empires combined to produce competition for even the most remote frontier areas. Before long the frontiers disappeared, divided and parceled out among states and colonial acquisitions.

Thus the frontier has *area;* it is a spatial phenomenon. The boundary, on the other hand, should be thought of as a vertical plane separating two states; it is *linear* on the ground. It appears on our maps as a line, but it also separates airspace above the ground and the rocks and resources below. Today, though, only one frontierlike area, Antarctica, remains (Fig. 6.6). There several states claim zones

FIG. 6.6 Spheres of influence in the Antarctic.

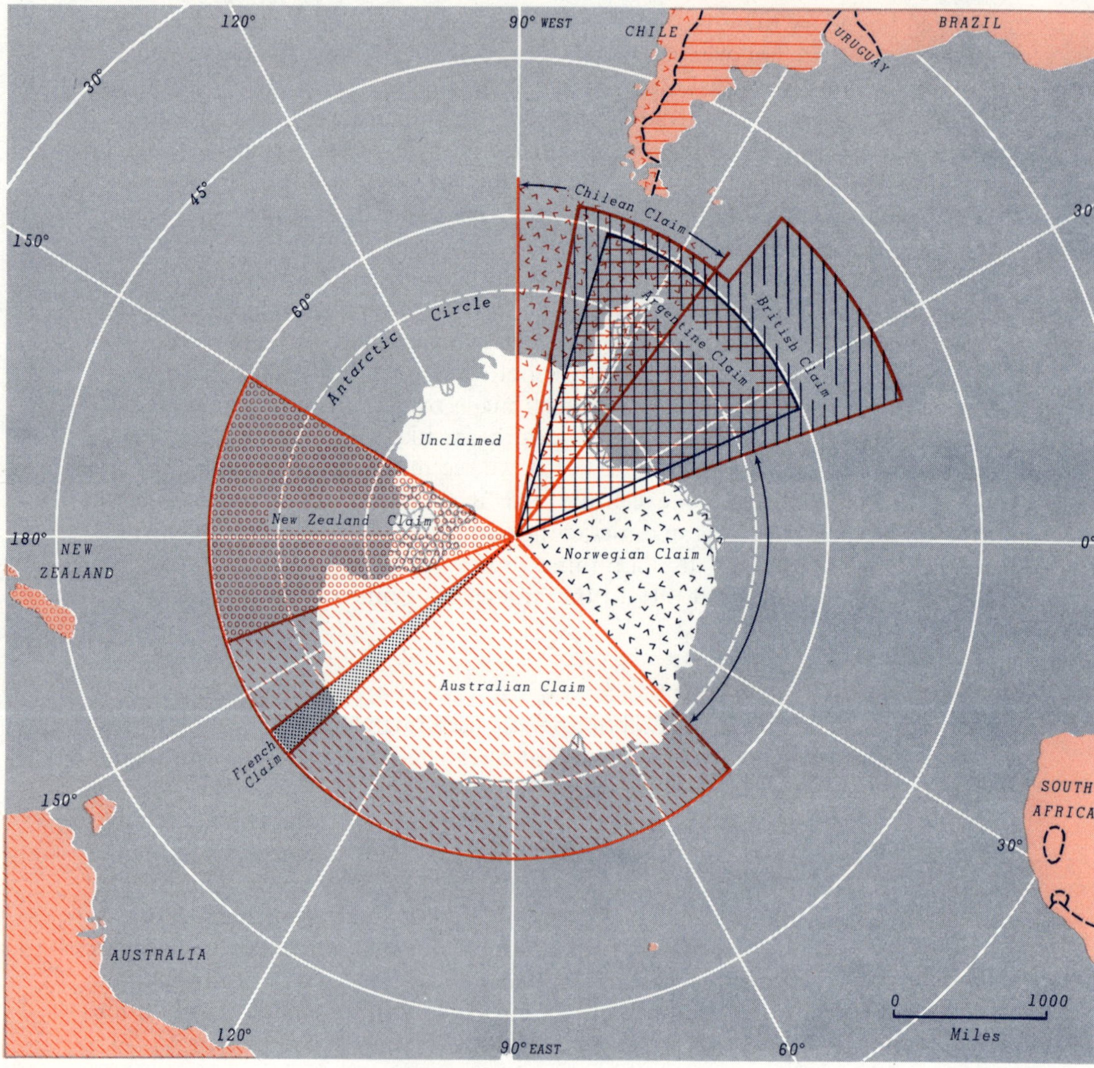

of preeminence, some of them overlapping (Fig. 6.6). But Antarctica is mainly a region of research and investigation, and the claims have not yet led to serious conflict. A major discovery of some critical resource could change that picture considerably, however.

Boundaries, then, are the limits of states. They mark the extent of the state's territory, the limit of jurisdiction. They separate state from state, but they are also the place where states make territorial contact. This dual function, of separation and association, has led geographers to investigate the ways boundaries have come about, how they function (and how those functions have changed in modern times), and how they might be usefully classified.

Boundary Evolution

In ideal circumstances, boundaries evolve through three stages. Imagine a frontier area, about to be divided between two states. First, agreement is reached on the rough positioning of the border. Then the exact location is established through the process of *definition*, whereby a treatylike, legal-sounding document is drawn up in which actual points in the landscape are described (or, where a straight-line boundary is involved, points of latitude and longitude). Next, cartographers, using large-scale maps and referring to the boundary line as defined, put the boundary on the map in a process called *delimitation*. And, if either or both of the states so desire, the boundary is actually marked on the ground by steel posts, concrete pillars, fences, or some other visible means, sometimes even a wall. That final stage is the *demarcation* of the boundary. All boundaries on the world map are by no means demarcated. There are thousands of miles where you could cross from one state into the other without ever knowing it. Demarcating a lengthy boundary in any way is expensive, and it is hardly worth it in inhospitable mountains, vast deserts, frigid polar lands, or other places where there is virtually no permanent human population.

Peoples have sought to demarcate their territories for thousands of years. The Romans built walls in northern Britain to mark the limits of their domain, the Chinese Wall is still one of the wonders of the world, the pre-Inca Peruvian Indians used stone lines to confirm their territorial claims. More often, linear physiographic features such as rivers and mountain chains became trespass lines in practice, then boundaries in fact. Until as recently as the Second World War, boundaries were viewed as viable lines of defense. The French in the late 1930s fortified their northeastern boundary with Germany, called it the Maginot Line, and hoped that it would withstand the German onslaught. Modern warfare has more or less eliminated this function of the boundary, except in places where guerilla warfare occurs. There, a river or a mountain range may still present some strategic advantages. But the concept of the boundary as the state's ultimate and practical line of defense no longer applies.

In the Old World especially, but also in Middle and South America and in Africa, boundaries seem to zigzag endlessly and, it would seem, pointlessly. Why

States and communities have for many centuries built walls to confirm their boundaries, to keep outsiders out and to defend their domain and to keep would-be emigrants confined to their homelands. The Great Wall of China is more than 2000 years old, and it served all these purposes as China's northern boundary. The Berlin Wall, shown dividing one of the city's streets down the middle, also serves as a boundary, but mainly to stop East Germans from leaving their country for what appears to them to be a better life with greater opportunity in the West.

are all boundaries not like those between the United States and its neighbors, along a major river, through the middle of lakes, along a line of latitude? The answers are many. From Roman times on, Europe's cultural and political patterns have changed almost continuously. In places (such as along the Pyrenees between Spain and France) a fairly stable and rather permanent boundary evolved, but elsewhere states and nations fought over territory time and time again, and boundaries were defined, altered, defined once more, wiped out, reestablished. Even boundaries in mountainous areas and along rivers were subject to pressure; in the mountains there would be argument over the sources of river water and control over watersheds, and in the rivers there would be conflict over who would own the banks, the navigable channel, the river mouth. There is a piece of historical geography attached to almost every one of those bends and turns you see boundaries make on maps. And yet today, after all those centuries of adjustment and readjustment, there are still Italian-speaking people in France, French speakers in Switzerland, Dutch speakers in Belgium, Hungarians in Rumania, Germans in Poland.

If that is the situation in crowded and culturally complex Europe, what about the Americas, where the population numbers are so much less, or Africa, where external powers could lay down boundaries with impunity? In the Americas, as the daily press will confirm, the appetite of states for their neighbors' territory is far from satisfied. Numerous Middle and South American states consider sections of their boundaries temporary and subject to future redefinition. Venezuela has a potential dispute with Guyana, Guyana has a claim in Surinam, and Guatemala has thinly veiled hopes for the absorption of a whole country, Belize. Panama wants to acquire the Canal Zone and wipe out the United States-held corridor that renders it a fragmented state. And in Africa the colonial powers negotiated with each other and in only very few places actually contested their empires' borders, but they created some boundaries that may prove to be untenable in the future. African states have been busy trying to organize and survive following the colonial period, but as they gain strength it is likely that they will begin to contest the borders with which they were endowed. Already the Somali Republic has indicated that it wants the boundary with Ethiopia redefined (there are hundreds of thousands of Somali in Ethiopia). Uganda has expressed a desire to acquire a land corridor to the sea through northern Tanzania.

And so we can watch the very adjustments that result in the tortuous maze we see on the map. Israel acquires a small piece of Syria called the Golan Heights. Pakistan and India struggle over a snowy, mountainous area in Kashmir. China redefines its boundary with Burma and starts an argument with India over the Simla Conference, at which many decades ago the China-India border was defined. Cold, desolate country can provoke some hot, intense confrontations.

Functions

If boundaries today (with a few exceptions) no longer serve in a defensive context, how do they serve the modern state system? Boundaries may not be trespass lines any longer, but they do serve as the symbol of state inviolability. They are a factor in the development of that intangible emotional bond of national consciousness, especially if contrasts between societies and cultures on opposite sides are strong. On the front pages of newspapers, on schoolbooks, on stamps, even on flags, states display maps of their national territories. A real or even an imagined threat to the boundary can be used to arouse national feeling. But there are many states whose relationships with their neighbors across the boundary are such that the boundary hardly functions as a separator. People move across with only a minimum of control, and some even commute across daily. Social and economic conditions on the two sides are so similar that the boundary seems an unnecessary impediment. The United States-Canadian boundary comes to mind, and the boundaries between the Netherlands and Belgium, Norway and Sweden, Austria and Switzerland.

Even these boundaries (note that they are *within*, not *between* culture realms) still have important functions. One of these is obvious enough. The whole state system, with all its parts—education, taxation, law, conscription, and so forth—is built within this framework. The boundary marks the limit of state jurisdiction in all these and countless more spheres. The boundary signifies the limit of sovereignty; weakening its role implies yielding of some of that sovereignty. As we shall see, a number of states in the postwar period have proved themselves willing to do just that: to reduce the divisive functions of the boundary in order to join with other states in economic, political, and strategic blocs.

Even though boundaries no longer provide the protection of fortification to states, they still serve to shield the state in another practical way, against illegal traffic in contraband goods. To prevent or inhibit such movement, boundaries are sometimes fenced and policed; at sea a cushion of territorial waters is similarly guarded. States that want to prevent the movement of people into or out of their territories may create a no-man's-land through which it is difficult to cross undetected.

This brings us to the most important function of a boundary in modern times, its role in commerce and economics. A state can erect a tariff wall against goods that try to compete on the domestic market, and that wall coincides with the state's political boundary. Such a tariff wall makes foreign goods expensive and helps local producers sell theirs at an advantage. But there is a price, which, of course, is retaliation. The local producers may find that they face the same obstacle when they export their goods to the country where the foreign goods are made. And so states use their boundaries and the markets they enclose to compete and negotiate. Sometimes a group of states discovers that there is much to be gained *not* by creating economic barriers at their mutual boundaries, but by dropping many of them and joining in an economic union. After the Second World War this happened in Western Europe where six states, the "Inner Six," France, the Federal Republic of Germany, Italy, Belgium, the Netherlands, and Luxembourg, formed the European Economic Community. This bloc became known as the *Common Market*. Later, other states (Denmark, Ireland, even the United Kingdom), seeing the advantages involved, joined. But in so doing, all those states had to give up some of their sovereignty, some of their power to make decisions inside their borders, in order to share in the resultant benefits.

Thus we see boundaries functioning in many different ways: some states are still strengthening and confirming their borders (notably those states that have recently become independent); some still seek to move their boundaries outward to acquire part of their neighbors' territory; and other states can afford the luxury of reducing the divisiveness and absoluteness of their borders in order to participate in and open ways to a wider, international economic and political sphere.

Boundary Classes

The world politico-geographical map, with its maze of lines, does not seem especially susceptible to order and classification until we combine what we see with what we know. If we examine the boundary framework as it relates to some of the subjects we have considered in previous chapters, we note that political boundaries do have recurrent properties that permit classification. For example, if we lift the political boundaries from a map (by tracing them) and then superimpose the whole framework on a map of the physical features of the world, we find that boundaries in many places coincide with linear features of nature. Boundaries lie along the crests of mountain ranges (as between Spain and France, Argentina and Chile, Czechoslovakia and the German Democratic Republic). They also lie along rivers, for example between the United States and Mexico (Rio Grande), Zambia and Rhodesia (Zambezi), Uruguay and Argentina (Uruguay). These are *physiographic-political* boundaries, and many of them are holdovers from the time when natural barriers formed suitable trespass lines.

Our physical map does not help us with long stretches of other boundaries, however, so we superimpose our framework on a series of cultural maps (from Chapter Four) of languages, religions, and so forth. Again, we soon detect coincidences. Between India and Pakistan the political boundary follows a religious break in the cultural landscape. The boundaries between Hungary and its neighbors, and

---

MORPHOLOGY

In the context of boundary *morphology*, it is perhaps reasonable to remind ourselves of just what morphology means. We see this word in geomorphology; it is often placed opposite *genetics*, implying something static against something dynamic. The *morph* part comes from the Greek, and it means "having a certain form." Morphology, thus, is the study of forms or features for their own sake, without necessary reference to the processes whereby these forms came about. In boundary morphology, therefore, we look at the boundaries as they are, straight or bent or irregular, as a prelude to later study of genetics.

---

between Finland and the U.S.S.R., coincide with language changes. Some boundaries follow ethnic division lines, as between Israel and its Arab neighbors, Lesotho and some of the smaller African states and their neighbors, and between Southeast Asian countries. These boundaries are called *anthropogeographic* political boundaries, in reference to their cultural qualities.

Note that we are only comparing the existing boundary framework with present physiographic and cultural distributions; we are not using any information we might have relating to the development of the boundary or the landscape. We will use that information later. All we are doing here is viewing the situation as it is. And it is hard to miss still another kind of boundary, the boundary that does not coincide with either physical or cultural features, but that shows another notable characteristic: it is a straight line on the map (or slightly curved one, depending on the projection). We should be familiar with such boundaries. In both the north (with Canada) and the south (with Mexico) the United States has such borders. Africa has many miles of them, especially in the north but also between Kenya and Tanzania farther south. Such boundaries usually run through areas that are (or were) quite empty. Hundreds of miles of straight-line boundaries lie in the Sahara Desert, for example. These are *geometric* political boundaries, and they normally come about through treaty discussions. And speaking of functions, such straight-line boundaries, when delimited, obviously are expected to function very marginally. Who would expect them to separate dense populations in the Sahara Desert? But the convenience with which they were established can disappear quite quickly. Imagine a major oil find astride one of those Saharan borders, shared between competing states!

We have looked at boundaries in a factual, descriptive, morphological way. But we could usefully view them also in the context of their development, their genetics. Now we use some of our knowledge of the evolution of the cultural

A superimposed boundary: barbed wire cuts off a road in Germany. Beyond the barrier lies the German Democratic Republic, the Communist fragment of once-united Germany; the photographer stood in the German Federal Republic, or West Germany. Germany was divided among the victorious powers following the Second World War, and the dividing line became an ideological one. Scenes like this, with roads, railroads, and other means of communication severed abruptly, are abundant all along the East-West boundary.

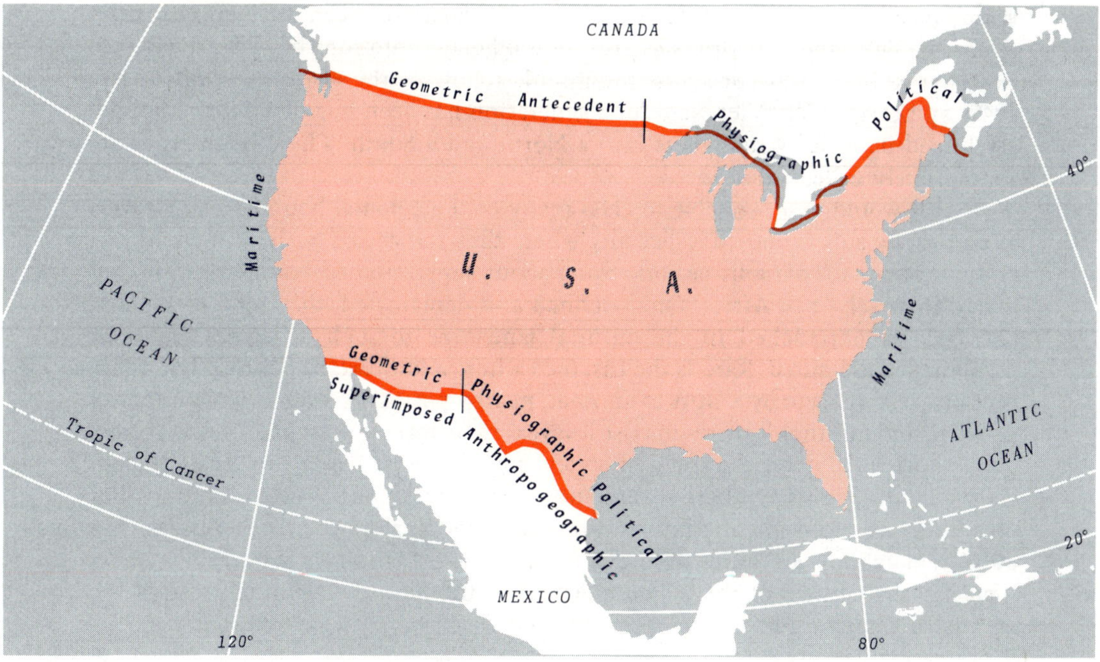

FIG. 6.7 Boundaries of the United States.

landscape through which the boundary runs and of the role of the boundary in that process. For example, we know that the geometric United States-Canada boundary west of the Great Lakes was defined by treaty long before the major elements of the cultural landscape as we see them today developed (Fig. 6.7). The boundary was there almost before any other modern cultural feature, and the evolution of the cultural landscape had to adjust to its existence. The boundary is *antecedent* ("prior," "preceding") to the landscape created by man. Other boundaries, however, were established during the "filling in" of an area. For instance, two states whose peoples moved into and occupied a frontier zone might agree on a boundary roughly coinciding with the areas claimed by their respective subjects. A sample case would be the India-Pakistan boundary, which was defined after the major elements of the cultural landscape in that area had existed for a very long time. When a boundary is created during or after the development of the human landscape, it is called *subsequent* ("after," "following"). But there is a still more extreme case, where the boundary results not from negotiation between adjacent, growing states, not from readjustment, but from the aftermath of war. In this case a boundary may be created by a foreign power that decides to carve up a formerly unified area or by two or more victorious allies who share the spoils of victory by claiming pieces of territory. We have several modern examples of this: Germany was split up into a Communist-dominated East and an Allied-administered

West (the controls have been relaxed, but the boundary between the two Germanies remains). The capital, Berlin, was also split, officially into four but effectively into two sectors. Korea was cut across the middle following the Korean war of the late 1940s and early 1950s. Viet Nam for a generation has been fragmented by a boundary forced upon it by foreigners into a North and a South. These boundaries are appropriately called *superimposed.*

It would be reasonable to class many of the colonial boundaries in Africa as essentially superimposed, especially where they cut across fairly unified ethnic, cultural regions. This tells us something about superimposed boundaries, in Germany, in Israel, or in Africa. Once defined and demarcated, they tend to remain, and become entrenched in the cultural landscape to such a degree that their permanence is assured. Rare is the instance where a long-extant political boundary is eliminated (except by a shift to another, nearby position). Such *relic* boundaries do exist, and cultural features may still reveal their former presence, but not often.

We shall defer the topic of states' boundaries at sea to our final chapter. States require limits to their sovereignty in the oceans as well as on land, but obviously the problems of administration, control, exploitation of resources, and even demarcation are quite different from those on land. We will discuss these topics in the context of the whole question of the nature and human use of the oceans in Chapter Ten.

CORE AREAS OF STATES

Every functioning state has a regional focus, a heartland, a *core area.* When we think of France, what comes to mind is the great city of Paris and its environs, the Paris Basin, for centuries the heart of the French Republic. In the United States we realize that the core area still lies in the east, including the great cities of megalopolis, and extending westward to the vicinity of Chicago. An American geographer, D. Whittlesey, wrote in his book *The Earth and the State* (1939) that many states typically "crystallized about a nuclear core that fostered integration. . . . in nearly all states the nuclear core is also the most populous part—its ecumene. [The core area is] the portion of the state that supports the densest and most extended population and has the closest mesh of transportation lines. . . . it is more richly endowed by nature than the rest of the state."

It is interesting that Dr. Whittlesey chose dense population and an intensive transport network to characterize the core area. Actually, it is one thing to form a general idea of the core area, but quite another to determine exactly what constitutes it. The biggest problem may be that of *scale.* Obviously we cannot expect a state such as Tanzania or Thailand to possess a core area as highly developed as France or the United States. On the other hand, there are territorially quite large states that simply have not developed core areas. Look at an atlas map of Mauritania, Central African Republic, Saudi Arabia, Mongolia. You will be hard put to find the population agglomeration and denser transport networks that would indicate the existence of a core area. The core area not only reflects the substantial population our nation-state definition required but also manifests organization.

Core areas, then, have the following qualities. They contain the largest cluster of population in the state, but the size of that cluster varies from state to state. In a country of 50 million people, we might expect to find 14 or even 20 million people in the core area. In a state that has only 3 million inhabitants, the core area might contain fewer than 1 million people. And, as maps of surface communications show, well-developed core areas are characterized by webs of railroads and roads that radiate outward to connect the heart of the state to more outlying areas. The interconnections are many within the core area itself, then become sparser with distance away from it. But again, we should keep the scale of things in mind. In a country where dirt roads and paths connect most of the towns and villages, a few miles of busy hard-surface, all-weather highway may be as important to the state as a whole network of roads and railroads is to a more developed state.

Core areas, too, are the most urbanized parts of states. Not only does the core area contain a large agglomeration of people; but it also constitutes a substantial part of the urbanized population. In the United States, cities within the core area include not only the megalopolis between Boston and Washington but also such places as Pittsburgh, Cleveland, Detroit, and Chicago (Fig. 6.8). In France, Paris, the largest city in that state by far, is the focus of the core area. In the Soviet Union, the core area centers on Moscow but also includes Leningrad, Gorky, and other major industrial centers. China's core area lies in the northeast and incorporates

FIG. 6.8 Core area of the United States.

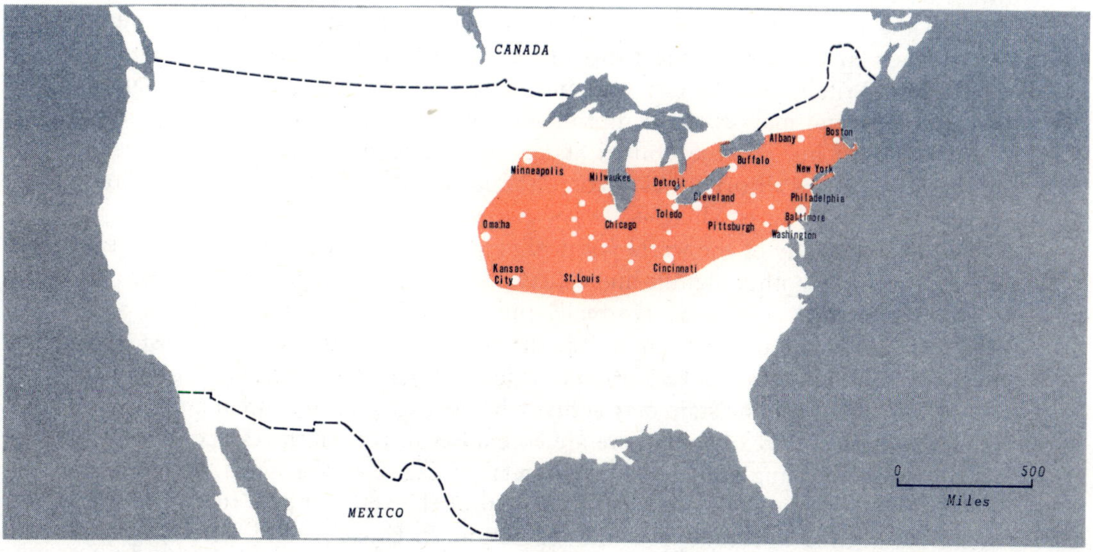

Peking, Shanghai, and so many other urban foci that the core area is far more highly urbanized than China as a whole.

This brings us to the capital city and its relationship to the core area. We shall view capital cities later in this chapter, but obviously the normal situation is for the capital city to lie within the core area, if indeed it is not the very focus of that region. In numerous states, France, the United Kingdom, Chile, Mexico, Egypt, and Thailand, for example, the capital city is also the largest city and therefore the obvious center of gravity of the core area as well as of the state as a whole. But there are cases where the capital city is not by any means the largest, as in the United States, Australia, South Africa, and Nigeria. Note that these states have something in common: they are federally organized, and their capital cities were either created or selected specifically to function as the states' governmental headquarters. In these cases the capital cities lie within the core area, but they are not the dominant foci. In a very few special cases the capital actually lies *outside* the core area. The outstanding example is Brazil, whose new (also federal) capital of Brasilia was created and located specifically to draw the attention of Brazilians away from their eastern, coastal core area and into the vast interior.

The core areas of states also are (or long were) the most productive regions of their states. Some core areas rose to prominence because they contained a combination of mineral resources that could sustain the forward march of the industrial revolution. Towns grew into cities, labor streamed in, and communication systems intensified. In other instances, core areas emerged on the basis of exceptionally favorable conditions for agriculture. Argentina's Pampa is such a case. Combinations between the two circumstances also occurred, as in South Africa where the major industrial heartland lies very near the most extensive and productive farming areas. And then, of course, there is the factor of self-perpetuation. Once in the lead, a core area can survive as the state's leading region even when other areas in the state prove to have equal or even greater advantages. London found itself rather far away from the focus of change when the industrial revolution transformed Britain, but the city was, and remained, the state's largest population agglomeration and market, the center of governmental offices and agencies, and a major transport hub. In the United States, people for many decades have been moving west, and California has become the largest state in terms of population, but the core area still is where it always was, in the east.

Our scrutiny of the world political map in connection with the core areas of states reveals another significant reality. Some states have not one but two or even three distinct core areas. Normally the situation is that one core area will clearly be dominant and the other(s) might be viewed as secondary or subsidiary (or emerging) core areas, as California obviously is. But there are instances where the core areas within one state may actually be nearly equal in content and importance. In Nigeria, for example, there are core areas in the southwest, centered on Ibadan and containing also the federal capital of Lagos, in the north, focused on Kano and constituting the Moslem stronghold of the state, and in the southeast, where Biafra tried to secede from the federation. In Ecuador, too, there is a core area on the coast and one in the Andean interior. The implications of such a situation are naturally serious. Competition between the peoples who form the majority in the state's multiple core areas can endanger the cohesion of the state system, as happened in Nigeria.

CAPITAL CITIES

The capital city is an essential element of the state. Some capitals may be so small that they do not deserve the term *city*, but large or small, states need a location where their governmental apparatus can be accommodated (Fig. 6.9). Quite a few states, in fact, began as cities and grew around urban centers that possessed nodality and attained strength and permanence. Such places might have been positioned on routes connecting areas that traded with each other, so they served as markets. Or they were located at points where a river could conveniently be crossed. Or the site was one of easy defense and safety. Whatever the origins of the capital city, it is now the urban focus, often the pride of the state, and the layout, architecture, and monuments and shrines reflect the things that are valued in history and the aspirations that exist for the future. The capital city is the center of national life, often the most cosmopolitan of cities. Frequently you hear the name of a capital city used to identify a state or a nation: "The response from Moscow is . . ."; or "Peking says that . . ."

Capital cities, therefore, perform tangible as well as intangible functions for the state. They must house the offices of government and administration and embassies and other diplomatic agencies. As the center of national attention, the capital city can act as a binding agent, a source of unity and coherence for the nation. In a large and internally varied state it must contain some substantial evidence of awareness of minority peoples or those removed by distance from the state's headquarters. At the same time the capital must be viewed as the source of power and authority. As in the case of boundaries, that function of actual strength provided by fortification (once a role of some capitals) has been replaced. The capital today is located most frequently in the economic heart of the country, from which much of the image of strength of the state emanates.

Capitals still have a function in this context, however. When we study a map of the capitals of the world, we note that some states are in the process of relocating their capitals. As we saw in the case of Brazil, such relocation may have entirely internal objectives. But in some other cases capitals are (and have been) used to confirm a state's presence in or near a contested or potentially contested area. A prominent case is Pakistan. For a long time, coastal Karachi was the state's headquarters, but that city lies far from the real heart of the state, the Punjab. The city of Lahore in many respects exceeds Karachi as a cultural, national focus. But when Pakistan decided to move its capital from Karachi, it did not choose Lahore; rather, a new capital was planned farther into the northern interior and nearer Kashmir, the area Pakistan disputes with India. Here is a case of what geographers called the *forward* capital. The city is used to further national interests

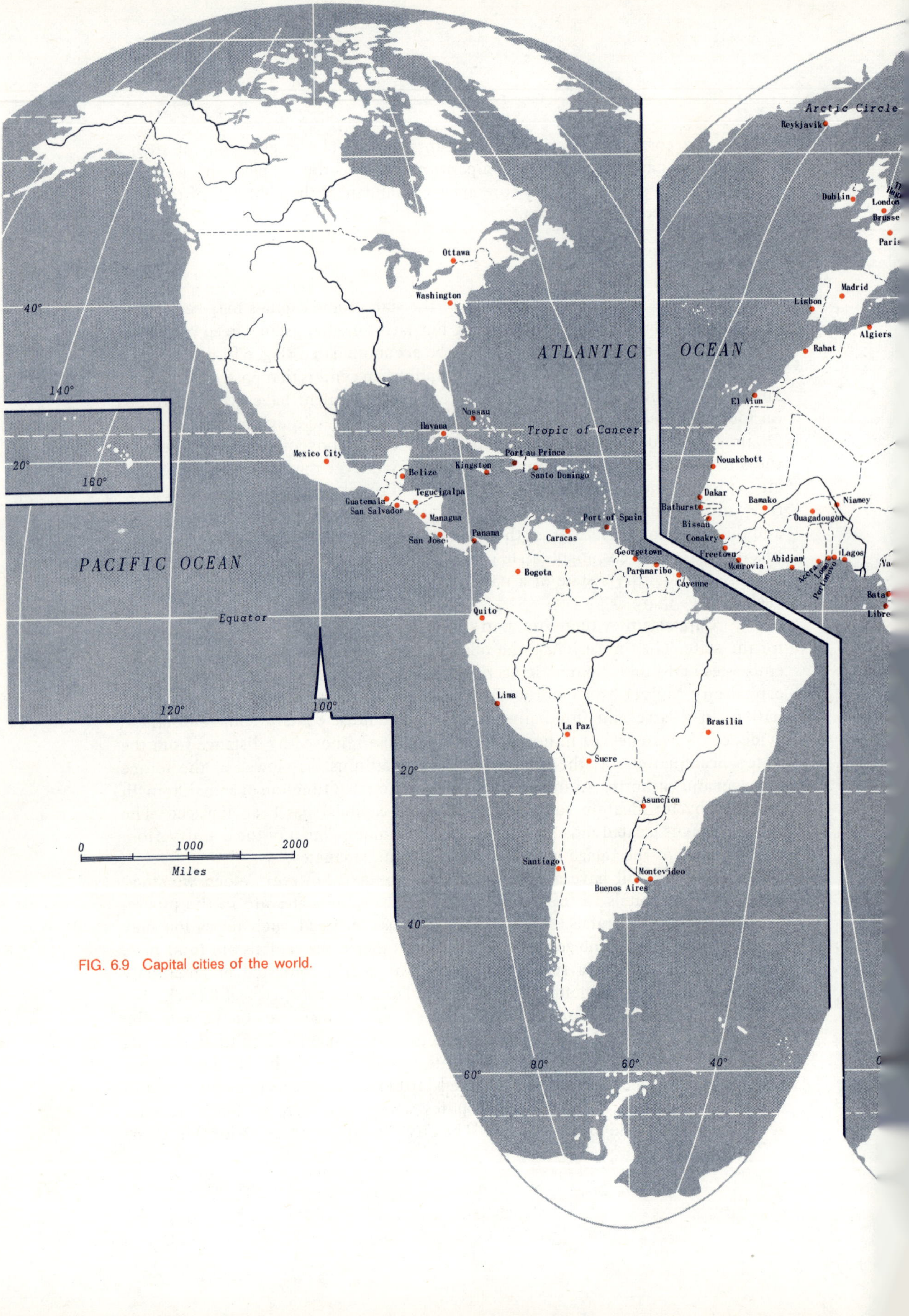

FIG. 6.9 Capital cities of the world.

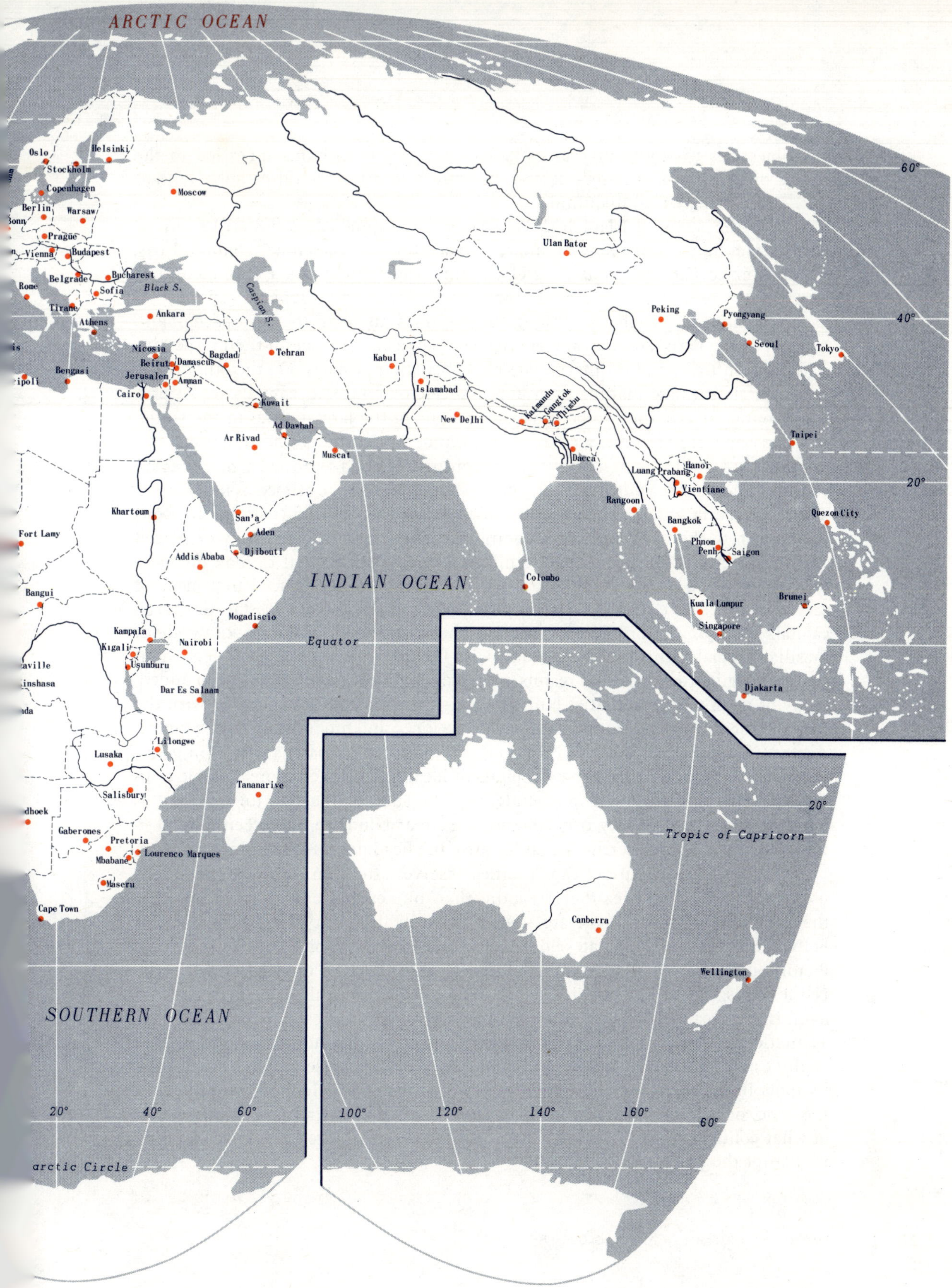
ARCTIC OCEAN
Oslo
Helsinki
Stockholm
Copenhagen
Moscow
Berlin
Warsaw
Bonn
Prague
Vienna
Budapest
Bucharest
Belgrade
Black S.
Rome
Sofia
Tirane
Caspian S.
Ankara
Athens
Nicosia
Bagdad
Tehran
Beirut
Damascus
Jerusalem
Amman
Bengasi
Cairo
Kuwait
Kabul
Islamabad
New Delhi
Katmandu
Gangtok
Thimbu
Ad Dawhah
Ar Rivad
Muscat
Dacca
Ulan Bator
Peking
Pyongyang
Seoul
Tokyo
Taipei
Hanoi
Luang Prabang
Vientiane
Rangoon
Bangkok
Phnom Penh
Saigon
Quezon City
Khartoum
San'a
Aden
Djibouti
Fort Lamy
Addis Ababa
INDIAN OCEAN
Colombo
Bangui
Mogadiscio
Kuala Lumpur
Brunei
Singapore
Kampala
Nairobi
Kigali
Equator
Usumburu
Kinshasa
Dar Es Salaam
Djakarta
Lilongwe
Lusaka
Salisbury
Tananarive
Gaberones
Pretoria
Mbabane
Lourenco Marques
Maseru
Tropic of Capricorn
Cape Town
Canberra
Wellington
SOUTHERN OCEAN
20°
40°
60°
100°
120°
140°
160°
60°
Antarctic Circle

and objectives in frontierlike areas. Somewhat similar circumstances led to the choice of Berlin as Germany's capital. Its eastern location would be an advantage in the face of Slavic competition.

If the name of Islamabad, Pakistan's new capital, does not strike you as awfully familiar, neither may Lilongwe, Nouakchott, or Gaberone. London, Paris, Rome, Mexico City are something else again. And therein lies a telling contrast between different classes of capital cities. On the one hand, there are capitals that have functioned as such for so long that we may identify them as *permanent.* Athens, Rome, Paris, London, and Peking have served as the foci for their respective states for centuries, and no alternatives challenge their primacy. On the other hand, new capital cities are making their appearance. Some famous capitals are surprisingly young: Tokyo, the capital of Japan and one of the largest cities in the world, replaced ancient Kyoto just a century ago. Canberra, Australia's headquarters, was founded and built during the twentieth century. Brasilia is an even more recent venture. These more youthful capital cities should perhaps be classed as *introduced.* Tokyo, Canberra, and Brasilia were all chosen deliberately to fulfill certain functions in addition to the usual ones of accommodating the government. Tokyo replaced an interior city when Japan's modernizers wanted their capital to look out over the sea, which was to become Japan's lifeline. Canberra was placed in a piece of carved-out federal territory as a compromise when Australia's coastal, territorial capitals vied for the privilege of housing the government. And we know about Brasilia's special spatial role. Now today, with literally dozens of newly independent states emerging from colonial realms, a whole list of new capitals has been added, many of them towns and cities built by foreign powers and not endowed with the required facilities. Some young states have developed and modernized their capitals: Accra (Ghana), Dar es Salaam (Tanzania), Abidjan (Ivory Coast) are cases in point. Others have founded wholly new capitals in locations deemed more suitable than the alternatives available at independence. The state of Malawi, for example, is building a new capital city from the ground up at Lilongwe, more centrally positioned to the elongated territory than Zomba, the headquarters the Europeans used.

One other feature of capital cities deserves attention. Again the map tells us that in some countries there is more than one capital city. This means that the functions of government are shared between different places. The map of Bolivia, for instance, designates both La Paz and Sucre as capitals. In South Africa, Pretoria is the administrative capital, Cape Town the seat of the legislature. In the Netherlands, the parliament sits in The Hague, but the royal palace is in Amsterdam. In Libya and Laos, too, this circumstance of *divided* capitals prevails. Such a situation reveals compromise, the rather inconvenient resolution of conflict. That conflict may be between royalty and popularly elected representation, it may have its roots in history involving a conflict between regionally defined groups within the state, or it may have religious overtones. A divided capital is a manifestation of what political geographers call centrifugal forces within the state, forces tending to disrupt the system.

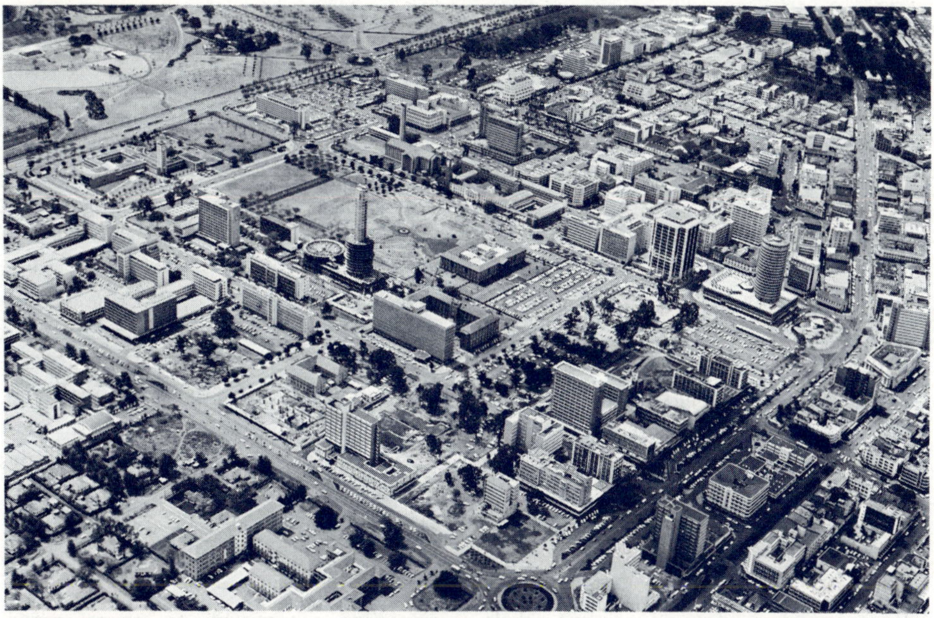

An air view of Nairobi, the modern, spacious capital of the new state of Kenya. In just three generations, Nairobi has become East Africa's largest and most modern city. It symbolizes the modern development of Kenya, the country's aspirations, and its comparative well-being.

THE EXPANSIONISM OF STATES

A German political geographer named Friedrich Ratzel (1844–1904) once suggested that states are like biological organisms, passing through stages of growth and decay. Just as an organism requires food, so the state needs space, territory. Any state, to retain its vigor and to continue to thrive, must have access to more space, the essential, life-giving force. The acquisition of frontier areas with their resources and populations infused states with renewed vigor, argued Ratzel. Boundaries he saw as dangerous straitjackets, capable of administering the death knell to the state.

Ratzel may not have intended it, but this *organic theory* of state evolution, purely an academic speculation, became a matter of practical politics in the hands of some of his students and successors. In Germany a school of *geopolitics* developed whose practitioners were political geographers less in the business of objective scholarship than involved in advising governments on the best way to acquire *lebensraum* ("life space") and to break real and imagined "encirclements" of the German state by powerful enemies.

Certainly those geopoliticians had ample evidence that the acquisition of more territory could do a state no harm. During Ratzel's lifetime Europe's colonial era reached its climax: Germany itself gained major colonies in Africa, Belgium secured the Congo, France and Britain were completing the acquisition of huge and far-flung empires, and the Netherlands had become rich from its East Indian dependency. Ratzel never said that the territory a state ought to absorb should be contiguous to that state, and it is not difficult to see why.

Let us not conclude, however, that the desire for more territory was or is a monopoly of colonial powers alone. Excuses always exist to justify expansionism. Israel seeks greater security, Pakistan wants control over vital river headwaters, Ethiopia needs an outlet to the sea, China declares old treaties invalid. And so Israel invades Arab territory, Pakistan advances into Kashmir, Ethiopia takes custody of Eritrea, China starts border conflicts with India. Still, the pattern of expansionism may be changing. Although Portugal, France, the United Kingdom, and the Netherlands still hold remnants of colonial empires, the days are over when a state could with impunity begin a campaign of territorial acquisition in a distant continent. Note that territorial claims just mentioned are all *contiguous* to the states involved. Perhaps a world conflict will some day arise during which states will again gain far-flung empires, but the present trend is toward the dissolution of such holdings and the disentanglement of colonial ties.

Colonialism

The colonial era is ending, but while it prevailed the world was transformed. Colonialism was the vehicle upon which the ideas and skills of Europe were carried to many parts of the globe. The flow of Europeans from Spain, Portugal, France, Britain, and the Netherlands to other parts of the world began as a trickle during the fifteenth century, then swelled to a climax during the nineteenth and early twentieth centuries. As we noted in Chapter Three, the European emigration varied in direction. Compared with the numbers who came to the Americas (after the timid beginnings of Spain and Portugal), the movement to Australia was insignificant. Still fewer Europeans went to Africa. But no matter how limited the numbers, the impact the Europeans made was enormous. On the heels of explorers and adventurers followed the chartered companies, commercial organizations that went out to exploit newly opened areas with the sanction and support of the home state's government. The administrations of those companies in effect became the governments of the European states' spheres of influence, and when the companies dissolved, the states themselves took direct control. During the eighteenth century, when Europe's colonial powers were revitalized by the fruits of the industrial revolution and technology soared ahead, the ill-defined spheres of influence were secured, confirmed, and sealed by treaty. Quite suddenly Africa, Southwest, South, and Southeast Asia acquired rigid boundary frameworks, and the last frontier areas were colonized and delimited. In the Americas, the nineteenth century brought a harbinger of what was to come. There the Europeans, in the majority unlike in Africa and Asia, shook off the political ties with the European states they no longer needed or wanted.

The European invasion did more than transform the map. It separated peoples with similar cultures and threw together peoples of different life-styles. It gave advantages to some indigenous groups but dislocated and destroyed others. Cities were founded and transport systems laid out. Land was taken from its occupants and turned into plantations and ranches. Taxation forced millions into the money economy. In religion, in education, in health standards the Europeans brought changes. Explosive population growth outraced whatever improved food distribution arrangements the Europeans introduced. And throughout the colonial world European minorities settled as permanent residents, not as temporary representatives of their states' governments, and began to take over powers formerly held in London, Paris, or Lisbon.

MOTIVES What drove the European states to seek distant empires, and what led Europeans to leave the comparative familiarity of their homes to venture into foreign areas? Strategic considerations ranked high. The European states were competing, and the advantages of foreign resources and markets could not be left to adversaries. The case of the Netherlands is symptomatic. Having acquired, virtually unobstructed, a rich and productive sphere of influence in what is today Indonesia, the Dutch found it necessary to protect and secure the route to the Indies by gaining a foothold at the southern tip of Africa. There they acquired a depend-

A picturesque village or town square in Germany or Switzerland? No, this is the building that long housed the German colonial administration in Dar es Salaam, a city that was formerly the headquarters for administration of German East Africa and is now the capital of Tanzania. This building (recently torn down in a modernization program) and several others were part of a small-scale German city on the shores of the Indian Ocean in East Africa, faithfully reproduced according to the home example.

ency almost by default, for Amsterdam was unable to stop the trek of Dutch settlers into the interior, the Cape of Good Hope. Later the British saw the advantages inherent in possessing Cape Town, and they captured the city. So began the sequence of events that brought South Africa a white population in conflict, a war over primacy (the Boer War), and a modern outpost of European culture still dominant over the nonwhite majority.

Some Europeans, then, went to the colonies as representatives of state governments in commerce or the armed forces, but others went on their own terms. "God, greed, and gold" sustained the Spanish in their penetration of the Americas, but the missionary zeal was not unique to the Iberians. In fact, missionaries were often the first to enter unknown areas, soon to be followed by traders, hunters, and others in search of a different kind of profit. And another kind of missionary zeal sustained people like Cecil Rhodes, Karl Peters, and De Brazza—a certainty that it would be in the best interests of the indigenous peoples, their home state, and the world at large if their national flags would fly over as much of the colonized world as possible. Rhodes spent huge sums of his own money sending pioneer parties into the southern African frontier to hoist the British flag. In 1973 his name was still on the map. Karl Peters furthered German interests in what is now mainland Tanzania (formerly Tanganyika, and before that, German East Africa). De Brazza secured the northern lower Congo area for France, and the capital of Congo still carries his name. Imbued with strong nationalist fervor, with boundless energy and luck, such individuals often transformed whole regions by their efforts and changed the course of history.

But many Europeans went to the New World, to Africa, Asia, and Australia not because they were so strongly attracted there, but because they had every reason to leave Europe itself. Hunger, epidemics, war, and religious persecution drove countless white men from their homes to distant places where, they hoped, better opportunities for a reasonable life might exist.

EMPIRES It was the economic motive that dominated, however, for the colonies offered raw materials in abundance, cheap labor to extract them, and markets on which to sell homemade goods. This was the lesson that Japan learned and applied so effectively and that thrust it into world prominence. European examples were many. The Netherlands, a small country in a corner of Northwest Europe, achieved a golden age of wealth and worldwide influence unprecedented and almost unimaginable. The Dutch reorganized their most profitable possession, Java, into one vast plantation. The Spanish gold finds in Mexico and the Andes were legendary. The French failed to oust the British in India, but they gained an enormous empire in Africa and another in Southeast Asia (Fig. 6.10). Germany obtained some small footholds in the Pacific, but its late colonial charge had success principally in Africa, where three large territories (including South West Africa, now held by South Africa) and one smaller one were secured. Belgium obtained its Congo via its King Leopold II, who had acquired it virtually as a private reserve.

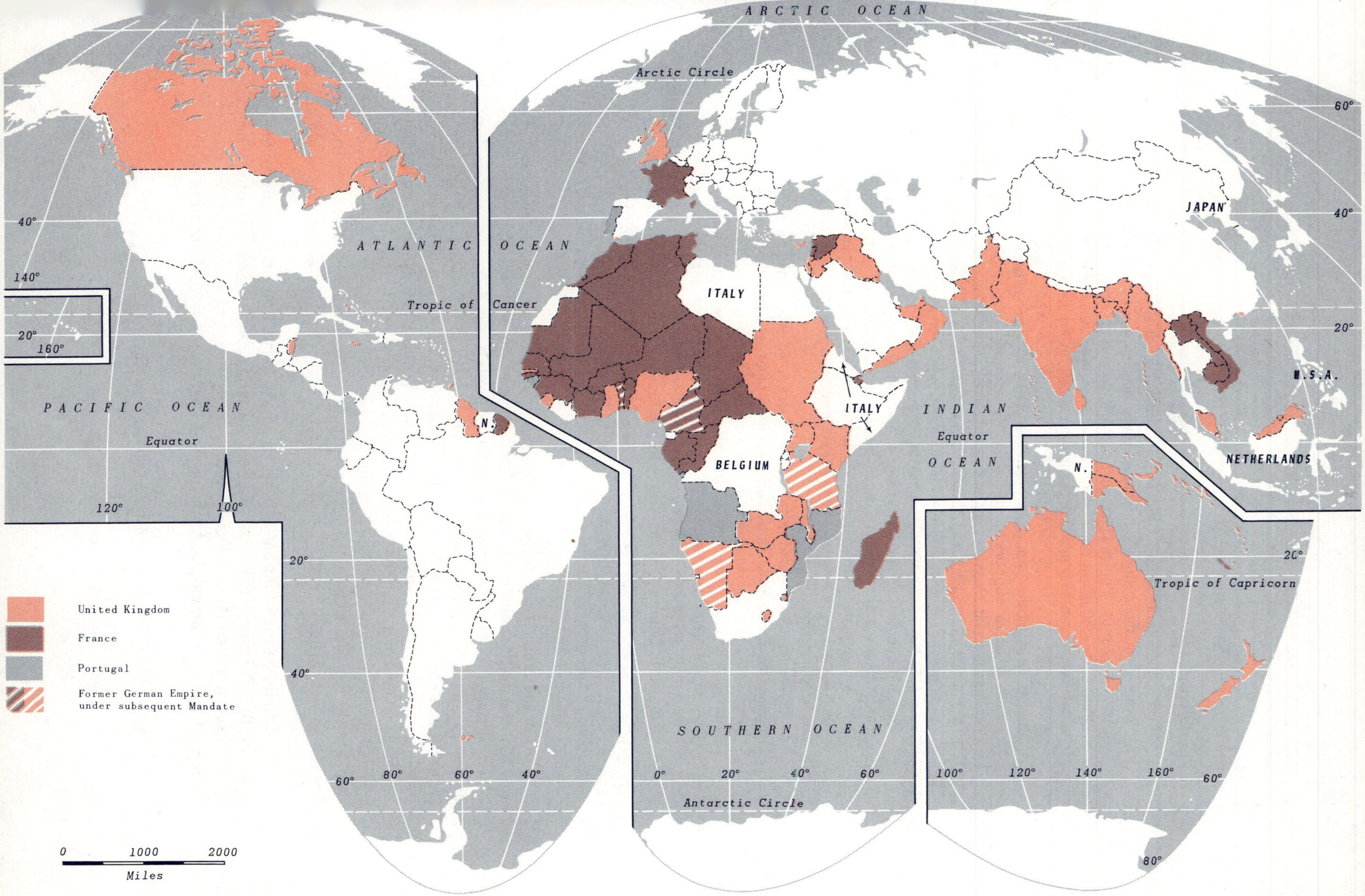

FIG. 6.10 Colonial realms and major spheres of influence of European powers.

IRREDENTISM

The political boundaries of the world's states do not always conform to "breaks" in the cultural landscape. There are Hungarians in Rumania, Turks in Cyprus, Somali in Ethiopia (among dozens of other instances). Often such a situation does not produce difficulties, but there are occasions when problems arise. Sometimes it is felt, in a particular country, that people who really form part of the nation, but who live across a boundary in another state, are mistreated there as a minority. The state begins to show support for its "nationals across the border" through propaganda, and pressure may be exerted on the neighboring government to improve its treatment of these prople. Assistance may be given to resistance movements, and radio broadcasts urge national unification. That may only be the beginning: ultimately the state may openly claim the land and people involved as historically and culturally a part of its domain and initiate a campaign for "reunification" and a boundary shift. Arms supply, sabotage, and similar provocations may form part of this effort. Such a policy by one national state toward a neighboring country is termed an *irredentist* policy.

Irredentism (the name comes from "Terra Irredenta," an area once claimed on these grounds by Italy) is a form of expansionism, and there are times when the expansion-minded state acts more in its own self-interest than in support of its mistreated nationals. Current cases include the Somali drive for a Greater Somali Republic that would include parts of Ethiopia, Kenya, and the Afars-Issas Territory, the Greek and Turkish interests in Cyprus, Eire's concerns for the fate of Catholics in embattled Northern Ireland, and Afghanistan's support for the peoples in Northwest Pakistan.

In the manner the colonial powers organized their colonial dependencies spatially and institutionally we can discern some of the qualities and aspirations of the European states themselves. France, for example, viewed its overseas domains as just that, *France d'Outre Mer.* Overseas France would be shaped as much like metropolitan France as possible. A French-acculturated elite was developed, many of whose number were educated in Paris itself. When times changed and independence approached, the French sought to leave their dependencies with as strong a French heritage as possible. The capitals, adorned with French architectural style, would be the foci of unitary states just as Paris dominates France. In the universities and schools, in music and poetry, in the highest of life-styles, France's impress would be permanent.

Britain, on the other hand, never sought to create a second England in the colonies. British institutions were exported, certainly, but so was flexibility. The British tried to duplicate an efficient and incorruptible civil service, one of their sources of pride, in the empire from India to Nigeria. As if in recognition of the regionalism of their own state (where Wales, Scotland, and Ulster have their own identities), the British left their colonies and protectorates with widely different systems of government. Ghana, for example, became a unitary state based substantially on the Westminster model. Nigeria, Uganda, Pakistan, and India were endowed with federal systems of diverse properties. Kenya, Zambia, and Ceylon became unitary states. There were failures as well as successes, but there was no attempt to apply the same formula everywhere.

Portugal is the only colonial power still to hold on to most of its original empire. Even in Indonesia, independent now for a generation, Portugal still holds its half of the island of Timor. Although it lost Goa to India, Portugal still retains its influence in huge Angola, in Moçambique, and in Portuguese Guinea in West Africa. African nationalists have the Portuguese under pressure in all three African territories, but the struggle has not succeeded in ousting the Europeans. The Portuguese have modified their relationships with their colonies repeatedly in recent decades. These are colonies in the classic sense, that is, they produce raw materials cheaply, provide cheap labor, and are ready markets. But the Portuguese retitled their dependencies "Overseas Provinces" in the early 1960s, thereby avoiding by technicality the scrutiny of the United Nations. In 1973 the territories became known as "Autonomous States" in the Portuguese community. But the essential qualities of life in the Portuguese dependencies changed little. Only at mid-century did practices of forced labor and compulsory cropping end; in the 1970s there was little sign of the multiracial partnership, the Brazil-type society the Portuguese indicate they wish to build.

Belgium saw its colonial task as a paternal one. It had no intention of creating an overseas Belgium in the former Congo; but Belgians did not envisage, until very near the end, that they would be yielding to an independent state in their huge equatorial African domain. The Congolese were treated more or less as children, to be brought slowly and equally to desirable levels of "civilization." It

Portugal's African dependencies contribute considerably to the national economy. Southern Moçambique, for example, serves as an outlet for both South Africa (through the port of Lourenço Marques) and Rhodesia (through Beira). The profits involved are reflected by this photo of modern, booming Lourenço Marques. The spacious bay is in the background, and the cranes and other equipment of the well-appointed port can be seen beyond the buildings.

would take generations. In the meantime, the Belgians organized the Congo spatially into six provinces, each of which consisted of a piece of the rain forest-filled Congo Basin and a section of the higher, surrounding plateau-rimland, the more productive area of the colony. The Belgians could not foresee that this scheme contained the seeds of a sort of neofeudalism. But when independence approached, the provincial capitals became centers of secessionist-minded African leaders, and the territory nearly split apart; the capital, Leopoldville, had hardly played a binding role throughout its existence and could not now. Even after African government was established at Kinshasa (as Leopoldville was renamed), the problems of regionalism persisted. Not until the whole colonial system was replaced by a new framework could effective administration begin.

The Congo is independent Zaïre now, and the colonial era recedes. Indigenous peoples are seeking to reestablish connectivity in their histories, to trace ties to the pre-European past. But the impact of Europeanization is lasting and, in some respects, probably permanent.

INTERSTATE COOPERATION

Earlier in this chapter we noted that some states voluntarily yield some of their sovereignty and independence of action in order to participate in a community, or bloc, of states. States that have joined in the North Atlantic Treaty Organization, for instance, must agree to permit foreign armed forces to be based on their territories. At one time this would have been unheard of except under unusual circumstances, but NATO was born of a joint desire for security following the end of the Second World War, when a new threat to Western Europe seemed to be looming from Moscow.

This is not to suggest that countries did not band together in earlier times. Indeed, the old Greek city-states did just that; later there was a Hanseatic League of cities with common interests; and always there were states that had allies in war and friends in peace, and whose actions and policies coincided strongly. But the period after the First World War suggested that a new phase of interstate cooperation might be at hand. The forerunner of the United Nations, the League of Nations, was the first worldwide venture of its kind, even if a number of states (including the United States) stayed out of it. Then, after the Second World War, interstate cooperation became a practice so widespread and in places so effective that a new era had clearly arrived (Fig. 6.11).

Interstate cooperation occurs in a variety of spheres. The twentieth century has been a time of international ideologies, growing cultural-regional awareness and assertion, and intensifying economic competition. States have joined together to further their shared political ideologies, their economic objectives, their power position in military-strategic terms. In 1973 there were more than 40 such organizations and others were under discussion.

Once again it was in Europe that the first major experiments in interstate cooperation were undertaken. Three small countries led the way, the Netherlands,

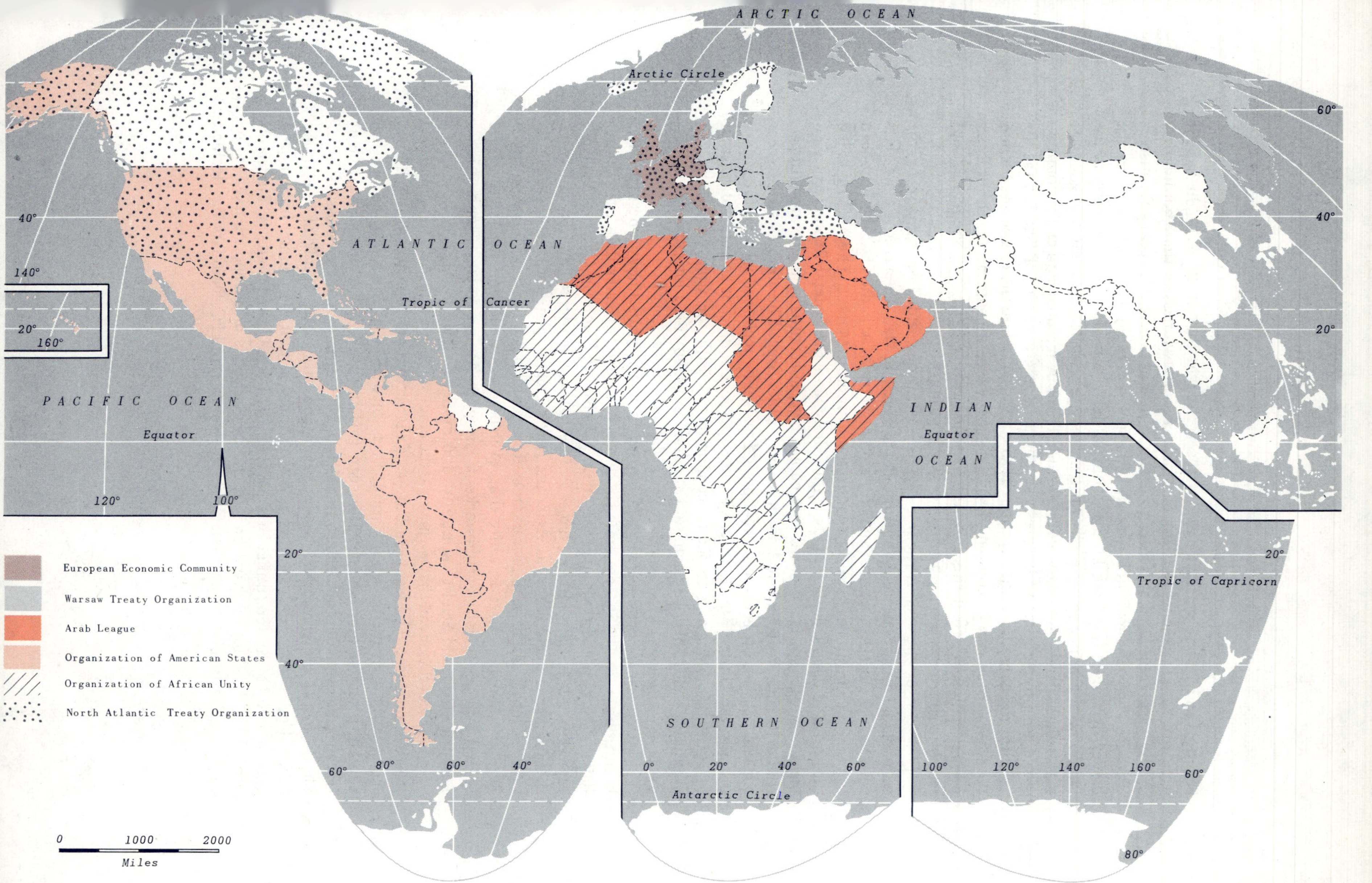

FIG. 6.11 Selected international unions, 1973.

Belgium, and Luxembourg. It had long been thought that all three might benefit from mutual agreements that would reduce the divisiveness of their boundaries; in any case, they have much in common. Northwest Belgium speaks Flemish, a language very close to Dutch. Luxembourg speaks French, like Belgium across the border. But more importantly, the countries have considerable complementarity. Dutch farm produce sells heavily on Belgian markets. Belgian industrial goods go to the Netherlands and Luxembourg. Would it not be reasonable to create common tariffs, to eliminate import licenses and quotas? Representatives of *Benelux* (as the organization came to be called) thought so, and even before the end of the Second World War they met in London to sign an agreement of cooperation. Other European countries watched this experiment with great interest, and soon there was talk of larger, more encompassing economic unions.

Out of the reconstruction of postwar Europe, aided so strongly by the infusion of Unites States support through the Marshall Plan, several economic blocs emerged. The most successful and consequential of these is the EEC, the European Economic Community, also known as the Common Market. The EEC included the three Benelux countries, plus France, Italy, and the Federal Republic of Germany. This group, as we noted in Chapter Five, substantially makes up Europe's core area, and so it became known as the "Inner Six." The union's success was considerable, and it became an economic force to be reckoned with in the world. Initially the United Kingdom stayed outside, for it could not accommodate itself to EEC rules and retain intact its economic ties with the Commonwealth. But eventually the United Kingdom sought entry, and following a long series of conferences, its admission became effective on January 1, 1973.

As anyone who has been to Europe in recent years knows, the realm has been transformed by its economic successes. Mobility is greater than ever before: Italians work in Amsterdam, Frenchmen go to live and work in Germany, Belgian capital invests in Italy. Moving across an international border to go to live and work in another member country is easier than ever. License plates bearing the exhortation "Europa!" can be seen on cars from Rotterdam to Rome. And indeed, Western Europe is moving closer to unity than ever before. In addition to EEC there are several other cooperative frameworks: a European Atomic Energy Community, a European Coal and Steel Community, and more.

Does this mean that we may see the emergence of a United States of Europe? It is still doubtful. The lesson of Europe is that economic cooperation is achieved more easily than political unification. Ever since 1948, a Council of Europe has existed, a sort of experimental European parliament. As an embryo government of a future, unified Europe, the council has barely begun to make progress comparable to that of the continent's economic unions. Still, there is talk of selecting a "capital of Europe," and the hope is often expressed that the council may indeed turn out to be the forerunner of a government for a United States of Europe. If something does not interrupt present trends, the states of Europe are likely to be drawn still more closely together. Other states may join EEC. Denmark has decided to apply, and Ireland is considering it. But Norway, given the opportunity, found its electorate opposed.

Although interstate cooperation (or *supranationalism*, as it is often called) has progressed quite far in Europe, other regions and realms of the world have also achieved much in this area, and Europe has no monopoly. Not surprisingly, states with much at stake in the world have floated unions in which they are the dominant member. Thus the United States plays a powerful role in the Organization of American States and in the Alliance for Progress, seen by some as an attempt to perpetuate United States influence in Latin America. Similarly the Soviet Union dominates the Warsaw Pact, its answer to NATO, and Comecon, the Eastern Europe-directed Council for Mutual Economic Assistance. But elsewhere shared objectives and joint aspirations have produced better-balanced unions, though sometimes still ineffective ones: the Arab League, a cultural-political union of 18 states in Africa and Southwest Asia, and the Organization of African Unity, a similar association of 41 black African states.

Some of the interstate groupings that now exist will undoubtedly fail and break up, and others will emerge. It is unlikely that the states of the world will march uninterruptedly toward amalgamation into larger, superstates. But interstate cooperation is occurring in so many areas and in so many different spheres of activity that the phenomenon has taken on permanence. "Seek ye first the political kingdom," said the late leader of Ghana, President Nkrumah, "and the rest shall be given." In the same speech he referred several times to his hopes for a United States of Africa. The time seems to be at hand for an idea that has taken hold everywhere.

THE STATE AND THE INDIVIDUAL

So far in this chapter, we have looked at states as they are constituted internally and function in their relationships with other states. But it is people who make up the state and people who guide the ship of state, people who make decisions on behalf of the nation. The people who negotiate the agreements and sign the treaties and who make the laws and modify the state system act on the basis of what they perceive as best for their country. They may or may not respond to expressions of public opinion; there was some doubt in the United Kingdom in 1972 that a referendum on the question of entry into the EEC would have produced approval.

What has just been said has to do with a phenomenon worth studying: human behavior in the context of the state, the nation, territory, boundaries—in short, human behavior in political geography. Some political geographers have researched in detail the voting behavior of particular sectors of the electorate, seeking to understand, for instance, the factors that influence minorities in their choices. Others have studied the way decisions are made, that is, the circumstances, inputs, and pressures behind significant politico-territorial moves. For example, when Brazil decided to move its capital from Rio de Janeiro to Brasilia, it was not really "Brazil" that decided; rather, it was a set of the state's representatives that made the decision. What was the record of that decision-making process? How much was known, could be predicted with some reliability? What were the intangibles? Similarly, the study of the decision-making processes that lead to a state's entry into a supranational union can have results of great importance.

Our behavior is based on the way we perceive reality and on the things we have learned as members of a culture. Everyone has seen those postcards sold in big states like Texas and California that show these states so large that they occupy almost the whole United States, with the rest of the states crowded and undersized along the narrow rim that remains. Such postcards are intended to be funny, but none of us has a precise image of our town, state, or world. When we are asked to draw a "mental map" of our country or some other part of the world, we will distort it in a way that may significantly reflect our misconceptions. In other words, we hold a model of our spatial environment involving notions of distance, direction, shape, accessibility, and so forth, and on the basis of this model, which is at variance with reality, we operate.

So, too, do the politicians who make decisions that guide the state. Complicated issues are simplified and thus distorted, and then decisions are made (and opinions swayed) accordingly. Brasilia was positioned a few hundred miles into the interior. Some Brazilians now wonder whether it was far enough, whether the distance involved is not simply a matter of inconvenience to the members of government rather than a true move into the frontier. When a state such as Pakistan places a forward capital in or near contested territory, it does so in the hope that its adversary will perceive the significance of its presence there. With intangibles such as these the possibility of miscalculation is obviously very great, and it is toward a reduction of such misjudgment that research is often oriented.

A related issue that has aroused a great deal of interest is expressed by the single word *territoriality.* In 1966, a book written by R. Ardrey appeared, entitled *The Territorial Imperative.* In it, the author argued that people, like animals, tend to defend their territories as a matter of natural instinct. He was not the first to express this view, but his forcefully written work brought focus to the debate. Ardrey, in saying that man is in effect a territorial animal, suggested that man's sense of territoriality as an instinctive drive is a greater determinant of his politico-territorial adjustment and organization than what he has learned through social and cultural evolution. He provided much support for this argument, but when you read his work you will find that there is a tendency to attribute to aggregates (groups and nations) behavior that has been seen to exist in individuals. This may not be altogether appropriate; we cannot really prove that a state behaves as a whole in a manner identical to the population that it contains. And this brings us back to our earlier point about representatives who guide states through their negotiations and decisions relating to other states. Do these people act as representatives, using learned skills, reflecting in some measure the history and prevailing social standards of the nation, or do they operate, say, in the context of a boundary question, from instinct?

Some political geographers now distinguish between the *instinctive* territoriality of animals and, perhaps, early man and the *cultural* territoriality that emerged when man began to organize himself into territorial states (expressed so powerfully by the German geopoliticians of the interwar period). But that argument, too, has its detractors. For example, if cultural territoriality becomes stronger as man compartmentalizes into the rigidity of nation-states, then presumably the sense of territorial inviolability simultaneously intensifies. But read what anthropologist A. Alland says about this in his book *The Human Imperative* (1972): if the nation-state is the epitome of fiercely defended territory, why draft laws, compulsory anthem-singing, pledge-reciting, and flag-waving? All these rituals point to weakness and a need to instill rather than innate feeling and strength.

The whole area of behavioralism as it relates to the nation-state of today is full of exciting possibilities for study and research. Anthropologists, psychologists, as well as geographers are contributing to the debate, a debate to which no geographer could be indifferent!

CHAPTER SEVEN

Agriculture and Society

SEVEN

A fundamental feature of every culture lies in the way food is produced, distributed, and consumed. It may be that an age of synthetic foods is approaching and that man will begin to lessen his dependence on the soil; but for the time being, practically everything we consume is the product of a process or a sequence of processes that begins in the earth itself. Farming has become the basis of existence all over the world.

Viewed in the perspective of all human history, farming actually is a rather recent innovation; its beginnings date back a mere 12,000 years, and perhaps less. Even today some societies survive much as man did before agriculture was developed, by hunting and by gathering what nature has to offer and sometimes by fishing. Peoples who subsist this way have been pushed into difficult environments by more powerful competitors (as the Bushmen were by the Bantu and the white man), and their survival seems to involve the weathering of one crisis after another. Drought is the worst enemy. It withers the vegetation, kills and drives off the wildlife, cuts off natural water supplies such as springs. Still the Bushmen, the Australian indigenous peoples, the Indians of southern Argentina, and several other groups in the Americas, Africa, and Asia manage to sustain themselves in the face of great odds. They do so by knowing and exploiting their environment exhaustively. Every seed, root, fruit, berry, and beetle is sought out and consumed. Hunting is done with poisoned spears, bows and arrows, clubs, and sticks. When the people do not depend on a particular water hole for their own water, they may poison it and follow the animals that have drunk there. But without agriculture and without the practice of storing food in preparation for future periods of shortage (very few hunting-gathering peoples do this), life is difficult. It was easier where the land was more productive, but most of the remaining hunting-gathering peoples have been driven from better-endowed areas into dry, cold, and less hospitable environments.

Even so, there were peoples who subsisted on hunting and fishing as well as gathering who had specialized to some extent in some area of production. The oak forests of parts of North America provided an abundant harvest, sometimes enough to last more than one year. Indians near the Pacific Ocean coast became adept salmon fishermen. The buffalo herds provided sustenance for centuries without being wiped out. In northern regions people followed the migrations of the caribou herds. The Eskimo and the Ainu (northern Japan) developed specialized fishing techniques.

From these ways of life we can deduce to some extent the means by which our distant predecessors survived. In Chapter Two we observed that the forest margins in Africa may have been a place of early settlement and community growth. There, perhaps, early man had the options of gathering in the forests when hunting yielded poor results and of hunting when the opportunities improved again. Possibly those forest-margin people could do what few hunting-gathering peoples could afford, become semisedentary and stay in one place for a length of time and create a more or less permanent settlement. That is one of the contributions the development of agriculture made: it permitted man to settle permanently in one location, with assurance that food would be available in seasons and years to come.

AGRICULTURAL BEGINNINGS

Man probably began to domesticate plants simultaneously in several different parts of the world. Scholars still are not certain of the sequence of events, but independent invention in areas as widely separated as the Middle East, Central and South America, and Southeast Asia seems plausible. The process itself, which happened between 10,000 and 20,000 years ago, was not a clear-cut "agricultural revolution," as it is sometimes called. It is more likely that man slowly came to realize that he could do more than simply harvest wild crops. Roots and tubers could be stuck in the ground and they would multiply to add to what the gathering practice produced. The first cultivation undoubtedly occurred in association with other means of survival, and it has been suggested that the fishing villages of ancient times, which afforded a measure of permanence of settlement, may have been the places near which the first agriculture was tried.

Wherever agriculture first developed, we should differentiate between seeding and the planting of root crops and cuttings. Transplanting of roots and plants from the forest probably came earliest. Possibly some root crops began to grow spontaneously from food remnants thrown on refuse heaps. Seeding involved more deliberate activity: the gathering of grains from plants, the preparation of a piece of land, then the sowing, followed perhaps by some weeding and general attention to the crop.

Where could agriculture be invented, more than 10,000 years ago? Again, we must speculate to some extent. In a book entitled *Agricultural Origins and Dispersals* (1952), Carl O. Sauer suggests that Southeast Asia may have been the place where the first farmers lived. He points out that the conditions of natural vegetation, climate, and soils must have been favorable for such a development, and believes that the knowledge about cultivation gained there spread out into China, India, even Africa. Other scholars have described the region of agricultural invention that existed in Middle and South America, where the first cultivation of roots and cuttings also was followed by the more sophisticated practice of seeding. The Middle East, both in North Africa and in Southwest Asia, was another leading region of plant domestication; here lies the famous Fertile Crescent, about which we will say more later. China's source area, discussed in Chapter Three, also was a region of early plant domestication. And although it took a long time for scholars to recognize it, Africa, too, shared in the process. In his book *Africa, Its Peoples and Their Culture History* (1959), George P. Murdock reports:

> *It has hitherto escaped attention . . . that agriculture was independently developed . . . by the Negroes of West Africa. This was, moreover, a genuine invention, not a borrowing from another people. Furthermore, the assemblage of cultivated plants ennobled from wild forms in Negro Africa ranks as one of the four major agricultural complexes evolved in the entire course of human history. Interestingly enough, the innovators have belonged to four distinct races. Along with the Caucasoids who developed the Southwest Asian complex, the Mongoloids who achieved the Southeast Asian complex, and the American Indians who elaborated the Middle American complex, we must now align the West African Negroes as one of mankind's leading creative benefactors.*

As Fig. 7.1 shows, Africa may have had more to offer even than that. Another area of significant plant domestication lay in the African great lakes region, and in other areas of South America and Eurasia as well. From these locations, knowledge of crops and cultivation techniques diffused to other parts of the world, and undoubtedly there was some interaction between certain of the early farming sites. It is thought that millet, a small-seed grain, was introduced to India from West Africa, and sorghum, another grain crop, from West Africa to China. The watermelon spread from West Africa, first to nearby regions but eventually all over the world. Corn (maize) spread from Middle America into North America; later, after the Portuguese brought it across the Atlantic, it became a staple in much of Africa. The banana came from Southeast Asia, as did a variety of yams. The process of dispersal was a slow one for thousands of years, but the worldwide communications set up with the expansion of Europe during the past 500 years accelerated it greatly.

Shifting Cultivation

The ancient farmers learned to plant crops, but they knew little about soils, fertilizing, or irrigation. It is likely that they had to abandon areas in subtropical zones after the soils became infertile and crops stopped growing. The farmers would move on to another parcel of land, clear the forest, and try again. This practice of *shifting cultivation*, like hunting and gathering, still goes on. In tropical areas, where the redness of the soil signifies heavy leaching but where luxuriant natural vegetation thrives, a plot of cleared soil will carry a good crop at least one time, perhaps two or three. But then the area is best left alone, to regenerate its natural vegetation and to replenish the nutrients lost during cultivation. Several years later the plot may yield a good harvest once again.

Shifting cultivation is a way of life for many more people than hunting and gathering. Between 150 and 200 million sustain themselves this way in Africa, Middle America, tropical South America, and in places even in Southeast Asia. At one time this was the chief form of agriculture in the inhabited world, just

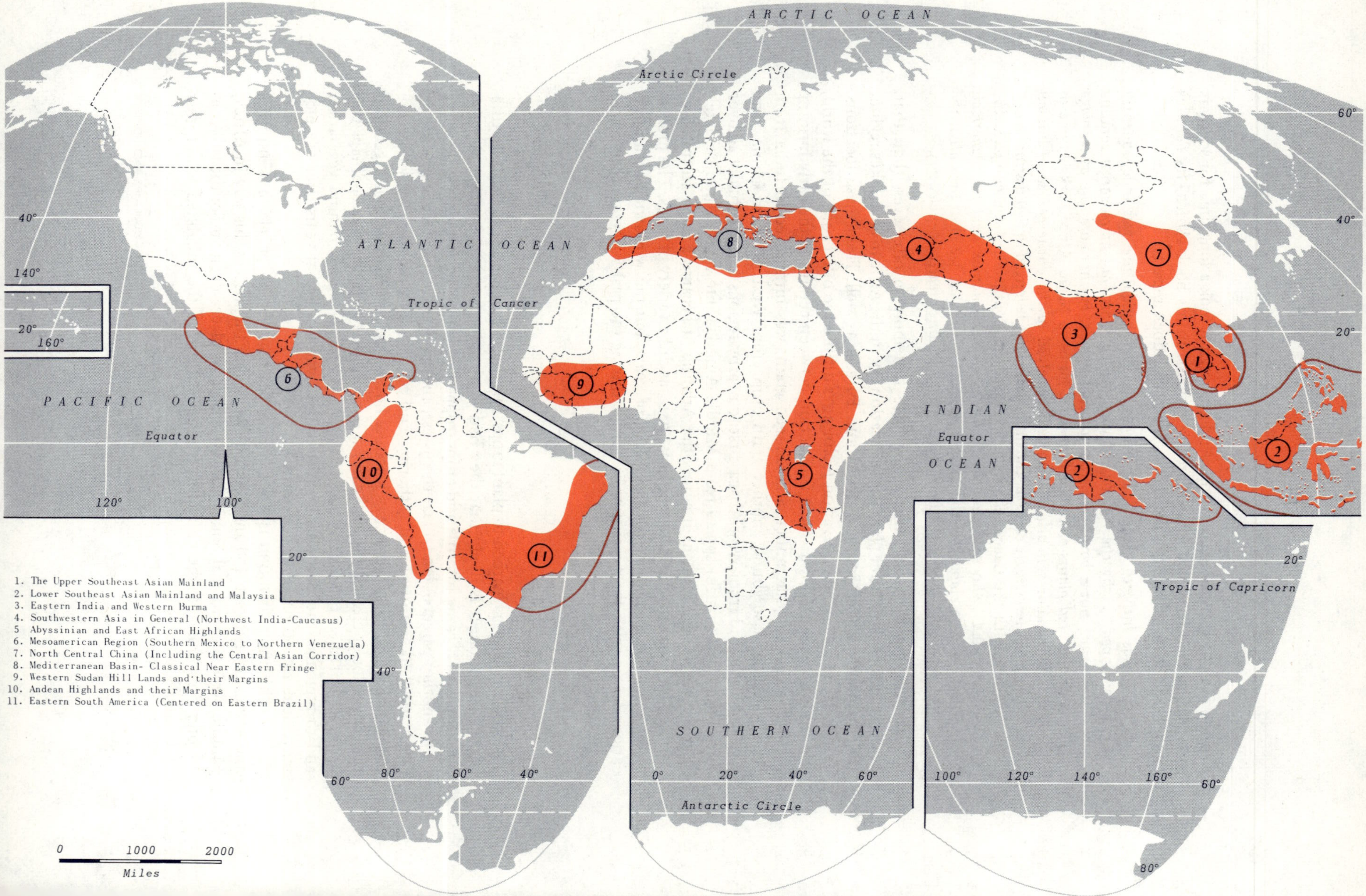
ARCTIC OCEAN
Arctic Circle
ATLANTIC OCEAN
Tropic of Cancer
PACIFIC OCEAN
Equator
INDIAN OCEAN
SOUTHERN OCEAN
Tropic of Capricorn
Antarctic Circle
1. The Upper Southeast Asian Mainland
2. Lower Southeast Asian Mainland and Malaysia
3. Eastern India and Western Burma
4. Southwestern Asia in General (Northwest India-Caucasus)
5 Abyssinian and East African Highlands
6. Mesoamerican Region (Southern Mexico to Northern Venezuela)
7. North Central China (Including the Central Asian Corridor)
8. Mediterranean Basin- Classical Near Eastern Fringe
9. Western Sudan Hill Lands and their Margins
10. Andean Highlands and their Margins
11. Eastern South America (Centered on Eastern Brazil)
0 1000 2000
Miles

as hunting and gathering once prevailed as modes of existence. The process involves a kind of natural rotation system in which areas of forest are used without being permanently destroyed. It does not require a nomadic existence of the farmers, because usually there is a central village and parcels of land in several directions are worked successively. When the village grows too large and the distances to workable areas become too great, a part of the village's population may move to found a new settlement some distance away. This implies, of course, that population densities in areas of shifting, "patch," agriculture cannot be very high; there has to be room. But as we noted in Chapter Three, high densities were rare in ancient times. It was through regions where shifting agriculture prevailed that crops and tools for farming were exchanged and dispersed. Wheat, probably first grown in Southwest Asia, thus was transmitted to China (and to Western Europe as well). Rice spread from Southeast Asia in all directions to Indonesia, China, and India. Until the Europeans connected the continents, certain crops remained confined to the Americas, corn, squash, potatoes, some types of beans, cassava, and peanuts, among others. When corn reached Africa, it soon spread across the savanna and became a major staple. The cassava also became a major food producer in Africa. Numerous crops did better in new environments than the ones in which they were first domesticated. With the spread of the potato to Europe, corn to Africa, wheat as far as Australia and Canada, the diffusion process begun by those ancient farmers was finally complete.

Animal Domestication

For hundreds of thousands of years man and his ancestors hunted wild animals. As communities grew larger, better organized, and more stable, animals became a part of the local scene, not to be hunted, but to be kept as pets or for some other (for example, ceremonial) purpose. Just as plant domestication still produces unanswered questions, the origins of animal domestication are a matter for speculation, but it is likely that animals attached themselves to human settlements as scavengers and even for protection and thus contributed to the idea that they might be tamed and kept. Any visitor to an African wildlife reserve can observe that when night falls, a permanent camp will be approached by certain species (gazelle, zebra, monkeys) that spend the night near and sometimes even within the camp's confines. With daybreak the animals wander off, but the dangers of the next nightfall bring them back. Similar behavior probably brought animals to the settlements of the ancient forest farmers. Hunters might bring back the young offspring of an animal killed in the field and raise it. Such events probably contributed to the concept of animal domestication.

Just when this happened is also a matter for guesswork. Some scholars believe that animal domestication came earlier than that of the first plants, but

FIG. 7.1 Regions of significant plant domestication. (After a map in J. Spencer and W. Thomas, *Cultural Geography*, Wiley, New York, 1969, p. 148.)

others put animal domestication as late as 8000 years ago. In any case, the goat, pig, and sheep became part of a rapidly growing array of domestic animals, and in captivity they changed considerably from the wild state. Archeological research indicates that when such animals as wild cattle are penned in a corral, they develop types different from the original. Protection from predators led to the survival of animals that would have been eliminated in the wild; inbreeding entrenched the modifications nature would have wiped out. It is thought that wild cattle may have been domesticated first for religious purposes, perhaps because the shape of their horns looked like the moon's crescent. Apparently cattle were strongly associated with religious ritual from the beginning, and, as we saw in Chapter Four, there are regions where they continue to hold a special position as holy animals. But how did cattle come to be used as suppliers of milk and as work animals in agriculture? Again, the origins may be in religion: cattle whose very young calves were taken away for sacrifice were milked to provide relief; cattle pulled sacred equipment on sled-platforms before they pulled plows.

The domestication of animals provided early man with another new source of food. Whether animal domestication or plant domestication started first in various parts of the world, soon societies were combining the two and thus lessening their dependence on a single food source.

Nomadism

Nevertheless, there were always peoples who could not keep herd animals (those living in tropical forests, for example) and peoples who could not cultivate (those in the dry steppes of northern Africa and Asia). We know how the forest peoples managed. How did the inhabitants of the drier areas sustain themselves? They did so, and some still do, by a system of pastoral nomadism. By this means, the peoples who preferred the dry country, those who actually saw their habitat become progressively drier through climatic change, and those who were ousted from better-watered areas could provide for themselves in regions formerly incapable of supporting human societies. Pastoral nomadism involves the almost constant movement of people and their animals, the animals in a cyclic pursuit of pastures and water, the people in pursuit of the animals. Today we see Indian peoples do this with the caribou in the far-northern latitudes; the Masai live this way with their cattle on the very equator in East Africa.

The Masai today occupy an area astride the Kenya-Tanzania boundary, only a remnant of their region of former centuries. Their habitat consists of the lowland of a rift valley and its adjacent sides and higher land. This is arid country, with a rainfall of 20 to 25 inches annually (an unreliable average), with long droughts and a parching sun. In scattered groups, the Masai follow their herds of cattle across these sparse grasslands, searching always for pasture and water holes. The Masai diet consists of blood, drawn from an incision made just below the cow's throat, milk, and sometimes the meat of goats and some sheep, which accompany the cattle herds.

A Masai shepherd and his cattle herd, in the Arusha District of Tanzania. The mountain in the left background is Mount Meru, an extinct volcano; smaller volcanic cones lie in the middle distance. The grasses in this picture are rather sparse, but the animals are in good condition.

Although the Masai are nomads, they do build settlements in an area to which they return seasonally when the rains break. When the drought comes they move upslope with their cattle, where it is cooler and where the moisture lingers longer (this is a variation of the process called *transhumance*). But when the rains begin to fall again and the dust is washed from the remaining grass and from the shrubs and trees, and the rift-valley grasses turn green once more, the people and their herds come back to their familiar surroundings and the small *enkangs* they left behind.

The Masai never sells his animals for beef. With diseases and food shortages attacking the stock, individual animals are normally emaciated; milk production is very limited from such animals. But what they lack in weight and productivity they make up in social value and prestige for their owners. The larger a man's herd, the higher his status. Anything must be done to preserve the numbers. And so the cycle goes year after year, the Masai's entire life centered on their herds.

Very few nomadic peoples are actually "pure" pastoralists. Even the Masai try a few patches of corn during the wet season when they inhabit their *enkangs.* Often attempts at cultivation are made because of a contraction of the grazing region and resulting problems of overgrazing and general dislocation. The Tuareg of the Sahara, for example, were pushed back by the invasion of both Arabs and Europeans

TRANSHUMANCE

The term *transhumance* is used to denote the seasonal migration of livestock, accompanied by their owners or hired herders, in pursuit of suitable pastures. One type of transhumance occurs in the European Alps, whose snows partly melt during the summer. When this happens, the herdsmen, and sometimes the whole farm family, take up living in cottages built specifically for this purpose near the snow line, high up the slopes. With the arrival of autumn, the cattle and goats and their keepers descend to the valleys and other low-lying areas, abandoning pastures and cottages to the winter's approaching snows. The next spring, when the melting snow exposes the pastures again, preparations begin for another move to the high slopes.

Transhumance, then, is a form of cyclic migration, but it is not confined to vertical movement. The Masai and their cattle also practice transhumance as they traverse the plains in search of pasture and water; their cycle of movement may not be as neat and planned as that of the Swiss, but it is recurrent nevertheless.

---

in North Africa. Once in control not only of the Saharan steppelands but also of trade routes and oases, the Tuareg have been reduced to some interior plateau and mountain areas and in some cases to tenant farming of small palm groves.

Pastoral nomadism is not confined to Africa, and it is not always based solely on cattle herds. When the herders acquired horses and camels to ride and with which to control their herds, they also attained new power and mobility with which to threaten sedentary societies. From China to Russia to North Africa, the nomadic riders of the plains emerged as feared invaders and ruthless victors. Sometimes they would attack and destroy cities, looting them and taking off again, but on occasion they took temporary political control.

SUBSISTENCE AGRICULTURE

We tend to think of agricultural geography in terms of cash cropping (that is, farming for sale and profit), plantations, ranches, mechanization, irrigation, the movement of farm products, marketing, exports and imports, and so on. When we associate certain crops with particular countries, these are usually cash crops: Brazilian coffee, Colombian tobacco, Egyptian cotton, Australian wool, and Argentinian beef, for example. But the fact is that a great number of the world's farmers are not involved in all this. Hundreds of millions of farmers use their plots of land in the first instance to grow enough food to survive, sometimes with only marginal success. Their chief objective is subsistence, not profit. Before we consider "modern" agriculture, therefore, we ought to have a look at the way so many farming families sustain themselves in these modern times, by subsistence agriculture (Fig. 7.2).

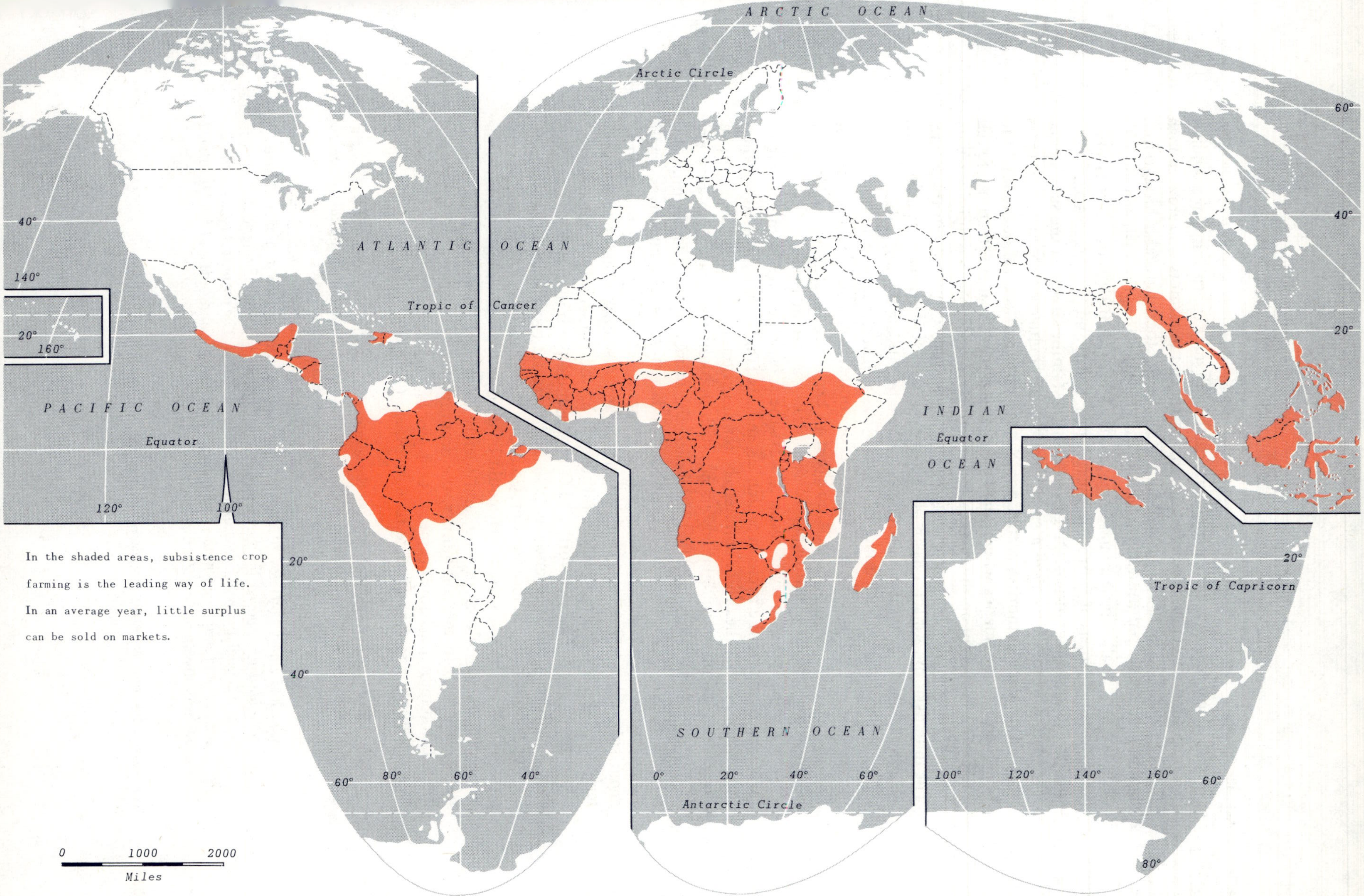

FIG. 7.2 World regions of primarily subsistence agriculture.

The term *subsistence agriculture* can lead to some confusion. Some scholars use it in the strictest sense of the words, that is, to refer to farmers who grow enough food to sustain themselves and their families, who use their natural habitat to find building materials and firewood, and who do not enter into the cash economy of their country at all. Those remaining societies where shifting agriculture is practiced, in remote areas of South America, Africa, South and Southeast Asia, would qualify under this definition. On the other hand, families living at the subsistence level, but who sometimes sell a small quantity of produce (perhaps to pay taxes imposed by some authority) would not. Yet the term *subsistence* ought surely to be appropriate to societies where small-plot farmers may be able sometimes to sell a few pounds of grain on the market, but where poor years threaten hunger, where poverty, indebtedness, stagnation, and sometimes even tenancy are ways of life. The Indians of the Amazon Basin, the sedentary farmers of Africa's savanna areas, villagers in India, peasants in Indonesia—all share subsistence not only as a way of life but also as a state of mind. Experience has taught the farmer and his father and grandfather before him that moments of comparative plenty will be paid for by times of scarcity, that there is no point in expecting progress when fertilizer is always too expensive and improved seed strains a distant rumor.

It is tempting to try to think of ways the subsistence agriculturalist might be helped to escape the cycle in which he and his family find themselves, when higher productivity, healthier children, better nutrition, longer life spans might be the rewards. European colonial powers put such thinking into practice in a variety of ways, by demanding taxes that compelled the farmer to raise some funds, by genuine assistance in the form of land-consolidation schemes, soil surveys, and social incentives, and by initiating forced cropping schemes. If you read a little about subsistence agriculture, you will find many a scholar wondering how "to tempt [subsistence farmers] into wanting cash by the availability of suitable consumer goods," as A. N. Duckworth and G. B. Masefield put it in their book *Farming Systems of the World* as recently as 1970. But there are other sides to subsistence farming we ought not to lose sight of. Changing those attitudes might have effects on the cohesion of the society that could be destructive. Subsistence brings with it the communal holding of land, and society is of need quite equal: surpluses are shared by all in the community, the accumulation of personal wealth is restricted, and individual advancement at the cost of the group as a whole is limited. As A. H. Bunting said in *Change in Agriculture* (1970):

> *To allocate the land or manage the seasonal migrations, and to survive through hardship and calamity, these societies have to be cohesive, communal and relatively little differentiated socially and economically: the chiefs, elders or elected headmen may be little richer than their fellows—to many of whom they are in addition linked by ties of relationship within the extended family. Mutual dependence, imposed by the environment and the state of the agricultural art, is maintained and reinforced by genetic relationships. The community is enclosed socially and may even tend to be isolated culturally. Landlords and feudal rulers are unknown; the cultivators are poor but free.*

Change the economic system, and there will be unpredictable and incalculable modifications in the society, which, indeed, may founder under the stress of "progress."

In your reading you will also note that it is customary to divide agricultural societies into groups, for example, "subsistence," "intermediate," and "developed" or, using different terms to express the same idea, "primitive," "traditional," and "modern." Actually, the world's many hundreds of societies lie along a continuum, and these divisions are for purposes of simplification and discussion and are not absolute or inarguable. This brings us back to the question we raised earlier. Just what is subsistence? What is the boundary between "primitive" and "traditional"? Is it the kind of ownership of land? The amount of produce that is consumed by the farmer compared to that sold on a market? There are no hard and fast rules, so the terms and their application become a matter of individual judgment. And that judgment is often debatable. Maps of Africa, for example, may show African land use to be all "traditional" or "tribal" except in South Africa, Rhodesia, and in the Kenya Highlands where whites dominate or dominated agriculture. The

Where soils are poor and rain unpredictable, people must spend their entire lives trying to eke out enough food to subsist. These people are often ravaged by disease as well as malnutrition, and daily life has few moments of pleasure. This is a village in Upper Volta, a country struck severely by the 1973 drought. The sun is hot. One cloudless day follows another. The woman holding the baby is going blind. There is nothing to do but wait and hope.

interpretation, then, is that when land use is Western, it is "modern" or "developed," and if not, it is "tribal" or "traditional." But if the map can show the white farms of Rhodesia and (formerly) Kenya as modern forms of agriculture and land use, then why not the cocoa farmers of Ghana or the cotton growers of Uganda? Their response to the profit motive is not very different from the wheat farmer's or the cattle rancher's or the Rhodesian tobacco grower's. It may be that, in addition to their products for sale, African and Asian "traditional" farmers also raise some crops for their own consumption. But so do farmers in Bulgaria and Ireland. When you run across such typologies, interpret them carefully.

All this is not to suggest that there is no transitional phase to be witnessed when subsistence land use changes to more intensive farming, cash cropping, and ultimately to the mechanized, almost automated kind of farming in which huge pieces of equipment, not people, do the work and actually touch the soil. However, the change involves more than the use of the land itself, and the stages should be viewed in this context. For example, land held communally by true subsistence-agriculture societies is parceled out to individuals when cash cropping becomes a major element in life. Social organization, too, ceases to be communal, and the system that ensured an equitable distribution of surpluses breaks down. The distribution of wealth becomes stratified, with poor and nearly destitute people at the bottom and rich landlords at the top. It is thought by scholars that the innovation of irrigation may have contributed to the first differentiation of subsistence society, but there is no doubt that the greatest number of farmers in "intermediate" societies find themselves in their present situation as a result of the changes brought to their part of the world by expanding, imperializing Europe. In the aftermath, it is a bit ironic to call these the "traditional" societies: they are caught between traditions destroyed when the European invasion came and traditions found inadequate when the Europeans sought to introduce their own.

THE PLANTATION

The Europeans brought the first large-scale agricultural profit-producing project to the non-Western world in the form of the plantation. The term *plantation* brings to mind a huge, productive, comparatively modern enterprise, where the work is done by local people and the owner and his family live in an elaborate, comfortable mansion of beauty and style. But such an impression does not apply to many plantations. Some plantations may be quite large but never produce much of anything. In fact, there is so much disagreement on just what a plantation is that no definition has ever satisfied everyone.

In that case, is it useful to know something about these enterprises? Indeed it is. Plantation agriculture transformed whole areas of the world from subsistence to cash cropping; it caused large areas of land to be taken from their indigenous users; it helped generate enforced migrations of millions of people for labor purposes; and it generated enormous wealth for some of the plantations' owners (companies as well as individuals). Today, the character of plantations is changing, but the institution survives.

What are the main qualities of a plantation? One criterion is size. Plantations often are quite extensive, much more so than farms. There is no precise lower limit, but some plantations in South America cover more than 250,000 acres, or 400 square miles. That is about one-third of the state of Rhode Island! Many plantations are smaller, but the average plantation is much larger than the average farm in the United States. This relates, of course, to the availability of great stretches of land when the Europeans penetrated Middle and South America, Africa, and Asia. Even when there were people living on the land, it was awarded to white men by royal decree or taken by conquest.

Plantations are owned by private individuals or by corporations (such as, for example, the United Fruit Company, which has large banana plantations in Middle America). In recent years, plantations formerly owned by whites have been bought or taken over by the governments of newly independent states. But those governments have recognized the economic value of the enterprise and have set up corporations to run them, as has been done in Kenya. In a few cases the plantations have been divided among peasants. This was a great experiment in Mexico, part of the revolution that began in 1910 and stipulated by the constitution of 1917. As of that date, landowners with more than 750 acres had their lands taken over for redistribution to Indian farming families. In other forms, large estates were also expropriated in the Soviet Union and in China and incorporated in collectives and communes. But all these are changes of the twentieth century. Originally the plantation was an individual or a corporate enterprise.

The administration of the plantation is still today predominantly in the hands of people of European extraction. It is, after all, the product of European culture. Large estates were plentiful in Europe during the days of colonial expansion, and the possession of land was a matter of great prestige. Thus Spanish invaders in South America would acquire land, as much of it as they could, even when it was unproductive land. In that way many of those unproductive *haciendas* we referred to came into existence. Whether productive or not, the plantation has a centralized management, run by the family that owns it or by a company through a representative. The work is done by hired labor or, in some places still, by tenants. In the early days, when the Europeans were fanning out across the world and establishing plantations, enough labor could not always be secured. In the southern part of North America, in Caribbean America, and in northern and eastern South America, the coming of the white man and the enslavement of the rather small Indian population led to such a rapid decline in the latter's numbers that the importation of black Africans began quite soon. The plantations needed abundant and cheap labor; later Asians from India also were brought to Caribbean America as indentured workers, a status just a shade less captive than that of the slave. The British thus conveyed tens of thousands of Indians from their Asian empire to America; they did the same in Africa, where Natal's lucrative sugar plantations

A banana plantation in Honduras, Middle America. This plantation is owned by United Brands, a division of United Fruit Company. Conveyor belts, having taken over from human carriers, move the plastic-covered banana bunches to the factory.

needed labor the Africans were unwilling to provide. The Dutch brought laborers from their East Indies (now Indonesia) to Surinam. The introduction of the plantation was a major factor in population migrations of world significance. And not surprisingly, plantations thrived in Southeast Asia where labor was plentiful and did not require supplementing.

Another hallmark of the plantation, especially when viewed against the background of prevailing land use nearby, is its specialization. Plantations emerged because there were lucrative markets, principally in Europe, for products that could be grown in warmer climates. Sugar was a prime example, and sugar plantations from Caribbean America to southern Africa transformed whole countrysides. Cotton, coffee, bananas, tea, tobacco—millions of acres of land were cleared and converted to their production. But the plantation would lie like an island of intensive cultivation in a sea of unchanged tradition. To this day, seemingly endless fields of sugarcane abut suddenly against land where patch agriculture is the way of life. And there have been times when those nearby traditional areas saw the plantation come and go. Specialization brings with it certain risks. For one, it reduces the options in time of crisis. Disease struck ruinously at the coffee plantations of Ceylon (Sri Lanka), Ecuador's coffee, Central America's bananas. Substitutes, too, have put plantations out of business, for example, sugar beets for sugarcane and synthetic rubber for natural rubber. Plantation monopolies sometimes proved to be temporary.

Two other qualities of the plantation are relevant. First, the methods and tools of farming tend to be more complex than those of the smaller farms. Although the history of availability of cheap labor has slowed the mechanization process, recent political conditions and related pressures have led plantation operators to turn to greater mechanization as a means of reducing dependence upon people. And second, plantations, unlike farms, often have equipment to process the produce

VON THÜNEN AND THE ISOLATED STATE

J. H. von Thünen (1783–1850) was a geographer of remarkable qualities. He lived during a time of far-reaching change in Europe. The population was growing, and people were moving to the towns and cities. Around these urban places, farmers produced the foodstuffs needed there in increasing amounts; Von Thünen was one of them. In 1810, he acquired a fairly large farming estate not far from the town of Rostock in the German state of Mecklenburg. As he farmed his lands, Von Thünen became interested in a subject that still concerns economic geographers, the effects of distance and transportation costs on the location of productive activity. For four decades, he kept meticulous records of all the transactions of his estate, and while doing so he began to publish books on the spatial structure of agriculture. His series, under the title *Der Isolierte Staat* [the isolated state], in some ways constitutes the foundations of spatial theory. Von Thünen's conclusions are still being discussed and debated today.

Von Thünen called his work *The Isolated State* because he wanted to establish, for purposes of analysis, a self-contained country, devoid of outside influences that would disturb the situation, a sort of regional laboratory within which he could identify the factors that affect the location of farms around a central urban area. He assumed that the soils and the climate were uniform, that there would be no river valleys or mountains to interrupt the flatness of the surface, and that a wasteland surrounded his circular "state" with its single, centrally positioned city. Von Thünen had the farmers do their own transporting; to get to the urban market they had to travel directly overland by oxcart. This is the same as assuming a network of radially converging roads of constant quality, and in such a situation, of course, transport costs would be directly proportional to distance.

In such a state, Von Thünen then asked himself, what would the spatial arrangement of agricultural activities be? He concluded that farm products would be grown in a series of concentric zones outward from the market center. Nearest to the city those crops would be grown that yielded the highest returns, for example, vegetables and dairy products. Farther away, you would find more durable goods such as potatoes and grains. And eventually, since transport costs to the city increased with distance, there would come a line beyond which it would be uneconomical to produce crops. There the wasteland would begin.

Von Thünen's model, then, incorporated five belts surrounding the market center; from the city outward, these would be first, a zone of intensive agriculture and dairying; second, a belt of forest for timber and firewood; third, a broad belt of increasingly extensive field crops; fourth, a zone of ranching and animal products; and fifth, the surrounding, unprofitable wasteland. If the location of the second zone, the forest, surprises you, remember that the forest still was of great importance in Von Thünen's time as a source of building materials and fuel. All that was about to change with the onslaught of the industrial revolution, but there are lots of towns and cities left in the world that are still essentially preindustrial. When Professor R. J. Horvath went to study Addis Ababa, Ethiopia, in this context he found a wide and continuous belt of eucalyptus forest surrounding that city, positioned more or less where Von Thünen would have predicted it to be and serving functions similar to those attributed to the forest belt of the isolated state. Horvath's 1969 article on this in the *Annals* of the Association of American Geographers makes fascinating reading.

Modern cities are still ringed by recognizable agricultural zones, though modern transportation, urban sprawl and suburbanization, and numerous other developments have disrupted patterns so much that they are sometimes almost unrecognizable. Decades ago you might have found economic geographers willing to discern Thünian patterns in the farmlands around, say, Chicago; these days they might talk about remnants of such patterns, but a concentric zonation still prevails around many smaller centers in America, and certainly in much of the rest of the world. Next time you fly into a city or travel in and out by road, make the trip even more interesting by trying to recognize the pattern!

prior to shipment. This makes the plantation more of an industry than the farm normally is.

These are some of the properties of plantations. Let us not complicate things by adding still another definition to the ones that already exist. Whether plantations must lie in the tropics or subtropics, whether they are run by foreigners using local labor, whether they must be farm industries that process as well as plant in order to qualify, plantations are clearly a discrete form of land use. There are many kinds of plantations, but those with the greatest impact were located in populated areas of the Americas, Africa, and Asia. Apart from generating enforced labor migrations, they occupied good agricultural land and pushed hundreds of thousands of people who might have remained subsistence farmers into the money economy. Others found themselves tenants on land they had always viewed as their own, forced to work off a debt incurred simply by living there. Plantations also had some positive impact, more so in recent decades. Besides providing an economic stimulus, they contributed to the development of improved health facilities, education, and local communications. Crops first introduced on plantations spread to surrounding areas. Agricultural techniques were learned and adopted.

For all their importance in the world's recent transformation, plantations do not cover large areas of the world and never did. They did come to extend over much of Cuba and Puerto Rico, coastal Northeast Brazil, lower areas of Venezuela and Colombia, and Middle America. In Africa, they developed in coastal Natal, parts of East Africa, and relatively small sections of West Africa from Senegal to Nigeria. In Asia, there are the tea plantations of Assam (India) and Sri Lanka (Ceylon), the sugar estates of Java and the Philippines, and rubber plantations of Malaya. But, as Fig. 7.3 shows, these are mere patches in the farming regions of the world.

GRAIN CROPS

The great majority of the world's people today are sustained chiefly by the major grain crops, the *cereals.* Along with two root crops, potatoes and cassava, the three leading cereals, rice, wheat, and corn, account for more than three-quarters of the food supply produced on this earth each year. Grain crops of lesser, local importance, such as sorghum, millet, barley, oats, and rye (among others), fruits (the banana is a staple in a few areas), vegetables, meats, and dairy products make up the remainder. In Chapter Three, we noted that all too many people do not, in their diets, enjoy a balance of these foodstuffs.

Rice

Probably about half the people of the world depend upon rice for their daily sustenance. Rice and wheat are among the "oldest" of the cereals, having been domesticated in Southeast and Southwest Asia respectively during the earliest beginnings of agriculture. The rice and wheat cultivated today, of course, are not

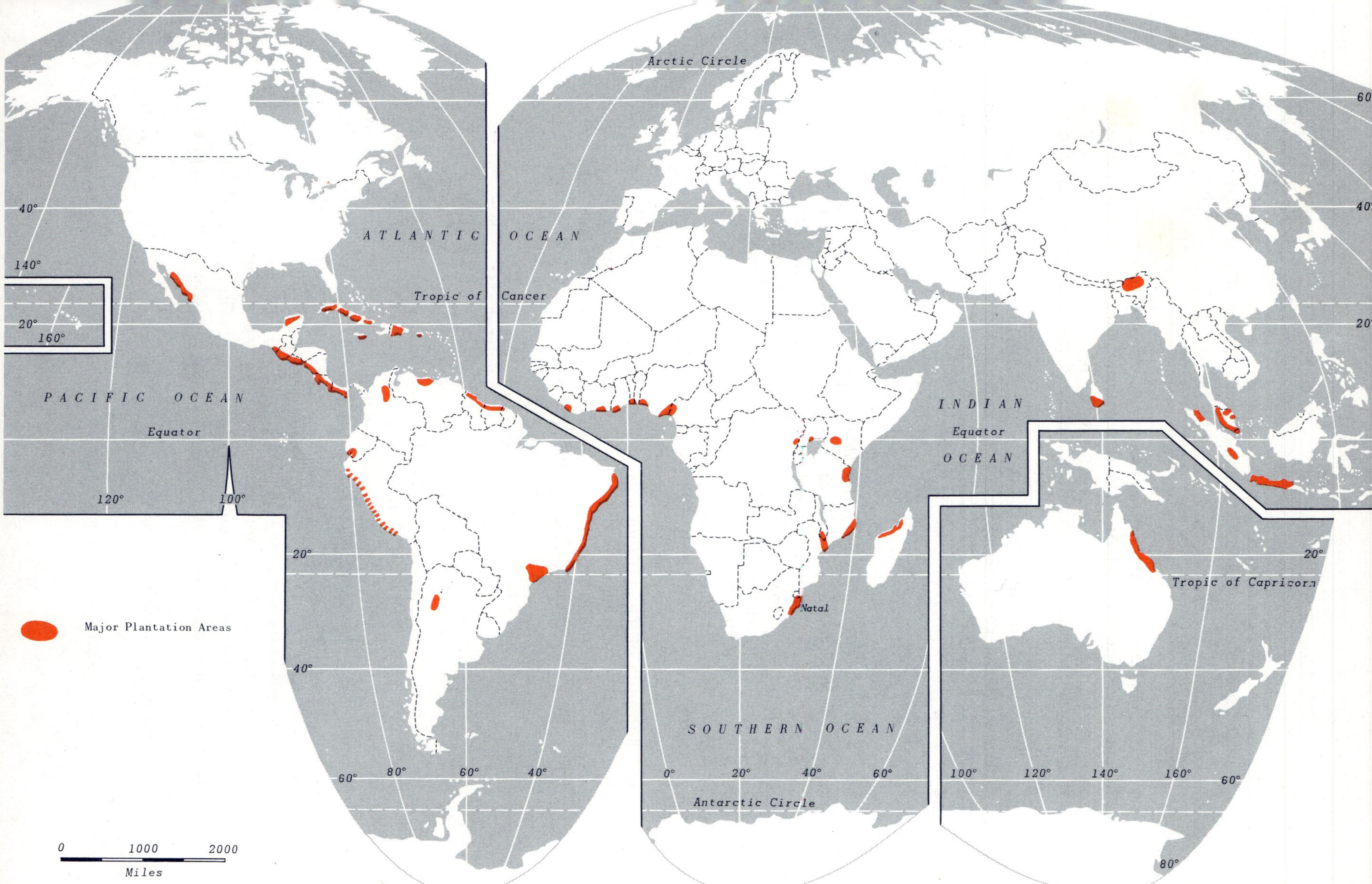

FIG. 7.3 Major plantation areas of the world.

the same kinds as those first grown thousands of years ago. The strains have been modified over time, and new ones are still being developed, such as the "miracle rice" of recent years that has boosted yields in many rice-growing countries enormously.

Compared to rice and wheat, corn is of less importance. It was domesticated in America and did not reach Africa and Eurasia until after the first European visits to the New World. Much corn is consumed indirectly, that is, it is used for livestock feed, and the products of the livestock, milk and meat, are then consumed by people.

The map showing present areas of substantial rice cultivation (Fig. 7.4) indicates that rice continues to be grown predominantly in and near the region where it was first domesticated. Compared to the areas where wheat and corn are grown (Figs. 7.5 and 7.6), the extent of the rice-producing areas seems rather limited. But the map should be viewed in the light of per-acre yields, which for rice are far greater than for wheat, nearly twice as high, in fact. Fast-maturing rice strains have been developed, and now two crops can be harvested on the same acre of land, with time left over for some vegetable growing. There are even places where three crops are harvested in a single year. This is done by planting the seed in special plots and then transplanting the seedlings to the paddies immediately after the previous harvest is completed. Although rice has been grown to latitudes 40 degrees, Fig. 7.4 reminds us that it is basically a low-latitude, warm-climate crop. Much of the region outlined on the map has a 12-month growing season, but the period shortens in northern China, Korea, and Japan.

The critical element in rice cultivation, however, is water. Rice needs standing water in which to mature, and that means, of course, that it requires level land. But much of the area shown as rice producing on Fig. 7.4 is by no means level land. The farmers have created the necessary environment by terracing hillslopes and controlling the flow of water. Whole countrysides have been transformed into giant staircaselike elevations, meticulously maintained and intensively farmed. The water in which the rice stands has two ancillary advantages: it keeps down the weeds that would damage the rice if it were grown in less moist areas, and fish are raised in it while the paddies are inundated.

A comparison of Figs. 7.4 and 3.1 suggests how large a percentage of the world's population is concentrated in the rice-producing region and how important rice is as food. Nearly 90 percent of the rice is grown in East, Southeast, and South Asia, and most of the people there, more than three-quarters of them, are busy producing it. This regionalism of production and consumption also occurs with the other grains. The rice-producing countries consume and trade with each other in rice; the countries where wheat is produced do the same with wheat. There are no large-scale exports of rice from Asia to, say, Europe or North America.

Rice is a very nourishing food, but despite Asia's enormous annual production, more is always needed. You will note that figures on world rice production vary with the source being consulted, but China undoubtedly is the world's leading producer. China has come a long way, during the last generation, toward self-

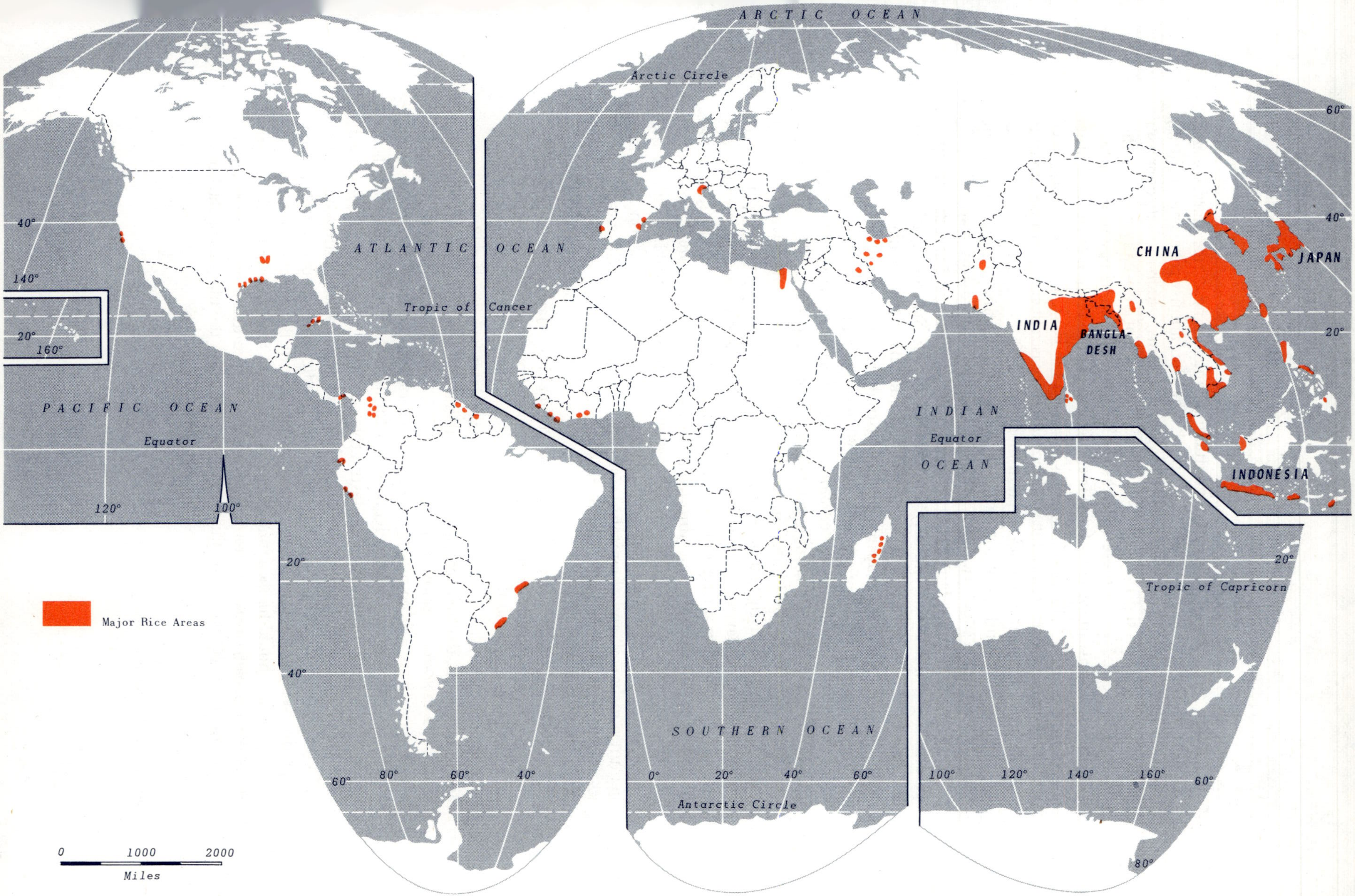

FIG. 7.4 World areas of rice cultivation.

sufficiency and adequacy in food production, and Chinese rice harvests may amount to over one-third of the world total. The Communist government during the early 1950s began to implement sweeping land reforms, which initially depressed production; later, in the 1960s, there were several years of unusually poor weather and bad harvests. Despite these setbacks the reforms seem to be paying off. A huge percentage of the Chinese labor force is engaged in the production of rice and other foodstuffs. Probably four of five Chinese work on the land. This has always been a characteristic of rice production. Its labor-intensity is reflected by those carefully maintained terraces and paddies. Even in modernizing, industrially powerful Japan, where only three workers in ten work on the land, this principle applies, and the farmers work their small, two-to-three acre plots by hand, row by row.

By world totals, Japan is not a major rice producer. It harvests perhaps 7 percent of the world total annually. But it is remarkable that the Japanese are able to produce enough food at home to feed three-quarters of the population, over 75 million people. Because of Japan's great industrial prowess we may overlook what Japanese farmers have accomplished. Over 90 percent of all farmland is under food crops (well over half of it for rice alone), and per-acre yields are the highest in the world by far. Japanese farmers get 80 bushels of rice per acre; Chinese farmers average under 60. In the United States the rice growers are happy with anything over 50 bushels, and in Southeast and South Asia 20 to 30 bushels per acre is considered a good harvest. The Japanese farmers have achieved their high returns from the soil by diligent irrigation (more than half of Japan's farmlands are irrigated in some way), by transplanting, through painstaking terracing (despite the difficult Japanese topography), multiple cropping wherever possible, notably south of the Kanto Plain, which lies just north of Tokyo, and intensive fertilization by animal manures, night soil, and chemicals. Japan still has to import about one-quarter of its annual food requirements, and in this respect it shares the problems of some other Asian countries.

Those problems are great in the Philippines, Malaysia, and Sri Lanka (Ceylon). They also exist in Indonesia, where food production has not been keeping up with population growth, and they are felt also in India, Bangladesh, and Pakistan. Although the last three countries are sometimes described as self-sufficient in rice and other cereals, that self-sufficiency is calculated on rather low per-person caloric intake. Of the whole region mapped on Fig. 7.4, only the countries of mainland Southeast Asia are able to produce a substantial rice surplus and export this commodity to their less fortunate neighbors. But for decades Southeast Asia has been involved in one way or another in destructive conflict. Dikes have been bombed, farm populations frightened away, stored harvests deliberately destroyed. In mid-1973 it remained to be seen whether the region would finally recover and approach its productive potential, currently materialized only in Burma and Thailand.

If that should happen, one of the beneficiaries would be India, along with Sri Lanka and Bangladesh. Although India produces as much as one-fifth of all the rice annually harvested, all too many people still live on the margins of starvation;

The cultivation of rice is still a job for many hands and long hours. On this commune in northern Kunming, China, some two dozen workers are turning over the soil and preparing the paddies for a new planting.

regional crop failures can still have disastrous results. India has bought grains from the United States and other countries, but the proximity of substantial annual surpluses in Southeast Asia forms an attractive alternative.

Wheat

Whereas rice is the subsistence crop of the developing countries of Asia, wheat is the staple of the developed world. This generalization can be carried too far because wheat is cultivated and consumed in India and China as well as in Europe and the U.S.S.R., the United States, Argentina, and Australia. But our map of wheat-producing regions (Fig. 7.5) reflects that situation quite strongly. First domesticated in Southwest Asia, wheat still grows on the shores of the Mediterranean from Syria to Morocco. But today the bulk of the world's wheat stands in the Ukraine, in the American Great Plains, in Argentina's Pampa, in Australia, and northern China. It is a distribution very different from that of rice.

This has to do, of course, with the many different varieties of wheat that have been developed since its domestication. Wheat is like a tall grass, and it was first cultivated consciously in a form that had grown wild. Experimentation over centuries has produced strains that can survive on minimal moisture, stay alive

FIG. 7.5 World areas of wheat cultivation.

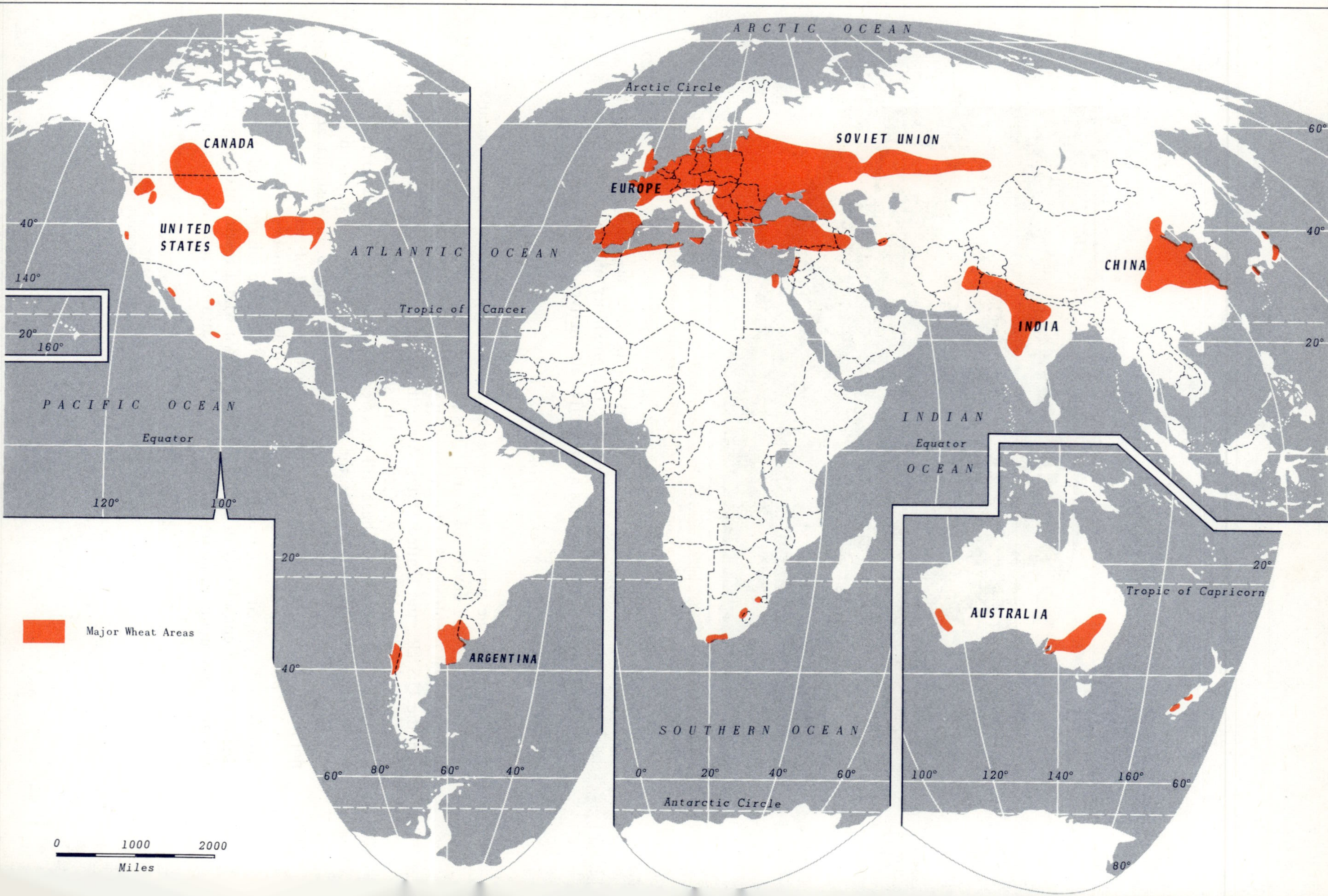

under snow to be harvested after the winter, mature in the shortest of growing seasons. But even wheat has its limits. Water, for example, is a friend of rice, but it can be an enemy of ripening wheat. There are places where wheat grows on as little as 10 inches of annual rainfall, but the rain must come during the early period of growth and then cease during the ripening period. Too much moisture, and too much heat, make the wheat vulnerable to diseases and the ravages of insects. Figure 7.5 shows wheat to be a higher-latitude crop than rice (and, in a few areas, higher-altitude, too). A comparison between Figs. 7.5 and 2.12 shows the versatility of wheat in terms of climatic conditions.

VARIETIES Wheat, like rice, grows in different varieties. In the case of rice the most common differentiation is between *lowland* rice, which has ample water and stands in flooded paddies, and *upland* rice, which must grow with less moisture on higher slopes. Wheat varieties are identified less on the basis of location of growth than on the seasonality of planting and harvesting. Thus *winter* wheat is sown during the fall, then lies dormant under the snows of winter, and ripens during the spring and early summer. *Spring* wheat is sown where the winter cold is so severe that the seedlings would be killed. Hence the seeds are planted as early in the spring as possible, and the crop matures and ripens by the middle of the summer or slightly later. This is the wheat that can be produced where growing seasons are quite short. Winter wheat is the most common variety cultivated, by about two to one (in terms of acreage) over spring wheat.

Varieties of rice and wheat are also tied to the way they are used for consumption. Thus there is *brown* rice, so called because the kernel retains its coat of nutritious, brown-colored, oily bran, and *polished* rice, whose bran coat has been removed (along with its valuable minerals and vitamins) to expose the familiar white grain. In the case of wheat, there is a distinction between *hard* wheat, grown in the drier zones and used mainly for bread, and *soft* wheat, a product of moister climates and used in pastry crusts. All this accurately suggests that wheat, like rice, is virtually all consumed directly by people (that is, it is not used for animal feed). In the Western world bread is made from it; in eastern Asia, noodles and dumplings.

Whereas rice is the crop of small paddies and individual attention, wheat is the crop of huge fields and mechanized production. True, wheat is grown in India and China as rice is, on small plots and with much individual attention; but in the Soviet Union, the world's leading producer, in the United States, in Canada, Australia, and Argentina wheat is grown on a large scale, with machines doing the seeding and harvesting. The last four countries mentioned are also the leading exporters of wheat; despite its enormous production, perhaps 30 percent of the world's total, the Soviet Union frequently must buy wheat from other countries. In a great effort to expand home production, the Soviets during and after the 1950s opened up some areas only marginally suitable for wheat growing (the "virgin lands" scheme under Khruschev's term of office was a prime example), and although there were some successes, a price was paid in yield variability. In some years the wheat did well, but in other years it did not. The 1972 harvest was a notable failure, and the U.S.S.R. was forced to buy large quantities of wheat from the United States and Canada.

The United States, which produces winter wheat in the Great Plains and in the Northwest and spring wheat in the Dakotas (as Fig. 7.5 shows, the spring wheat zone extends into the Canadian Plains), harvests about 15 percent of the world's wheat. It could produce even more. Most of the wheat is grown on commercial grain farms of several hundred acres, sometimes more than a thousand. Especially in the Great Plains, the whole operation has been mechanized to the point that farmers may do their own planting, but the harvesting is done by professionals who own huge combines for threshing the wheat. These professional harvesters work their way northward as the wheat ripens, starting in Texas in May and finishing up in North Dakota in September. Some even work the Canadian wheat farms.

For the wheat-growing countries outside Eurasia, this is big business. Wheat accounts for 6 percent of United States exports by value and for 12 to 14 percent in Canada (the two neighbors must compete for the same markets with their product). In Australia, where mechanization has gone so far that the output per

Wheat fields near Moscow, Idaho. The rolling hill country is clothed in vast stands of this grain. Similar sights can be seen in the Ukraine and Argentina's Pampa (although the land is generally somewhat flatter there), in the Great Plains, and in Australia.

man is about twice that in the United States, wheat exports are increasing in value and account for more than 10 percent of annual export revenues. Argentina, too, sells wheat on world markets, and its receipts amount to some 7 percent of export revenues per year. From these details it is clear that wheat trade is quite different from rice trade. Rice is mostly consumed in the country and maybe in the very village where it is produced. Probably less than 5 percent of annual world production is traded externally, and then this trade is almost exclusively within the South and East Asian rice-dependent region. Much of it goes by small boat down rivers and along coasts. Wheat, on the other hand, moves in giant cargo ships across oceans from distant producing areas to (mainly Eurasian) consumers. As much as 15 percent of world production of wheat in a given year may enter such international trade, in which Australia sells to Europe, the United States to the Soviet Union, Canada to China. These are just examples. All the producers compete to sell to the buyers. The United States *and* Canada sell wheat to India, Japan, and the U.S.S.R., among major purchasers.

If you look further into this business of international trade in grains, you will note some apparent contradictions. China, for instance, may have a deficit of rice and import wheat, yet China is listed as an *exporter* of rice as well! Such a situation reflects a governmental decision that food exports to certain countries are in the national interest, no matter what the situation is at home. The capacity to export to a friend or ally or to undercut a competitor on a nearby market is deemed to be essential for national strategy. When North Viet Nam's food supplies were reduced during the later stages of the Indochina war, China sent replacements, though China was itself in need of expensive grain imports. In another arena, when the Soviet Union found itself needing huge quantities of wheat in 1972, the United States was able to oblige, but there were calls in Congress for strings to be attached to the sales. It was argued that the United States should insist that the U.S.S.R.'s Jewish community receive better treatment than it allegedly had if United States wheat was to bail out Soviet planners, farmers, and consumers. Some observers believe that some Soviet concessions did indeed accompany the big wheat deal. Food and power are closely linked; food is a potent weapon.

Corn (Maize)

Corn, or maize, as it is called in other parts of the world, was domesticated by American Indian peoples in Middle or Andean America or in both and not known to other peoples until the sixteenth century. It did not take long for the crop to be tried elsewhere, once the Europeans brought it to Africa and Asia, and today corn grows in practically all parts of the world. It stands on subsistence plots in Indonesia, the Philippines, China, and India; on commercial grain farms in South Africa and Argentina; on farms from Turkey, Spain, and Brazil to the countries where it may have been cultivated first, Mexico, Guatemala, Colombia, and Peru. The world's leading area of corn production is still in the Americas, the famous United States Corn Belt (Fig. 7.6).

FIG. 7.6 World areas of corn (maize) cultivation.

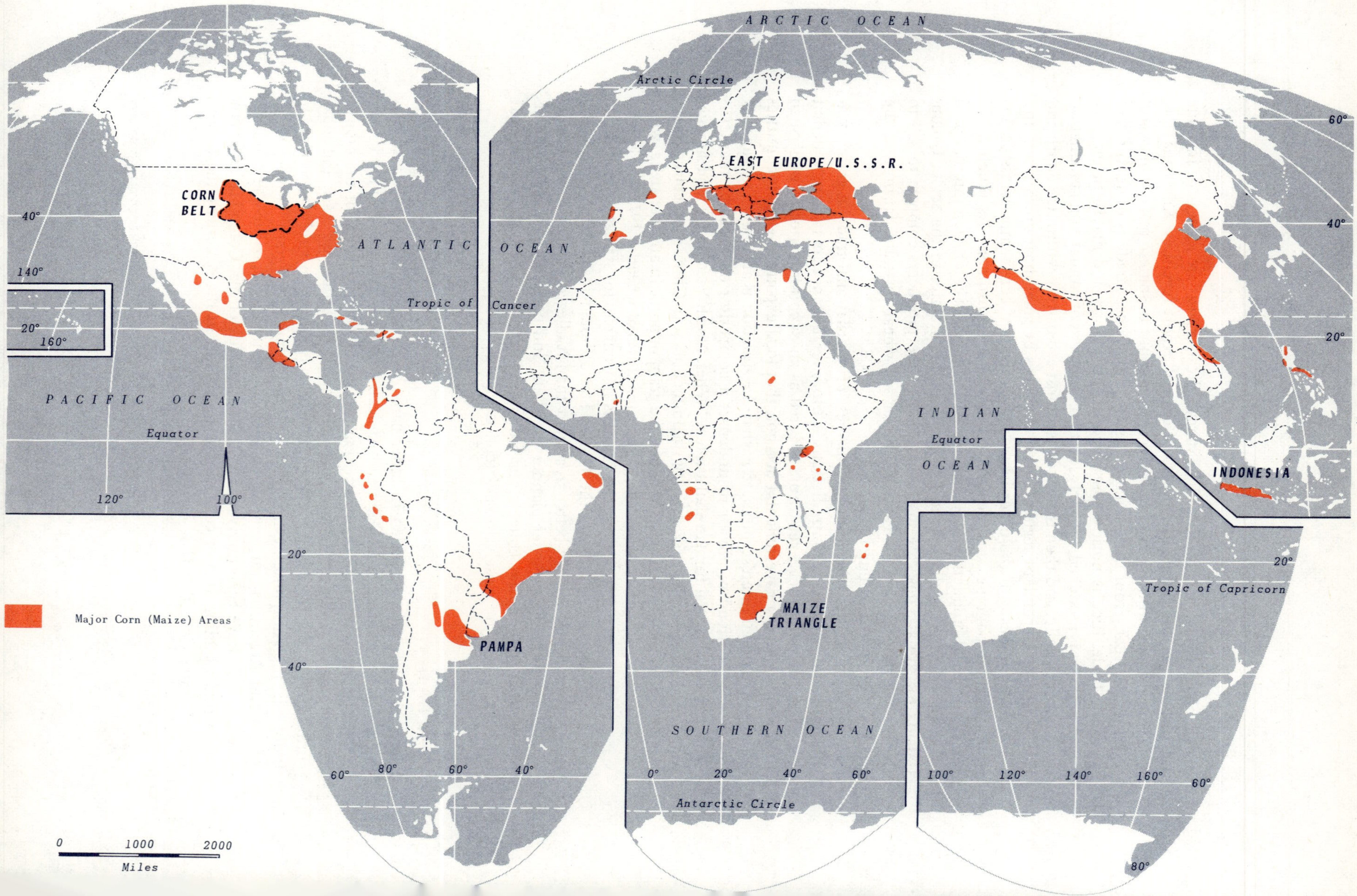

This underscores a quality of corn as a crop. It can be grown under many different soil and climate conditions. Corn can grow in dry as well as moist climes, in valleys and on high mountain slopes, in growing seasons of less than three months. Our map shows the principal areas of cultivation, but corn can be found as a subsistence crop in large areas of Africa, Middle and South America, and Eurasia as well. Corn was, in fact, the early "subsistence" crop of the white man as the North American interior was entered and occupied. Later, scientists developed more productive strains of corn, and one of these, hybrid corn, proved a revolutionary advance in corn cultivation and production. The soils of the United States Corn Belt give yields twice as high as those of Europe or South Africa because of the productivity of the new strains.

The United States produces more than half of all the corn annually harvested, but unlike wheat and rice this corn is used primarily to feed livestock, mainly cattle and hogs, and poultry. As much as 90 percent of the corn harvest consumed domestically is used this way; the remainder goes into corn bread, breakfast cereals, cornmeal, and margarine and related products. Some corn is consumed as a vegetable. In general, however, people tend to eat rice and wheat in preference to corn, and corn is consumed indirectly through the livestock and its products.

Corn is produced in the United States in areas other than the Corn Belt, as Fig. 7.6 shows. Natural conditions are most favorable in the humid-continental climate of the Corn Belt with its prairie and chernozem soils and rather level land, but corn is also grown on farms in the South and East. Those acreages are principally devoted to the feeding of animals. The Corn Belt itself is not a region devoted exclusively to corn growing. Wheat, oats, soybeans, and other crops are grown in addition to corn.

International trade in corn focuses upon Europe and Japan. From the United States, Argentina, and South Africa corn is exported to the United Kingdom, West Germany, the Netherlands, and Japan. There is also some local trade. The United States exports some corn to Mexico, South Africa supplies Botswana, Lesotho, Namibia (South West Africa), and Rhodesia when necessary. But very little corn enters the world market, a smaller percentage even than rice.

Other Cereals

Although rice, wheat, and corn are the modern world's leading cereal grains, other grain crops exist and may have local importance. *Barley*, for example, is grown on all the continents, and it is a crop of even greater versatility than corn. Barley can mature under a shorter growing season, on poorer soils, and with less water than any other cereal. It serves as livestock feed, but is also used in the manufacture of liquors and beer. The crop is grown substantially in the United States-Canadian Plains, in central Mexico and interior, high-altitude Peru, in Argentina, in Europe, the U.S.S.R., Southwest Asia, India, and East Asia. The United States, China, U.S.S.R., and Europe produce the bulk of the annual harvest, and the import pattern is not unlike that of corn, with Western European countries and Japan in the lead. *Millet* and *sorghum* are grains grown by subsistence farmers from West Africa to India to China, but also cultivated in the U.S.S.R. (millet) and the United States (sorghum). These grains can grow in the most adverse environments and survive with very little attention. Millet is like a grassy wheat; sorghum resembles a variety of corn. In the United States, sorghum is grown as livestock feed, but elsewhere these crops are grown primarily for human consumption. *Rye* and *oats* also are grown for human as well as animal food. Rye is second only to wheat as a bread grain in Europe and the Soviet Union, and oats is second to corn as a fodder crop.

ROOT CROPS

Two of the three major grain crops we have discussed were domesticated in the Old World. The most important root crop, the *white potato,* was being grown in the Americas when the Europeans appeared here. It was cultivated in the high-altitude valleys of Middle America and the Andes, and the white men took it back to Europe, where it became a staple from Ireland to Russia. The potato thrives in the cool, rather moist conditions prevailing in the North European Lowland, a region that extends from Britain through the Netherlands, Belgium, and northern France, the two Germanys, Poland, and Russia. In those countries the potato brought a new and dependable supply of food, and the population growth that Europe experienced during the eighteenth and nineteenth centuries was sustained by this agricultural contribution from America. The potato provides starches, proteins, iron, and some vitamins; in millions of European households the standard dinner includes a vegetable, some meat, poultry, or fish, and a hefty helping of potatoes covered with a gravy or sauce. Livestock, too, are fed on potatoes, and the crop also yields alcohol, starch, and other raw materials. In terms of overall productive capacity, the potato is the only crop that can match rice, the most nutritious of the cereals.

The potato spread also to other parts of the world (ironically it came to North America *via* Europe!), but Europe and the U.S.S.R. produce and consume the great majority of the annual harvest (the Soviets contributing perhaps as much as one-third). The crop continues to gain. In India, China, Japan, and Middle and South America consumption is rising as the hardy potato, given some care and attention, brings substantial rewards.

Another American agricultural invention is the *sweet potato,* which was domesticated in South America's Andean region and, like the white potato, was carried to many parts of the world. Similar to the sweet potato is the *yam,* a root crop first cultivated in Southeast Asia and grown in many tropical and near-tropical areas today. The yam has a higher starch content and contains somewhat less sugar than the sweet potato. It supplements diets in India and China, among other countries.

Still another root crop native to America is *cassava* (also called *manioc*), a versatile plant that grows in many different environments with little attention.

Cassava grows in the heat and humidity of the interior Amazon Basin, in the drier country of the Central African Republic, on the monsoon coast of India. Brazil remains the leading producer, with Indonesia a distant second, but substantial cassava production also occurs in Nigeria and other West African countries, tropical Africa, and Southeast Asia. Cassava makes a good substitute where climatic conditions prevent the cultivation and maturing of a grain crop. After cassava is harvested it is usually dried and consumed in the form of a bread or porridge.

In terms of their overall role in feeding the world's people, the root crops are no match for the cereals. Still, they play an important role, the potato as a staple in the diets of many millions of Europeans and Soviets and cassava as the basis of subsistence for Africans, Asians, and South Americans.

OTHER FOOD CROPS

Diets in which cereals or root crops form staples are supplemented by other specialized products of agriculture. Spices, sugar, beverages such as tea and coffee, fruits, and vegetables are but a few among many products that diversify and accompany mealtimes in various parts of the world. Some of the combinations can be regionally identified quite easily, such as Indian curries, Indonesian *rijsttafel* (literally, "rice table," a variety of dishes served on beds of rice), Mexican tortillas, and so on. Meats, poultry, and fish bring further balance to many diets.

Fruits and Vegetables

Wild fruits have been used to diversify diets for many thousands of years, but the intensive cultivation of citrus fruits, grapes, apples, and the like, is a rather modern phenomenon. Orange groves, vineyards, and apple orchards reflect a higher-than-average standard of living, for farmland so used cannot be given over to staples. The distribution of the significant fruit-producing areas, and the location of the commercial vegetable gardens reveal a close association with the markets of developed countries. This does not mean that all fruit-growing areas lie in the Western world. Citrus fruits, for example, are grown in such widely dispersed places as Brazil, Moçambique, Malaysia, and Indonesia, as well as California and Florida. What it does mean is that the growing regions must be positioned to take advantage of the lowest possible transport costs to the wealthier markets. The vegetable farms (other than subsistence plots scattered all over the inhabited world) also lie within easiest reach of large markets, in this case, urban markets. Although large cities in the developing world also have nearby vegetable gardens, truck farming (as this kind of agriculture is appropriately called) is typical especially of the areas near European and American cities.

Sugar

All kinds of justifications exist for focusing briefly on sugar alone. An enormous quantity of this commodity is consumed in our diets. More than 15 percent of the calories consumed by Western peoples are in the form of sugar. Although the figure is somewhat lower elsewhere, the total amount of sugar consumed annually is still huge. Production has been between 70 and 75 million metric tons in recent years, not much less than half the annual rice harvest. Unlike rice, though, there really is no substitute for sugar. And sugar produces more energy than any cereal.

Sugar comes from two quite different plants, sugarcane and the sugar beet. Pictures of sugarcane harvesting usually show rows of people, bent over, chopping the cane by hand. Why should that be so, when much sugarcane is grown on big, otherwise modern, plantations? The fact is that sugarcane does not lend itself well to mechanical harvesting, and many hands are needed for the task. It was sugarcane, planted by the white man in the Caribbean area, that generated the first flow of slave labor from Africa, and other labor migrations later. Europe was the prime market for this lucrative commodity, but eventually the Europeans produced a competitor, the sugar beet. During the nineteenth century the production of sugar from sugar beets began to overtake that from cane, but, as current figures indicate, sugarcane regained its old position during the twentieth century and now accounts for more than 70 percent of the commercial world sugar crop each year (Fig. 7.7). Today Europe and the Soviet Union produce more than 25 percent of the world harvest of sugar from beets.

The production of sugar has expanded enormously during the past half century, especially cane sugar. Whole national economies (such as that of Cuba) have come to depend upon revenues from the sugar harvest, and countries such as India and Brazil count heavily on such income as well. But the markets are in the technologically developed countries of the middle latitudes, where local farmers grow sugar beets but are unable to fulfill the demand. To protect the domestic sugar farmers and to bring some order to the international flow of sugar to world's markets, it has become necessary to fix tariffs and quotas. This has in turn produced almost continuous international wrangling. The circumstances are such that the major importers can threaten the very survival of the economies of some of the producers. When the political disagreement between the United States and Castro Cuba arose, the United States cut off its imports of Cuban sugar. Though the Cuban export trade was cushioned by alternatives in the Soviet sphere and Canada, this was a staggering blow. As the map suggests, Cuba's most advantageous market is not half a world away in Eurasia, but next door in North America.

Luxury Crops

So far we have discussed crops that are essential to calorie intake and diets, producers of energy and bulk, ensurers of survival. But man needs and wants diversity as well, and many acres of farmland that could sustain subsistence food crops are given over to what we may debatably call the "luxury crops." Many of these exist, but tea and coffee rank among the most important. *Coffee,* first domesticated in Northeast Africa (in the region of present-day Ethiopia), today thrives in Middle and South America. Brazil, Colombia, Mexico, the Central American republics, and

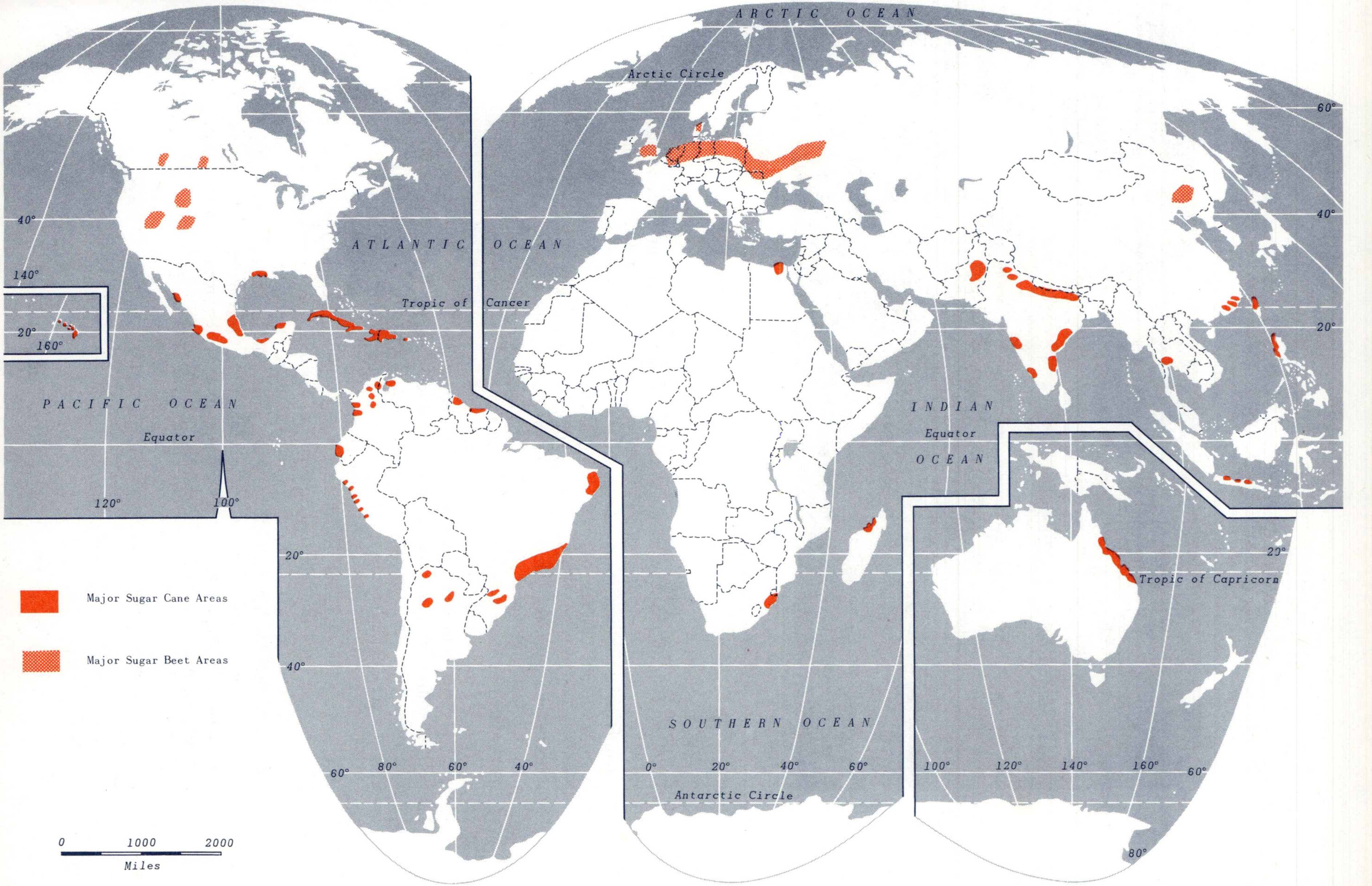

FIG. 7.7 World areas of sugarcane and sugar-beet cultivation.

The cultivation and harvesting of sugar does not lend itself well to mechanization. In Cuba at harvesttime, all hands are called out to the fields to help with the job, a vital one to the national economy. The photograph shows a group of workers cutting and stacking the cane. Both women and men do the work.

Caribbean countries together produce and export about 70 percent of the world's annual harvest, from both huge coffee *fazendas* such as those of Brazil and small, carefully hand-cultivated farms. Africa still produces coffee, in Angola, Ethiopia, and countries of the west coast for example, and there are coffee farms even in Indonesia and India. But the heart of the coffee industry lies in America, and the United States buys more than half of all the coffee sold on the world markets annually. Western Europe imports most of the remainder. Compared to coffee, *tea* is consumed in lesser quantities by weight and in greater amounts in the areas where it is grown: India, China, Sri Lanka (Ceylon), and Japan. Whereas coffee is the beverage cultivated and consumed dominantly in the Americas, tea is the Eurasian equivalent: it goes from the Asian producing areas to the United Kingdom and Europe. Tea also is a rather recent addition to Western diets. It was grown in China perhaps 2000 years ago, but became popular in Europe only during the nineteenth century. The colonial powers, chiefly the British, established enormous tea plantations in Asia, and the full-scale flow to the European markets began. It has continued after the decolonization period. Another kind of beverage crop is *cacao* (the beverage itself is *cocoa*), which went the reverse route from coffee. First grown in tropical South America, it was tried successfully in Africa, and today West African countries, mainly Ghana, Nigeria, and Ivory Coast, produce the bulk of it. Cacao also is used for making chocolate, a product very popular in the major importing regions, Western Europe and the United States.

An ancient, time-honored luxury crop is the *grape,* the source of the world's wines. The grape was cultivated in France during Roman times; France today produces a harvest amounting to one-fifth to one-quarter of the world total, depending on the conditions of maturation and harvest. Italy is a close competitor in terms of volume, and the harvests of Spain and the United States (California and New York) are increasing in volume. Substantial production also takes place in Portugal and Algeria; in Turkey, Greece and Eastern European countries, and the U.S.S.R.; in Chile and Argentina; in the Cape Province of South Africa; and in Australia. If this set of countries reminds you of the distribution pattern of Mediterranean climatic regions discussed in Chapter Two that is a relevant observation. Most (though not all) of the vineyards lie in or near those areas.

INDUSTRIAL CROPS

Among the crops that produce industrial raw materials, too, it is possible to differentiate between necessities and luxuries. *Tobacco,* for example, would fit right into the category we just discussed. It occupies enormous acreages of soil in many parts of the world, notably in the technologically developed nations where much of the consumption also takes place. A great number of varieties of tobacco can be cultivated under very diverse conditions of soil and climate, and so tobacco grows above 50 degrees North in Russia, in the St. Lawrence valley in Canada, in the southeastern United States, and in Brazil on the very equator. Such cultivation the world over is oriented toward local markets, for smoking is a worldwide habit. But there is also a great deal of international trade, for tobacco varieties can be combined to produce desired blends. Thus the United States, the world's leading producer, exports to Western Europe and imports from Turkey, Southeast Europe, and Caribbean America. China and India, on the other hand, are the next most productive countries but consume virtually all their home-grown tobacco at home.

Fibers

More essential to our existence, though less so in these days of synthetics than in the past, are the plant fibers. This group of crops includes cotton, jute, flax, hemp, sisal, and several other fibers, some of them cultivated for thousands of years. Among these, cotton stands out, accounting by weight for more production than all other fibers combined.

Cotton is grown in many areas of the world, but the major areas of production are relatively few: the Cotton Belt of the United States, Northeast China, Soviet Central Asia, and central peninsular India. But cotton fields of significance also exist in Mexico, in eastern and northeastern Brazil, in Egypt's Nile Delta, and in the Punjab region shared by Pakistan and India. Literally dozens of other cotton-growing areas could be identified, and in those countries the product is grown for the domestic market, as it is in China, a leading producer.

Cotton cultivation expanded greatly during the nineteenth century when the industrial revolution produced machines for cotton ginning, spinning, and weaving that multiplied productive capacity, brought prices down, and put cotton goods in reach of the masses. As they did with tea and sugar, the colonial powers laid out large-scale cotton plantations, sometimes under irrigation (for example, the famed Gezira scheme, in the triangle between the White and the Blue Nile in the Sudan). Even today, with the competition from synthetics such as nylon and rayon, most of the cotton fields' production still finds buyers. Cotton goods made at home are cheaper than synthetics imported from overseas. And the developed countries have not stopped buying cotton either. Japan, the United Kingdom, and Western Europe still import large quantities of the fiber. But in the case of cotton the former colonial countries have to contend with competition from the United States as an exporter. The United States sends cotton to Japan and Western Europe. Egypt's cotton harvest also goes to Europe, and such countries as Sudan, Uganda, and Turkey also compete for that market.

Cotton has spread all over the world, but *jute*, a much rougher fiber used for making burlap bags, underlays for carpets and similar products, is still grown near the place where it was first cultivated, in Northeast India and Bangladesh. There, in the moist, warm delta region of the Ganges and Brahmaputra river systems, jute thrives. Some minor production of jute also occurs in scattered areas of Southeast Asia and a few places in Brazil's Amazon Basin, but the great bulk, perhaps 90 percent, still comes from Bengal and Bangladesh. Bangladesh exports almost all of its jute production, but India uses most of its own harvest at home, and even imports some from its neighbor.

The other fibers, such as *flax*, used in especially fine textiles, *hemp*, used in string and thread, *kapok*, with which many pillows and life jackets are filled, and *sisal*, which is made into rough twine, find markets where the major fibers sell the best, in Europe, Japan, and the United States. Flax and hemp are grown in the Soviet Union and Europe primarily, and their international movement is small. The production of kapok, sisal, henequen, and some of the other minor

---

COTTON STAPLE

Cotton is often classified as short, medium, or long *staple*. This refers to the average length of the actual lint, or fiber, as it comes from the cotton bolls. To determine the class of the cotton, a set of fibers is held between thumb and forefinger and pulled for straightening. A selection is then taken for measuring. Short-staple cotton has fibers less than an inch long; long-staple fibers are in excess of 1.2 inches approximately. Length of the fiber is important because it plays a role in the strength of the thread that can be made from the cotton. Long-staple cotton is the most highly prized.

Cotton grows successfully in many parts of the world, and the crop can be harvested mechanically, as is being done here in the Tajik Republic, U.S.S.R.

fibers may not seem large, but to the low-latitude countries where they are produced their importance may be considerable. Sisal, for instance, brings in more than 10 percent of all of Tanzania's export revenues.

Rubber

Initially, rubber was a substance not cultivated but gathered, collected from rubber-producing trees that stood (among many other tree species) in equatorial rain forests, mainly the forests of the Amazon Basin. Those were the days, around the turn of the present century, when the town of Manaus on the Amazon River experienced an incredible rubber boom, and the river itself was an artery carrying one of the world's most desired commodities.

The Amazonian rubber boom was rather short-lived, however. Ways were sought to create rubber-tree plantations, where every tree, not just some among many, would produce rubber, where trees could be given attention, and where collecting the rubber would be more efficient and easier. Seedlings of Brazilian rubber trees were planted elsewhere, and in Southeast Asia they did especially well, so well, in fact, that two decades after the height of the Amazonian rubber boom, just under 90 percent came from the new plantations in Malaya, the (then) Netherlands East Indies, and neighboring countries, and no longer from Brazil.

As time went on, more and more uses for rubber were found, and the demand grew continuously. The advent of the automobile was an enormous boost, and still today most of the rubber produced is used up in vehicle tires. Then the Second World War forcefully brought the need for alternative sources for rubber to United States attention, for Japan occupied much of Southeast Asia. This stimulated the production of synthetic rubber, and although plantation rubber came back to some extent after the war, synthetic rubber has been in the lead ever since. In 1970, world production totaled just under 8 million metric tons, 5 million of it synthetic (nearly half of that produced in the United States) and less than 3 million natural (90 percent of which came from Southeast Asia).

One question that presents itself is this: why were those rubber plantations placed in Southeast Asia and not in the Amazon Basin itself where the trees were known to grow or in the Congo Basin, also an early source of wild natural rubber? The answer has less to do with trees than with people. In the Southeast Asian countries a large supply of cheap labor was available, larger than in Amazon South America or Congo Africa, and the rubber tapping requires the constant attention of a lot of workers. Indeed, some rubber plantations *were* established in Africa, namely in Liberia, on a large scale by the Firestone Company. And lately efforts have been made to try the plantation system along the Amazon in the heart of Brazil.

Rubber is needed in enormous amounts where there are many millions of cars on the roads. The United States is the world's greatest consumer of natural rubber, the largest producer *and* consumer of synthetic rubber, far ahead of second-place Western Europe.

ANIMAL HUSBANDRY

Earlier in this chapter we dwelled briefly on nomadic pastoralism as a means of subsistence. We now look at the keeping of livestock in a very different way. Commercial animal husbandry is a far cry from nomadic herding, but it may involve the same animals. Cattle are the mainstay of life for the Masai of East Africa and the rancher of Argentina. Sheep sustain nomadic people in Soviet Asia and wool exporters in New Zealand. The differences lie in the nature and scale of operations and the disposition of the livestock's products.

Compared with the cereals and sugar, the products of domesticated animals contribute in only a minor way toward the fulfillment of man's energy requirement (dietary diversification is another matter). One acre, planted to a cereal such as wheat, will generate three times more calories than when it is allotted to livestock for meat and dairy products. But milk and its products, meat, eggs, and animal fats do contribute importantly to balanced diets and general nutrition and, like plant crops, domesticated animals also generate industrial raw materials including wool, leather, and specialized products such as feathers, ivory, and even hair.

Dairying

The dairy industry is a bit like the vegetable-garden industry. It is oriented primarily toward the urban areas and other zones of dense population in the more developed, richer countries. The primary product, of course, is milk, and milk and cream are perishable. The dairy industry began before refrigeration techniques became as efficient and widespread as they are today, and so the dairy farmer's optimum location was always as near to the market as possible, that is, near cities and towns. Dairying as an industry began in Europe, where rich, green pastures and cattle feed from other sources were available in abundance and where markets grew by leaps and bounds. There the distribution of dairy cattle shows a marked concentration from northern and northwestern France and the United Kingdom through Belgium and the Netherlands, West Germany, and Denmark into southern Sweden.

In the United States and Canada, the dairying industry similarly shows a marked orientation and proximity to the core area of dense population eastward from Minnesota to the Atlantic coast. This region is sometimes called the Dairy Belt, and traveling through it you would see not only numerous cattle in grassland pastures, but also fields of hay, grown to serve as cattle feed during the winter season. You would note that there are also other kinds of farming in the Dairy Belt, but near cities and towns the dairy farm prevails. Stop even in small towns and check the phone book, and you will find that there is at least one, perhaps several local dairies. Make a call and discover that those dairies are supplied by nearby farmers on a regular, contracted basis. Then stop in a supermarket and check the price of milk. It will confirm that dairy products are luxury items and remind you that it takes a prosperous economy to sustain such an industry.

When we check an atlas map showing the world distribution of the dairy industry, it seems at first to contradict what we have just reported. Far away from the European and North American markets, New Zealand and Australia generate dairy products and export them in quantity. How can those distant industries compete? There are several reasons. First, despite its large production, Europe requires more dairy products than it can make. The United Kingdom especially, but France and other mainland countries, too, need imports of dairy products. Second, refrigerator ships make it possible to send perishables over long distances. Third, although milk may not be a practical export, milk can be used to manufacture other dairy products (butter, cheese, powdered milk) that can withstand distant shipment and delays to market. And finally, we should not forget that Australia and New Zealand have domestic markets comparable in terms of purchasing power (although not size) to the United States and Europe.

All this should not suggest that the consumption of dairy products is a monopoly of the wealthy mid-latitude countries. The dairy industry is nowhere as well developed as it is there, but in other parts of the world people also consume milk when they can obtain it, and dairy belts have sprung up around major urban centers from Santiago to Seoul. Milk is a prominent item in Masai diets; in India, where there are so many cattle, milk is about the only substantial benefit from that overpopulation. Milk is a vital dietary ingredient, especially for young children, and hundreds of millions of people do not consume nearly enough of it because it is too expensive to obtain. Goat's milk serves as a substitute in some areas, and condensed and powdered milk stand on the shelves of village stores from Bolivia to Botswana to Burma. But the ample supply to which we are accustomed is an exception and a privilege in this world.

Beef Cattle

The map of the world distribution of cattle (Fig. 7.8) might give the misleading impression that India, with its enormous population, is best endowed in this respect. India has millions of cattle, but the religious prohibition on eating meat prevents the use of this big herd for consumption. Other developing countries with no such prohibitions still do not consume much meat. Beef is an expensive commodity. To import meat would not be economically feasible, and to allot acreages to pasture and fodder crops when rice or other grain staples are urgently needed would be impractical. And so beef is another item of comparative luxury. About half the people on this earth in the mid-1970s still eat very little meat, and when they do it is meat of goats or some other animal other than cattle, or perhaps fish on occasion.

Our map indicates that cattle are widely distributed around the world (it represents both dairy herds and beef cattle). The population is thin in the very high and very low latitudes. The former are too cold (not just for cattle, but also for the cultivation of fodder crops), and the latter's hot, tropical conditions are not conducive to the animals or to the development of pastures. Beef cattle do best in the middle latitudes, and the map shows major concentrations in the central United States, in Europe and the Soviet Union, in Argentina and Uruguay, and in Australia and New Zealand. But there are substantial herds also in Southeast Brazil, in West, South, and Northeast Africa, and in eastern China.

As always, it is important to differentiate between production and export figures. In terms of beef and veal production (veal is the meat of a calf normally not more than 12 weeks old), the United States is far ahead of any other country, annually producing about one-third of the world total. Here the Corn Belt plays its role, for there the feed is grown on which millions of beef cattle are fattened. This production is virtually all consumed at home. The smaller farms, where the farmer mixes the raising of some beef cattle with the cultivation of several grain and other crops (*diversified* farming, as it is called) may send their animals to nearby markets. The larger ranches of the Great Plains, where beef-cattle raising is a huge operation, rail or truck cattle to centers such as Kansas City, from where the meat is distributed to the national market. By world standards, American meat consumption is enormous, though the *per capita* intake of meat is somewhat greater still in one or two other countries.

When it comes to exports, it is Argentina in the lead, with about one-third by weight of all the meat traded internationally each year. The grasslands and favorable climate of the Argentine Pampa, with its capacity to produce alfalfa as a fodder crop, have come to sustain a very large herd. Argentina's human population is not so substantial (27 million), and so the country has a major surplus of beef to export. Those exports have been going to the United Kingdom, several mainland European countries, and even to the United States. Behind Argentina, Australia ranks as the next most productive beef-exporting country. Australia also has a great deal of pastureland, a large cattle herd, and a small domestic market. New Zealand also is a beef-exporting country, and for many years meat from Australia and New

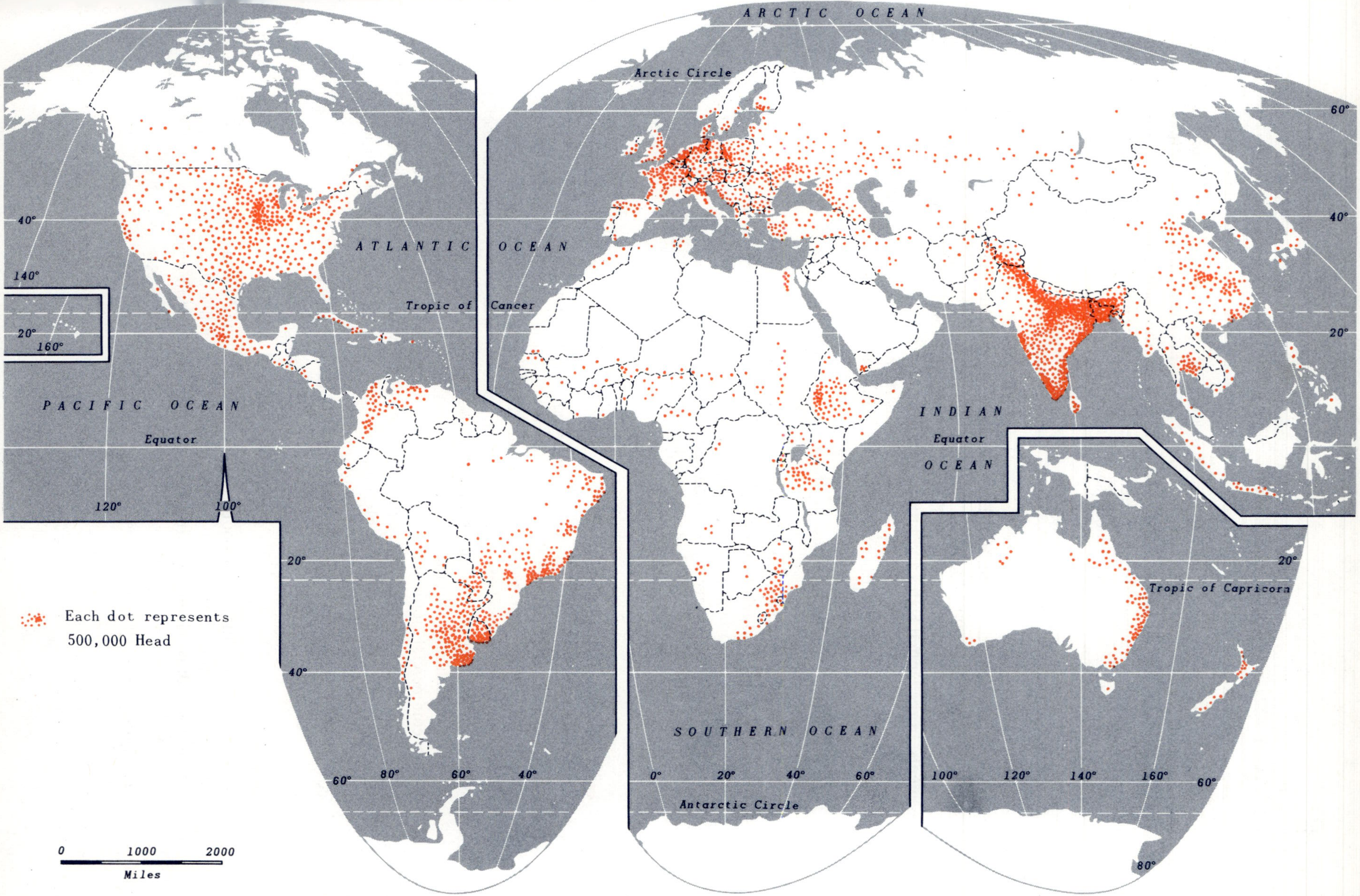

FIG. 7.8 Distribution of cattle in the world.

Zealand has been going to the United Kingdom, the natural trading partner in Europe. In early 1973 it was too early to tell what effect the entry of the United Kingdom into the Common Market would have on the trade connections with sterling-bloc, Commonwealth countries. (The sterling-bloc countries are those whose currencies are connected to the British pound sterling, thus facilitating trade and exchange.)

Hogs

Hogs, source of pork, bacon, and lard, are a most useful alternative where meat sources are hard to come by, and even where they are not. How many Americans start their day with bacon and eggs? Hogs can survive in widely different environments. In the United States and Europe they are raised commercially, and fed corn and potato-derived feed. But in many other parts of the world they are simply scavengers, surviving on leftovers and garbage. Thousands of Chinese villages would be incomplete without their hog population (China is probably the world's leading hog producer), and hogs abound in Brazil, Mexico, and other South and Middle American countries.

Hogs (or pigs, or swine) could serve even more effectively to diversify diets if religious taboos (chiefly Moslem) did not exclude them from large areas of the world. There are no hogs in any number in North Africa and Southwest Asia, or in Southeast Asia, and very few in India and Africa. Almost anomalously, the major production (other than in China) is in the already meat-rich Western countries, which trade among themselves in ham, pork, bacon, and lard.

Sheep

Increasingly, sheep are raised not only for wool, but also for their meat, mutton. Sheep are able to live off land that could not sustain cattle, and subsistence farmers raise them not only for meat and fiber but also for milk (and the cheese made from it), and even for skin, used for leather. Later we will view sheep from the standpoint of their production of industrial raw materials, but we tend to think of these animals as wool producers alone, which is not appropriate. Crossbreeding of the commercial herds has helped diversify the production from sheep. For example, in England sheep were raised principally for mutton until the breed was mixed with the wool-producing Merino. Now the new variety not only has good meat, but also can be profitably sheared during its lifetime for its much-improved coat of wool. Figure 7.9 shows the world distribution of sheep.

Goats

No livestock can survive on poorer land than goats: they manage to climb the steepest slopes after the smallest tuft of vegetation; they live in large numbers in the most arid of populated country. From Africa to southern Europe to the Middle

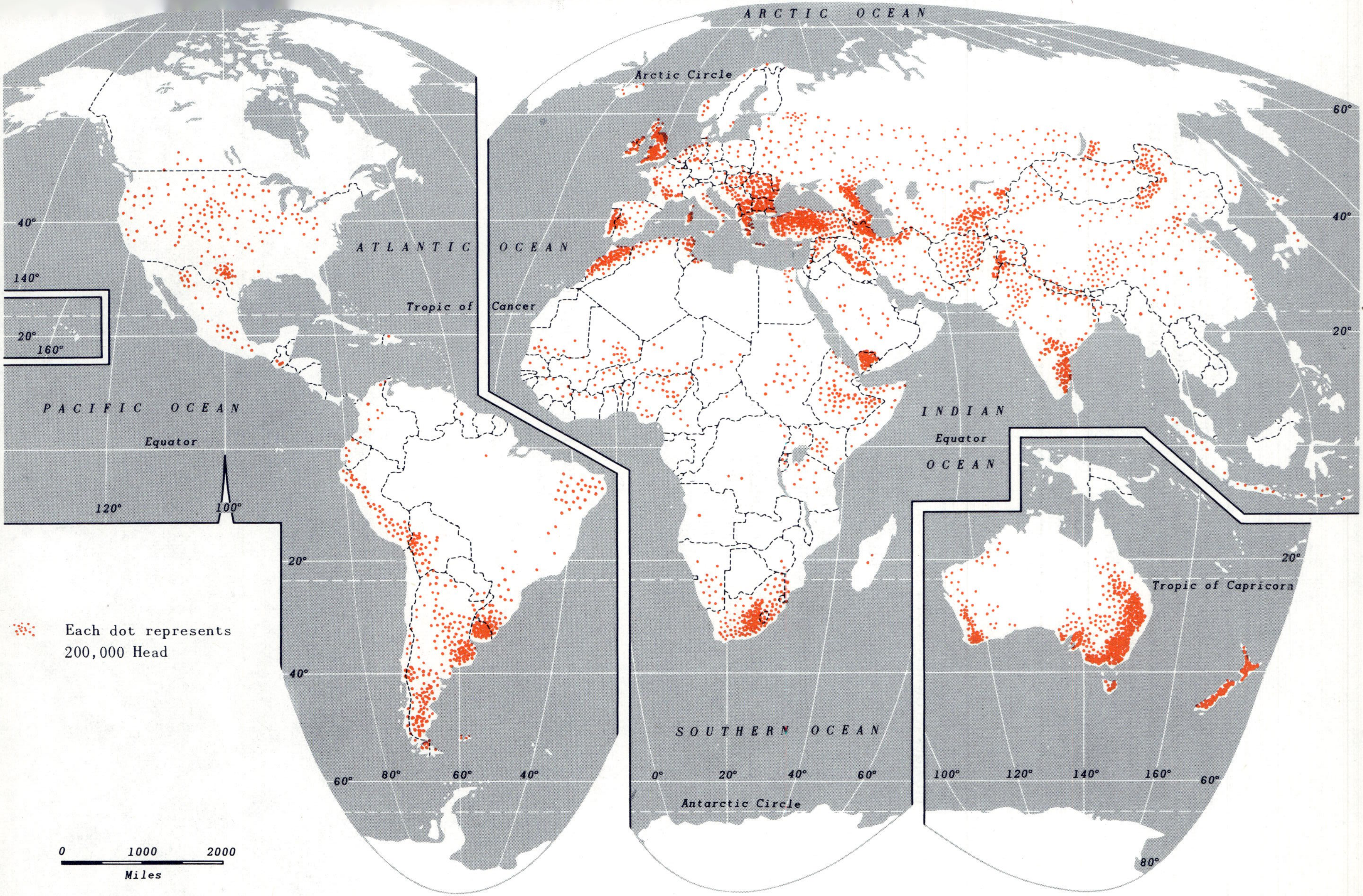

FIG. 7.9 Distribution of sheep in the world.

East to southern Asia the goat provides meat, milk, skin, and even hair to its owners. Goats constitute the final alternative for farmers in densely populated rural areas, where there is little grazing land to be set aside for livestock and where poverty prohibits purchasing meat. And goats even provide valuable industrial raw materials. Specialized breeds produce the ingredients of which mohair coats and cashmere sweaters are made. If there is any negative aspect to the omnipresence of goats in poorer areas, it is that by their numbers and their persistence in search of food they do contribute to soil erosion and land denudation.

Poultry

Poultry may well be the oldest of all livestock domesticated by man, and it, too, provides its owners with a set of valuable commodities in return for little attention or land. Chickens live as scavengers but can also be fed systematically. In terms of weight, they generate the best rate of return per pound of feed expended. They produce meat and eggs, crucial in diets where proteins are badly lacking.

In the technologically advanced countries, the poultry industry is as thoroughly organized as that of cattle or hogs, and the local market consumes enormous quantities of poultry products. Breeding has created specialized strains, some noted for their eggs, others for the meat. Such experimental breeding has also been aimed at better meat production in other poultry such as turkeys and ducks. In the developing world these refinements are not enjoyed. Still, it is difficult to imagine an African village without its share of a few goats and a flock of chickens, livestock that demands very little and returns a great deal.

INDUSTRIAL PRODUCTS

Man keeps livestock principally to produce food, but, as in the case of plant crops, some important industrial products also come from this source. One contribution is not in the form of a tangible product at all, but in the form of power. Livestock is used in many parts of the world for pulling plows, wagons, and other equipment,

It would be difficult to overestimate the importance of the water buffalo as a beast of all sorts of burdens in Asia. On a farm in Cambodia, two water buffalo and a plow turn the soil prior to the planting of a new rice crop, doing the work of dozens of people.

for carrying people and commodities, and various additional kinds of work. The horse is the most widely used draft animal, but mules, donkeys, buffalos, oxen, camels, even elephants perform work tasks for people. These animals may be adapted to function especially well in particular environments, as the water buffalo does in the rice paddies of Asia and the camel in the deserts of Africa and Arabia.

When it comes to industrial raw materials produced by livestock, we find that the range of these goods is not nearly as large as that of the foodstuffs they make. Hides and skins are valued products of livestock in subsistence as well as commercial economies (they are, for example, significant exports of Argentina and valuable trading items for Saharan nomadic peoples). Ostrich feathers, karakul (a breed of sheep) pelts, certain kinds of bone, and hair (mainly from goats) are specialized and localized livestock products. But in terms of world importance, only two items, wool and silk, stand out prominently.

Wool

Notwithstanding the growth in the use of synthetic fibers and the increasing use of cotton, wool continues to be a popular product. Sweaters, coats, blankets, and other wool textiles are in demand in the cold-winter climes of North America, Europe, and the Soviet Union, but the sheep that produce the wool are raised in places far away from those markets. The great sheep herds roam the ranges of Australia, New Zealand, Argentina, and South Africa. The wool produced there is bought by the textile makers of the United Kingdom, Western Europe, Japan, the United States, and the Soviet Union, by now a familiar set of importers. The United States and the Soviet Union both produce a considerable amount of wool themselves, but not enough to export any significant amounts; their home markets are too great. The Southern Hemisphere countries that produce the wool for export all have relatively small populations, a consequently small domestic market, and the sort of climate that does not call for warm, woolen garments.

Sheep are widely distributed across the world, though, and wool as well as meat production is rising also in countries that do not export those products, for example, China (Fig. 7.9). As we noted earlier, sheep are hardy animals that can live relatively well where cattle would starve, and subsistence farmers in South America, Africa, and Asia benefit from this versatility and the animal's productivity of fiber, meat, and milk and by-products.

Silk

Somehow it does not seem appropriate to classify a silkworm as livestock, but it is an animal nevertheless, an animal that produces a fiber and in this case, a high-priced one. Silk was an ancient product of East Asia, and across the Asian continent it made its way via the "Silk Route" to Southwest Asian markets and into Europe. Later it became a prize in the European penetration of China. And still today, Japan and China produce the bulk of the world's raw silk, harvested from silkworm colonies tended carefully by hand as the Chinese and Japanese farmer can. An enormous amount of labor goes into the production of silk; almost none is produced in the United States and very little in Europe, though some comes from France.

Raw silk must be converted into the finished product, and this *is* done in the United States, Western Europe, and the Soviet Union. Japan manufactures more of its own silk materials now, and France also makes the textile, but the United States continues to be the major manufacturer. The silk market over the years has had its fluctuations, and it has been highly speculative. Silk garments such as kimonos bring high prices, but the market is limited and the demand greatly variable.

What might be our chief conclusion from our overview of world agriculture? A prominent one surely must be that the range and variety of products with which we are familiar is an anomaly in this world of simple diets. A worldwide network of commerce in farm products is oriented to that one-fifth of the world's population living in the United States and Canada, Europe, and the U.S.S.R., and not everyone even in those areas benefits from the flow: there are malnourished and poorly dressed people from Alabama to Armenia. The array of produce we casually pass by in an American supermarket multiplies by a hundred times what the average peasant in Peru, in Brazil, in Bangladesh, in China ever sees, much less gets to choose from. We live in an island of prosperity and diversity and have mobilized a whole world to our sustenance and comfort.

CHAPTER EIGHT

Industrialization and Development

EIGHT

The title of this chapter might suggest that industrial growth and national development are synonymous. Certainly the "developed" countries have the big industries, and industrialization commensurate with state resources, capacities, and needs is an essential ingredient of development. But industry is not the sole criterion of development. There are other routes along which a state can progress. In some developed countries national well-being is based less on heavy industry than on agriculture, less on steel mills than on farms.

DEVELOPED AND UNDERDEVELOPED COUNTRIES

Reading about the economic progress of states, you will frequently run across the terms *developed countries* (DCs) and *underdeveloped countries* (UDCs). Sometimes the term *developing* is used in preference to *underdeveloped* because people in these countries have expressed a dislike for the latter word. Whatever the terms employed, though, one thing is clear: these are subjective, imprecise concepts. No universally applicable, generally approved criteria or indices exist according to which we may classify one country as developed and another as underdeveloped or developing. Everyone who thinks about the problems of development seems to have a personal definition, a set of impressions of what an underdeveloped country is. And just about everyone would agree that certain countries are indeed UDCs: Peru, Guatemala, Tanzania, Egypt, Pakistan, Thailand, for example. Similarly, it would not be too difficult to get agreement on a list of DCs, which would include the United States, the United Kingdom, France, West Germany, Sweden, Australia, and New Zealand. But here is a set of countries that produces no such unanimity: Brazil, Mexico, Bulgaria, Turkey, South Africa, and Chile. Some economic geographers would say that states such as those in this last group are at the "take-off" point, about to join the DCs. Others argue that although these states may be developing more rapidly than some others, they are being left behind ever further by the powerful DCs such as the United States and the Soviet Union and by Japan. You can read about the question of development and come away feeling optimistic—or profoundly depressed. Here is how the biologists Paul and Anne Ehrlich differentiate between countries in *Population, Resources, Environment* (1970):

> *Look for a moment at the situation in those nations that most of us prefer to label with the euphemism "underdeveloped," but which might just as accurately be described as "hungry." In general underdeveloped countries (UDCs)*

differ from developed countries (DCs) in a number of ways. UDCs are not industrialized. They tend to have inefficient, usually subsistence agricultural systems, extremely low gross national products and per capita incomes, high illiteracy rates, and incredibly high rates of population growth. . . . most of these countries will never, under conceivable circumstances, be "developed" in the sense in which the United States is today. They could quite accurately be called "never-to-be-developed" countries. . . .

Not an awfully bright picture, but a rather accurate one, unfortunately. We may optimistically speak of developing or "emergent" countries when, in fact, little or no economic progress is being made in many of them. Populations continue to grow, but life expectancies do not, and all too many people will know hunger and malnourishment as a prominent reality of life. The cycle of poverty is not easily broken. The DCs are not immune from the problems of poverty and hunger, either. In the United States a "war on poverty" became a theme of government spending in the 1960s.

The Ehrlichs' just-quoted statement includes some issues we should examine more closely. The UDCs are said to have an extremely low *gross national product*, or GNP. The GNP is calculated as the total value of all goods and services produced by a country during a certain period, usually a year. This value is measured at prevailing market prices in order to make it possible to attach a worth to all the many different items that are produced. When the GNPs of many countries of the world are compared, they are all converted to one currency (say, the dollar), again at prevailing rates of exchange. Thus there are some pitfalls both in the calculation and the interpretation of GNP data. Monetary rates of conversion may not be realistic (witness recent devaluations of the dollar). The purchasing power of the equivalent of, say, \$100 in Nigeria may be far different from what \$100 is worth in the United States. Nevertheless, GNP statistics tell much of the story of underdevelopment. Reduced to per capita figures, we quickly get an idea of the contrasts between DCs and UDCs. The International Bank for Reconstruction and Development, which annually reports on such matters, showed the 1970 per capita GNP in the United States to be about \$4,400, in Canada, \$2,700, in the United Kingdom, nearly \$2,000, but in some 80 UDCs it averaged less than \$250. Of course, we really should be careful with these GNP figures. Record keeping in underdeveloped countries is not very good, and a lot of those statistics on which conclusions are based began as estimates and guesses.

This is especially true where farming is of the subsistence variety, another characteristic of underdeveloped countries. The quality of a national economy can be effectively gauged by measuring the percentage of the labor force that is engaged in agriculture. In the previous chapter we noted how that percentage has declined in the United States over the past century and how large a proportion of the workers in India and China are still engaged in growing crops to feed themselves and the small percentage of people not living directly off the land. Such differences in the kind of work done by the labor force give insights into the contrasts between DCs and UDCs.

Before the arrival of the Europeans, who transformed economies all over the world, and before the industrial revolution, there were well-developed, widely distributed, and very productive handcraft industries in many parts of the world, from Middle America to India to Japan. In most areas the European invasion damaged or destroyed this industrial base, as happened in India, for instance. But in Japan it survived and became the foundation for the country's later, self-generated modernization. An ancient handcraft industry, the weaving of cloth, characteristically of brilliant colors, is shown in this photograph taken in Guatemala.

In Chapter Three we discussed another of the criteria mentioned by the Ehrlichs: the high population growth rates that prevail in so many of the developing countries. Coupled with the shorter life expectancies, this means that most of the people in the UDCs are young people. In Middle and South America in the early 1970s nearly *half* the total population was under fifteen years old; in the Western countries of North America and Europe the figure is under 30 percent. Most of these youngsters are not producers, but *are* consumers, thus constituting an enormous burden and cost.

This brings us to the criterion that interests us most in this chapter, the differing levels of industrialization of DCs and UDCs. In the quotation above, the authors say that the UDCs are not industrialized and imply that the DCs are. This, too, needs to be amplified. There is some industry in all countries; it is in the

CRITERIA FOR DEVELOPMENT

What distinguishes a "developed" economy from an underdeveloped one? Obviously it is necessary to compare countries on the basis of certain measures; the question cannot be answered simply by subjective judgment. No country is totally developed; no economy is completely undeveloped. We are comparing *degrees* of development when we identify DCs and UDCs. Our division into developed and underdeveloped economies is arbitrary, and the dividing line is always a topic of debate. There is also the problem of data. Statistics for many countries are inadequate, unreliable, or unavailable.

The following list of measures is normally used to gauge levels of economic development:

1. NATIONAL PRODUCT PER PERSON This is determined by dividing the sum of all incomes achieved by a country's citizens by the total population. Figures for all countries are then converted to a single currency for purposes of comparison. In DCs the level is as high as $2,000 and over; in some UDCs it is as low as $100.

2. OCCUPATIONAL STRUCTURE OF THE LABOR FORCE This is given as the percentages of workers in various sectors of the economy. A high percentage of laborers engaged in the production of food staples, for example, signals low development levels.

3. PRODUCTIVITY PER WORKER This is the sum of production over a period of a year by the total number of persons engaged in productive activity (that is, the labor force).

4. CONSUMPTION OF ENERGY SUPPLY PER PERSON Apart from noncommercial sources, how much electric and other power is used in a country? High-quantity use reflects greater development, for example, but the data must be viewed to some extent in the context of climate.

5. TRANSPORT AND COMMUNICATION FACILITIES PER INDIVIDUAL This measure reduces railroad, road, airlines, vehicle numbers, and so forth, to a per capita index.

6. CONSUMPTION OF METALS PER PERSON A strong indicator of development levels is the quantity of iron and steel, copper, aluminum, and other metals used by a population during a given year.

7. RATES A number of additional measures are employed, including literacy rates, caloric intakes per person, percentage of family income spent on food, and amount of savings per person.

kinds of industries and their prevalence that the differences between DCs and UDCs lie. Among indicators used to gauge the level of development of a country (in addition to those already mentioned) are the productivity per employee in the labor force; the consumption of power, such as electricity, per person; the quantity of metal used; and the transport facilities that have been developed. These indicators also reflect a country's industrialization because the output per worker goes up when industry becomes increasingly mechanized, commercial energy is consumed in huge quantities by industries, metals play a major role in industry, and transport networks reflect the exchange character of an economy.

Small wonder, then, that some countries whose governments seek to accelerate their economic development have tried to go the route of massive industrialization. In the Soviet Union, planners poured great amounts of the country's resources into a determined program to industrialize. UDCs from Indonesia to Egypt want their own steel mills and national airlines as symbols of progress. But not all countries have made this decision. The governments of some UDCs realize that the criteria by which development is measured reflect not only those particular attributes but also the transformation of a whole society. A domestic steel mill or some other major industry will do little to speed that transformation, and it remains nothing more than an alien anomaly, a costly producer of misleading statistics. A good example of a different approach is the case of Tanzania where, after independence, the *first* national order of business was to change the country's agriculture, to help farmers living a life of inadequacy under subsistence to gain toward adequacy, to diffuse improved farming techniques, fertilizers, and tools (not necessarily tractors, but better hand hoes, for instance), and to encourage and support cooperatives. Few of the results would appear as indices on the list of development criteria, but Tanzanians know that there has been progress.

We are dealing here with numerous uncertainties, debatable statistics, arguable criteria, so Fig. 8.1 must be viewed with caution. We already know that no absolute line of separation between DCs and UDCs can be drawn. Creating a sort of middle category, as we did on page 344, is merely an admission that we cannot be sure of our interpretations. And our map fails to show that numerous countries include different societies in which contrasting levels of development prevail. South Africa is a case often used as an example. Its statistics merit inclusion in the middle category, but it is the white minority of less than 4 out of 20 million that generates the high incomes, consumes the bulk of the energy, develops and uses the transport networks. Black South Africa is another world in terms of wages and incomes, literacy, nutrition, and so on. We do not even have to go all the way to South Africa to learn of different income averages and related qualities of plural societies.

Our map deals with national units, and generalizes internal regional contrasts away. As such, it is a rather optimistic representation of world levels of economic development. We have taken into account the factors of change (how much a country has developed since about the beginning of this century) and resources and the limitations they impose on development in some countries. The

FIG. 8.1 Developed and underdeveloped countries of the world.

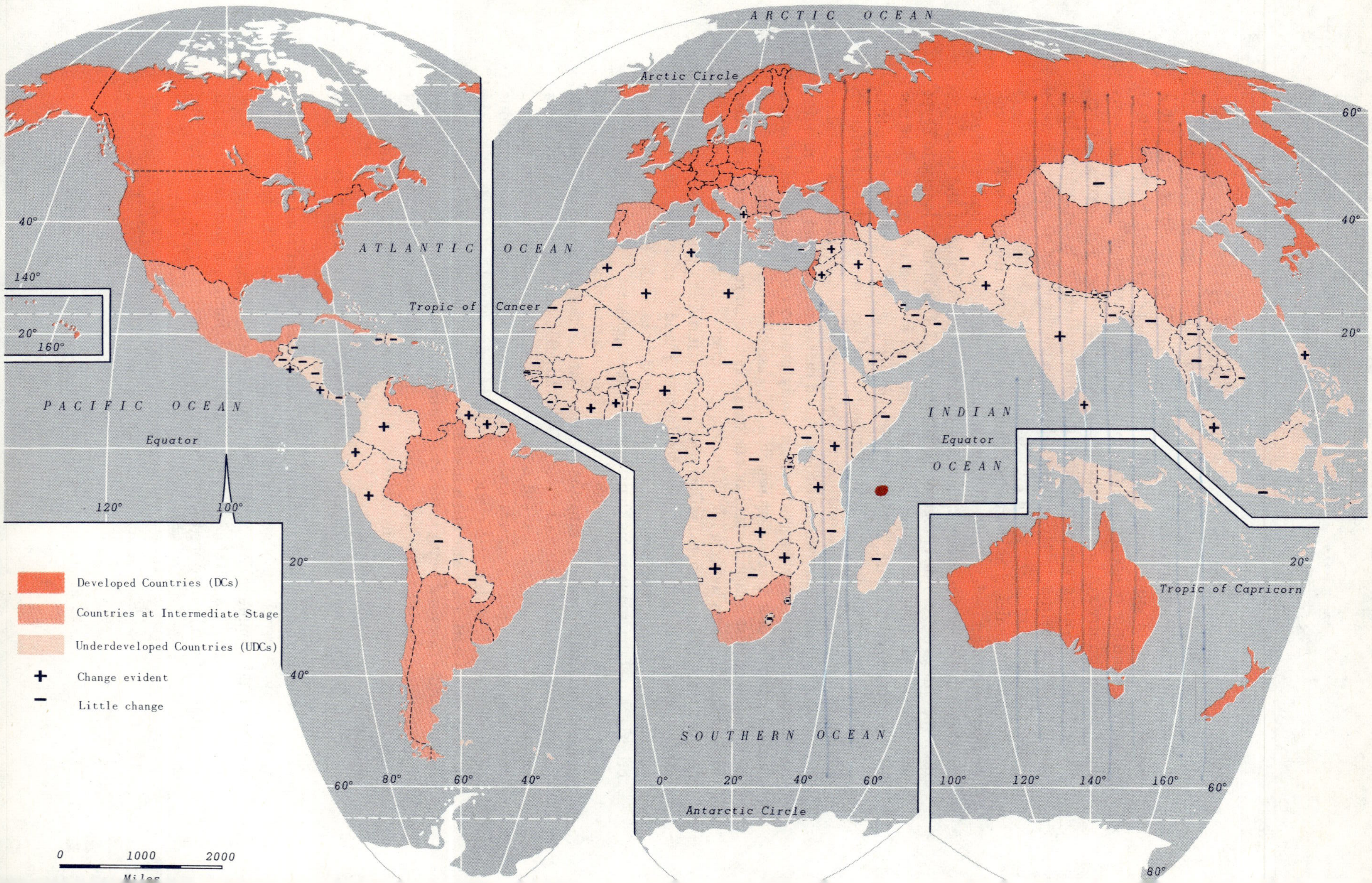

Soviet Union in 1900 was a stagnating, underdeveloped country. Today it could by some criteria still be ranked as intermediate, but three generations have produced a veritable transformation. Japan, too, may still carry some of the symptoms of developing states, but its rate of growth is evidence to support its inclusion as a developed country today. Similarly, China is emerging from underdevelopment into the intermediate category, as are Mexico and Brazil; in southern South America, Chile, Argentina, and Uruguay might be classed by some as developed countries. A comparison between Figs. 8.1 and 3.1, however, produces the conclusion that more than 70 percent of the world's population inhabits countries that are either UDCs or, at best, in the intermediate stage of economic development.

MAN'S INDUSTRIES

We shall return to the theme of national development later, but first we ought to become familiar with some geographic aspects of industrialization. Industrial activity takes place in certain places; it does not occur elsewhere. What determines where industries develop? The industrial revolution centered in a particular region, then diffused to other areas. Why? Today, several discrete areas of the world are outstanding industrial zones. Where do they lie? And in view of the enormous variety of man's industrial activities, can some sort of order be recognized?

Let us take the last question first. The making of steel is an industry, but so is fishing. The fishing industry is Iceland's very means of survival, more important even to that nation than our steel industry is to us. But somehow fishing, the extraction of fish from the sea, is not the same kind of industry as the manufacture of a product such as steel rails. Fishermen catch what exists in nature. Steel makers mix raw materials and come up with something artificial, new. And what about the transportation industry? It produces no product at all, but moves people and goods from place to place. Still it is also an industry!

Thus three classes of industry can be established. The *primary* industries are extractive, as fishing is. Other primary industries include forestry, mining, quarrying, and hunting. Though in a different subgroup, the various forms of agriculture discussed in Chapter Seven are also primary industries. The *secondary* industries are the manufacturing industries. Raw materials are transformed into products ranging from plastic toys to railroad cars. When we speak of "industrial" nations, we really mean the manufacturing nations. When the Ehrlichs wrote of the UDCs as not being industrialized, they had in mind the virtual absence of secondary industries. Finally, there are the *tertiary* industries, involving the production not of goods, but of services. Teaching is such an industry, as is transportation. There is such a thing as the medical industry. Dentists, bankers, bus drivers, and postmen all operate in this sector.

Primary Industries

The primary industries include all those economic activities that involve the extraction of materials from their natural sources, the cultivation of crops on the earth's soils, and animal husbandry. In the present discussion we will not focus upon the last two of these activities, which we discussed earlier, but we should look at the extractive industries. In many of the UDCs, the extractive industries have greater importance than secondary or tertiary industries, and raw materials are sold untreated or only partially modified to overseas consumers.

FORESTRY The forests have been among man's oldest allies, providing food, shelter, building materials, firewood, and even clothing. As we reported earlier, man may have been enabled to live in communities first on forest margins. And still today, the forests of the world provide important resources for human use. Building materials, paper, plywood, and specialized woods such as balsa and teak come from forests. It is not surprising that the earth today no longer carries the forest cover it once did. It is estimated that about half of the earth's land area was covered with forests when man's exploitation began, and today the figure is near 25 percent. Some areas have been completely denuded or very nearly so. China is such an area, and Mediterranean and Western Europe are also largely deforested. Next time you take a trip through Michigan and the eastern Great Lakes area, look around you and imagine that, not very long ago, your path would have led through mile after mile of dense, green forests.

Still, a great deal of forest survives, much of it saved from exploitation by inaccessibility and the costs of distance to would-be consumers. Figure 1.13 represents the present situation in generalized form, except that it does not take note of major areas of deforestation. Compare it with Fig. 8.2, which shows the existing forested areas approximately. Note that there are three major zones: (1) the tropical, broadleaf, evergreen rain forests, (2) the mid-latitude, deciduous and mixed broadleaf forests, which mix with and give way to (3) the high-latitude, narrow- or needle-leaf, evergreen conifers. A bit unfortunately, the terms *softwood* and *hardwood* have come into use without actually referring to the quality of the wood itself. The softwoods are the conifers, the trees with needlelike or scalelike leaves; the hardwoods are the trees with broad leaves, deciduous in middle latitudes but evergreen in the warmest regions of the world, such as the Amazon and Congo (Zaïre) basins. You will see these terms used in statistics reporting annual lumber volumes, but the so-called softwoods may actually be tougher and harder (Douglas fir, for example) than some hardwoods.

The world's forests are exploited by man in many different ways. In the savanna of Africa and in the hill country of Middle America you will find trees being cut for charcoal making. In the forests of Ecuador and Burma respectively there are stands of light balsa and heavy, hard ironwood, so valuable that they can bear the cost of selective extraction. Mahogany from America and teak from Asia are valued by furniture makers. But most of the tropical forests stand virtually untouched, protected by their inaccessibility, the cost of shipping heavy logs great distances, and the variety of their composition. The tropical forest in a given area may be made up of several dozen species. To seek out and cut only one or two of these is not practical under most circumstances. Only very valuable woods or

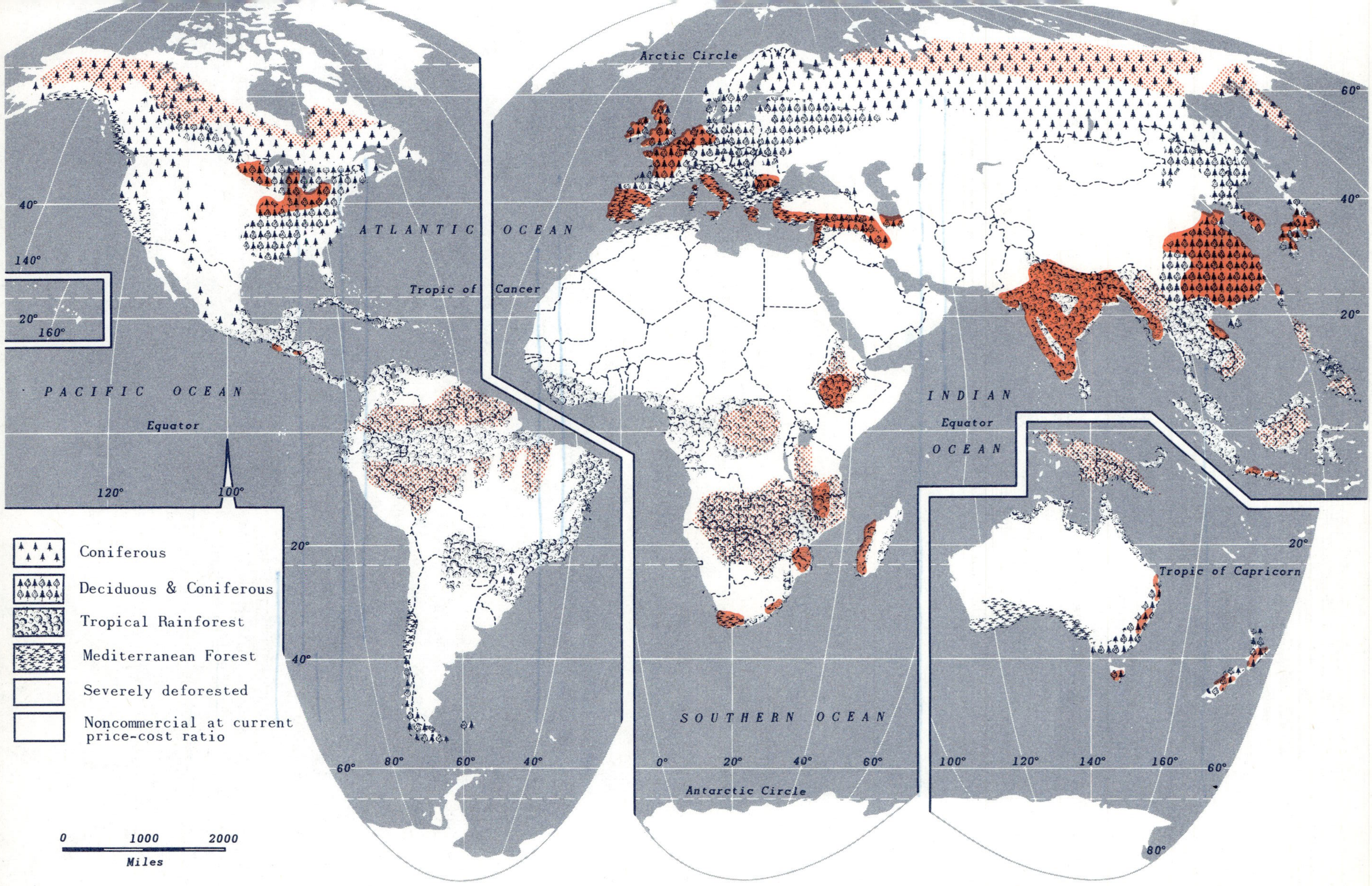

FIG. 8.2 Major forests of the world and areas of severe deforestation. (After a map in R. S. Thoman, E. C. Conkling, and M. H. Yeates, *The Geography of Economic Activity*, McGraw-Hill, New York, 1968, p. 377.)

favorable locations produce a lumbering industry. In Africa there is some forest exploitation for lumber in Gabon, near the Atlantic coast. In South and Southeast Asia, too, there is some lumbering along rivers and near coasts.

The large-scale lumbering industry, however, centers on the softwood and mixed mid-latitude forests and upon the coniferous forests farther north. These forests are desirable for several reasons. They lie near major markets and their composition (especially in the softwoods) is far less varied than the tropical forests. In the United States, lumbering began in the Northeast with the beginning of white settlement, and as western and southern frontiers were penetrated, the loggers followed. During the nineteenth century enormous forested areas were exhausted and cut over; where the lumbermen did not do the job the farmers did, for the trees stood in the way of cropping on fertile soils. Early during the twentieth century the lumber boom hit the Northwest, a boom that still continues. Though far from

---

OTHER FOREST PRODUCTS

In addition to wood in its various forms, the forests of the world still yield a large number of other products of great utility. Among these are:

Cork, taken from the cork oak, which grows in Mediterranean areas, and which has a wide variety of uses because of its low weight and its capacity to contain liquids while permitting air to penetrate. It serves as insulation against sound, as flotation material in life preservers, as bottle stoppers, and in many other ways.

Bamboo, which in some areas of the world, notably Southeast Asia, is used in hundreds of ways, as building material, in the manufacture of containers, mats, furniture, even in road and bridge building.

Resin, which comes in a variety of types and is an important element in the paint industry. The pine forests of the southern United States produce the world's largest volume of resins and turpentine.

Gums such as gum arabic, a base for chewing gum, which come from trees in drier parts of the world and are used in the manufacture of glues, paints, and even medicines. Gums are distinguished from *latexes* such as chicle, another chewing-gum base.

Tannin in various forms, which is used in quantity by the leather industry; hides and skins are treated with tannin extract to prevent their deterioration and to make them soft.

Quinine which was at first a forest product, drawn from the cinchona tree; its chief value lies in its role in the prevention and treatment of malaria. Cinchona plantations have developed.

Wine and other products, which come from the palm trees of various areas of the world. In southern Nigeria the oil palm forms the very basis of life, furnishing building materials, raw materials for marketable products such as baskets, hats, and so forth, oils for cosmetic manufactures, and wine for drinking.

The cutting down of forests for building materials and fuel has long been practiced; some areas of the world were virtually denuded in the process. Man has learned to plant forests to replenish the acreages lost, but the loss still exceeds the replacements. The area shown is an afforestation in British Columbia, Canada, and the cutover area has been reforested by the logging company with small seedlings.

the eastern markets in North America, the Pacific coast loggers sell their lumber easily as prices rise with increasing demand, not only in the United States but also as far away as Australia and Japan. In 1973 the Japanese offered to buy lumber at prices far beyond those prevailing, and if the normalization of relations with China continues, another foreign buyer for Pacific lumber should appear on the market.

Lumber no longer leads by value among forest products. Pulpwood has taken over. Pulpwood supports the paper industry, and pulpwood also is a raw material in the manufacture of some synthetics (rayon, for example), wallboard, and even explosives. Anyone who has been reading daily and Sunday newspapers over the past years will have noted part of the reason for the rapid increase in demand for pulpwood. The papers are getting bigger and bigger. In 1969, the latest year for which accurate statistics are currently available, the value of processed wood produced (sawn wood, railway sleepers, and so forth) was $16.3 billion, but for pulpwood it was $20.7 billion. In distant third place was the category for panel products, such as plywood, wall panels, and veneers, accounting for $6.0 billion by value.

The overall, world picture of forestry shows that the Soviet Union is presently the leading exploiter and producer, and as the map shows, the U.S.S.R. has enormous reserves of forest. But the United States is by far the largest per capita consumer. Characteristically, the Soviet Union exports more of the raw material and the United States exports more of the finished product. Countries of Western Europe are also major importers, much of the import coming from the coniferous forests of Scandinavia and Finland. It is noteworthy that one of the countries we mapped as "developed" on Fig. 8.1, Finland, has exports based mainly on the primary industry of forestry: paper brings in some 27 percent of the export revenues, timber 16 percent, and wood pulp 13 percent. Sweden, too, depends heavily on forest-product exports, but no one who looks at Fig. 8.2 will be surprised that Canada is the world's leading exporter of these commodities, most of the production going to its southern neighbor. North America, Europe, and the Soviet Union together cut more than half of the world's trees each year.

As with the production of cereals for subsistence, it is possible only to estimate how much wood is annually consumed the world over. In Middle and South American, African, and many Asian countries much wood is cut and used for fuel or locally as building material. As the mid-latitude and northern forests of North America and Eurasia are exploited, it may be that the tropical woods will enter the trade picture more than has been the case. Certainly both prices and demand are rising rapidly.

FISHING Fishing, with hunting, is another of man's oldest occupations. And, like the products of the forests, the products of the oceans continue to sustain people. But where ancient man combed the shallows and beaches for shellfish and later began to fish with spears, hooks, and nets, contemporary fishermen ply whole oceans, search out the fish with electronic gear, and make their catches with mechanical equipment capable of harvesting huge areas very efficiently.

The dangers of overfishing and the exhaustion of the world's fishing grounds is a topic we shall pursue in our final chapter. But it is clear that effective allocations of fishing zones and limitations on the catch are needed: in spite of increases in annual fish production, this is an endangered resource. The world map induces a false optimism. There seems to be so much ocean and surely so much fish that the supply must be almost limitless. But reality is different. In the first place, those vast oceans include areas almost as devoid of fish as there are deserts on land devoid of people. Fish are part of what has been called a *food web.* Smaller fish live on plankton, drifting plant and animal material composed of minute organisms. Somewhat larger fish live on the smaller fishes, and still larger fish depend on the medium-sized and smaller varieties. Break this hierarchy somewhere, by fishing out the small or medium-sized fish, and the whole balance is destroyed.

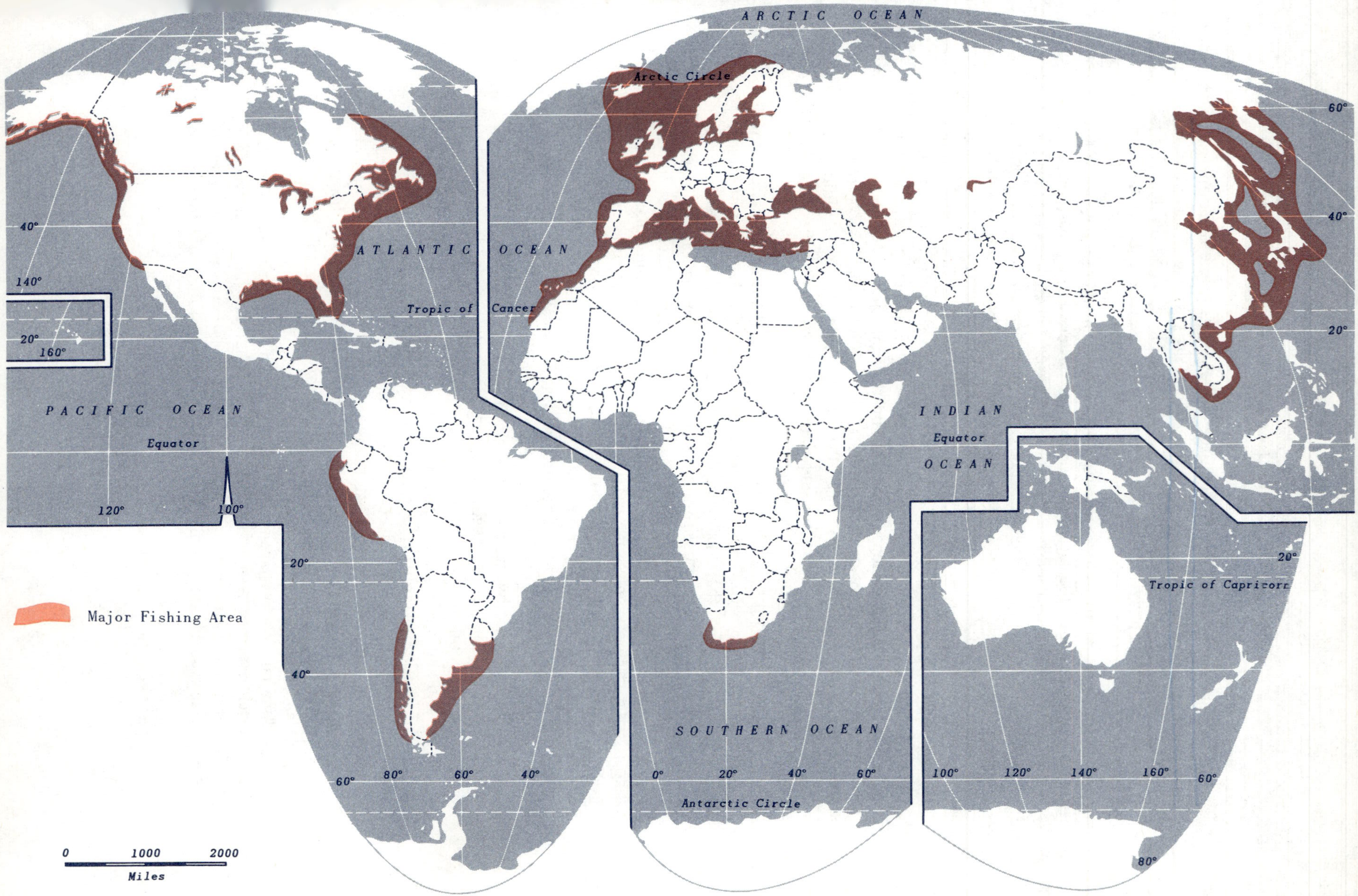

FIG. 8.3 Major world fishing areas.

As it happens, plankton develops best in the shallow parts of the sea, principally the areas adjacent to the continents, the continental shelves. Thus, concentrations of fish exist on and near these shelf areas. Another look at our world map shows that the continental shelves are not even remotely as extensive as the oceans (Fig. 8.3).

The most substantial continental shelf areas (generally at a depth of less than 600 feet or about 100 fathoms) lie off northwestern Europe, eastern North America, and eastern and southeastern Asia. Narrower shelves do exist elsewhere, for example, the Agulhas Bank off the southern tip of Africa and the shallow waters off part of New Zealand. For centuries, peoples living in coastal countries near these shelf areas have fished in the adjacent waters, but in the period after the Second World War, fishing fleets began to appear on fishing grounds thousands of miles from home. In response, countries (especially the UDCs) began to expand their territorial waters, the adjacent zone of water over which they insist on having sovereignty. And some developed countries did the same, notably New Zealand, which in 1969 decided to claim 12 miles of coastal, territorial waters, thereby chasing off the Japanese, who had been fishing right up to the 3-mile limit, and Iceland, which announced that it would claim waters within 50 miles from land as exclusive fishing domain.

Japan is the world's most active fishing nation today. If you check the annual United Nations Food and Agriculture Organization's *Yearbook of Fishery Statistics,* you will note that Japan is near the top of the list of producers in terms of tonnage, and in a class by itself in terms of value, well in the lead. By tonnage, Peru is the leader, but Peru's catch is mainly in the form of anchovies for the fish-meal industry. Peru's waters also yield sardines, mackerel, and tuna, however. Still, it is Japan that stands out, its large fleets competing for the catch almost everywhere in the world, pursuing marine life from herrings to whales. Meat-poor Japan consumes more fish per person than any other country, as much as 80 pounds, which constitutes more than three-quarters of all the animal protein intake. Thus Japan's fishing industry is a national necessity, supplemented by a growing fish-farming industry in freshwater ponds and rice paddies on land. Importantly, the bulk of Japan's fish harvest is caught in waters lying within a few dozen miles of Japan itself: the country looks out over one of the world's richest fishing grounds. There, a warm and a cold ocean current meet (see Chapter Ten) and a rich fish fauna results, including sardines, herring, mackerel, and tuna in the warmer waters and cod, halibut, and salmon in the cooler, more northern seas. Along Japan's coasts there are more than 4000 fishing villages, and tens of thousands of boats, most of them small, ply the waters offshore to bring home catches that are distributed locally. In fact, the larger tuna boats bring back a harvest too valuable to consume at home. Japan even exports canned fish.

China ranks below Japan in annual fish catch, but actual figures for the Chinese catch have not been available. The Chinese harvest the seas actively for the same reason the Japanese do. They need more food, especially the protein element made available through fish. The Asian continental shelf off China is one

WORLD'S LEADING FISHING NATIONS (1970)

BY TONNAGE (Thousands Metric)			BY VALUE (Thousands U.S. Dollars)		
1.	Peru	12,613	1.	Japan (1969)	2,357,700
2.	Japan	9,309	2.	U.S.S.R.	n/a
3.	U.S.S.R.	7,252	3.	United States (1968)	496,900
4.	Norway	2,980	4.	Philippines	433,900
5.	United States	2,714	5.	Spain	378,000
6.	India	1,746	6.	India	341,300
7.	Thailand	1,595	7.	France	286,700
8.	South Africa	1,519	8.	Pakistan	238,300
9.	Spain	1,497	9.	Italy	231,700
10.	Canada	1,378	10.	Thailand	218,600

Source: Britannica *Book of the Year*, Encyclopaedia Britannica, Chicago, 1973, 300.

of the world's richer fishing grounds, and the Chinese catch is made almost entirely by small boats based at coastal fishing villages. China's modern fleet cannot match Japan's.

The Soviet Union, high among the world's top five fishing nations, contests the northern Pacific Ocean with Japan and the United States. Its wide-ranging fleets, modern and massive, frequently find themselves in conflict in other parts of the world as well. Floating factories process on the spot the harvest brought in by schools of trawlers, trawlers that sometimes drag their nets within sight of foreign shores. Even in the Atlantic Ocean, in the fishing zones on which fishermen from the United States Northeast depend, the Soviets trawl. In recent years the Soviet catch has been more than double that of the United States, the next-ranking producer. The U.S.S.R. benefited enormously when, after the end of the Second World War, it gained primacy in the rich Northwest Pacific fishing grounds where Japan had formerly prevailed.

The United States fishing industry is really two industries, not counting the inland, freshwater fisheries. The East Coast fisheries extend from Maine to Texas, and they account for about 70 percent of the product (somewhat more by tonnage, a little less by value). The cooler waters of the north produce halibut and herring, among other fish, and, of course, lobsters, clams, and oysters. From the warmer southern waters come red snapper and flounder. The Gulf of Mexico yields a valuable shrimp harvest. On the other side of the country, the Pacific Coast fisheries are based primarily on tuna and salmon catches, the salmon being especially productive in Alaska and the tuna along the California coast. The salmon industry has been going through a difficult period, involving international conflict over quotas, river pollution (salmon come to the fresh water of rivers to spawn), and river deterioration because of dam building and lumbering operations.

These five countries are the world's leading fishing nations, but we should

A fleet of fishing boats pursues schools of fish located by radar on the mother ship (at the right of the photograph). Such fleets can roam the oceans, going thousands of miles from their home bases. With their electronic devices to locate the catch and sophisticated equipment to bring it aboard, they endanger the fish fauna and increase the risk of overfishing in world fishing grounds. The mother ship is a floating factory as well as a supply ship to the smaller boats, and the fish that is caught is processed and packaged, canned, or refrigerated, all at sea.

remember that fishing plays an extremely important role in countries with more modest catches as well, in fact, a much more important role in the overall economy even than in Japan. Norway, the United Kingdom, Iceland, Denmark, the Netherlands, and France form a cluster of such countries, but fishing is of consequence also in South Africa, India, Canada, Thailand, and Spain, as the table (page 353) shows. Note that the list begins to include an increasing number of UDCs. The harvest lies open for all to secure. The pressure on the seas is growing.

MINING Mining is the third of the primary, extractive industries we discuss here, and when we remind ourselves that whole periods of human cultural development are named after the chief substance mined and used (the Stone Age, the Bronze Age, the Iron Age), we realize that this, too, is an old occupation. But even today, when the variety of minerals and fuels brought to the surface is so great, there are not many more people engaged in mining than there are in forestry and fishing, perhaps 1 percent of the labor force. Machines, pumps, and pipelines do the work with increasing frequency and effectiveness. True, miners still go down the mine shafts of Pennsylvania, Peru, and Poland, but cars, wheels, and conveyor belts have replaced countless human hands.

Mining has to take place where the resources are. In Chapter One we looked at the haphazard distribution of the known, major mineral resources of this earth; there is not a single country, no matter how large, that has a complete range and volume within its boundaries sufficient to sustain all its industries. Thus there is much trade in minerals and mineral fuels, and the search is always on for new reserves. Human choice in this context is limited to the decision on which among several alternative sites to exploit first, and as the highest-quality reserves are worked out, the lower-grade ores are brought into production. Improved refining techniques are making it possible to profitably mine ores so impure that they were thought to be useless just a few decades ago. The industrialized, developed countries seek out and buy ores and fuels wherever their purchase is warranted by demand and cost considerations. Not only do they buy the ores, but they also invest heavily in the equipment needed on-site to mine them. Thus Japan, in buying iron ores from a rich deposit in Swaziland, southern Africa, spent money on machinery and even on a railroad to link the field with the nearest port, in this case Lourenço Marques, the capital of Moçambique. United States investments span the globe, from the oil fields of the Persian Gulf area to the diamonds of South Africa. As with the foodstuffs and other raw materials we discussed in Chapter Seven, the flow is from the UDCs to the technologically advanced countries of North America and Eurasia.

IRON ORE Iron and steel form the basis of so many of the manufacturing industries (which we shall discuss shortly) that it is true to say that without them, or without other metals such as aluminum and copper, the modernization we have witnessed in the DCs could not have been achieved. The industrial revolution could not have taken the course it did, and the world would be a very different place. The technological developments of the past two centuries have affected even the places most remote from the foci of change.

Iron ore happens to be one of the most widespread of our resources: in the early 1970s some 70 countries were reporting a production of this commodity. Some of the discovered ores are very pure, containing up to 70 percent iron; others a mere 25 percent or even less. Figure 8.4 shows the distribution of major world iron ores. Many factors enter into the decision to mine or not to mine at a given time. Accessibility of the ores, the grade of the ore, the size of the deposit, distance to markets, even the political situation in a country will affect the decision to select one ore deposit over another for exploitation. Figure 8.4 actually shows the known *reserves* of iron ore; so far as can be determined India appears to have the largest supply available, perhaps with as much as one-quarter of all reserves. The Soviet Union, the United States, and Brazil also have large iron-ore reserves, but the picture still changes. Australia recently proved to have larger reserves than first thought; in Africa several countries also have new, proven deposits.

FIG. 8.4 World iron-ore deposits. (After a map in *Man's Domain*, McGraw-Hill, New York, 1968, p. 22. Base Map © copyright, General Drafting Co., Inc.)

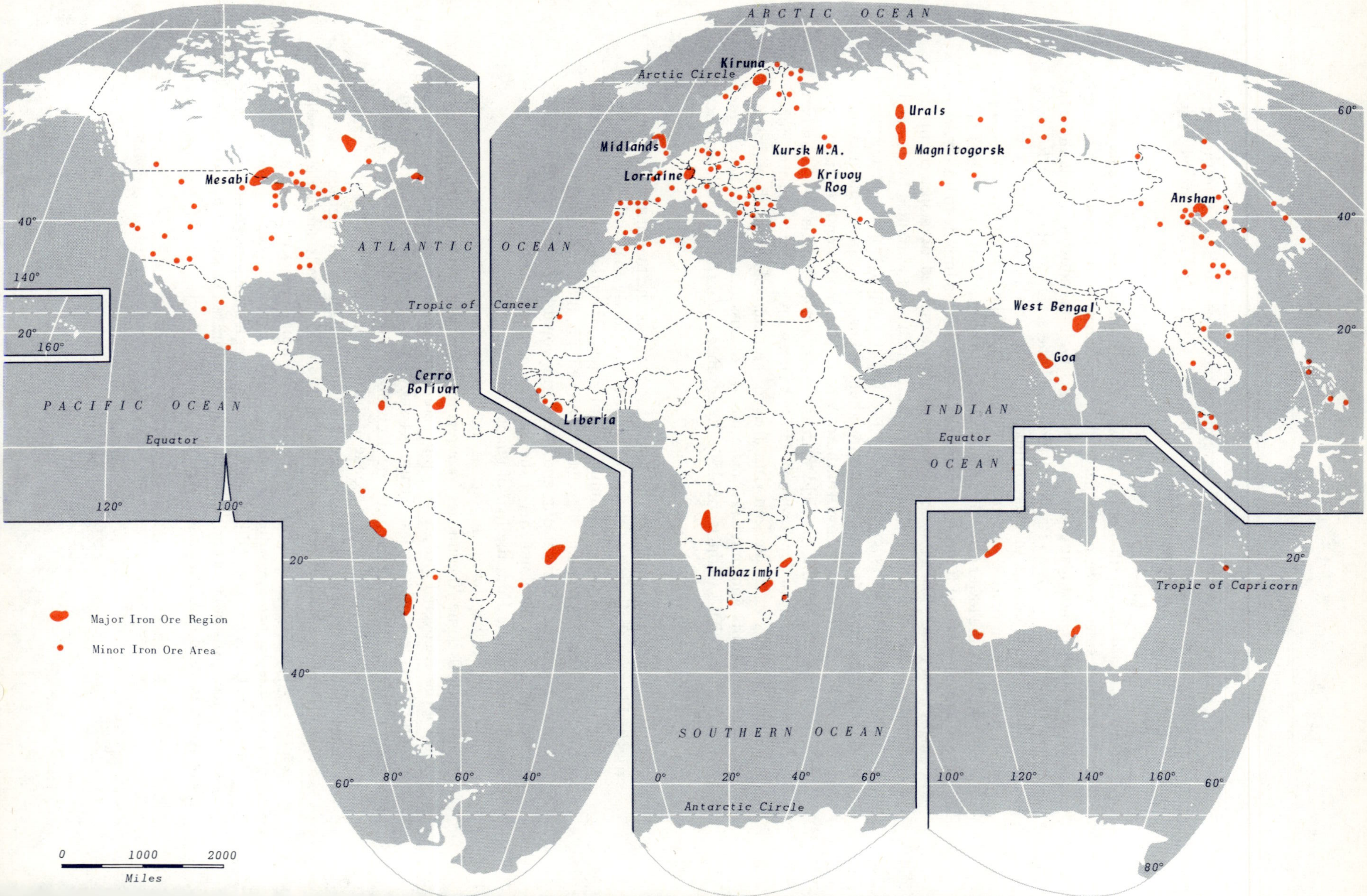

The production of iron ore is presently dominated by the Soviet Union, which mines nearly twice as much iron ore (some 100 million metric tons) as the United States (55 million). Canada (25) ranks next, and Australia has risen from sixth to fourth place, passing Sweden and France. When it comes to the production of steel, however, another picture emerges. Here the United States in 1971 still led the U.S.S.R., and third place was held by Japan, which is way down on the list of iron-ore producers. This underscores the heavy international trade in iron ore, in which the steel-producing countries participate intensively. There was a time when major steel producers could depend mainly on their home supplies, but no longer. The United Kingdom (where the industrial revolution first took hold), West Germany, Japan, and even the United States import enormous amounts of iron ore. Among these countries, only the United States could possibly do without those imports (it is cheaper for the United States to import foreign ores than to exploit some local deposits); the United Kingdom, West Germany, and Japan have steel industries that are quite dependent on foreign iron ores. So the international traffic in iron ores is likely to increase over the years as the industries of the developed countries grow.

THE INDUSTRIAL REVOLUTION

Fishing the oceans and seas, exploiting the forests, and extracting minerals and rocks from the ground are, as we have said, old industries of man, ancient pursuits that went on many thousands of years ago. Still today, men in canoes pursue seals, men with spears hunt in forests, and desert people dig up salt for sale on distant markets. But after countless centuries of slow change industry took a great leap forward in the last half of the eighteenth century and the decades thereafter. First it was a localized phenomenon, centered in Britain, but soon ripples and then shock waves radiated out into Europe and into much of the rest of the world. The great transformation had begun.

This is not to suggest that there were no substantial industries at all by the second half of the eighteenth century or that the industrial revolution generated them. On the contrary, the roots of the revolution lay in the search for improvements in the machinery of Britain's textile industry. British merchants were importing fibers by the ton, and there was not enough labor to keep pace with the growing trade. Machines capable of larger production were needed, better spinning and weaving equipment. They were invented, but, as before, they were driven by water, the old source of power. Then Watt (along with others) invented and refined his steam engine (1765–1788) and adapted it for various uses. About the same time, it was realized that coal could be transformed into coke and that coke was a superior substitute for charcoal in the smelting of iron.

These momentous innovations had immediate effect. The power loom revolutionized the weaving industry. Freed from their dependence on dwindling wood supplies from the remaining forests, iron smelters could now be concentrated near the British coal fields, the same fields that supplied fuel for the new textile mills. And one invention led to another, each innovation made to serve some particular industry also serving other industries. Pumps could keep water out of flood-prone mines. Engines could move power looms as well as locomotives and ships. As for the capital required, there was plenty available for investment. Britain had been getting rich from its overseas empire for many years.

The industrial revolution also affected transportation and communications. The first railroad in England was opened in 1825. In 1830 the city of Manchester was connected to the port of Liverpool, and in the next several decades thousands of miles of rail were laid. Ocean shipping also entered a new age as the first steam-powered vessel crossed the Atlantic in 1819. England enjoyed even greater advantages than those with which it entered the period of the industrial revolution. England held a monopoly over products that were in world demand, and it alone possessed the skills necessary to make the machines that manufactured them. Europe and America wanted railroads and locomotives; England had the know-how, the experience, and the capital. Soon the fruits of the industrial revolution were being exported, and British influence around the world reached a peak.

Meanwhile, the spatial pattern of modern, industrial Europe began to take shape. In Britain, industrial regions, densely populated and heavily urbanized, developed in the "Black Belt," near the coalfields. The largest complex was (and still is) positioned in north-central England. In mainland Europe, a belt of major coalfields extends from west to east, roughly along the southern margins of the North European Lowland (see Fig. 1.16), due eastward from southern England across northern France and southern Belgium, the Netherlands, Germany (the Ruhr), western Bohemia in Czechoslovakia, and Silesia in Poland. Iron ore exists in a broadly similar belt, and the industrial map of Europe reflects the resulting concentrations of economic activity (Fig. 8.5). But nowhere on the mainland are the coalfields, the iron ores, and the coastal ports located in such close proximity as they are in Britain.

Industry: Location Factors

In the primary industries, production has to take place where the resources are. The timber, the fish, and the ores and fuels leave no choice; it is necessary to come and get them where they are. But we noted just a moment ago that textile industries and other industrial plants began to position themselves in clusters as Europe's industrial revolution gained momentum and that iron smelters, among these others, were positioned near the coalfields of the Black Belt. This means that the iron ores had to be brought to the coalfields, a reality that illustrates an important aspect of the secondary industries, the manufacturing industries. Decisions have to be made about where to locate them. Weight and bulk have a lot to do with such decisions. Surely it is sensible to refine an ore or to saw off useless waste from tree trunks practically at the site of production and *then* ship the raw material for final processing into finished products, rather than ship them long distances first. In the case of iron ore and coal, both needed in large quantities to make steel, the alternatives are (1) to take the iron ore to the coalfields, (2) to take the coal

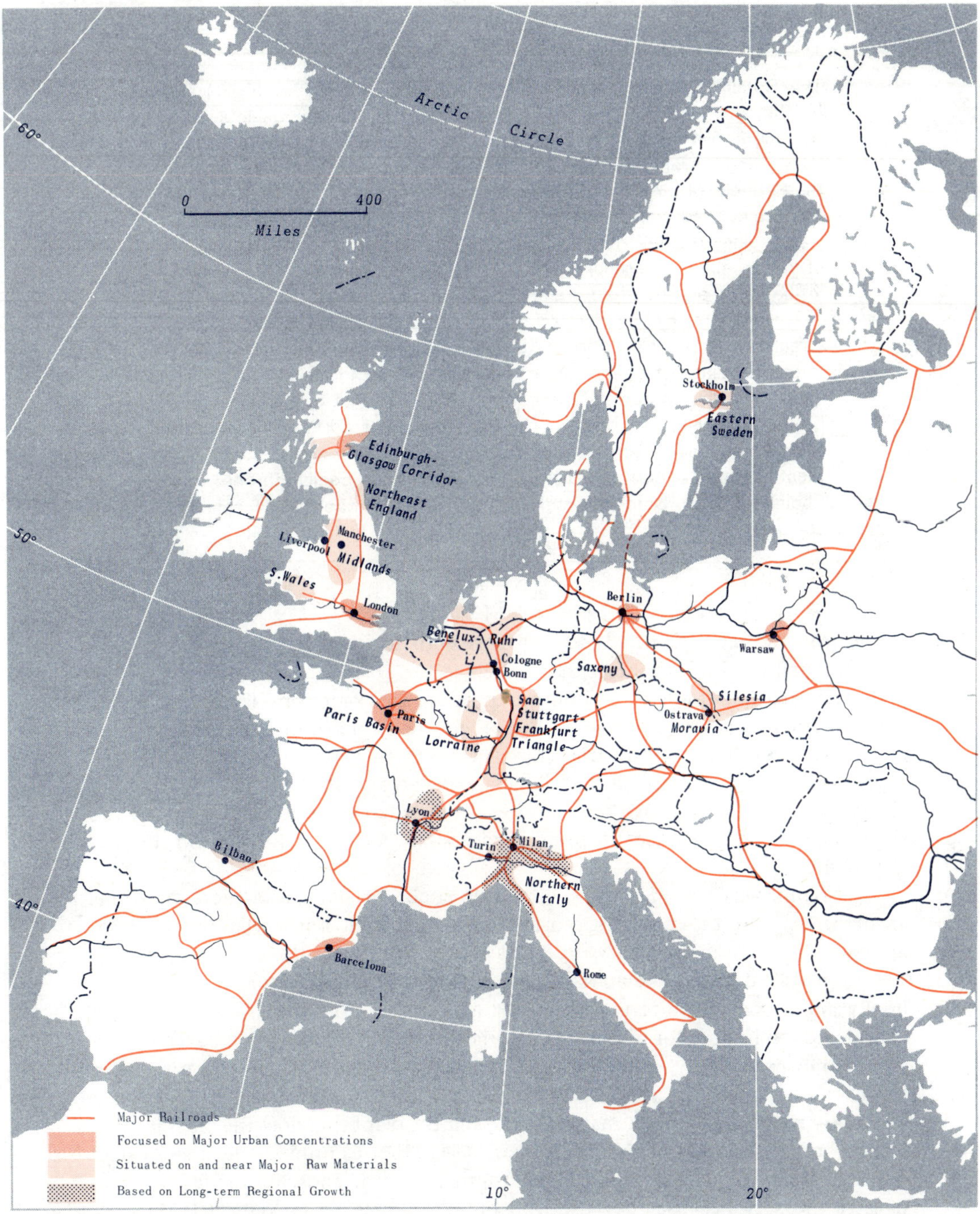

FIG. 8.5 Major industrial regions of Europe.

THE BRITISH ISLES

The British Isles consist of two large islands, Britain (the larger) and Ireland, surrounded by several thousand smaller islands and islets. They became known as the British Isles during the lengthy period of British domination not only over Britain itself, but over Ireland as well. Ireland, the Republic of Eire, did not achieve total sovereignty until 1921. But the Irish did not gain control over all of Ireland; a northern corner of the island was retained by the British and placed under London's control. The island of Ireland, then, is partitioned into two political entities, the Irish Republic in the south and Northern Ireland in the north.

The island of Britain also has political divisions. Sometimes Britain or Great Britain is referred to as "England," but this is incorrect. England is only the largest of the three political areas on the island of Britain, and it was the center of power from where the rest of the Isles were brought under unified control. The English conquered Wales during the Middle Ages, and Scotland was first tied to England early in the seventeenth century, when a Scottish king ascended the English throne. Thus England, Wales, Scotland, and Northern Ireland together became the United Kingdom.

to the iron-ore deposits, or (3) to transport both to some intermediate location. Most frequently, the iron ore is transported to the coalfields, sometimes after the partial elimination of waste and impurities. There are exceptions. For example, coal is transported in large quantities to the French Lorraine industrial region, where coal is scarce. When an industrial complex develops near coalfields and the coal supplies become exhausted, it may be less expensive to start importing coal from elsewhere than it is to move the factories. Even when both coal and iron ore are shipped over large distances to some intermediate location, the iron ore usually travels the farthest.

Numerous considerations enter into such decisions, and these questions are of prime interest to economic geographers, who want to know what the processes are whereby economic activity organizes and adjusts spatially. Obviously the *resources* involved, the raw materials, play a major role. In Chapter One we looked briefly at the way taconite is now mined as a useful iron ore, after having been considered too impure for profitable exploitation. Still taconite would be too expensive to transport to, say, the Chicago-Gary steel mills, so it is refined on the spot before it is shipped in a much purer state. Technically that refining process is a secondary industry, a kind of manufacture, and the plant that does the job is located relative to the situation of the raw materials. Another example is the coastal steel industry of the United States along the Eastern Seaboard. Those industries are there in large measure because they use iron ore shipped in from Venezuela, Liberia, and other overseas mines. Rather than transfer them from the arriving ships onto trains and railing them inland, the ores are used right where they arrive, practically at the point of unloading. So in this case far-away ore deposits had much to do with the location of industry in the United States.

A second factor in industrial location is the supply of *labor*. Even in this day of automated assembly lines and computerized processing, the available skills of a substantial labor force remain important criteria in the location of manufacturing plants. Over decades and generations a certain cumulative effect has manifested itself, whereby the people in certain areas have become known for their particular talents. The Swiss make fine instruments, specialized textiles are made (still) in Britain, cameras in Japan. The book-publishing industry in the United States is concentrated in New York, with a secondary center in San Francisco. In these places the several skills needed to transform a manuscript into an attractively designed book are available. A new industry would be likely to consider strongly making use of the available capacities and locating in or near these centers.

It may not be skills, but the cheapness of labor that will attract some industries. In the United States, there are areas where wages are lower than they are elsewhere, and unless other factors of industrial location are of overriding importance, a manufacturing concern may locate where it can pay labor less.

Another very obvious influence in the location of industry is the *market*. Obvious though the influence may be, it is not so easy to understand. In the developed countries, much depends upon the transport networks as well as upon the location of large population concentrations. A company is likely to establish itself in or near the largest city, where it has the biggest market nearby, knowing that good communications radiate outward into the rest of the country and that distribution there will be very efficient. Industries that produce perishable goods (dairies, for example) will locate as near to their markets as they can. Some industries make products specifically designed to be used by other industries, for instance, certain equipment for automobiles. It would not be sensible for a factory making speedometers for cars to position itself awfully far from Detroit. But there is still more to the issue. When a commodity is very heavy and bulky, the proximity of the market is far more significant than when it is smaller and lighter and, as is likely, more valuable per unit of weight. Even if the Swiss sell more watches in the United States and Canada than in Europe, it is unlikely that they will move their factories to New York. On the other hand, an American automobile manufacturer planning to sell a small car on the European market may decide to build the factory in Europe rather than to build the cars here.

Another factor in the location of industry is the availability of an *energy* supply. This used to be much more important than it is now. Those British textile mills, while they depended on water rushing down hillsides to drive the looms, had little alternative in the context of location. But these days power comes from different sources and can be transmitted by way of power lines over long distances without much loss. Industries are thus able to locate primarily on the basis of considerations other than power, except when the industry needs really exceptionally large amounts of energy, as, for example, aluminum-refining plants do. Those industries are positioned near sites where abundant energy is created, as is the case at hydroelectric plants.

The role of power supply should not be understated, however. In the DCs, dense transmission networks bring power to most areas, but in the UDCs, electrification still is an elusive goal for whole regions. The electrification program undertaken in the Soviet Union during the present century has been one of the prideful accomplishments for that country. Elsewhere, hydroelectric projects and power lines have the same aura of modernization and progress that causes UDCs to pour huge investments into the face-lifting of capital cities and the maintenance of a money-losing national airline. The hope always is that the cheap electricity, once available, will attract industries. Those anticipations are not always realized.

These are major factors of industrial location. They are interrelated; they do not operate independently. The location of one or several plants in a given area will influence others to locate there as well, so that a kind of clustering effect occurs. Without the already-established manufacturing plants, those other industries might have positioned themselves elsewhere. There is also the factor of political stability and receptiveness to investment. Industries are frightened away if there are signs of uncertainty in the political future of a country or if a government is planning to nationalize industries owned by foreigners. Policies of taxation play a role, too, and some countries try to attract industries by offering huge tax exemptions over a fairly long period. A few industries are located where they are because of environmental conditions. For example, the film industry has been so strongly concentrated in southern California because there was a time when there was a large number of clear, cloudless skies there each year.

INDUSTRIAL REGIONS OF THE WORLD

For reasons we have just noted, manufacturing industries tend to cluster in certain areas favorable to their development. It is not always possible to discern just why a certain assemblage of industries has developed in some region. We may understand that raw materials are brought in at low cost, but there may be other factors involved that are not so clear or obvious. History has to do with it sometimes. Plants converted from one kind of power to another but stayed where they were, and then the clustering process took over. Perhaps national planning was the deciding factor. In the Soviet Union some industries were built to produce goods that could have been bought more cheaply than they could be made at home. In the 1930s the U.S.S.R. located plants far from a potential threat (Germany in the West) so that they could continue to function even in case of war and invasion.

Whatever the underlying causes, and they are many and complex, a small number of major industrial regions exist today whose factories produce a large proportion of the entire world's industrial output. Take note that we are now considering the *secondary* industries, the industries that manufacture goods from raw materials made available in the primary industries. It is hardly necessary to enumerate these industries, for when we view the major regions we will have frequent occasion to refer to specific manufactures. There are four such regions: (1) western and central Europe, (2) the United States and Canada, (3) the core area and eastward extension of the Soviet Union, and (4) the Japan-China industrial concentration.

European Manufacturing Region

It is appropriate to begin with Europe, for here the industrial revolution commenced. Compare Figs. 8.5 and 5.2, and you will find the industrial regions of Europe to lie within the European core area; in fact, in large measure they *constitute* that heartland. Also note that we are speaking of more than one industrial area. Europe's industrial region consists of a number of districts, from the British Midlands in the west to Silesia in the east.

Nowhere is the principle of functional specialization better illustrated than in the industries of Britain. Wool garment-producing cities lie east of the Pennines, the mountain backbone of the island, centered on Leeds and Bradford. The textile industries of Manchester, on the other side of the range, produce cotton goods. Birmingham, now with $2\frac{1}{2}$ million inhabitants, and its neighbors concentrate on the manufacture of steel and metal products, including automobiles, motorcycles, and airplanes. The Nottingham area specializes in hosiery; Leicester adds boots and shoes to that field. In northeast England, shipbuilding and the manufacture of chemicals are the major large-scale industries, along with iron and steel production and, of course, coal mining. Coal became an important British export early, and with this resource's seaside locations in northeast England, Scotland, and Wales it could be shipped directly to almost any part of the world. In Wales coal mining is supplemented by shipbuilding and the production of iron and steel. The same complex of industries is based on the coalfields of Scotland, where Glasgow (2 million) has become one of Britain's major industrial cities. It is a measure of the impact of the industrial revolution that today four-fifths of the population of Scotland is concentrated in the area identified on Fig. 8.5 as the Edinburgh-Glasgow Corridor.

Britain, at the vanguard of the industrial revolution, has not always kept up with modern developments. Plants that at one time were the epitome of industrial modernization still operate today, but they are comparatively inefficient, expensive to run, slow, and wasteful. With the relative decline of the industrial cities of the Midlands and northern England, there is a tendency for industrial enterprises that can afford it and are able to do so to relocate near the old focal point of Britain, London. Here, still, is the greatest domestic market of the British Isles, and increasingly that market is important to local manufacturers as the competition for markets from elsewhere intensifies. All this reflects the decreasing importance of coal in the energy-supply picture (which is changing toward nuclear power), the desire to start afresh with up-to-date machinery, and a realization that

London, in addition to forming a huge domestic market, is also a good port through which to import raw materials. So London, too, shows up on Fig. 8.5 as an industrial district within the European region.

Paris does also. When the industrial revolution came to the mainland, Paris was already continental Europe's greatest city, but Paris did not (as London did not) have coal or iron deposits in its immediate vicinity. Even so, Paris was the largest local market for manufactures produced for hundreds of miles around, and when a railroad system was added to the existing road and waterway connections, the city's centrality was strengthened still further. As in the case of London, Paris itself began to attract major industries, and the city, long a center for the manufacture of luxury items (jewelry, perfumes, fashions), experienced great industrial expansion in such areas as automobile manufacturing and assembly, the metal industries, and the production of chemicals. With a ready labor force, an ideal position for the distribution of finished products, the presence of governmental agencies, a nearby ocean port (Le Havre or Rouen can serve), and France's largest domestic market, the development of Paris as a major industrial center is no accident.

But, as our map in Chapter One (Fig. 1.16) showed us, Europe's coal deposits lie in a belt across northern France, Belgium, and western Germany to southern Poland, and along this zone developed mainland Europe's concentrations of heavy industry. Before its fragmentation into an East and a West Germany, three such industrial districts lay in the German state: the Ruhr, based on the Westphalia coalfield, the Saxony area near the boundary with Czechoslovakia (now in East Germany), and Silesia (now in Poland). Among these, the Ruhr has become the greatest industrial complex of Europe. After Britain, Germany is Europe's top coal producer; it is Europe's first iron and steel producer, bar none.

The Ruhr, named after a small tributary of the Rhine River, provides us with an illustration of the advantages of a combination of high-quality resources, good accessibility, and a position near large markets of high purchasing power. The coal of the Ruhr is of excellent quality for steel-making purposes. The iron ore locally available was soon exhausted, but ores could be shipped in from overseas with only a single transshipment. Long before the Ruhr emerged as Germany's major complex of heavy industries, cities had been growing nearby along the Rhine, including Köln (Cologne) and the present capital of West Germany, Bonn. The surrounding agricultural area was productive and densely populated. After the early 1870s the Ruhr expanded rapidly and increasingly received the benefit of government support in the form of subsidies and tariff protection. The chief manufactures made here are iron and steel (and products forged from these, such as railroad equipment, vehicles, and machinery), textiles, and chemicals, with an almost endless variety of associated products.

While the Ruhr region specializes in heavy industries, Saxony (now in the German Democratic Republic, or East Germany) is skill- and quality-oriented. Today this region, which includes such famous cities as Leipzig (printing and publishing), Dresden (ceramics), and Karl-Marx-Stadt (textiles), benefits from nearby coalfields, but even before the industrial revolution made its impact here these places were substantial manufacturing centers.

Farther eastward, the industrial district of Silesia also was first developed by the Germans, although it now lies in Poland and extends into Czechoslovakia. A major industrial district is developing in southern and southwestern Poland, based upon local, high-quality coal resources and lesser iron ores, which are, however, supplemented by imports from the Soviet Ukraine. Czechoslovakia shares the raw materials with Poland; they lie astride the gap between the Erzgebirge (the Ore Mountains) and the Carpathians. In Czechoslovakia this particular industrial district is called Moravia, and the city of Ostrava, with its metallurgical and chemical industries, lies at the center.

Our map shows other industrial districts in Europe, notably those focused on large cities such as Berlin and Warsaw, and the rapidly emerging northern Italian district. Even so, the map is but a generalization of Europe's total industrial strength. We have highlighted only the really outstanding areas. Other noteworthy concentrations of industry exist in eastern Sweden, around Stockholm (noted for paper, wood products, textiles, precision instruments), in northern and eastern Spain, and in east-central France, where the focus is on the Lyon textile complex. These industrial areas, just districts in Europe's massive industrial structure, would stand out as major concentrations in other parts of the world.

North American Manufacturing Region

Europe's industrial prowess notwithstanding, the North American industrial complex has no rival in the world today. Sustained by a wide array of natural resources and spurred by networks of natural as well as man-made transport networks, remote from the devastation wrought by wars in other industrial regions, and positioned on the doorstep of the world's richest market, North American manufacturing developed rapidly and successfully. Ample capital, mass production, specialization, and diversification mark the growth of this region.

The bulk of American manufacturing (in the United States and Canada) is concentrated in the region delimited on Fig. 8.6, a region that extends from the Eastern Seaboard to Minneapolis and from Montreal to St. Louis. Manufacturing in North America began in *New England* as early as colonial times, but New England is not especially rich in mineral resources. This oldest manufacturing district still produces leather goods, textiles, metal goods (hardware, kitchen goods), and other light manufactures. It is one of those manufacturing districts where skills are generated, the kinds of skills that contribute importantly to balance in industry. But, as we know, other areas have overtaken New England in terms of productivity and diversity. As white settlement in North America marched westward, manufacturing spread as well. And farther west is where some great mineral deposits, fuels as well as ores, were discovered.

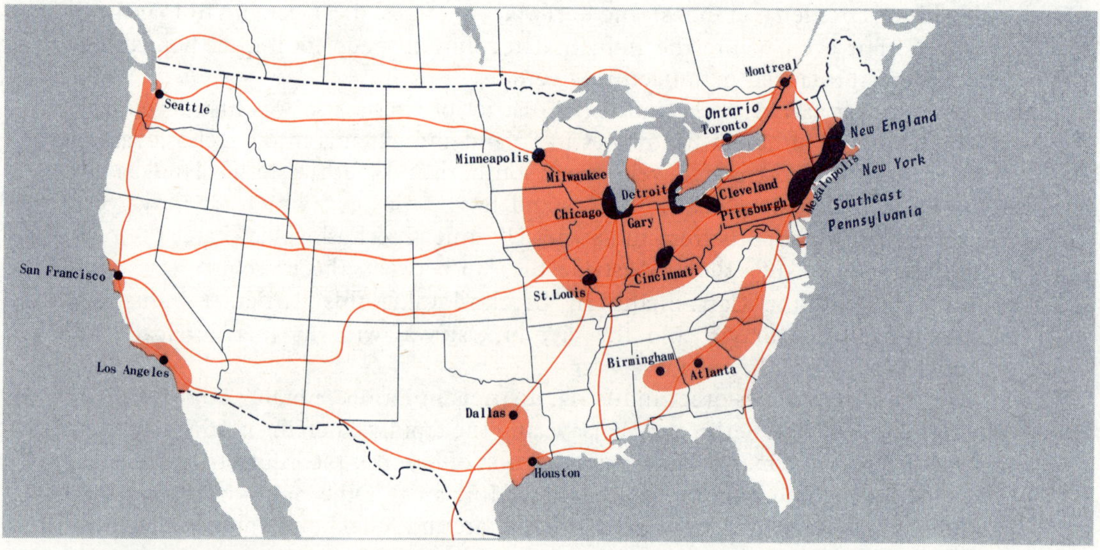

FIG. 8.6 Industrial regions of North America.

Another district of rather early importance centers on metropolitan *New York*, at the very heart of megalopolis and the locus today of tens of thousands of industrial establishments. Again, New York and its immediate environs are not especially well endowed with mineral resources, but we have here a situation not altogether different from the one we encountered in the cases of London and Paris. The agglomeration process plays a major role, New York being the country's largest, most concentrated market (and lying at the heart of a larger market still), the place where a huge skilled and semiskilled labor force exists, the focus of an intensive transport network, one of the world's greatest ports and break-of-bulk locations. New York is also a center for business and finance; numerous companies with nationwide tentacles have their head offices, their management in the city. As for the character of its manufacturing, the New York district, like New England but on a more massive scale, is an area of predominantly light industries. Clothing is made here, and books, metal goods, various kinds of food. There is also some treatment of incoming commodities from overseas, for example, the refining of petroleum and sugar.

Still farther south along the Eastern Seaboard, the southern end of megalopolis stands out as a district of heavier manufacturing. This district centers on *Southeast Pennsylvania*, specifically Philadelphia, but it extends on to Baltimore. As we noted earlier, iron ores shipped all the way from South America and Canada are smelted right on the waterfront in "tidewater" mills; in the past the district's raw materials came from local sources. In addition to steel mills, there are chemical industries, textile factories, and lighter manufacturing plants. A large market is clustered here, a substantial labor force, and some of the major ports on the Eastern Seaboard; and an excellent transport network serves the district.

Looking farther west, we can discern a rather well-defined industrial district in interior *New York State*, extending approximately from Albany, on the Hudson River, and Buffalo, on the shore of Lake Erie. As in England, much local specialization occurs in this district, and a wide variety of manufactures is produced. Some familiar city names and products here: Rochester for its cameras and optical products, Albany for textiles, Syracuse for shoes, Buffalo for its flour mills and steel-making plants. The old Erie Canal used to serve this district, dug early in the nineteenth century to connect the East Coast to the (then) far interior.

On the other side of Lake Erie (that is, on the northern and western side) lies Canada's *Ontario* district, really a continuation of the Albany-Buffalo industrial belt. Toronto is the focus of this district, which is Canada's industrial heartland and which extends all the way to Windsor, opposite Detroit. Toronto and Hamilton manufacture steel, and various locales in the region manufacture products ranging from heavy machinery to clothing to canned foods. Figure 8.6 shows Montreal also to lie in an area of manufacturing. The Montreal-Ottawa area, along the St. Lawrence River, is no match for the Ontario district, but it has one big advantage: cheap hydroelectric power. This, we know, attracts such industries as aluminum refining and papermaking; other industries have followed suit and now flour mills, sugar refineries, and textile plants have located there.

Returning to the United States, Fig. 8.6 shows several major industrial districts to lie west of New York State, including those we might identify as Pittsburgh-Cleveland, Detroit-southeast Michigan, Chicago-Gary-Milwaukee, and, on a lesser scale, those focused upon Minneapolis, St. Louis, Cincinnati, and a large number of smaller but still substantial urban centers. Collectively we might identify these as the United States *Interior* industrial districts, but to try to identify the really huge variety of products created here would be pointless. Many of them are common knowledge. Who thinks of Pittsburgh without thinking at the same time of its huge steel mills, or of Detroit without the image of automobiles, or of Gary (Indiana) and Chicago without miles of furnaces and mills and Lake Michigan barges bringing loads of iron ore from Lake Superior shores? This is where United States industrial power really transforms the landscape, where Appalachian coal and Mesabi iron ore are converted and autos, bulldozers, harvesters, armored cars, and tanks roll off the assembly lines. Computers, refrigerators, record players, toys, cornflakes, pills—these and thousands of other products pour from the factories of the Midwest.

Figure 8.6 reminds us, though, that this industrial heartland we have been discussing is not the only region of significant development in the United States. A region we might call the *Southern* district extends from the vicinity of Birmingham, Alabama, to Richmond, Virginia. Birmingham is a name associated with iron and steel in this country as well as Britain, and local raw materials have sustained this industry here for many years. Atlanta, rapidly becoming the South's urban focus, has a growing industrial base. Cotton and tobacco rank high in this district's array of manufactures, and, as we noted, timber is an important product also. On the Gulf of Mexico, the *Texas* district is emerging, centered on major urban areas of Houston and Dallas-Fort Worth. The oil fields generate a major petrochemical industry here, and meat-packing and flour-milling industries are also substantial. We know Houston from space-flight reports emanating from there ("This is mission control . . ."), and the new aerospace industry employs thousands of people in this industrial district as well. Denver, also outside the major industrial region of the United States, shares substantially in the growth of this industry. And finally, our map shows three manufacturing areas lying on the west coast, in the Los Angeles, San Francisco, and Seattle environs. These *West Coast* districts respond to the West's growing markets; the particular physical environments have boosted food-processing industries (citrus products in southern California, wines in the vicinity of San Francisco). Many of the aircraft you may have traveled on are built in the Seattle area, where the aerospace industry is of major proportions. Still, these large cities with their substantial industrial complexes are but outliers of the North American industrial region, the most productive world region of its kind.

Soviet Manufacturing Region

As one would expect in so vast a country, the Soviet Union's manufacturing districts are widely dispersed. Two of them, the districts around Novosibirsk and Karaganda in Soviet Asia, do not appear on our map (Fig. 8.7), which focuses on the western part of the U.S.S.R. Although those distant districts show signs of developing into important centers, they are so far dwarfed by the districts shown on our map: the area around Moscow, the Ukraine industrial district, the Volga district, and the district that lies astride the Urals. These lie in what some geographers would term "European Russia," still the most populous region of the country and demarcated to the east by the Ural Mountains.

The Moscow district, often called the *Central Industrial Region,* is the Soviet Union's oldest industrial district. It is not especially well endowed with natural resources, but it has other advantages. With the present distribution of population, Moscow has great centrality. Roads and railroads radiate in all directions, to the Ukraine in the south, to Minsk, Belorussia, and Europe in the west, to Leningrad and the Baltic coast in the northwest, to Gorky and the Ural district in the east, and to the cities and waterways of the Volga in the southeast. A canal links the city to this, the U.S.S.R.'s most-used navigable river. Moscow is the focus of an area that includes some 50 million people (more than one-fifth of the country's total population), many of them concentrated in such major cities as Gorky, the automobile-producing "Soviet Detroit" (over 1 million), Yaroslavl (½ million), the tire-producing center, Ivanovo, the "Soviet Manchester" with its textile industries, and Tula, the mining and metallurgical center of the south.

The comparative paucity of natural resources within the Central Industrial Region is counterbalanced by this excellent transport system, which facilitates the

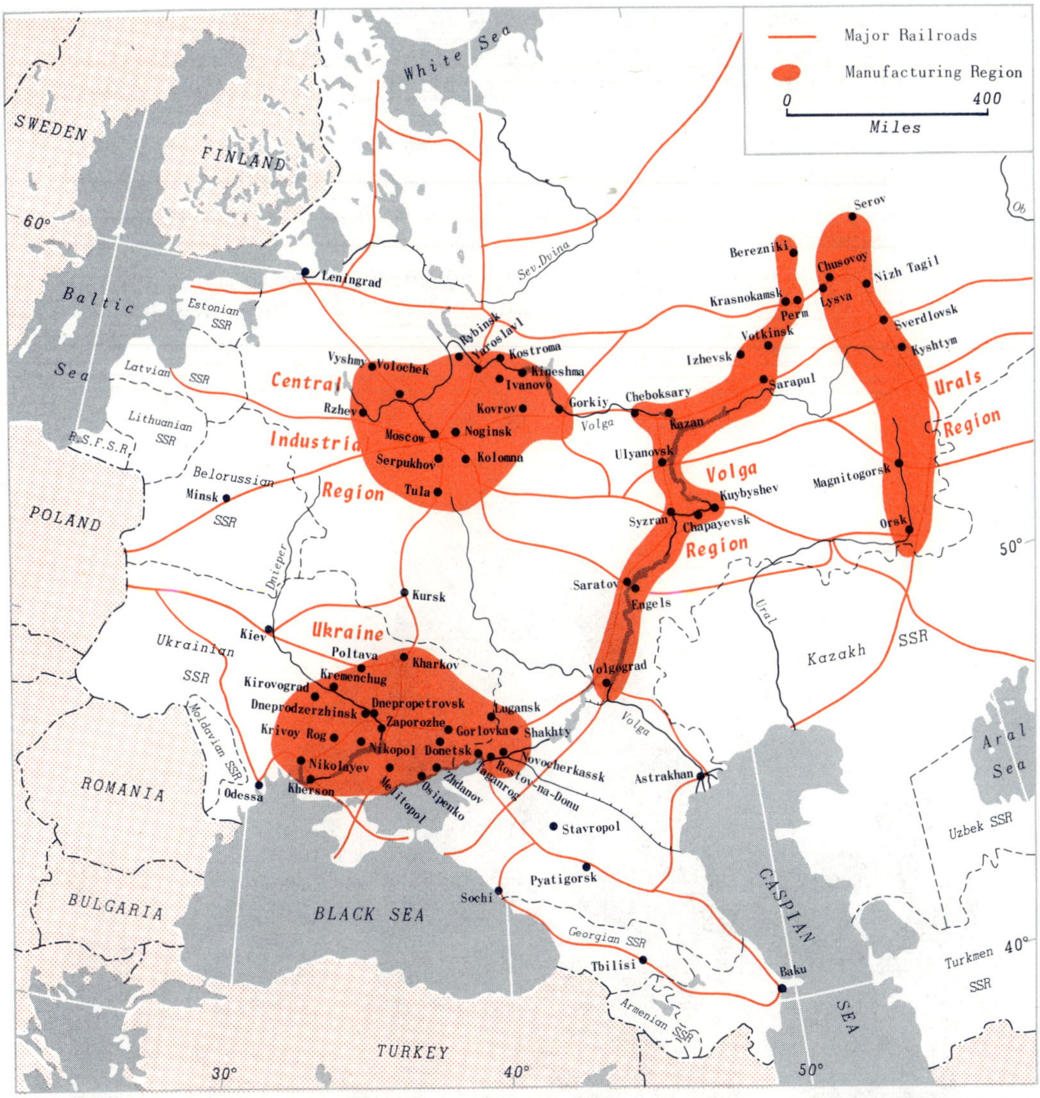

FIG. 8.7 Industrial regions of the Soviet Union.

inflow of a wide range of commodities, gas from Stavropol, petroleum from the Volga fields, and electricity from Volga River dams for energy, and metals, wood, wool, cotton, flax, and leather (among many other items) for processing. The major flow, other than that of food, is of these materials into the Central Industrial Region and of finished products being distributed to the consumers. Thus the manufactures of the district are marked, as are those of the London and Paris areas, by high values in relation to bulk. Moscow has always been the leading center of skilled labor in the Soviet Union, and the available skills are those that, as Russia found out after Western Europe had, do not develop quickly and cannot be imparted overnight to peasant labor freed of farm work. Primacy, time, and place have favored Moscow and the Central Industrial Region.

In the *Ukraine* the Soviet Union has one of those regions where major deposits of industrial resources, fuels and ores, lie fairly close together. The Ukraine began to emerge as a major region of heavy industry toward the end of the nineteenth century, and one major reason for this was the Donets Basin, one of the world's most productive coalfields. This area, known as *Donbas* for short, lies north of the city of Rostov (Fig. 8.7). In the early decades this Donbas field produced

Leningrad, the Soviet Union's second city, lies on the northern margin of the Central Industrial Region, overshadowed by Moscow. Still, Leningrad has a substantial industrial base, and it is an important Soviet shipbuilding center.

over 90 percent of all the coal mined in the country. Other fields have since been opened up, but even today, three-quarters of a century later, the Donets Basin alone still accounts for between one-quarter and one-third of the total Soviet coal output. And the coal is of high grade, some of it is even anthracite, and most of it is suitable for coking purposes.

What makes the Ukraine unique in the Soviet Union is the location, less than 200 miles from all this Donbas coal, of the Krivoy Rog iron ores. The quality of this ore is very high, and although the best ores have been worked out, the field still produces. Major metallurgical industries arose on both the Donets coalfield and near the Krivoy Rog iron ores: Donetsk and its satellite, Makeyevka, dominate on the Donets side of the district (this is the "Soviet Pittsburgh"), while Dnepopetrovsk is the chief center in the cluster of cities on the Krivoy Rog side. In one way or another all the major cities located nearby have benefited from the fortuitous juxtaposition of minerals in the southern Ukraine: Rostov, Volgograd, and Kharkov, and even Odessa and Kiev farther away. Like the Ruhr, the Ukraine industrial district lies in an area of dense population (and hence, available labor), adequate transportation, good agricultural productivity, and large local markets. And when the better ores were exhausted, somewhat lower-grade ores on the basis of which production could be maintained proved to be available. The biggest development in this context is the discovery of the so-called Kursk magnetic anomaly, positioned north of the Ukraine district and south of the Central Industrial Region (Fig. 8.7), and thus able to serve both districts. The Kursk magnetic anomaly is a large iron-ore deposit whose exploitation had been delayed by technical problems. Mining began in 1959 and the ores have given the metallurgical industries from Tula to Rostov a new lease on life.

Two other, elongated industrial districts appear on Fig. 8.7. Their elongation relates to their uniting physiographic feature: the Volga River and the Ural Mountains. The *Volga* region has witnessed major development in the decades starting during the Second World War. With the Ukraine and the Moscow area threatened by the German armies, whole industrial plants were dismantled and reassembled in Volga cities, protected from the war by distance. Kuybyshev was even the Soviet capital for a time. Then, after the war, the progress set in motion earlier continued. A series of dams were constructed on the Volga, and electric power became plentiful. Oil and natural gas proved to exist in quantities larger than anywhere else in the U.S.S.R. Canals linked the Volga to Moscow and the Don River, making the importing of raw materials easy. The cities along the river Volga, spaced at remarkably regular intervals, developed specializations: Kazan is known for its leathers and furs, Volgograd for its metallurgy, Saratov for chemicals, Kirov for wood products, and Kuybyshev for its huge oil refineries. This district's contribution to the Soviet economy is still increasing rapidly.

The last major industrial district on our map is the *Urals* region, the easternmost district of major importance. This area, too, developed rapidly during the Second World War. But there has been nothing artificial in its growth. The Urals yield an enormous variety of metallic ores including iron, copper, nickel, chromite, bauxite, and many more. The only serious problem is coal. There is not enough of it, and what there is does not have the required quality. So coal is railed in all the way from the Kuznetsk Basin, a thousand miles to the east. In the cities of the Urals district, metals, metal goods, and machinery are produced in great quantities. With the Siberian centers, the Urals district now produces as much as half of all the iron and steel made in the entire Soviet Union. It is a sign of the eastward march of the country's center of gravity and a part of the national planners' grand design.

Japan-China Industrial Region

Japan at present is the world's third industrial power, and if current rates of growth continue, Japan will be in second place (after the United States but surpassing the U.S.S.R.) within a few years. This is all the more remarkable when it is realized that Japan has a very limited domestic natural-resource base. Much of what Japan manufactures is made out of raw materials bought all over the world and shipped to that small archipelago off the East Asian coast.

In Chapter Five we noted how Japan rose from comparative obscurity, from an underdeveloped country to a modern industrial power in just one century. But we should not forget that Japan, prior to its modernization era, did have a domestic manufacturing industry. During the first half of the nineteenth century manufacturing in Japan, of the handcraft variety, was widespread. In home industries and community workshops the Japanese produced silk and cotton-textile manufactures, porcelain, wood products, and metal goods. During those days the small domestic resources were enough (Japan does have some ores) to supply the local industries. Power came from human arms and legs and from wheels driven (as in England) by water. The chief source of fuel was charcoal. Thus there was an industry, there was a manufacturing-oriented labor force, and there were known manufacturing skills.

Japan's push toward industrial modernization began with the integration and enlargement of these community and home workshops, the export to world markets of light manufactures, and the adoption of the techniques generated by the industrial revolution. There was some coal, and Japan's topography produced numerous sites for hydroelectric power development. As industry grew and the Kyushu coalfields became worked out, it was realized that raw materials would have to come from overseas. The age of the factory had arrived. The coastal cities were in the best position to receive the raw materials and to convert them to finished products, and the regional industrial pattern shown in Fig. 8.8 took shape.

Figure 8.8 reveals that the largest district of manufacturing in Japan is the *Kanto Plain,* which contains not only Tokyo (the capital and largest city) and Yokohama, its major port, but also several smaller but substantial industrial towns such as Kawasaki and Chiba. The Kanto Plain and its giant conurbation produces between one-fourth and one-fifth of the country's industrial output. Tokyo is the leading steel producer, with ores coming from as far away as the Philippines, Malaya,

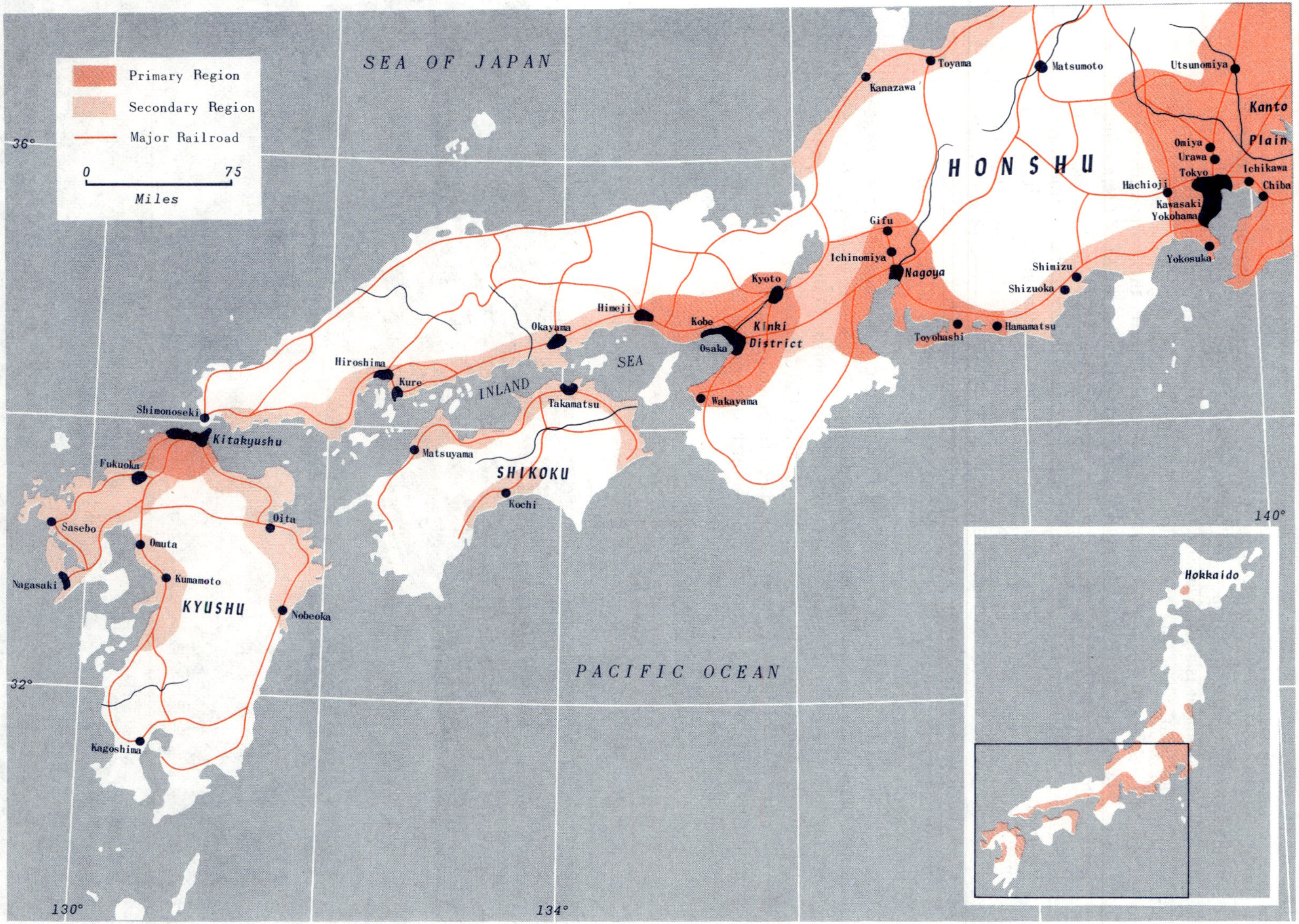

FIG. 8.8 Industrial regions of Japan.

India, and even Africa. Although hydroelectric power is brought in from the interior mountain sites, coal is shipped to Tokyo from a new field in Hokkaido (the northernmost major island of Japan) and from Australia and America. Petroleum comes from Southwest Asia and Indonesia. The Japanese are able to pay for all this by selling enormous quantities of their manufactures on the world's markets.

The second-ranking manufacturing district in Japan is the *Kobe-Osaka* area, with Kyoto as an outlier (Kyoto is the ancient Japanese capital, still a significant light-manufacturing center). This district, long an area of diversified manufacturing, has been developing heavier industries in recent decades and now produces not only textiles and chemicals but also iron and steel, machinery, ships, and airplanes. The Kobe-Osaka district, as Fig. 8.8 shows, lies at the eastern end of the Inland Sea. The Inland Sea in many ways is the heart of Japan. Overlooking its waters are dozens of large and small cities, and crisscross traffic is as busy as on any of the world's major rivers.

Between the Kanto Plain and the Kinki district (the general name for the Kobe-Osaka-Kyoto triangle) lies the *Nagoya* district, where a wide variety of textiles are made, and where heavy industries are also developing. Many of the Japanese automobiles you see on the world's roads are built at Nagoya.

At the opposite end of the Inland Sea from Kobe-Osaka lies a fourth major industrial district, called *Kitakyushu*, a conurbation of five north Kyushu cities.

Most of the raw materials that are the bases of Japan's huge industrial machine are brought in from overseas, and so the country's industrial complex is oriented toward the coast. Here the foundries of the Sumitomo Metal Industries can be seen positioned at the shoreline, which enables the large supply ships to deposit their cargoes directly into the factory.

Here were Japan's first coal mines, and here the first steel mills of modernizing Japan were built, which for many years remained the largest the country had. Look for this district to develop even faster than it has, should the trade normalization with China continue. No place in Japan is located in a better position to do business with mainland Asia. Presently, heavy industries dominate, with shipbuilding and iron and steel making in the lead, supplemented by a major chemical industry and numerous lighter manufacturing plants.

Only one manufacturing district in Japan, in the area depicted by our map, lies outside the belt extending from Tokyo in the east to Kitakyushu in the west. This is the district centered on *Toyama,* on the Sea of Japan. The advantage here is cheap electricity from nearby hydroelectric stations, and the cluster of industries reflects it. Paper manufacturing, chemical industries, and textile plants have located here. Of course, our map gives but an inadequate picture of the variety and range of industries that exist throughout Japan, many of them oriented to the local (but significant) market. Thousands of manufacturing plants operate in the cities and towns other than those on the map, even in the cold, northern island of Hokkaido. As we did in the case of Europe, North America, and the Soviet Union, we focused here on the really outstanding manufacturing regions of Japan.

CHINA China has more than eight times the number of people of Japan, but its industries cannot match those of its island neighbor. Still, China too is on the move, and we ought to look at the places where its progress in industry is most evident. In the past, China's industries were established and maintained by foreigners, first the Europeans, who built factories in the coastal cities to process goods to be sold at home, and then the Japanese, who invaded Korea and established a sphere of influence in Manchuria, China's northern province (Fig. 8.9). As in the case of the Soviet Union but on a smaller scale, some dispersal of industries into the interior was occasioned by the Second World War. Chungking and Chengtu, for example, grew substantially during this period. But the major industrial development of China has occurred during the modern, Communist period. When the Communist planners took over in Peking, one of their priorities was to develop as rapidly as possible China's own resources and industries.

China is a very large country, and it is likely that some of its natural resources have yet to be discovered. Even so, China has already proved to possess a substantial domestic resource base, one that is far superior to Japan's, for example. In terms of coal, there is hardly a limit on industrialization in China. The quality is good, the quantity enormous, and many of the deposits are near the surface and easily mined. China's iron ores are not so productive and are generally of rather low grade, but new finds are frequently made, and the picture has improved steadily.

China's industrial heartland extends from Manchuria to the Lower Yangtze, from the vicinity of Shenyang to the environs of Shanghai. As Fig. 8.9 shows, this industrial region is not contiguous, but consists of three major and several minor districts. *Southern Manchuria* has become China's industrial heartland, and the northeast's largest city, Shenyang (formerly Mukden), with a population of over

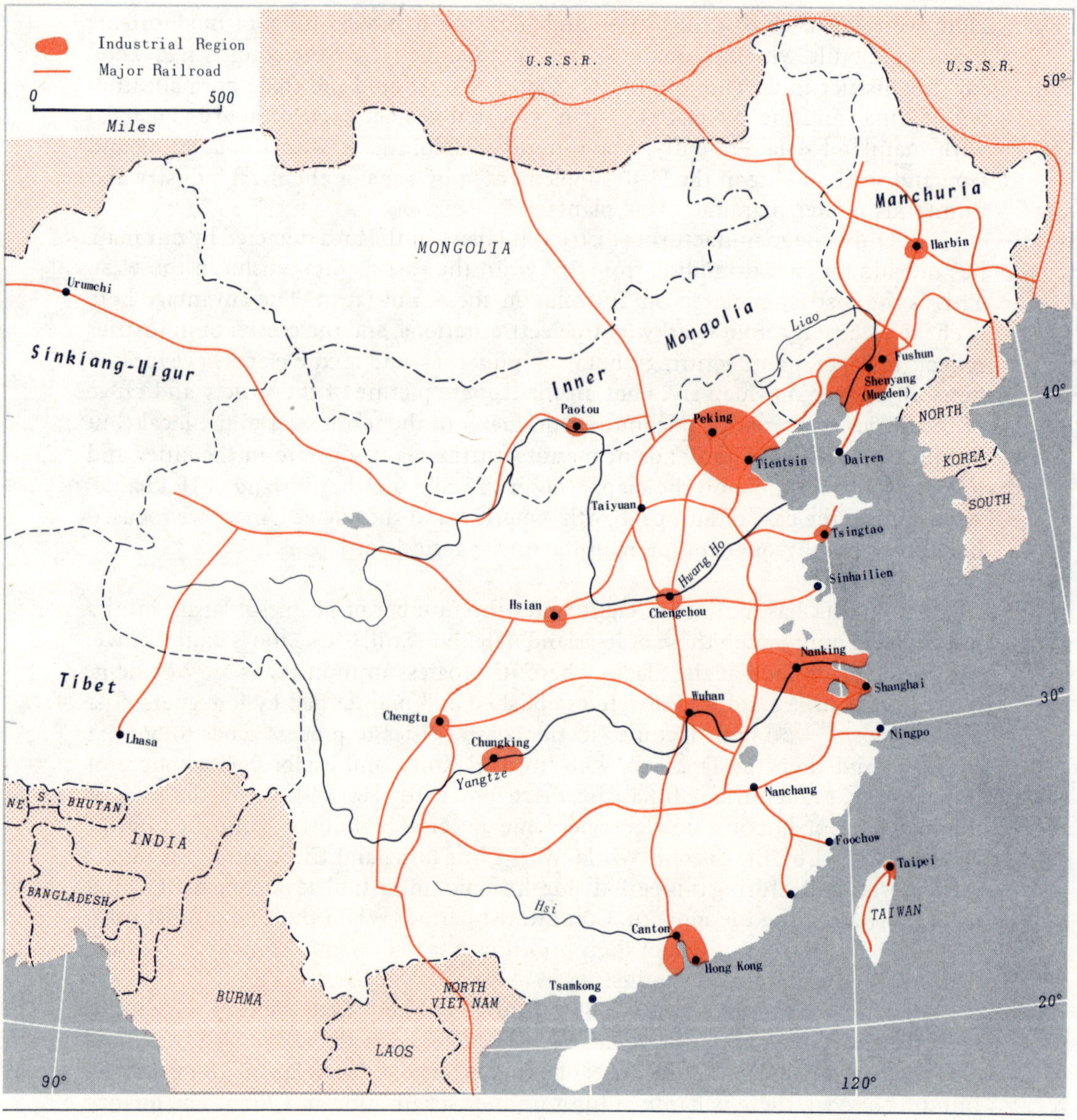

FIG. 8.9 Industrial regions of China.

5 million, has emerged as a "Chinese Pittsburgh." This is based on Manchuria's coal and iron deposits, which lie in the basin of the Liao River. There is coal within 100 miles to the west of Shenyang and about 30 miles to the east, at Fushun, as well; to the southwest, just 60 miles away, there is iron ore at Anshan. Since the iron ore is of rather low grade, the coal is hauled from Fushun to the iron reserves.

Under the circumstances this is the cheapest way to make iron and steel. Anshan has become China's leading iron- and steel-producing center, but Shenyang has the largest, most diverse overall industrial complex, with metallurgy, machine making, and other large engineering works. Before long the city of Harbin will become part of a contiguous Manchurian industrial region. Already, Harbin with $2\frac{1}{2}$ million residents has plants making farm machinery, textile factories, food-processing plants, and light manufacturing.

The *North China* district, including the capital, Peking, the port of Tientsin, and the city of Tanghsien (among others), also benefits from nearby coalfields. As one might expect, in the productive basin of the Hwang River, food processing and textile making rank next to the heavy industries. Still farther to the south lies the third district in China's industrial heartland, the *Yangtze* district. On our map the Yangtze district divides into two areas, one centered on Shanghai, China's largest industrial city, and the other focused upon Wuhan, astride the Yangtze several hundred miles upstream. In fact, still farther upstream along the Yangtze River lies Chungking, the city that grew so strongly during wartime. Whether we view this Yangtze district as one region or as three, this too is a Chinese industrial heartland, if not in terms of iron and steel production alone, then in terms of its diversification of production and its local specializations. Railroad cars, ships, books, foods, chemicals—an endless variety of products comes from the Yangtze district.

Industrial growth by agglomeration and in response to the expansion of urban markets has generated manufacturing centers in and near the Canton-Hong Kong area and in China's larger interior cities. The fortuitous wide dispersal of the country's raw materials places many such cities near natural resources, stimulating industrial development and broadening China's industrial base.

Industrialization Elsewhere

Compared to the regions we have discussed so far, industrial development in other parts of the world is minor (Fig. 8.10). Industrialization in *India* is yet in an initial phase of development. India's industrial core area lies on the margins of the Chota Nagpur upland. There is good coking coal in West Bengal (though not as much as India would wish), and high-grade iron ore in the same general area. Jamshedpur has become the center of this manufacturing district, positioned fortuitously on the doorstep of the great Ganges population cluster and not far from Calcutta, the great urban focus here. Unlike its Southwest Asian neighbors, India does not have large oil fields (some oil is being produced in Assam), and although there are good hydroelectric sites in many parts of India, their development is expensive and the rate at which they can be developed is necessarily slow. Still, with the coal situation deteriorating to the point where India must consider importing this commodity and the expense of moving locally available coal over great distances to India's widely dispersed cities, the Indian government is investing quite heavily in hydroelectric projects, especially those near major cities (Bombay and Madras, for instance,

FIG. 8.10 Smaller, developing industrial regions of the world.

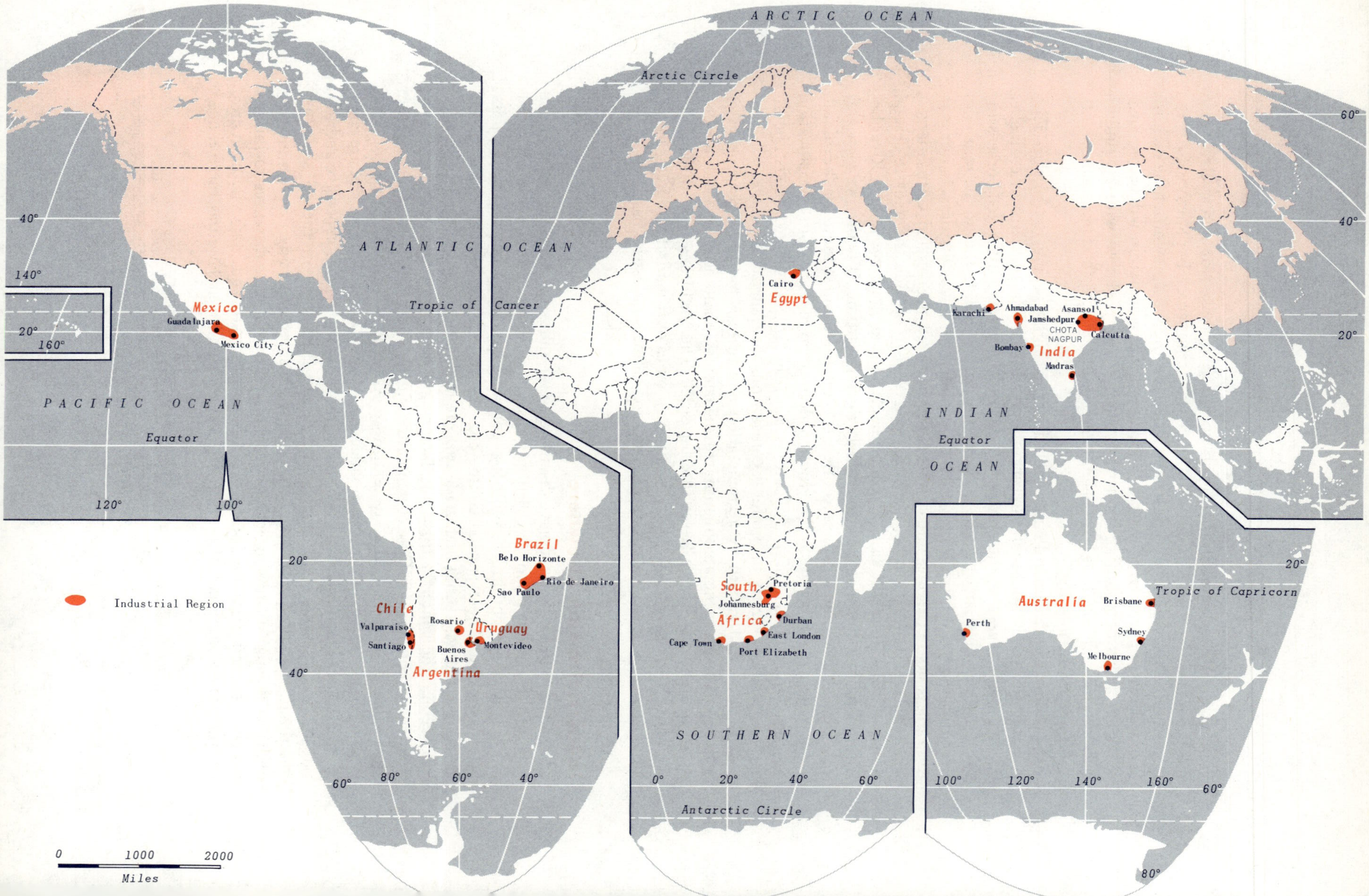

are served by hydroelectric power) and those where dam construction leads not only to improved power supply but also to new irrigation areas and improved flood control.

In addition to iron and steel industries in Jamshedpur, India has a major cotton-manufacturing industry, centered on Bombay and Ahmadabad (India is the world's second-ranking textile exporter, after Japan), and has, in Calcutta, a major center of diversified manufacturing. Calcutta has railway assembly plants, food-processing factories, jute mills, chemical industries, printing plants, and a host of other industries mainly oriented toward the domestic market.

This last factor, orientation and response to local markets, is an important one to consider in any discussion of world manufacturing. In developed countries such as Australia and New Zealand and in developing countries such as Brazil and Nigeria, manufacturing is encouraged primarily *not* because it is hoped that the products of the factories will immediately sell well on the world market. The hope is to reduce the country's dependence upon foreign manufactures, which are expensive to buy. Light manufacturing, then, dominates in those areas, and tends to concentrate in and near the largest local markets, the urban centers. In Australia, manufacturing centers on Sydney, Melbourne, Brisbane, and Perth; production for the domestic market includes the making of clothing, shoes, foods and beverages, and similar industries. Australia, of course, does have a substantial export trade in certain foodstuffs (dairy products, meats) and wool, and some of the country's manufacturing involves the processing and packing of those goods.

In Africa, a growing industrial complex lies in the southern African interior, in the southern Transvaal and the northern Orange Free State. Johannesburg lies at the center of this industrial district of some 4 million people, which includes the ironworks and steelworks near Pretoria and steel mills in the Orange Free State. This complex had its origins in the Witwatersrand goldfields and subsequently benefited from the presence of other, industrial resources nearby. Lesser industrial districts have developed around Cape Town and Durban, and at Port Elizabeth, all on the South African coast. South Africa is blessed with a wide range of mineral resources, and its industrial strength is considerable, far ahead of any other African country. In second place might be Egypt, which boasts a steel mill built about 20 miles south of Cairo where iron ore from the country's southern reserve is smelted with coking coal imported by way of the Nile. Egypt has also developed a rather substantial textile industry, based upon its agricultural output principally of cotton.

In South America, substantial industrial development is taking place in few areas, and the pattern of light manufactures for local markets, concentrated in the major cities, prevails. The exceptions are in the Rio de Janeiro-São Paulo district of Brazil, where some heavy industries including steel making have been established (Brazil has coking coal and iron ore among its domestic resources), in the core area of Argentina and neighboring Uruguay (mainly light manufacturing), and in central Chile, where industries are likewise serving the local market (some Chilean plants process mineral ores prior to export).

TERTIARY INDUSTRIES

Our discussion of industrialization would be incomplete without a look at a set of industries that do not (as the primary and secondary industries just discussed do) generate an actual, tangible product. These are the tertiary industries, including services such as transportation and communication, financial services (banking is such an industry), retailing, recreation, even government. Here the measure is not the quantity of a commodity produced, but the effectiveness with which the particular service has been performed.

Transportation

Few maps of the world are as significant as Fig. 8.11, which shows the distribution of railroads, their network densities—and their absence over large parts of the inhabited world. In the Northern Hemisphere, the two regions with the most intensive network are Europe, especially Western Europe, and the United States, especially the eastern United States. The evolution of these networks went hand in hand with the industrialization of those regions, and it is significant now to see tentacles of the Soviet system reaching out into interior Soviet Asia, with interconnections developing between the major arteries. We are seeing here the deliberate policy of national integration of the Soviets' distant eastern frontier into the state's framework. The colonial powers recognized the effectiveness of railroads in establishing and confirming spheres of influence, and many of those railroads you see in Southeast and South Asia and Africa were built by European powers for multiple purposes. In the case of India, the railroad network seems more intensive than would appear to be commensurate with that country's level of industrialization, but in crowded India the railroads have long played an extraordinarily important role in moving large numbers of people between the major cities. Africa's railroad pattern, on the other hand, looks quite different. Note how many rail lines commence in the interior and run to the coast, without any interconnections. This reflects the economic situation of the continent, long an exporter of raw materials hauled from the interior to ocean ports and then shipped to the major manufacturing regions of the developed world. In Africa, South Africa is the sole exception to this situation. South Africa, we noted, is also Africa's most developed country industrially.

Australia's railroad system is yet incipient, with just a single artery extending from east coast to the west and north-south connections existing only on the eastern seaboard. In Australia as well as Argentina, the railroad system (in Argentina it is a true network) reveals the character of development, still peripheral in the former but strongly clustered and central in the latter.

Paths, roads, highways, waterways, airways, even pipelines are part of the transport complex. And under the rubric of "communications" as a tertiary industry we should even add mail service, telephone service, radio, and televison. The transport and communications industries face constant change as modernization dates existing facilities. This is happening to the railroads, notably in the United

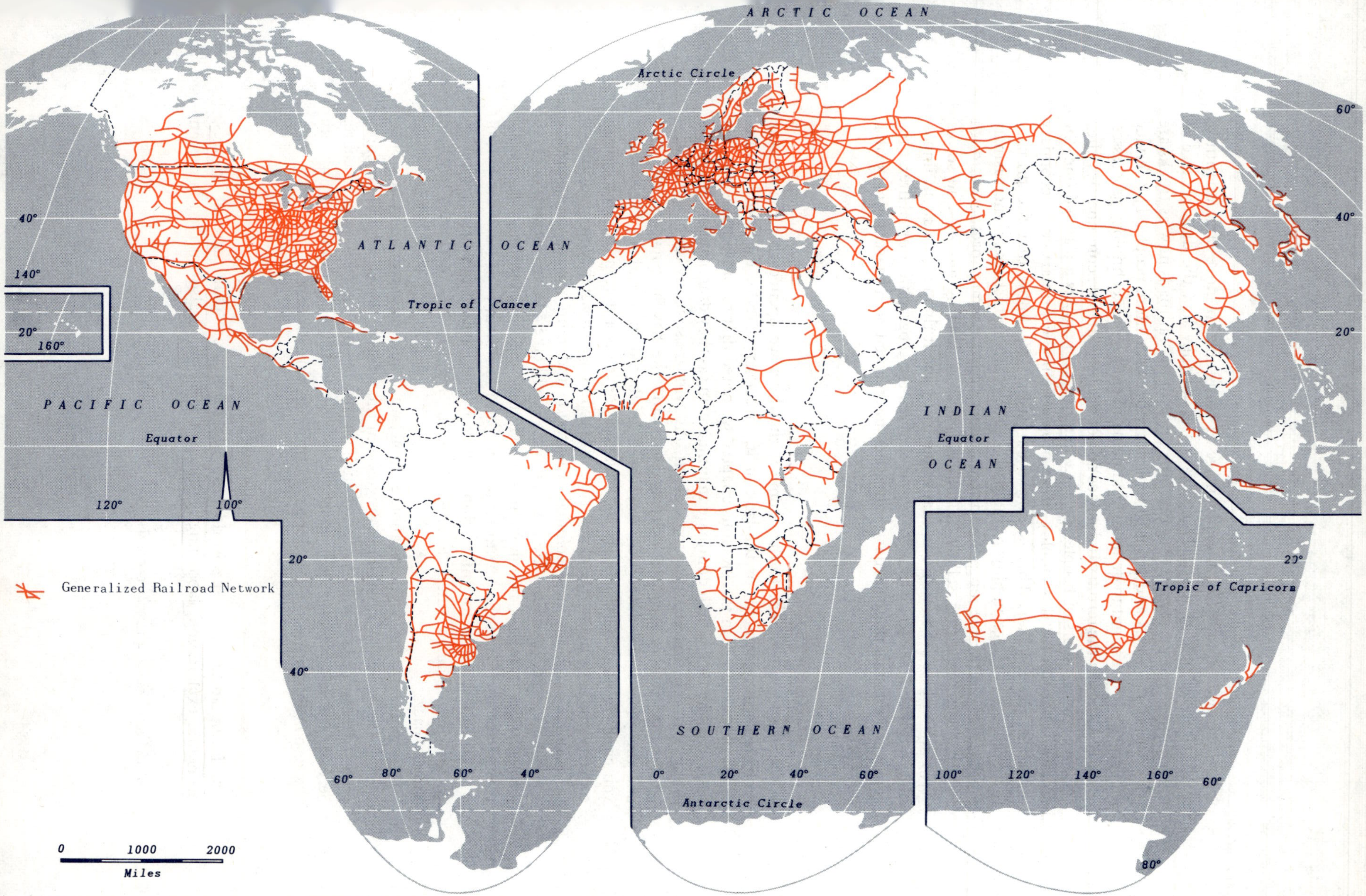

FIG. 8.11 The world railroad network. (After a map in *Man's Domain*, McGraw-Hill, New York, 1968, p. 12. Base Map © copyright, General Drafting Co., Inc.)

States; railroads always had the advantage of being able to move large quantities of goods and people at relatively modest cost, but they had the disadvantage of inflexibility. That disadvantage is catching up with the railroads now as trucks are taking away the goods and buses and planes move ever more people. The service industries must respond to demand, and demand changes as technology progresses.

Although the tertiary industries do not produce tangible commodities, they employ large numbers of people. Except where the economy is a planned one, the tertiary service industries employ *more* people than the primary and secondary industries put together. This is not so surprising. We know that the primary industries are not large employers and that the secondary industries are increasingly mechanized. So the big job is to bring producer and consumer together, through advertising, transporting, selling, installing, servicing, lending money, and so on. The relationships are so intricate that generalizing about them is just about impossible, but if you take some time to browse through the geographic journal literature, you will find that these tertiary industries have posed some fascinating spatial problems.

DEVELOPMENT?

We return for a moment to the theme with which we opened this chapter, the problem of development, or the lack of it, in certain parts of the world. Again, in this discussion as in the previous chapter, we noted that the advantages all seem to lie with the DCs, which dominate manufacturing as they dominate other spheres of life. Why has this situation come about, and why are so many of the advantages concentrated where they are? Why the have and the have-not nations, the rich and the poor? Those differences have been attributed to just about every factor you could think of. Climate has been blamed for it (as a determinant of human capacity), as have innate racial ability, cultural heritage (resistance to innovations, for example), and, of course, colonial exploitation and, more recently, neocolonialism. And indeed, the causes are many, though it will take a lot of research to prove any of these factors as major villains. Obviously, the distribution of accessible natural resources and the territorial fortunes and misfortunes of states are part of it. The opportunities for interaction and exchange have always been greater in certain areas than in others. Isolation from the mainstreams of change has held peoples back.

But before we go on to another topic, we should raise two questions. First, there is the question of whether "progress" in the terms to which we have become accustomed is the kind of progress to which the developing world will want to aspire. Second, and less philosophical, more urgent, is the question of what, if any, are the responsibilities of the developed countries in this context.

Progress

The Europeanization of the world brought with it more than material changes, economic exploitation, labor migrations, and taxation. It also injected European philosophies of government, religion, justice, and other values into non-European societies. One of these values is the concept of progress, the kind of progress that is reflected by productive factories, busy shopping centers, skyscrapered cities and by longer, healthier lives, improved education, and greater personal security. Government leaders of UDCs seek to emulate Western countries; the Western standards of progress have pretty well become universal objectives. Such UDC governments note that there are developed countries that, like the Netherlands and Denmark, have achieved a great deal without really substantial mineral resources, and they note the rise of powerful Japan, a non-Western industrial country with economic tentacles all over the world. And so, urged on by Western economists, they pour their scarce capital into industrialization programs, their expectations high, even when that capital may be direly needed elsewhere in the domestic economy. The UDCs note those statistics that say that in the DCs, more labor is in industry and

Home workshops, community industries, and handcrafts are important to any society seeking to progress economically. Nurturing and supporting these industries may be more efficacious for a country than establishing large, modern industrial plants, plants that do not use much labor and that may damage the domestic economy by putting some of the home industries out of business. This old loom makes textiles as it did long ago: human hands and attention go into the making of multicolored cloth.

less in agriculture. So the answer must be to generate manufacturing, to put more of the labor force to work in factories. But what actually happens? Modern industry is not labor-intensive; it simply does not require all that much labor. And those new plants may do damage at home. They may put the home workshops, the community industries, the handcrafts, out of business. So no matter what economists have written about subordinating the interests of farming to manufacturing to get to the so-called take-off point, the UDCs were misguided. Societies based largely on variations of the subsistence-agriculture theme need industries subordinated to farming. They need fertilizers and tools. They need to nurture the workshop industries.

That rather puts a dent in Western notions of progress. Now, let us briefly consider the role of those countries that have the resources, the factories, the markets, and the money. Should we simply say that that's the way life is and watch the rich countries get richer, the poor more populous? Perhaps not. These days, we frequently hear commentaries about how the underdeveloped countries are now free and should be able to succeed on their own merits, the colonial yoke having been removed. But it is not quite that simple. Thrust into a world system of exchange and capital flow, the outlines of which we have just reviewed, these UDCs find themselves falling further and further behind, not because they will not reform or because they resist "progress," but because the character of world development has been that the DCs, already way ahead, will make the gap still bigger. The advantages the areas of early industrialization, the areas where agglomeration took place and continued even after the resources began to run out, had over other areas are now the advantages of the DCs over the UDCs. And when the colonial yoke was lifted, all the UDCs got control over was the domestic economy, not the rules of the international system in which they now have to compete. The buyers decide on the quotas of sugar the UDCs may sell to the wealthy market. The countries that can afford it make decisions about aid, not the recipients, except perhaps by toeing an ideological line. And so the DCs, if they are serious about the role the UDCs are to play in preserving a stable and secure world, must respond to the UDCs' demands that the discriminatory system of international commerce be changed, that the UDCs' exports be given a greater chance and greater returns. We, here in the comfort of the well-fed, well-housed world, should afford it. Your grandchildren, could they do so, would demand it.

CHAPTER NINE

The Urbanizing World

NINE

Throughout the world, in India, in the Netherlands, in Brazil, and in North America, people are moving to the cities. Villages are growing into towns, towns into cities, and cities are coalescing into megacities, of which the megalopolis on the Eastern Seaboard of the United States, which extends from Boston to Washington, is an outstanding example. This modern age is indeed the age of urbanization, and the phenomenon is by no means confined to the developed countries. Wherever they are, in Kenya, in Japan, or in New Zealand, the cities attract the people like magnets, and the rural-to-urban migration flow goes on. But the cities often are incapable of accommodating all the new arrivals, as the slums and ghettos that afflict nearly all large urban areas attest. So there is good reason for studying the character of urbanization as a process and the nature of towns and cities. As we noted in Chapter Three, many of the most serious problems of life on this earth—deteriorating environments, the dehumanizing effect of life in overcrowded, congested areas, the breakdown of culture and stability, rising crime—are symptoms of the city especially and problems that remain unsolved.

It is not that urbanization is a completely modern phenomenon. There were cities and towns thousands of years ago, and they, too, had their problems. What *is* new, though, is the greater and still growing *proportion* of populations living in cities. The ratio of urban to rural inhabitants changed rather little for many centuries, but then the onslaught of the industrial revolution came. The population increase accelerated, and the image of the city as a place of jobs and opportunities, combined with the often grinding poverty of rural areas, led to a mushrooming of cities that is unprecedented in the history of the world.

PREINDUSTRIAL CITIES

The industrial revolution and the Europeanization of the colonial world brought great increases in rates of urbanization; industrialization was accompanied by the clustering of ever larger numbers of people in the urban-based labor force. But cities existed long before the industrial revolution began. The phenomenon of urbanization may have made its appearance as long as 6000 years ago, probably in Mesopotamia. It is likely that the first sustained agricultural surpluses were the key. When farming no longer required the constant, daily work of every able person in a society, there was time for other activities. These activities no doubt included early efforts at writing and record keeping, arithmetic, politics, the arts, and religion. It was also possible to consider and plan better ways of defending the community against enemies, to make better tools and weapons, and to store and distribute food. Such developments required the building of structures other than those intended for shelter alone. They were accompanied also by political and social changes. Society was beginning to become more complex. The first stage of the urban revolution had begun.

It is sometimes difficult to tell which modifications in the social order of those early days were causes of urbanism and which were caused *by* urbanism. The earliest towns probably experienced the same growth factors that much later sustained the modern rise of such cities as Paris and London. As the principal centers, crossroads, markets, places of authority, and religious headquarters, they drew the talents, the trade, and the travelers from far around. Where should the beginnings of metallurgy have been based but in such incipient cities? Where would a traveler, a tradesman, a priest, or a pilgrim rest before continuing the journey? And so the city had to have facilities, the kinds of facilities you would not find in farm villages, buildings to house the visitor, to accommodate the protectors, to store the food, to treat raw materials, to worship. Yes, cities grew and drew attention at a price. Their contents of food and wealth spread their reputations far and wide, and there soon were enemies who realized that a takeover of the city would mean power, prestige, and riches. So the early cities built defenses, walls, moats, and the like, and maintained armed forces. Those forces came in handy when it was necessary to exact tribute and taxes from the surrounding rural region and to subdue rebellious elements in the city's sphere, too.

How large were the ancient cities? There is no certainty about this because it is impossible to conclude from excavated ruins the number of people contained in each dwelling unit. But it seems that, by modern standards, the ancient cities were not large. The cities of Mesopotamia and those of the Nile valley may have had between 10,000 and 15,000 inhabitants after nearly two millenia of urban growth in those areas. That appears to have been the maximum size that could be sustained by existing systems of food distribution and social organization. So these urban places were but islands, exceptions in an overwhelmingly rural society. Urbanized societies such as we know did not emerge until several thousand years later.

Cities did grow larger, of course, as time went on. Each advance, every additional element in the culture was reflected in some way in the city. Greece's city-states had capitals much larger than the old Egyptian cities, and Rome grew larger still. Rome may have had more than 300,000 residents at its zenith as the capital of the Roman Empire. Later Constantinople, heir to the Roman Empire, probably reached the same size. This second phase in the urban revolution began approximately 600 B.C., about the time of the rise of ancient Greece, and continued until the fall of the Roman Empire. Following the collapse of Rome, Europe's trade connections were destroyed, Moslem power gained footholds in Mediterranean areas, and Europe was forced to survive mainly on its own agricultural merits, which were not enough to sustain the urban vitality Roman power had generated. Cities survived, but they stopped growing; some even declined in size.

In what way did Greece and Rome bring new stimuli to city growth? The answer lies in trade and transportation, in intensified exchanges of commodities, and in improved agricultural yields. Farming was better organized, tools were better, crops had been developed for particular environments. It was feasible to import foodstuffs from across the Mediterranean to sustain Greek and Roman cities. Rome benefited from its government's tight control over the empire of which it was the center and from which produce was requisitioned. Rome's manufacturers (of metalware, jewelry, pottery, and so forth) sent their wares by ship to distant colonies in North Africa, and the ships came back carrying grains and other foodstuffs.

Cities emerged in other parts of the world as well, however, and we should not assume that Southwest Asia and Europe had a monopoly over urbanization.

An overview of the ruins of Persepolis, one of the ancient cities of Southwest Asia. Known for its magnificent stonework and its tall columns (several still stand, at right), Persepolis was a focus for a large region nearly 2500 years ago. The city was plundered and severely damaged by Alexander the Great's invaders and went into decline. The rectangular area that forms the foreground in this photograph is all that remains of the Hall of One Hundred Columns (whose bases can be seen quite clearly). These columns were as tall as those at right, and the entire structure was covered by an imposing roof.

ZIMBABWE

From the Cape of Good Hope to the equator and beyond, Africans for centuries have built their houses of wattle and daub, with roofs made of thatch. Some homesteads stand alone; elsewhere there are hamlet- and village-sized clusters. The traveler in rural Africa soon becomes familiar with a countryside patterned by farm fields and dotted with gray dwellings.

On the Rhodesian plateau, not far south of Fort Victoria, stands a group of structures so stunning in their contrast to this common scene that the first European invaders who saw them must have stared in disbelief. In a valley and high on a hillside stand buildings whose walls are made of stone, carefully hewn and expertly laid; those walls are higher than the roof of the highest traditional house in Africa. In the valley, an oval-shaped building stands today amid the ruins of many smaller ones; this single building, with 30-foot high, 17-foot thick walls, is alone larger than many a whole African village. Decorative art work finishes the tops of the walls, and conical towers and rounded entryways attest to the technical capacity of the ancient builders. On the crest of the nearby hill, sections of wall connect granite boulders, turning a natural protection into a military fortification. This is the great Zimbabwe, surely one of the wonders of the world.

When was Zimbabwe built and who built it? Research has produced the following conclusions: Zimbabwe was built and rebuilt by at least three successive occupants; the foundations were laid as long as a thousand years ago, perhaps longer; at least some of Zimbabwe's structures had ceremonial significance; and the people of Zimbabwe traded, directly or indirectly, with distant lands including the Arab world and even China. They may have done so via the now-defunct port of Sofala on the coast of Moçambique.

Still there are numerous questions. No mortar was used by the ancient builders, and every piece of granite was carefully hewn to fit into a precise pattern. How was the technique learned? Where was the rock quarried? (The workings have never been found.) How was the great structure planned? Was it drawn by architects or built ad hoc? Was the oval-shaped building in the valley a temple (as it has been called) or something else? Was the acropolis, as the fortlike structure on the hill is known, really a fortification or not? How many people lived here permanently? Infrared photography has revealed the presence of circular foundations of African traditional houses that stood both within the walls of Zimbabwe and outside of them.

Historical geographers believe that Zimbabwe once lay at the center of a large and powerful African state. Today, Zimbabwe is a focus of ideological strength for Rhodesia's black peoples: when Rhodesia attains majority rule, they say, the country will be renamed after its enigmatic, ancient city.

In the West African savanna belt, between the tropical forests of Africa's west coast and the drier interior and north lay cities and towns that found themselves on the trade routes between forest and desert. Timbuktu, about A.D. 1350, was a city of major consequence, a seat of government, a university town, a market, a religious focus. Farther to the east in Africa, in the upper valley of the Nile, Meroe had become the "Birmingham of Africa," the place where metallurgy, especially the smelting of iron and the making of weapons, was a major industry. Approximately contemporaneous to the Mesopotamian-Egyptian urbanization, cities were emerging in the core area of ancient China, the valleys of the Hwang and Wei rivers. Cities arose in the Indus valley (now Pakistan), where Mohenjo Daro and Harappa grew large, in Persia, whose Persepolis was the focus of a major culture, and in southern Africa, where Zimbabwe still leaves many questions unanswered. And in the Americas, too, urban growth took place. In Yucatan sizable ceremonial centers developed; on the Mexican plateau much larger urban centers emerged. Mexico's largest pre-Columbian city, Teotihuacan (near present-day Mexico City), may have had over 100,000 residents nearly a thousand years ago. Like Rome, Teotihuacan benefited from tight control over a substantial, tribute-producing empire and water transportation. Though not on the scale of the Romans, the Aztecs moved their goods on the lakes and connecting canals of their heartland by boat.

While some of these other urban places flourished, Europe's urbanization in post-Roman times had been brought virtually to a stop. Damascus, Bagdad, Cairo, and Constantinople were the cities of vitality and growing technology. In Europe's cities there often was stagnation, poverty, and disease. But then came the Christian counterthrust to the Moslem successes. First the invaders were halted, and then the Crusades were begun. The colonial, imperial era of Europe was about to begin, and old trade routes were reopened. In the dormant cities, the effects were quickly recorded. The pulse of commerce quickened, there was work, and the numbers began to grow again. Paris, Amsterdam, Antwerp, Lisbon, Venice, Naples, and many other cities and towns were revived. It is important to keep a perspective on the magnitude we are talking about. London by the middle of the fifteenth century had perhaps 80,000 inhabitants, Paris, 120,000. And while colonies were secured and riches poured into the continent, London around the beginning of the nineteenth century still had fewer than 1 million residents, and Paris only 670,000.

What were these preindustrial cities like as places to live and work? Well, the adage of the "good old days" hardly applies. If today's cities are no bargain for many residents, preindustrial Ghent, Manchester, or Warsaw were not either. With better weaponry and the invention of gunpowder, cities faced threats not confronted before. Walls and moats no longer could withstand enemies' armies, so cities developed into veritable fortifications just at a time when they were also to accommodate growing numbers of people. Those fortifications could not simply be moved outward; once built, they more or less marked the confines of the urban center. Thus the only way people could be contained was by building not outward but upward, and so the four- and five-story tenements began to appear. Eventually these were built even higher, and by the seventeenth century the European cities were slum-ridden, dirty, unsanitary, and depressing. Epidemics, disastrous fires, much crime, and social dislocation prevailed. Yes, the picturesque, four-story merchants' homes overlooked the canals of Amsterdam and the parks of London. But the people who lived in them were the fortunate few who were in a position to manipulate the labor force of the city and to control the lucrative overseas commerce. For the ordinary people, the overcrowded cities more often than not were no place to be. Many decided just that when the chance came to leave for America, Australia, and other parts of the world.

NON-WESTERN CITIES

The changes that have come to urban life throughout much of the world as a result of the forces of European dominance and industrialization sometimes lead us to forget that the urban norm we are used to, represented perhaps by our hometown, or by our image of New York, San Francisco, or London, has its exceptions. As we will note later in this chapter, we can recognize certain recurrent features in the layout and structure of American and Western cities, and we make generalizations about these features, sometimes believing that we have discovered some universally operative rules of urban development. How accustomed we are to cities that have downtown areas, suburbs, and parks. Do not all cities everywhere have these characteristics? The answer is no.

Despite the nearly 200 years that have passed since the beginnings of the industrial revolution made their first impact on Britain's Midland towns, and the globe-circling imprint of Europe on the non-European world, and the worldwide adoption of the technologies emanating from the industrial revolution, there are still cities that remain essentially preindustrial in character, and many more in which the impact of the industrial era is limited. In these, most of the people still travel on foot (the bicycle has made inroads, though) or by horse and cart or by camel; houses, hostels, workshops, and other special-function buildings are all mixed up, so that it would be difficult and maybe impossible to find an urban region such as a downtown area. There is little vertical development in these cities, and from a vantage point overlooking the city it appears as a sea of roofs, all looking much the same, crowded closely together and separated by narrow, winding paths and lanes suitable for walking, but not for vehicular traffic. Of course, roadways for cars have penetrated most cities, but in preindustrial cities and in transitional urban centers there is a marked lack of internal interconnections for vehicular traffic.

Let us examine what a downtown area (or, to use the technical term, the *central business district*, or CBD) looks like in a non-Western city. We are familiar with it as the place where there are tall buildings, department stores and shops, offices, and banks; often the CBD is quite well defined and marked off by streets whose names are familiar to just about everyone living in the city. In the CBD you find many pedestrians on the sidewalks, and often congested street traffic; and here the land costs more per square foot than anywhere else in the city. In the non-Western world there are cities with just such CBDs, but when you find them (as in Nairobi, Kenya, for example), you can be sure they are colonial cities, laid out and built by Europeans. In the true non-Western city, for example in Ibadan, Nigeria, or in Lahore, Pakistan, you may find a weakly defined central business district or perhaps two such districts, one the traditional and the other the modern. India's cities have not one, but several such centers, one of which may be showing signs of CBD development. In India's smaller cities and towns, the CBD generally is still traditional, with the court, the religious shrine, the moneylender, the clothing retailer, and houses of the wealthy all crowded near the busiest part of town, thereby breaking all kinds of "rules" we derive from studying our Western downtown areas. In Africa, the cities and towns of Nigeria's Yoruba region similarly retain much of their traditional identity. The Yoruba people have a long history of urbanization; there was a time when they lived in fortified towns and spread out over their farmland in the daytime to do their work and retreated for protection to the towns

View over a section of a non-Western city: Kano, Nigeria. Kano was a major market and trade center before many Western cities were founded. After the European invasion of Africa, Kano was endowed with some of the attributes of modern times, a railroad link, motor roads, and Western-type living areas, but most of the city still looks like this, and the rules of urban life and the use of land are not those obtaining in Western cities.

at night. Yoruba cities of substantial size, such as Ibadan, have accommodated the invasion of modernization by developing dual business districts, the traditional market center with its economic, social, and political functions surviving some distance away from the newer, modern CBD. Other non-Western cities, from Kumasi (Ghana) to Colombo (Sri Lanka), have CBDs that, compared to those of Western cities, are still developing.

We in America are familiar with a zone of transition around our city's central business district, where one might find some light manufacturing, automobile sales, and other establishments that need rather too much space to locate within the CBD itself. This zone is also missing in the non-Western city; instead, the areas near the traditional *and* the developing modern CBDs tend to be very densely populated. Many people go to work on foot or by bicycle, and proximity to the market and business district is an advantage. Again, the traditional character of the non-Western population cluster comes through. These areas on the doorstep of the CBD or market are often still laid out not as streets, not linear, but in circular groupings of houses, facing the home of a chief, headman, or religious leader. Most non-Western cities contain a wide variety of ethnic clusters, and these tend to be associated with particular areas of the urban region. New migrants to the city will look for the settlement of their own kind and will seek to stay there, no matter how severe the shortage of space or how poor the facilities. Thus these areas near the CBD have become some of the world's worst slums, but eradicating them and substituting high-rise apartments has a major negative consequence: it may improve the physical facilities, but it tears at the fabric of society, already frayed by urban life.

Thus the near-downtown residential zones of the non-Western cities are not being deserted for the suburbs. On the contrary, they are becoming ever more crowded. So the very heart of the non-Western city is quite different from that of the Western city, and the question is: will the non-Western city eventually begin to respond to the laws of urban development that are operative here in America? The way incipient CBDs are emerging would seem to suggest an affirmative answer, but we ought to be cautious about generalizing too far in this direction. In the first place, Western cities have a different heritage. They inherited their land-value structure, the high-cost CBD, from preindustrial cities that are quite unlike non-Western preindustrial cities. Second, the Western city has emerged and has differentiated regionally and specialized functionally on the basis of the distribution and circulation of considerable wealth, wealth that is simply not available in non-Western cities and is not likely to be for a very long time. And third, there are deeply rooted differences in attitudes, toward ownership and property, toward the family and schooling, toward older people, for instance, between Western and non-Western societies. Those, too, will be reflected in the urban scene. The non-Western city will be recognizably distinct for a long time to come, and perhaps forever. We should be mindful of this as we resume our study of the urban development of Europe and the rest of the Western world.

THE URBANIZING WORLD

The impact of the industrial revolution and its associated effects upon Europe's cities came full force during the first half of the nineteenth century. As we noted previously, European cities about 1800 were not especially large by modern standards. London had well under a million people, Paris somewhat more than a half million. Large cities were not especially numerous either. There were perhaps 20 cities in Europe with more than 100,000 inhabitants, and they contained fewer than 4 percent of the continent's population. Urbanization ratios of those days compare with those of the UDCs today.

Even so, the cities of Europe were no longer dormant; they were productive, crowded, and growing. The rules by which cities thrived (or failed) were the rules of commerce and finance; it was also necessary to be able to keep enemies at bay. Thus the cities' lasting advantages were crucial ones in this era of mercantilism. These advantages can be summed up in two words, *site* and *situation*.

Site

The site of a city or town can have a substantial role in its development. The term refers to the actual physical qualities of the place the city occupies, whether it lies in a confining valley, on a piece of coastal plain, on the edge of a plateau, or perhaps on an island. In colonializing Europe, the protective or defensive qualities of a city's site often mattered a great deal. The changing technology of warfare, following the invention of gunpowder, rendered some cities vulnerable; others found sites where natural protection could be supplemented by effective fortifications. A famous example of this is Venice, whose small onshore settlement at the head of the Adriatic grew into a city of wealth and splendor. The city could have continued to develop on the mainland, but there was little natural protection to be found on that site. So the rich merchants built their warehouses, their mansions, and their monuments to the accumulated wealth on a group of islands offshore, thereby deterring attackers from land. Another example of the role of site in the emergence of a major city is Paris. The capital of France was founded on an island in the middle of the Seine River, again a place where defense was facilitated and also a place where the Seine could be crossed more easily than elsewhere in the vicinity. Exactly when settlement on this *Ile de la Cité* began is not known, but it probably happened even in pre-Roman times. Paris functioned as a Roman outpost for many centuries, and later the Roman regional administration was replaced by other authority, always making use of the defensible qualities of the island city. Eventually the rather small island proved too limited for Paris's growth, and the city sprawled onto the nearby river banks. But authority remained centered on the island, and potential adversaries knew that there was no point in challenging the outskirts if there were no prospects of conquering the *Ile de la Cité*.

Today, we tend to look at city sites from other viewpoints. The site of a city may severely hamper urban expansion and interconnection; marshes, swampy

Paris, where the river Seine divides to accommodate the *Ile de la Cité*. The picture was taken from the top of the Notre Dame Cathedral, one of the city's landmarks, and shows the river, now confined between artificial banks and bridged in numerous places, and the enormous urban sprawl of France's capital.

areas, valleys, escarpments, shallow harbor waters, limited dockage, and numerous other problems can confront the growing urban center. But in the days when European cities were hoarding wealth, when pirates from the sea and marauders from land preyed on them, defensibility was of prime concern.

Situation

More important still than the city's site is its situation. By situation is meant the position of the city with reference to nearby and surrounding areas of productive capacity, the extent of its *hinterland,* the location of competing towns, in other words, the greater regional framework in which the city finds itself, its *relative* location. Our previous example, Paris, combines advantages of site with a most fortuitous situation. The city lies near the center of a large and prosperous farming region, and as a growing market and distribution point, its focality increased continuously (Fig. 9.1). The Seine River, a navigable waterway, is joined near Paris by several tributaries, all of them navigable as well, which thus connect Paris to various parts of the Paris basin, and to areas beyond. To the west, the Seine River runs into the English Channel, and although today's ocean-going ships cannot reach Paris, the smaller vessels of the past could; and today only one transshipment, at

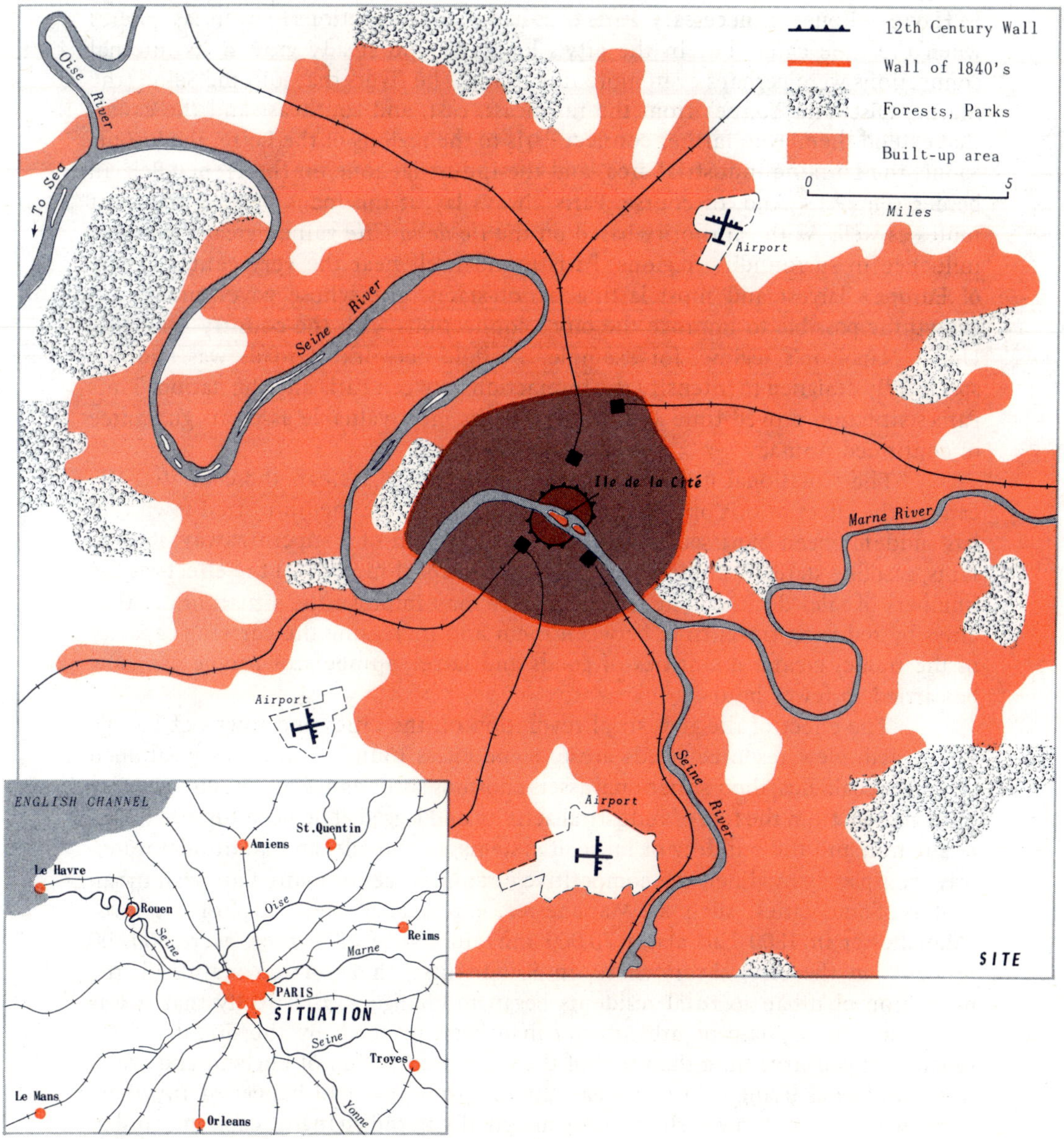

FIG. 9.1 Site and situation of Paris, France.

Le Havre or Rouen, is necessary. Paris thus had good connections to overseas spheres when the time came, but in the city's longevity and steady growth its internal connections were perhaps even more important. The rivers that join the Seine (the Marne, Oise, and Yonne) from the northeast, east, and southeast and the canals that extend them even farther connect Paris to the regions of the Loire, the Rhône-Saône, the Lorraine industrial area, and the industrial zone on the French-Belgian border (Fig. 9.1). And, of course, Paris always lay at the focus of land transport routes as well. As the authority based on the *Ile de la Cité* gained greater political influence in surrounding regions, Paris found itself near the very center of one of Europe's largest and most lasting nation-states, one whose governments did everything possible to improve the city's connections with the country generally. Under Napoleon's regime, for example, a whole network of roads was laid out specifically designed to intensify the interaction between Paris and the nation. Thus Paris's situation proved time and again to be its most valuable asset, its guarantee of continued primacy.

The situational qualities of a city can change, as can those of site. We can see the effect of this on the formerly busy center of a medium-sized town that has suddenly been bypassed through the construction of a superhighway. Restaurants, gasoline stations, and other commercial establishments feel the effects as the numbers of sales decline. The same highway may improve the situation of other cities by reducing travel time between them and increasing the carrying capacity of the traffic; greater quantities of goods and larger numbers of people can now be carried in fewer hours.

The effect of the industrial revolution on the cities and towns of Europe was varied. New resource needs arose. Some cities found themselves positioned very near coalfields, their situational assets enormously enhanced. Other towns were more remote from the foci of industrial activity and lagged. The need for intensified circulation and the building of railroad systems, new roads, and additional waterways completely modified the competitive circumstances of many European urban centers. Some cities, such as Manchester, experienced truly explosive growth (Manchester in 1800 had about 70,000 inhabitants; in 1850 there were 300,000 residents in the city). Everywhere, in Britain first, then on the mainland, the proportion of urban to rural residents began to change so drastically that it was clear that a new phase of urbanization had been reached. By the middle of the nineteenth century, more than half of the entire population of England and Wales was classified as living in urban areas, the first time that had happened anywhere in the world. Even if towns shared only marginally in the changing economic order, they could not escape the impact of the population explosion. Whole new industrial towns emerged. As the European mainland began to sustain the impact of the industrial revolution, the urbanizing phenomenon occurred there, too. Germany was the first mainland country to record a larger urban-based than rural population. If you could have taken a low-level airplane flight over those mushrooming European industrial towns, you would have seen three prominent sights: acres of factories, with a forest of smokestacks belching an enormous amount of pollution into the air; miles of gray, dirty, endlessly monotonous slums; and huge railway installations and yards carrying raw materials and finished products.

The urbanizing trend set in motion during the nineteenth century continues today. London, the United Kingdom's capital and leading city, now houses nearly 20 percent of the total British population (when the whole metropolitan area is taken into consideration); Paris, Rome, Moscow, Berlin, and other places that not long ago counted their populations in hundreds of thousands have become cities of millions. And cities outside Europe have overtaken their European counterparts in size: Tokyo is generally acknowledged to be the world's largest city, with 8.8 million residents according to the 1970 census but with 22 million people in the conurbation of which Tokyo is the heart. That accounts for more than 20 percent of Japan's population. Shanghai, in China, also is larger than any European city, with perhaps 10 million people in the city proper and its immediate environs. New York also ranks ahead of Europe's largest city (London). And in the underdeveloped countries of South Asia, Latin America, and Africa there are cities today of dimensions that would have seemed absolutely unimaginable just a century ago. Calcutta's urban area counts perhaps 7 million people; Bombay has 6 million residents. Of Brazil's less than 100 million people, no fewer than 12 million live in that country's two largest cities, São Paulo and Rio de Janeiro. Cairo, whose nearly 7 million inhabitants (immediate environs included) constitute one-fifth of the country's population, is among the world's 10 largest cities as of 1973.

The industrial revolution stimulated the growth of many urban places in England, but the accommodations that were built to house the labor force were mostly inadequate, with small rooms, poor facilities, and much overcrowding. The dreary, bleak urban scenes that emerged in those days are well represented by this view.

Many of these large world cities, Cairo, Bombay, Calcutta, developed for reasons different from those that stimulated the fast growth of such places as Manchester and Liverpool, Glasgow and Stuttgart. Calcutta, Bombay, and Cairo are not major industrial centers. Their emergence is testimony to the explosive population growth made possible by the technological advances made in those grimy European cities of the nineteenth century. Among the by-products of the industrial revolution were innovations in public hygiene and sanitation. It became possible to construct five- and even eight-story apartment houses with built-in sewerage systems and pipes to bring fresh water to every floor. Cities planned and laid out disposal systems and water treatment plants, and urban life became more tolerable and safer. Along with the results of progress in medical research and services, those innovations were exported to the colonial world. We considered the results quantitatively in Chapter Three; we can also read those results on the map, the map of urban places in the non-Western world.

It is important to remember, however, that although there are large and populous cities in the non-Western world, the *proportion* of urban to rural residents, the true measure of urbanization, is changing very slowly there. India's cities have mushroomed, true, but its total population has skyrocketed too. When it comes to the urban ratio, things are only slightly different from what they were before world population began to grow so rapidly. The most significant change occurred in Japan, but Japan already had an urban tradition before the European cities began their nineteenth-century growth. In 1800 Tokyo already was larger than London and probably was the largest urban center in the world. There are all sorts of problems in getting data to measure urban growth around the world. Different countries view in different terms the meaning of *urban*, and statistics are sometimes difficult to reconcile. But we can estimate that under 3 percent of the people of the world in 1800 were living in cities and towns over 20,000; in 1900 it may have been around 9 percent; in 1970, perhaps 35 percent. The bulk of that increase comes from the countries of the Western world and Japan. Compare the world figures with those for Britain, where they were 9 percent (1800), 62 percent (1900), and 85 percent (1970)!

CHANGING ACTIVITIES In our discussion on the previous pages, we have been referring to industrialization as a chief stimulus for the growth of cities and the whole phenomenon of urbanization. But have you noticed that when you go to a big city these days, to New York, or to London, or Paris, there seem to be huge numbers of people who very obviously are not industrial workers? And indeed, the employment picture is these cities is changing. The European city on the eve of the industrial revolution was a place of commerce and what we would now call light manufacturing. Many of the inhabitants of cities and towns were engaged in these activities. Then, with the advent of the industrial revolution, cities became factory complexes, with large industrial labor forces. But as industry evolved, it required fewer hands. Machines were created that could do the work more quickly and in larger volume. In the meantime, people were earning more than they had in the past, more than would cover the bare essentials of life. So when the numbers of people employed by the industries began to decline, there was other, service, work to be done. Transportation, retailing, banking, entertainment, government services, health services, and other industries we identified in Chapter Eight as tertiary activities became a major source of employment. In the United States, the service industries now employ not much less than two-thirds of the labor force. Many cities have become principally service centers, though others retain their primarily industrial base. We will discuss this question of the functions of cities in greater detail shortly.

DEFINITION AND HIERARCHY

So far in this chapter, we have been referring to *towns, villages,* and *cities* without defining those terms. When we mention the population of a large city, we may differentiate between the inhabitants of the "city proper" and those in the entire "metropolitan area." All this suggests a good deal of imprecision, and that is indeed the situation. Cities sprawl far beyond their municipal boundaries, and one observer's large village is another's small town. Even when we get away from specifics and speak only in terms of "urban" and "rural" populations, there are problems of definition. In Canada, all settlements with more than 1000 people are classified as urban. In the United States, incorporated settlements numbering over 2500 inhabitants have long been taken to be urban, and smaller ones as rural. In India, places having more than 5000 residents are classed as urban, but in Japan the lower limit is 30,000! When we base comparisons on such diverse criteria, we must be very much aware of the varying standards. The problem is more than one of size, too. In the United Kingdom, for example, "urban" is a question of status rather than dimension; communities are accorded such a status through application and government approval. In Italy, "urban" is considered to be a matter of employment. More than 50 percent of the economically productive population must be engaged in nonagricultural pursuits for a settlement to be so classified.

Settlement and Area

Another aspect of this issue has to do with the significance of a settlement in the context of the region it serves. A village-sized community may be lost among numerous others of similar and larger sizes in a populous region (such as, say, eastern Pennsylvania or New York State). But just such a settlement, positioned in the west-central United States, in the Great Plains or the mountain states, may be the prime nodal point for many miles around.

Various efforts are made to bring greater precision to the situation, to make comparisons (not only between countries, but also between successive censuses in the United States) more reliable. When a city grows beyond municipal boundaries established long ago there is not much point in comparing the 1960 and 1970 census of the city proper to identify growth patterns. Most of the growth is taking place outside. Thus the concept of the "urbanized area" is used to count all the people living not only within the city proper, but also in the contiguous built-up area, until the density declines beyond a certain level. You will also see references to the "standard metropolitan statistical area" (S.M.S.A.), which is a county-based unit that includes a central city or coalescing cities counting at least 50,000 inhabitants. But counties must satisfy certain criteria to be included in S.M.S.A.'s, and when an urban agglomeration fills two counties and just edges into a third, that third county may not yet qualify. So the S.M.S.A. population figure is usually less useful for comparative purposes than that of the urbanized area. The problem of definition in large conurbations is by no means unique to the United States. Many countries report dual figures for their large cities. Brazil's 1970 census, for example, gives the population of the city of Rio de Janeiro as 4,252,009, but the population of the metropolitan area of Rio as 7,094,211.

This brings up another, related problem: the dependability of census taking and reporting in the UDCs. The taking of a census is an expensive and technically complicated task, one requiring skills, resources, and communication capacities many countries simply do not have. This should be kept in mind whenever figures on populations in major cities, urban growth rates, and other data from the UDCs are compared with those from other sources. The census results from the developed countries are not always reliable either, witness, for instance, the doubt cast on the accuracy of the United States census of 1960 in so far as it reported on black America.

Hierarchy

Is the steplike differentiation implied by the terms *hamlet, village, town, city,* and *metropolis* a reflection of reality, or is the continuity too tight to allot these words any real utility? Obviously population size alone will get us nowhere. If we put all those settlements in the United States on a piece of graph paper, from the smallest to the largest, there would be a near-continuous string of sizes, with real breaks showing up only in the very largest categories. There would not be a gap between

---

THRESHOLD

Everyone knows what a threshold in a house is: on one side you are in the hallway; on the other, you are in the room. In geography, *threshold* is a term that denotes an important concept relating to the economic activities in urban places. It has to do with the size of the market, or the number of consumers, required to support any certain function in an urban place. For example, it takes far fewer consumers to support a grocery store than a supermarket. Thus the *threshold requirements* (or conditions of entry) of grocery stores are far lower than those of supermarkets.

---

the villages and the towns. But we could look at those places another way, not in terms of size, but in terms of the functions they perform and make available to their surrounding service areas.

Thus a *hamlet*, a settlement of fewer than 100 people usually, has the smallest number of services, probably fewer than 10. There might be a general store, a gas station, a coffee shop. But you would not find a post office, a church, a grocery. For those more specialized services you would have to go to a *village*. In a village in, say, the Wisconsin dairy region, you could expect perhaps 60 or 70 establishments, including a couple of gas stations, a restaurant and one or two bars, a farm elevator, two groceries. Obviously, such a village would draw customers from a much wider area than the hamlet. In fact, people from the hamlet have to go to the village for some services. Again, a *town* of, say, 2500 inhabitants has a still greater variety of functions and services. Here we find a greater degree of specialization, and people would come for a doctor, dentist, dry cleaner, various particular stores (hardware, furniture, appliance), a bank. Since such services are not available in either the hamlets or the villages, the towns have greater service areas (or hinterlands) still. Thus, rather than merely counting heads, we might define a town as a place where a certain assemblage of services is available, and whose hinterland includes the service areas of the hamlets and villages.

This hierarchy can be carried further to include cities, metropolises, and still larger conurbations; the advantage is that we do not rank settlements merely according to their population, but rather reflect their strength as places of trade and commerce. Hence the concept of *threshold* emerges. For an urban place to be a town, it must have the functions of a village *plus* certain functions that are not provided by villages. This approach to the determination of a hierarchy of urban places takes into account something that mere population numbers do not, namely, the interaction between the urban center and its trade area. In fact, we might think about addressing the problem the other way: rather than measuring the available goods and services in the urban centers themselves, the surrounding service areas are identified, and the *economic reach* of the cities and towns is determined. The ranking is then established on the basis of this economic reach, which generates a measure of *centrality*.

FUNCTIONS AND CLASSES

How often have you said, "My hometown is . . . a university town" (or a port city, or a mining town, or a farm village)? In doing so you try to summarize the nature of your hometown, to define what *kind* of an urban place it is. You are, in effect, classifying it with other places of the same type, disregarding size. There are mining towns large and small, industrial centers ranging from a few thousand to several million inhabitants. What these places share is not the same rank in the hierarchy we just discussed, but the same assemblage of functions. So it is on this basis that geographers try to classify urban places.

At first, classifying cities seems quite easy. You could jot down a list quite quickly. There are resort towns, retirement villages, ports, and so forth. But when we look into the matter somewhat more intensively, we find that port cities also have stores (and thus retail functions), factories (and thus industries), universities, and numerous other functions. This means that port cities are *multifunctional*; that is, they possess more than one sphere of economic activity, among which the transportation function is the salient one. But what about a city that does not have such a dominant function as a port city has (or a resort town or a mining town)? There are many cities, such as Denver, Kansas City, and Memphis, where no single function stands out to typify the place. On what basis can such cities be included in a generally applicable system of classification?

One approach lies in the activities of the labor force in cities. Where do people work? The census gathers such information; it determines the employment of the labor force in the major cities and reports the number of workers engaged in each economic activity. In 1943, Professor C. D. Harris published an article in the *Geographical Review* entitled "A Functional Classification of Cities in the United States," in which he used these data to develop a system according to which all United States cities could be classified. Thus he recognized two types of *manufacturing* cities: the M′ type, in which the total employment in manufacturing must be at least 74 percent of the total employment in retailing, wholesaling, and manufacturing combined, and the M type, in which that figure is reduced to 60 percent. In addition to these two classes of manufacturing cities, Harris recognized eight other classes. *Retail* centers, in which employment in retailing is at least half of the total number of people employed in retailing, wholesaling, and manufacturing combined, and *wholesale* cities, in which employment in wholesaling is unusually strong (and also quantified), are two major groups. Others in the classification are *transportation* centers, in which transportation and communication industries employ at least 11 percent of the labor force, *diversified* cities, where the data yield no clear-cut, paramount function, *mining* towns, where more than 15 percent of the workers are employed by mining industries, *university* cities, whose chief industry is constituted by the campus of a large educational institution whose enrollment (or combined enrollments, when there is more than one such institution) makes up at least 25 percent of the city's population, *resort-retirement* towns, and *political* cities. The last class of cities would include Washington, D.C., and a number of state capitals.

Not surprisingly, many geographers have tried to modify the Harris classification, either by adding new classes or by suggesting changes in the percentages used by Harris to establish his admittedly somewhat arbitrary criteria. But despite these reconsiderations, the pattern shown on Fig. 9.2 stayed much the same and is still quite current. Manufacturing cities of both classes (M′ and M) are heavily concentrated in the northeastern United States and in the North American core area generally. Retailing centers, on the other hand, tend to lie outside this manufacturing belt and, with wholesaling cities, are concentrated in a wide zone that extends north-south across the country from North Dakota to Texas. Diversified cities show a stronger concentration in the eastern half of the country; trans-

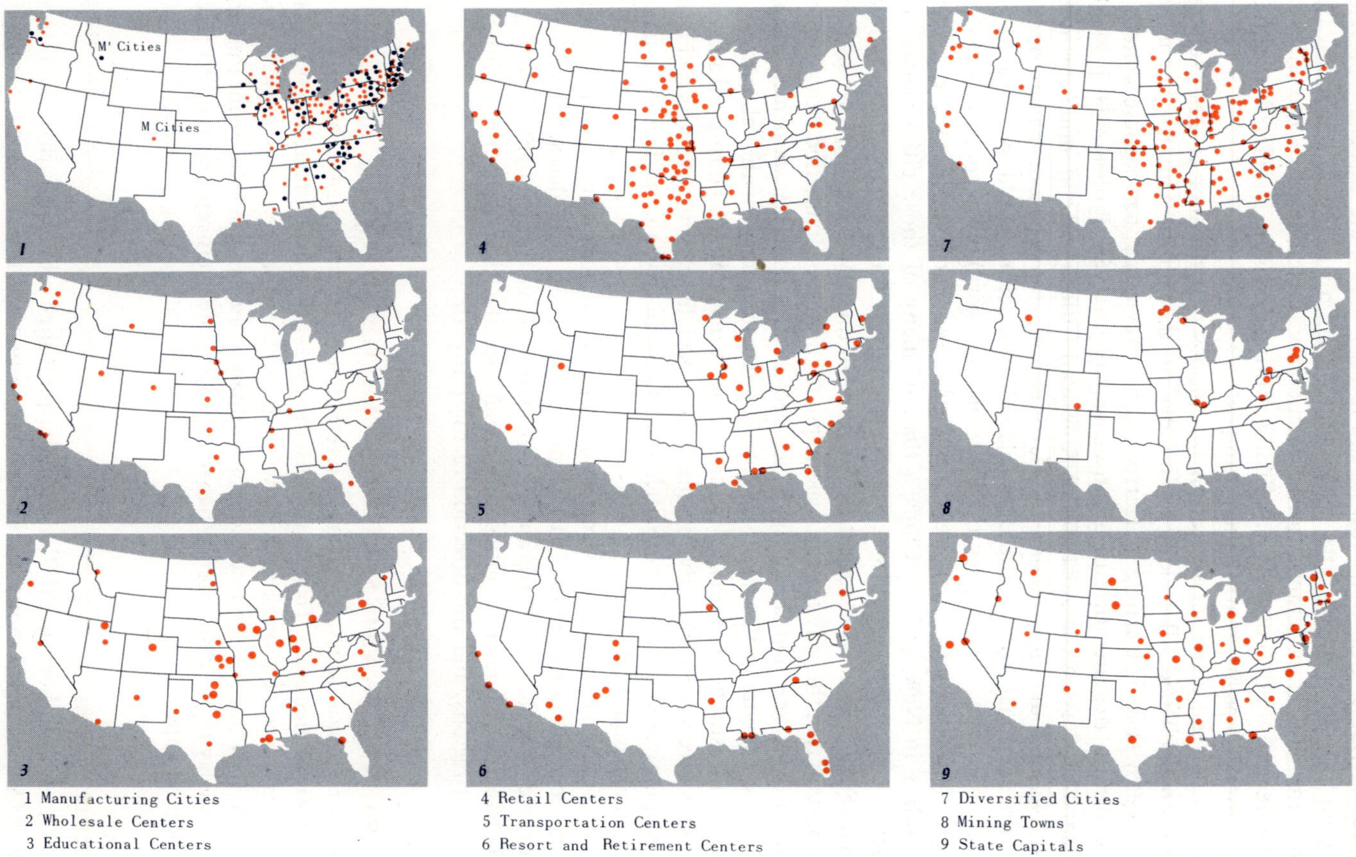

FIG. 9.2 Functional classification of cities in the United States. (After a map in C. D. Harris, "A Functional Classification of Cities in the United States," *The Geographical Review*, 33 (1943), p. 88.)

portation centers lie both along the coast and along major rivers and in places where railroads converge. University towns, as the map shows, are especially prevalent in the midsection of the country, where many small towns are dominated by large universities. Resort-retirement towns, predictably, concentrate in the southern, warmer sections of the country. And finally, mining towns are associated (of course) with mineral deposits, mainly the coalfields.

Since the Harris classification was proposed, a number of other schemes have been published, all seeking to improve on that effort. This has not proved to be an easy task. The article by Harris, quoted previously, has become something of a classic in the literature of urban geography. In 1955, Professor H. J. Nelson published "A Service Classification of American Cities" in *Economic Geography.* There, Nelson sought to overcome the Harris classification's lack of accommodation for multifunctional cities. In the Harris classification, cities would specialize in only one economic sphere (except in the diversified category), even though it was very evident that other functions were also important. For example, there is more wholesaling in New York or Chicago than in many of those cities classified as wholesale cities, but you will not find New York or Chicago mapped as wholesale cities on Fig. 9.2. That is because other functions of these two great cities are similarly so important that the only feasible class to put them in was diversified cities. Nelson, trying to overcome that difficulty, developed a more complex statistical formula based on employment figures also, expressed as percentages of the labor force, in which he measured how far cities exceeded the average calculated from all available data. This gave a more objective evaluation of specialization than Harris had done; Chicago, for example, could score high as a wholesale city, a manufacturing center, *and* a transportation center.

Is this search for a satisfactory classification system really worth the effort? Urban geographers feel it is. Apart from their effect in stimulating a continuing debate over specifics, the Harris and Nelson systems have thrown light on the nature (and changing character) of cities' functional specialization, their interaction with associated regions, and the direction of their growth.

DISTRIBUTION AND SPATIAL RELATIONSHIPS

In the foregoing discussion we have referred several times to the hinterlands of cities, their service areas. To such regions, the cities are *central places;* even a village is a central place to a small area. How, then, do these service areas relate to each other? Do they overlap? Do towns of approximately the same size lie about the same distance away from each other? What rules govern the arrangement of urban places on the landscape?

Central Places

In 1933, a geographer named W. Christaller published a book entitled *The Central Places of Southern Germany.* In this volume, which was not translated into English until 1966, Christaller laid the groundwork for central-place theory. He addressed himself to those questions we just raised, trying to develop a model that would show us where central places in a hierarchy (hamlets, villages, towns, cities) would lie with respect to one another. In his effort to discover the laws that govern this distribution, Christaller began with a set of assumptions. The surface would be flat, without physiographic barriers, and with soils of equal, unvarying fertility. He also assumed an even distribution of population and purchasing power and a uniform transport system permitting direct travel toward each urban center. Finally, Christaller assumed that a constant maximum distance or range for the sale of any item produced in a town would exist in all directions from the town.

Christaller's idea was to calculate the nature of the central-place system that would develop under such idealized circumstances, and then to compare that model with the real world situation, explaining variations and exceptions. Some places, he realized, are more central than others. The central functions of larger towns will cover regions *within* which smaller places with lesser central functions exist. What was needed, Christaller reasoned, was a means to calculate the degree of centrality of various urban places. In order to do this, Christaller identified *central goods and services* as those provided only at a central place and made available to the people of the surrounding region, as opposed to those services that might be available anywhere, without focus, and those that are produced for distant, even foreign, markets and are thus of no relevance to the local consumers. Next came the question of the range of sale of such central goods and services, the distance people would be willing to travel to acquire them. The limit would lie halfway between one central place and the next town where the same product is manufactured and sold at the same price, for under the assumptions Christaller used, a person would not go 51 miles to one place to buy an item if it were possible to go only 49 miles to another.

Each central place, therefore, has a complementary region, within which the town has a monopoly on the sale of certain goods, since it alone can provide such goods at a given price. From what we have just said, it would seem that such complementary regions would be circular in shape. But when we construct the model on that basis, problems are encountered. Either the circles are adjoining and leave unserved areas or they overlap, and when they do, there is no monopoly situation any longer. These two possibilities are shown in Fig. 9.3. The solution is a system of hexagonal regions, as shown in the same sketch.

Now Christaller proceeded to derive his model for the spatial arrangement of central places. To illustrate the result, let us take a specific case. Assume that a certain central place provides 100 different kinds of goods. We rank them in order of importance from 1 to 100. Thus the product ranked number 1 will have the largest "limit" and that ranked number 100, the smallest. If we now placed three central places, each producing 100 central goods, say, 100 miles away from each other (Christaller called these G places), the following would happen. The large hexagonal region shown in the center of Fig. 9.4 as surrounding the city (a G place) marks the limit of the product ranked number 1. The product ranked

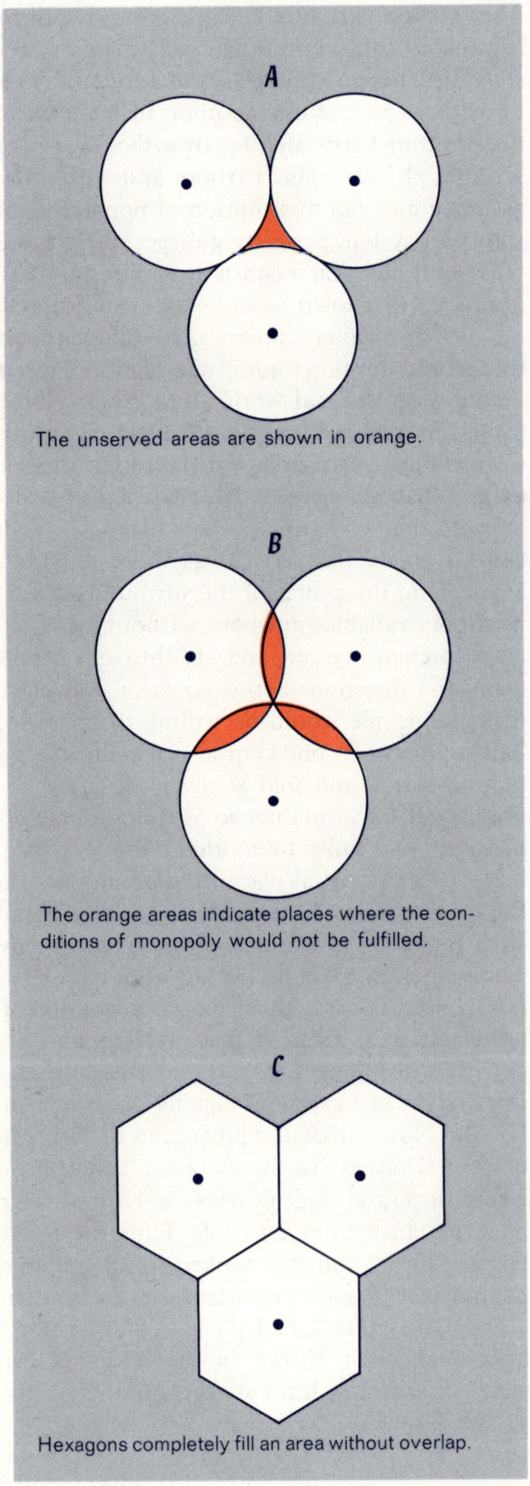

The unserved areas are shown in orange.

The orange areas indicate places where the conditions of monopoly would not be fulfilled.

Hexagons completely fill an area without overlap.

FIG. 9.3 Hexagonal and circular service areas.

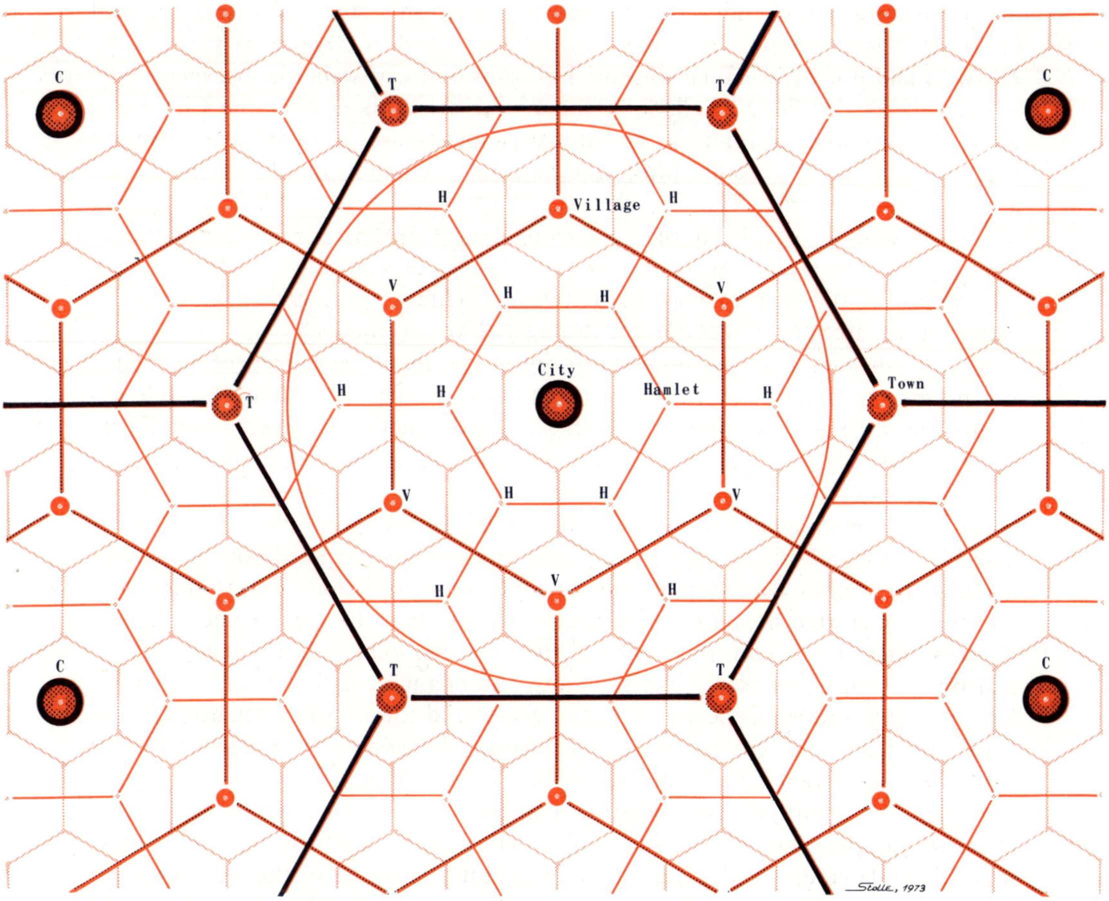

FIG. 9.4 The Christaller hierarchy of central places.

number 2 has a limit marked by a circle just inside the large hexagon and number 3 (not shown) has a limit slightly smaller again than that. Now when we draw circles for each of the 100 ranked products, we begin to leave an area outside of, say, circle 20 or 21 where there would be no service (all this is theoretical, of course). By the time we draw the circle for item ranked number 30, this unserviced area is large enough to accommodate its own hexagon and central place, and so a new network of places, B places, Christaller called them, is born (identified in Fig. 9.4 as towns). Now both G and B places provide goods ranked 31 and lower, and as we proceed and reach products ranked somewhere near 55 to 60, we find that the situation just described repeats itself again, and once more there is room for a system of central places of a still lower order (K places; villages in Fig. 9.4) and their own complementary regions. The ideal situation, based on Christaller's assumptions, is shown in Fig. 9.4.

Christaller was aware, of course, that many factors in the real world would operate to distort his model. But his book brought on a storm of discussion and debate that is still going on. Some geographers looked at large areas such as the Midwest and northern France and saw hexagons and Christaller landscapes everywhere; others saw none at all. Soon there were attempts to alter the Christaller model, to introduce different or additional elements, and to relate it to different parts of the world. Christaller himself joined the discussion; in 1950 he published a new volume entitled *The Foundations of Spatial Organization in Europe*, in which he said he had been right all along (pages 17–18): "When we connect the metropolitan areas with each other through lines, and draw such a network of systems on the map of Europe, it indeed becomes eminently clear how the metropolitan areas everywhere lie in hexagonal arrangements."

Christaller may have been overly optimistic about the validity of his model, but he set the pattern for future research on central places. Such research has focused on the nature and impact of city size on the model (a large city dominates smaller urban places, even for the same product or service), the effects of physical barriers and cultural differences, the cumulative, agglomerative effect of time, the failure of people to make rational decisions (some people *will* travel those 51 miles instead of 49), and other phenomena. After all, the hexagons need not be perfectly regular; they could be distorted, having some short and some longer sides, and maybe the distorted model comes closer to reality than the regular one. Changes in transportation and consumers' ability to travel ever longer distances, advertising and image making for places and products, and changing demands of the market have loosened the bonds between central places and their immediate hinterlands. Earlier we referred to the impact a superhighway bypass can have upon a service center whose main street formerly carried all that traffic. And that is only one possibility; competition may also stifle a small settlement and render it virtually dormant. On the other hand, a small town may find itself increasingly in the orbit of a nearby large city and may become a satellite, resulting in the opposite kind of change. Obviously, then, Christaller's model was only a beginning. But it was an important one. If you look at the periodical literature in geography these days, you will see numerous articles in which geographers are trying to unravel the complexities that Christaller, through his assumptions, circumvented. In the process, our understanding of the forces that influence the distribution of urban places and their growth and decline is strengthened.

World Distribution

A look at the world map of major urban areas (Fig. 9.5) quickly suggests that Christaller's proposed regularity is more the exception than the rule. Although many large cities do lie spaced well apart, many others have developed so close together that they are coalescing. In North America, that process is making a vast urban complex, a conurbation, of the Eastern Seaboard from Boston to Washington (the name "Bosnywash" has been suggested for this urban region). This enormous megalopolis, the greatest of its kind in the world, includes also New York, Newark,

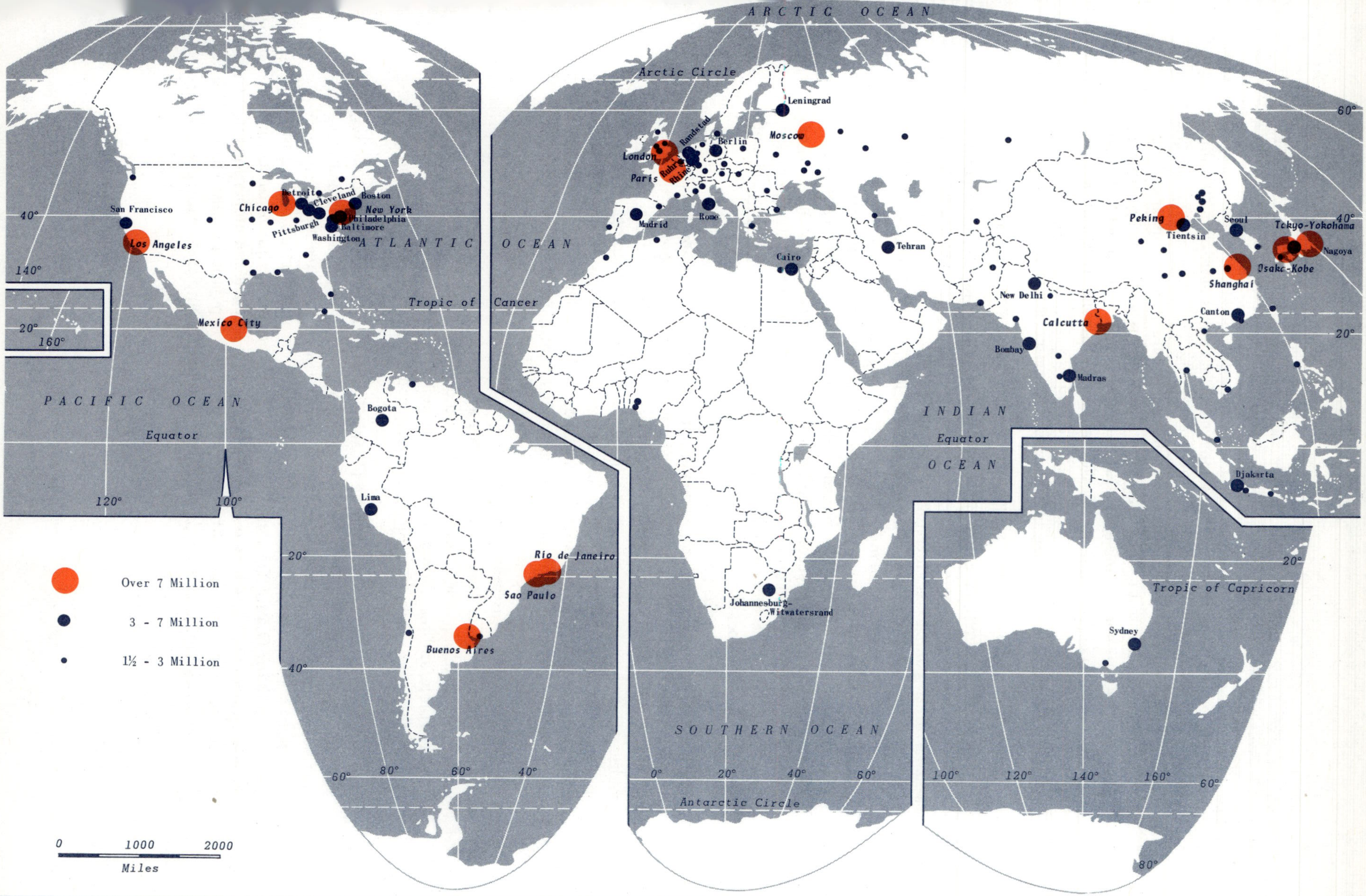

FIG. 9.5 World urban areas.

Philadelphia, Baltimore, and a host of smaller cities; it counts more than 40 million inhabitants and continues to grow, pushing suburbs into the countryside and absorbing towns and villages. Another such megalopolis, in a less advanced stage of development, lies from Detroit and environs to Pittsburgh and incorporates Cleveland, Akron, and Youngstown. Farther to the west, a Chicago-Milwaukee conurbation is emerging. And on the west coast, two urban clusters of megacity proportions are evolving: the largest, centered on Los Angeles, counts nearly 15 million residents and the other, of which San Francisco is the focus, about half that number.

In Europe, England's conurbations are approaching megalopolitan unification. London and environs count more than 13 million residents, and the cities of South Wales and the Midlands lie but a short distance away. On the mainland, a major urban complex is emerging in western West Germany, in the Ruhr and Rhine valley zone, including such cities as Düsseldorf, Essen, Frankfurt, Munich, and Stuttgart. In the Netherlands, planning for a triangular megalopolis (Amsterdam-Rotterdam-The Hague) is a matter of national priority; an attempt is being made to guide this megalopolis, already called "Randstad" (*rand* means "edge," "margin"), into a twenty-first century megacity-model, with parks, spacious housing, good communications and transportation, and optimal social services. Belgium, too, is experiencing the development of a coalescing urban complex, with Brussels and Antwerp the twin foci.

Other major urban agglomerations in Europe include the region centered on Paris, the less advanced but developing northern Italian zone, and the complex that extends from East Germany's Saxony to Poland's Silesia. Elsewhere there are major individual cities, such as Moscow, Leningrad, and Madrid, but not yet a true multicity urban region. Of course, we should keep the scale of things in Europe in perspective, and Fig. 9.5 is worth another look from that point of view. The whole urbanized area, from Britain's Midlands to West Germany's Ruhr-Rhine region, extends over an area not much larger than North America's megalopolis. Yet Europe's political fragmentation and cultural diversity lead us to identify discrete units within its huge urban core.

The only other area of the world where true megalopolitan development is occurring is Japan. The Tokyo-Yokohama and Osaka-Kobe conurbations rival those of Europe and the United States (the former had over 22 million inhabitants in 1970, the latter nearly 15 million), and they continue to grow by accretion and absorption.

This is not to suggest that the urbanizing trend that has generated Bosnywash and Randstad are not being experienced in other areas of the world. Johannesburg lies at the center of a smaller-scale megalopolis that includes several smaller cities along the Witwatersrand and eventually will incorporate Pretoria to the north and the northern Orange Free State to the south, for example. But urbanization outside Europe, the United States, and Japan has not reached the proportions it has in those areas. Even where a high percentage of the population is urban-based, the numbers are not large enough to create huge urban complexes: Australia has several large cities, but no megalopolis yet. In Argentina, Buenos Aires is a city of world proportions, but its nearest substantial neighbor, Rosario, lies nearly 200 miles away. The area between São Paulo and Rio de Janeiro, Brazil, does not have the urban development to support the evolution, in the near future, of a coalescing supercity. That mirrors the situation in the UDCs: large cities tend to stand alone, islands in a rural-dominated scene.

STRUCTURE

Cities are not simply disorganized, random accumulations of buildings and people; rather, they have *functional structures.* They are spatially organized to perform their functions as places of exchange, production, education, residence, and much more. If there are forces that govern the distribution of central places on the landscape, then certainly there are forces that affect the way cities are internally organized. One such force is the price of land, which is highest in the central city and tends to decline rather irregularly outward. So you would not look for a spacious residential area near the heart of a city or for the downtown area on a town's periphery!

Just by using these terms, *residential area, downtown,* we reveal our awareness of the existence of a regional structure in cities. Yes, cities have regions. When you refer to downtown, or to the suburbs, or to the municipal zoo, you are in fact referring to urban regions, where certain functions are predominantly performed (retailing, residing, recreation in the three cases just mentioned). All these urban regions, or zones, of course, lie adjacent to one another and together make up the total city. But how are they arranged? Is there any regularity, any recurrent pattern to the alignment of the various zones of the city, perhaps reflecting certain prevailing processes of urban growth? In other words, do the city's regions constitute the elements of an urban structure that can be recognized to exist in every urban center, perhaps with modifications related to such features as the city's particular site, its size, functional class, and so forth?

One way to go about this problem is to study the layout of a large number of cities, to compare the resulting maps, and to determine recurrent features. In the most general terms, it would soon become clear that cities have *central* zones, consisting mainly of the central business district, and *outer* zones, where sprawling suburbs and shopping centers lie. Between the central and outer zones lies the *middle* zone, an ill-defined, often rather mixed and disorganized region, where change is frequently evident, for example, in the deterioration of housing and the development of slums. This suggests that cities have a certain concentricity in their structure, and that impression was first formalized by Professor E. W. Burgess nearly 50 years ago when he studied the layout of Chicago. Although downtown Chicago lies near Lake Michigan, and the whole urban area is therefore a semicircle and not a full circle, Burgess suggested that his *concentric-zone theory* applied here as in most American cities. He recognized five concentric zones in the city rather than three, counting the innermost zone, the central business district (CBD). This,

The central business district of Chicago, Illinois, looking northeast toward Lake Michigan. The downtown area is marked by a great deal of vertical development, reflecting the high price of space here. Proximity to the lake also commands high land prices, and vertical development continues along the shore, north as well as south.

the heart of the city, has shops, offices, banks, theaters, and hotels; the larger the city, the more likely is the CBD to be divided into several districts of specialization, such as a theater district, a financial district, and perhaps even an area, near the edge of the CBD, of warehousing and wholesaling. The CBD is marked as region 1 on Fig. 9.6. It is the zone characterized by high land values and by maximization of floor space through the building of multistory skyscrapers. Transport lines converge on this zone, disgorging tens of thousands of workers into streets that bustle with traffic and pedestrians during peak hours, but may be nearly deserted late at night after the theaters have closed.

Surrounding the CBD, Burgess pointed out, lies a transitional area, a zone of residential deterioration (and the scene of urban renewal programs), marked also by the encroachment of business and light manufacturing. This is region 2 (Fig. 9.6), a zone of persistent urban blight these days, of tenements and slums, inadequate services, and deepening dislocation. We shall report on these problems later in this chapter. Still farther out from the CBD lies what Burgess called the "zone of working men's homes," a ring of closely built but adequate residences of the urban labor force (region 3). The next zone (4) consists of middle-class residences, suburban areas that are characterized by greater affluence and spaciousness. At this distance from the CBD of the city, local business districts (shopping centers) make their appearance. Now we have reached what we might call the "urban fringe," the commuters' zone (5), consisting of communities that are, in effect, dormitories for the CBD, where most of the economically active residents go to work. Here lie some of the urban area's highest-quality residential areas. Here, too, is the interdigitation of urban and rural areas, the leading edge of the advancing city. As we said earlier, small rural towns may be engulfed by the rising tide of suburbanization.

Burgess suggested that the zones thus identified are not static, but are mobile and encroach on each other. As a city grows, it is postulated, the whole system expands outward, so that CBD functions invade region 2, the characteristics of zone 2 begin to affect zone 3, and so forth. Zone 5 will lie ever farther from the central city. This, certainly, is something we can see occurring in many cities, but the Burgess theory also has its shortcomings when it is put to the test of real situations. For example, it does not account adequately for areas of heavy industry in the city; the industry mentioned as part of zone 2 is of the light-manufacturing variety. Certainly heavy industrial areas do not form a ring surrounding the entire city, but anyone who knows an American city with big industrial plants knows these to be concentrated in one or two places, related mostly to transport lines. And Burgess's concept of outward mobility also needs its qualifications. It may apply to residential areas, but many other urban elements, such as railroad yards and stations, port facilities, and large manufacturing plants tend to become entrenched

FIG. 9.6 The concentric theory of urban growth.

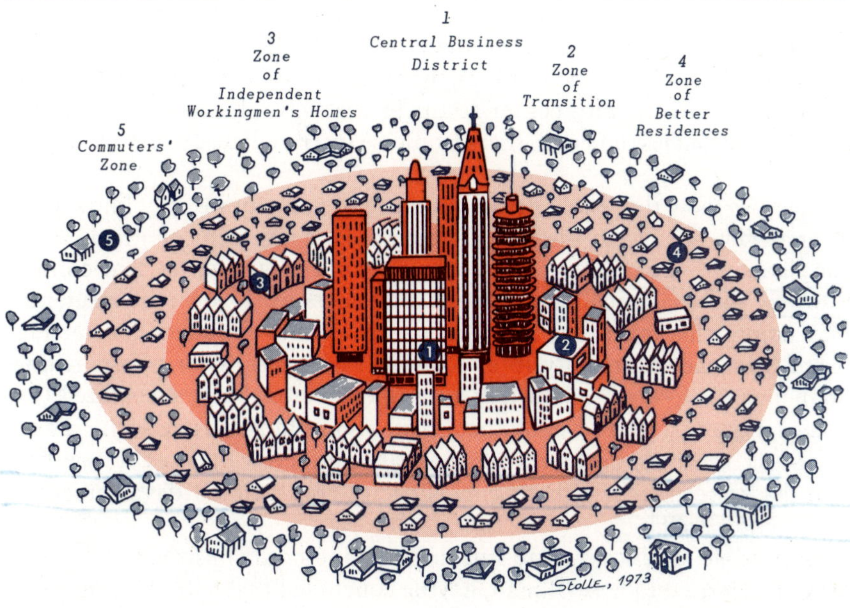

in the urban landscape and will resist this locational change. In fact, as Chicago itself has shown, this is true even of certain residential areas, especially when they have a strong cultural-national flavor. On the other hand, the experiences with airports tend to confirm Burgess's idea. Chicago's Midway Airport was an edge-of-town field when he did his study, and it grew into one of the nation's busiest. Then urban sprawl overtook it, congestion on the ground and in the air became acute, and more distant O'Hare Airport was built. Next time you fly into Chicago, note what is happening to all that open space there used to be around O'Hare. There will soon be some decisions to be made involving an airport still farther out to be added to the existing facilities or perhaps out in Lake Michigan on an artificial island.

But is such concentricity as Burgess proposed really the norm? When you travel into a city via a major artery, you will note that certain functions tend to cluster along that major roadway. Rarely do you see expensive homes along truck-congested, four-lane highways; rather you see factories, auto showrooms, and the like. This means that arterial routes tend to interrupt Burgess's concentric scheme, so much so that it is possible to argue that the major structural features of a city are pie-shaped rather than concentric. And this is precisely what economist H. Hoyt concluded in his study of urban layout (*The Structure and Growth of Residential Neighborhoods in American Cities,* 1939). He proposed a *sector theory* of urban structure, based on data related specifically to residential land use. Hoyt showed that much could be learned from an analysis of one aspect of residential life, the rent people pay. Following Burgess's theory, we would conclude that rent consistently increases with distance from the CBD, since the quality of housing is lowest in zone 2 and highest in zones 4 and 5. Hoyt, however, argued that a low-rent area could extend all the way from the center of a city to its outer edge. Similarly, high-rent sectors could be recognized in the cities he studied, flanked by intermediate-rent areas. Thus, a high-rent sector might extend along a very fast, convenient, uncluttered route into the city center (train or highway), but where communications were somewhat less efficient, there would be intermediate-quality residential areas.

When we view the whole urban structure of any city we may know fairly well, we may well be able to discern a pie-shaped arrangement of urban zones. Industrial plants may be concentrated along one or several arteries that connect the city to raw material supplies and hinterland markets. A low-rent residential sector may adjoin this industrial complex; a high-quality residential sector may fan outward from the central city in another part of town. Major transport lines may be bunched so closely together that a transport sector can be recognized (Fig. 9.7). A major university and several schools can also form a discrete sector, as our map suggests. As is so often the case when there are conflicting theories, it is perhaps possible to discern elements of both concentricity and sectorial development in the city you know well.

It is possible to conceptualize still another kind of urban structure. In 1945, Professors C. D. Harris and E. L. Ullman, having concluded that both the concentric

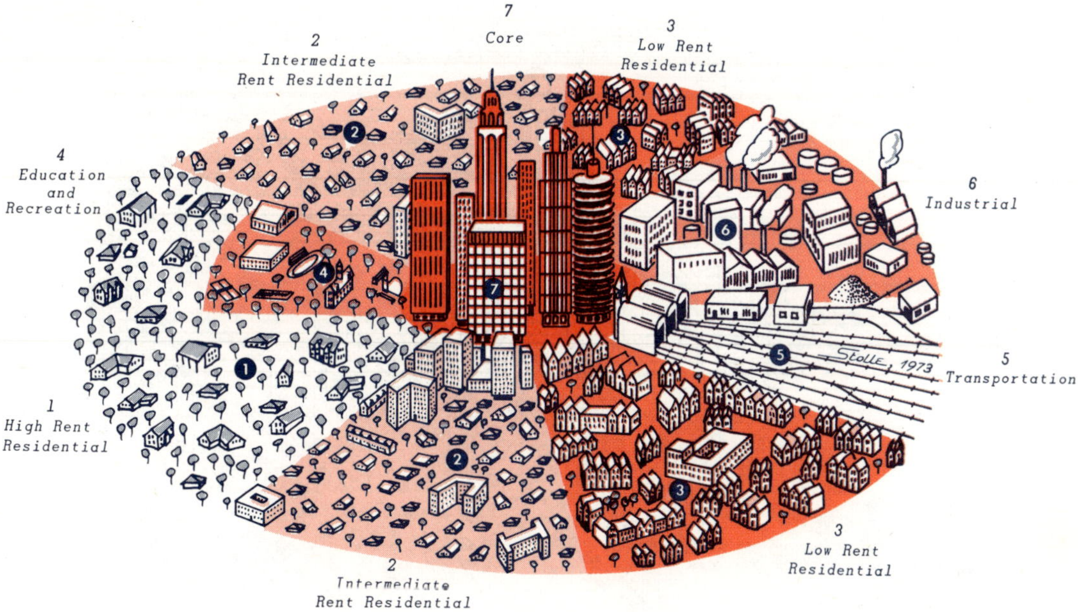

FIG. 9.7 The sector theory of urban development.

and sector theories left unsolved problems, proposed in an article entitled "The Nature of Cities" a theory that holds that cities consist of a number of discrete areas that form and evolve around separate nuclei. Once again, the idea arises from something we can observe every day as we live in or visit cities: the CBD is losing its dominant position as the undisputed nucleus of the city, and it is getting competition from other, vigorously growing business centers that constitute secondary nuclei. Thus the *multiple-nuclei theory* proposes that a city, especially a large city, may grow *not* around just one focus, but around several nuclei. These several foci include not only the CBD itself and subsidiary business areas such as large shopping centers, but any other element, a university, a port, a government complex, that can stimulate urban growth and attract other functions. In their diagram (Fig. 9.8), the authors suggest schematically what such a city might look like. Note that there is none of the regularity of the Burgess and Hoyt models. A city that comes to mind immediately, of course, is Los Angeles, but nucleated development of this sort is occurring in most of the world's larger cities today.

But even the multiple-nuclei theory fails to account for all the spatial properties of the modern city, and it shares with its two predecessors the problem that it is perhaps an excessive oversimplification of the situation in today's megacities. Still, the next time you spend some time in a large urban area, perhaps the place where your university or college is located, try to discern the elements of urban structure and relate them to the models we have just discussed (perhaps using a fold-out street map). You are likely to find the exercise quite interesting,

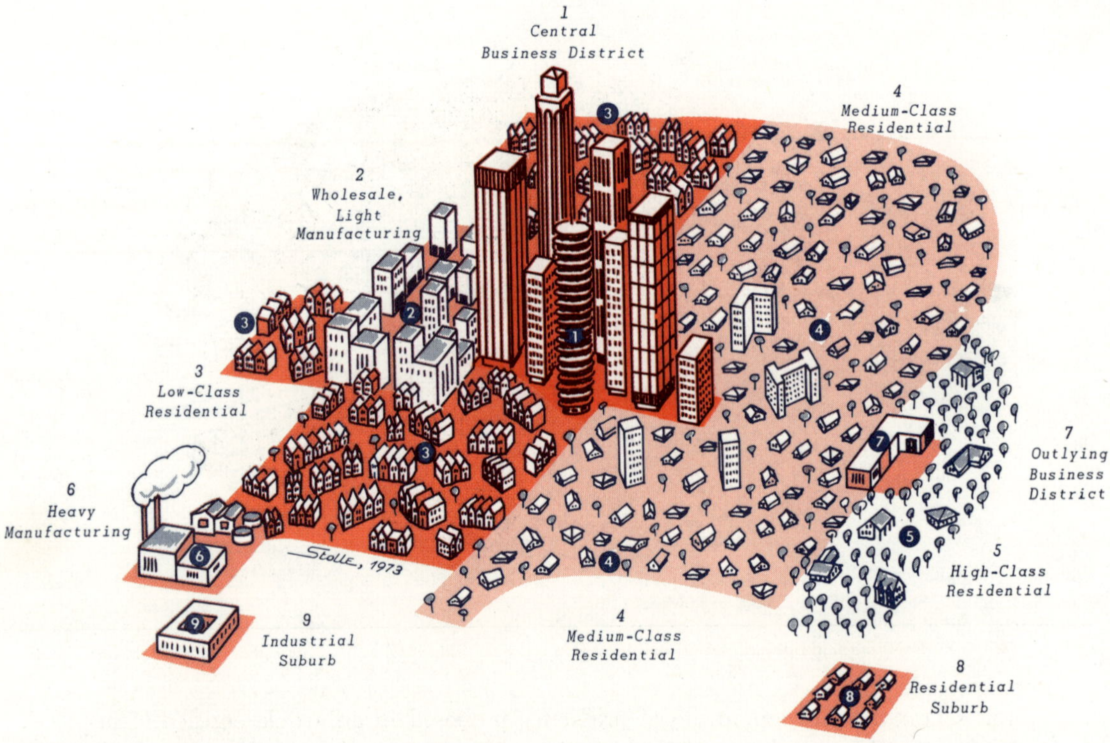

FIG. 9.8 The Harris-Ullman multiple-nuclei concept.

and perhaps you can produce your own theory of urban structure, integrating the theories already proposed. At the very least, you will get to know the layout of your urban environment much better!

PROBLEMS

In the long-distance perspective of time, historians may some day come to refer to our twentieth century as the beginning of the age of modern urbanization, the time when true urbanized societies made their appearance for the first time, and the problems of life in megacities and megalopolitan regions emerged in full force. Burgess, when he wrote about his concentric theory, viewed cities as great integrating forces, as assimilators of diverse people, as generators of American ways of life. But Burgess wrote in the 1920s, and cities over the past half-century have changed a great deal. Burgess based his views on urban America of the late nineteenth and early twentieth centuries, when European immigrants were streaming to America's cities and when assimilation seemed to be a matter of integrating Germans, Poles, and other Europeans into American life-styles. There was a major in-migration to the cities to come, however, and it started gaining momentum during the years of the First World War, when the European flow was cut off and the urban industries continued to expand and require labor. Now a campaign began to recruit black laborers from the South, to which blacks responded by leaving the South by the hundreds of thousands, mostly to the cities of the American core area. This flow continued after the war and increased during the 1920s; after a decline during the depression years of the 1930s it resumed its upward climb. At the turn of the century, only 10 percent of the black American population lived outside the South; in 1970, it was nearly 50 percent. This means that the black population in America became urbanized even more rapidly than the white population did, for the vast majority of those who went to the industrialized North came from rural areas in the South. Today, about 90 percent of black Americans living in rural areas remain in the South; of those in the North, over 90 percent live in the cities.

What happened when these black Americans came to the cities of the North? They usually settled in one or perhaps two areas of the city, clustering near where black people already were living, mostly in areas where rents were low. Crowding soon became a problem, and the high densities had two immediate consequences, the expansion of the black residential areas in the city and their

One of the "multiple nuclei": a shopping center away from the downtown area of an American city. This is the Northland Shopping Center, Columbus, Ohio, which opened about a decade ago with 60 stores, restaurants, and a theater. As the photograph shows, parking is ample and, quite unlike the situation downtown, free. A shopping center such as this employs several thousand people who, in turn, will tend to seek living quarters near it. Thus both customers and residents are drawn toward a nucleus other than the central business district.

deterioration. Black people could not simply move into any part of town and buy a house; most could not afford it, and those who might be able to purchase a modest home found that their choices were limited to few areas. A pattern of residential segregation evolved, quite similar to that with which the black people had been acquainted in the South. It is, we know, a persistent feature of the American city even now. The city has not been the assimilator some anticipated that it would be.

As the in-migration of black residents swelled the population of the inner cities, it generated an out-migration of white residents, and eventually the suburbs began to sprawl beyond the municipal boundaries, producing administratively independent communities. This has had several implications. First, the flight of the white inhabitants of the interior city reduced the opportunities for constructive contact between the cultures of the city. Second, those white residents who moved beyond the city's boundaries took with them their substantial taxable incomes. The city suffered the loss of the taxes. Many businesses followed their customers out of the city as well, and the city proved still less able to provide for its remaining inhabitants the services that were already under the pressure of numbers.

In Chapter Three we viewed some of the effects of urban congestion and a deteriorating environment upon the fabric of culture and the general condition of society. The inner city in America has become a place of crisis, of tenements, dirty air, polluted waterways, violence, decay. Harlem, New York, typifies the situation: there, nearly 3 million black Americans are crowded into apartment buildings, mostly six stories high, that were built as walk-ups during the late nineteenth and early twentieth centuries. Most of those buildings are old, unsanitary, worn out; yet their apartments are overfilled with people who must live with this discomfort and who cannot escape the vicious cycle that put them there. So crowded is Harlem that if all of Manhattan Island were as densely peopled, it would contain the entire population of the United States. Urban renewal, tried in so many places these days, is not always the answer. In the poverty of ghetto and slum life, the neighborhood, social structure, and life-style are nevertheless something of value. The construction of high rises and the relocation of the people may so disrupt that sense of community that the liabilities ultimately outweigh the assets, and people feel themselves to be no better off than they were before.

Spend a night in the downtown area of an American city, and compare your experience with a similar one in Europe, in Paris, Amsterdam, or Hamburg. In those European cities, the streets are busy late at night, lights are on in windows above shopping streets, in apartment buildings. In the American city, the streets are quiet and dark. Except for some activity in theater districts and in such areas as Atlanta's Underground or Chicago's Near North Side, the streets are dormant, buildings dark. Why the difference? One major difference lies in the continuing residential function of the downtown area in many European cities. The skyscraper has made its appearance, and some European cities are taking on American aspects architecturally (central Rotterdam, for example), but where the war did not destroy the inner city old patterns of mixed retailing and residential living still prevail. Thus downtown stays alive, and its continuity is reflected by old establishments, always dependent on local, neighborhood patronage, that still survive. What is happening in the American city may not be symptomatic of Western cities in general, but a regional phenomenon born of the special circumstances of American society. By building exclusive, expensive high-rise apartments near the central city (Chicago's Marina Towers, for instance), attempts are now being made to lure the better-off residents back into the downtown area. But the rising prices of land have driven away the bulk of those who might have remained, the broad population base the heart of the city needs. So the downtown region lies surrounded by the sickest sector of the urban complex, whose diseases spill over into its unsafe, deserted, concrete canyons.

If this seems an overly pessimistic appraisal, it may be because you have not experienced the circumstances of life in the cities' inner areas. The most positive, optimistic descriptions of the condition of the inner cities are written in

To lure the suburbanite back to the central city: modern high rises, offering the most up-to-date facilities, in-house parking, and often magnificent views over the urban scene. Proximity to work and entertainment areas and the elimination of daily commuting make living here attractive. These are the Marina Towers, Chicago, overlooking the Chicago River and positioned on the northern edge of the central city.

the libraries of comfortable suburban homes and in some government offices. The kind of furor that was unleashed in America's cities during the 1960s did not come about accidentally—and the causes have not been eliminated.

Three prominent problems beset the American city as it struggles to combat its ills: government, environment, and communication. Declining revenues and increasing dissatisfaction among the black population combine to render city administrations impotent; schools are overcrowded, services get worse instead of better, housing projects lag, law and order break down. Not only do black urban residents face residential and occupational discrimination; but they find that poverty also means political weakness. They are unable to stop their living space from being torn apart for the construction of expressways, along which the suburban white worker travels back and forth to his downtown job every day. Despite all efforts to reverse the prevailing trend, the crisis in which the large cities find themselves is deepening still. Government is enmeshed in endless complications; it is divided and duplicated, and coordination is lacking. Planning on an integrated, overall basis, with long-range objectives such as those of Holland's Randstad, is practically impossible, for there is little cooperation in multigovernmental urban regions.

In the process, it is not just the city's buildings that degenerate. As we noted

With the advent of the automobile and the mushrooming of its numbers, expressways have been cut through urban areas to facilitate movement into and out of central cities. Undoubtedly, these expressways have moved the traffic, but they have also broken up the urban area and separated people. They have often become barriers cutting residential zones off from each other. When routes for new expressways are selected, the people most likely to be dislodged are those who live in poor areas. They have the least power to influence the decisions involved.

in Chapter Two, urban areas where industries and other plants (power generators, for example) cluster suffer from environmental deterioration. The natural climate is modified, the air is less pure, atmospheric circulation fails, and noise and smell and ugliness assail the senses. The city's waterways are fouled by municipal, industrial, and even recreational wastes. Water shortages and power failures occur. Land for recreation, parks, waterfronts, and playgrounds is being taken over for business and financial gain. Zoning boards yield to pressure and the urban scene becomes a patchwork of inconsistencies.

Inevitably all this has an impact on the city's accessibility, its internal circulation. To function adequately, to be a reasonable place to live and work, a city must have good accessibility. The city provides advantages to various functions through permitting them to cluster closely together. But that advantage disappears when the system of contact and communication breaks down. We have seen this happen. Larger numbers of people in far-away suburbs need ever wider, multilane expressways. The expressways are built, but somehow the volume of automobiles seems to increase faster than the carrying capacity of the roads. A four-lane expressway is so congested that one six lanes becomes necessary; no sooner are the six lanes put in service when they are just as congested as the four lanes were a short time ago. In America's widely sprawling cities, mass-transit systems cannot be expected to solve such problems altogether either. In such places as New York and Chicago the subways and trains carry thousands of commuters, but the highways are still terribly overcrowded.

It is the mushrooming use of the automobile and the resultant personal mobility of large numbers of the population that has contributed greatly to the problems now confronted by the cities. Some cities have already experimented with schemes that would ban cars from certain downtown areas (the mayor of New York closed Fifth Avenue to traffic some years ago, just for one day, and made that bustling street a shopping mall for pedestrians only); parking problems have led to drastic tow-away measures. These measures, though, are mere stopgap answers when much more comprehensive action is needed. In the northeastern megalopolis, the outer fringes of cities are beginning to meet, and traffic congestion focused on outlying nuclei is approaching that of the central city itself. No one foresaw the impact the increased use of the personal automobile would have upon the spatial characteristics of our cities, and the implementation of potential solutions is difficult for reasons we noted earlier, principally among them, integrated planning.

Our cities are the foci of our culture, still the centers of production, innovation, administration, the arts. But our cities also mirror our society's major problems. People who need to work in the cities but who can afford it want to live as far away from them as they can. People who cannot afford to live away from the city must face the realities of a deteriorating environment. The cities assimilated East and West Europeans into the American life-style, but they failed to achieve the same with their latest immigrants, black Americans. They fragmented spatially, and thus came to reflect society as a whole. Ours at present remains a broken society, whose healing must start where the problem is most acute, in the great urban areas of America.

CHAPTER TEN

The Scramble for the Oceans

TEN

We have discussed nine topics in geography to this point, all of them dealing with phenomena that exist on the land surfaces of our globe. That is reasonable enough: practically all of the world's population lives on solid ground. The continents, however, constitute less than 30 percent of the total surface of our planet. The other 70-plus percent is under water. We should be better informed about the oceans than we are and more aware of the growing problems relating to the marine world that must be faced and solved.

What are these problems? We have referred to some of them in passing. The technological revolution is bringing the oceans increasingly within human reach. Not long ago we were catching fish by means of small, wooden boats and nets pulled in by hand. Now we are locating fish schools by radar, harvesting them mechanically, and processing them in floating factories. Resources buried in the ocean floor, always beyond reach and even undiscovered, are being found and exploited from man-made islands. The population explosion has also generated a pollution explosion as millions of tons of waste pour into the oceans, which in places are taking on the characteristics of an open sewer. And now the question of the territorial sea is giving trouble, too. Coastal countries that were long satisfied with a narrow strip of sovereignty over adjacent waters are demanding rights over ever wider zones. The oceans always were open, free, available for all to use. They are in danger of losing that quality.

THE OCEAN BASINS

The ocean basins at this point in the earth's geologic history are so full of water that they in fact overflow periodically. Today the edges of the continents are submerged by salt water. We know this to be the case because continental coastal plains (such as that of the southeastern United States) continue under the ocean's surface to a depth of between 500 and 600 feet (100 fathoms). Only after reaching this depth does the terrain fall off more steeply to the true ocean bottom. As we noted in Chapter One, this shallow, submerged surface that extends from the actual coastline to about 100 fathoms is called the *continental shelf,* and the submarine scarp that leads down to deeper ocean floors is the *continental slope.* The continental slope, thus, is the real geologic margin of the continents, and the coastlines as we see them today reveal the degree of inundation of our landmasses by the ocean's waters. If the oceans yielded those 600 feet back to us, the land surface of the earth would increase by nearly 5 percent, to about 35 percent of the total. With population pressure what it is (and what it will become), this is an unfortunate time to lose nearly 5 percent of the continents to the seas!

We know what continental landscapes look like, but we still have only glimpses of the ocean floor seascape. Deep down, darkness permanently prevails, and cameras lowered on cables can make only flash photos, not the best way to get a view of the lay of the land. A few researchers have gone down in submarinelike research vessels fitted with lights and small windows, but even they saw just a fragment of the vastness of the ocean's seascape. How, then, can we know much about the configuration of the ocean basins? The answer is another product of modern technology, the depth recorder, which operates by sending sound impulses

FIG. 10.1 The continental shelf and ocean floor off the northeastern United States. (Modified from B. C. Heezen and M. Tharp, *Physiographic Diagram of the North Atlantic Ocean,* Geological Society of North America, Boulder, Colo., 1968.)

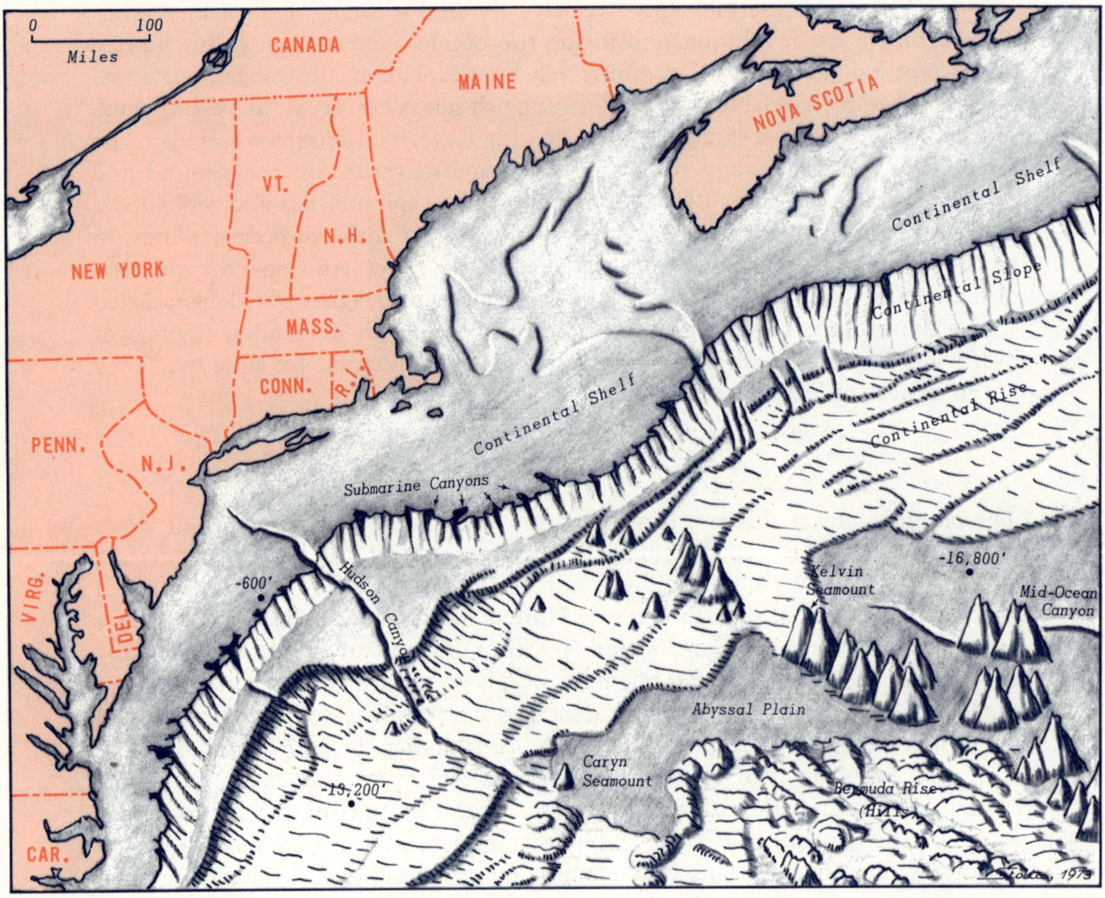

to the bottom and recording their echoes. As a ship traverses the ocean, its depth recorder creates a profile of the seascape over which it travels. Since the Second World War, many thousands of ocean-floor profiles have thus been made, and from these it has been possible to construct maps and diagrams of the submerged topography. Figure 10.1 is such a diagram. It shows the marine topography off the northeastern part of North America. Note how the continental shelf extends far beyond the coast and is then cut off by a rather steep scarp that in turn leads to a much deeper part of the Atlantic Ocean's basin. The real situation is not quite as dramatic as the diagram suggests, of course. Such models always require a certain amount of vertical exaggeration. But the continental slope is quite well-defined nevertheless.

The Continental Shelf

The continental shelf is a part of the continent adjacent to which it lies. It is, therefore, geologically similar, and it will have resources such as those also found below the land surface of the continent, fuels such as oil, metals, and other minerals. Narrow continental shelves are just a few miles in width, but in some areas they extend for more than 200 miles offshore (Fig. 10.2). The continental shelf generally has a smooth surface, although deep, canyonlike trenches have been cut through it in places, apparently made by currents of muddy water that flow downslope like submarine rivers. Man has learned how to overcome the technical difficulties posed by mineral exploitation below several hundred feet of water, and on the nearly level surface of the continental shelf stand oil platforms whose drills penetrate far into the *subsoil,* as the rock material of which the shelf is made is called. Not only the subsoil but also the overlying waters have economic potential. As we noted in Chapter Eight, most of the fish caught in the oceans are taken from relatively shallow waters, and the ocean is shallowest over the continental shelves. Note the correlation between Figs. 8.3 and 10.2 in this connection. It is not surprising, then, that coastal states have begun to lay claim to both the subsoil of the continental shelf and the waters overlying it, or the *superjacent waters,* to use the legal term. Even states that do not have the technology to exploit the resources that might exist in the subsoil are making such moves; in any case, they do not want foreign powers to put oil rigs in the sea within sight of their coasts. If there are going to be foreign companies, the coastal states want to be able to control their activities and to sell exploitation rights.

As Fig. 10.2 shows, the world's continental shelf areas are markedly concentrated in several areas. Off eastern Asia, from the Bering Strait, along the coast of China and Indochina, and into Indonesia's archipelago lie the most extensive of these. Australia has substantial shelf areas, especially in the north, and New Zealand also has shelf zones to the north of both of its major islands. On the opposite side of Eurasia, northwestern Europe has large continental shelves. North America has an extensive continental shelf along nearly its entire east coast and along much of its west coast as well. South America's continental shelves are best

FIG. 10.2 Continental shelf areas of the world.

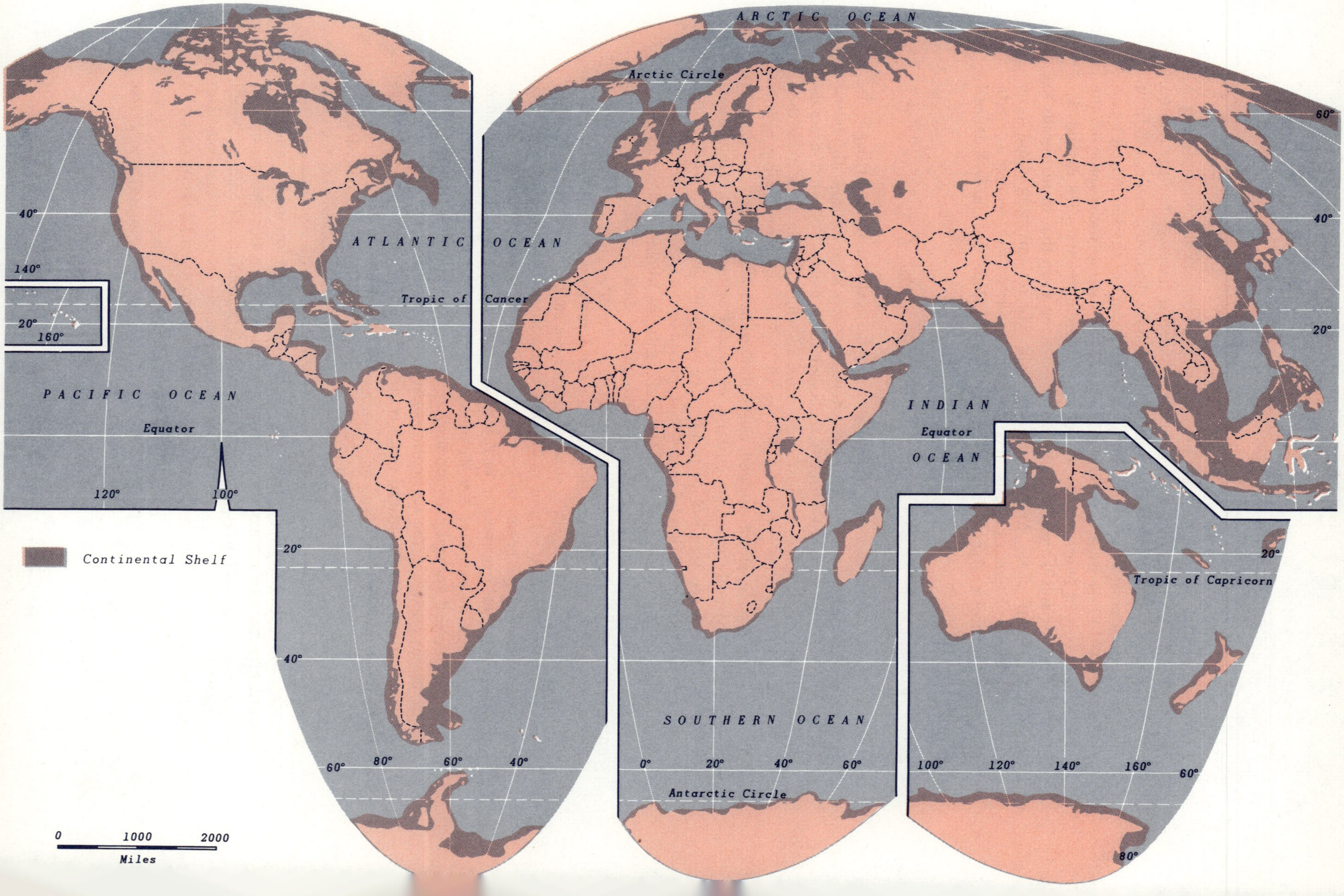

developed off the coast of Argentina and southern Brazil. There is a narrow shelf along southern Chile, Peru, and western Colombia, and also along part of the continent's north coast. Per mile of coastline, Africa is least endowed with continental shelves, which has to do with its position in Gondwana (see Chapter One). There is some shelf development along the west coast and off the southern tip of the continent, and in the Mediterranean Sea.

The Ocean Floor

The "real" ocean basins, then, begin where the continental slope drops off several thousand feet (3000-10,000) to the ocean floor. Much of the ocean floor lies between 10,000 and 20,000 feet below sea level, but the surface is not nearly as flat as that of the continental shelves. There are hills and mountains, ridges and valleys. Some of the submarine hills and mountains are high enough to protrude above the water surface as islands (Bermuda is an example). There are also remarkably flat areas, called *abyssal plains*, which probably result from the deposition of silt and mud and beneath which rougher terrain lies buried.

Two other features of the ocean floor have much significance. For many years the Mid-Atlantic Ridge was known to geographers as a lengthy, ocean-spanning mountain chain extending north-south in the Atlantic, about midway between the opposite coasts. There the ocean floor rises to depths varying from about 6000 to 9000 feet. What was not realized until more recently is that the Mid-Atlantic Ridge constitutes only a part of a ridge system that extends around southern Africa into the Indian Ocean and on into the Pacific (Fig. 10.3). Now the Mid-Atlantic Ridge is recognized only as a part of this worldwide *Mid-Oceanic Ridge System*.

The other feature displayed by the ocean floors has the reverse characteristic. It is not a ridge, but a series of very deep *trenches*. These are elongated, narrow troughs whose depths reach 25,000 to 30,000 feet and even more. They are particularly well developed in the western Pacific Ocean, but there are major trenches also off the coast of western South America, off the Pacific coast of Middle America, and in the Caribbean (Fig. 10.3). For a very long time, the relationships between the ocean-floor ridges and trenches went unrecognized, but now the sea-floor spreading hypothesis of continental drift makes the association apparent. The concept of *plate tectonics* holds that the earth's crust, ocean floors as well as continents, consists of several huge segments that float on the mantle below. Where those plates separate, a fissure known as a *spreading center* opens between them. Up through this fracture wells the molten material from the mantle, and it solidifies into a ridge, such as the Mid-Atlantic Ridge. Where the plates meet and push against each other, one is forced downward under the other, and in the process the trenches are created. This is why such submarine mountain ranges as the Mid-Atlantic Ridge do not resemble the folded linear ranges of the continents, such as the Appalachians and the Himalayas. The ocean-floor mountains are crystalline mountains; the continental mountains contain sedimentary rocks. Thus the oceans, such as the Atlantic of today, began as spreading centers, and first looked

FIG. 10.3 Mid-ocean ridges and ocean trenches. (Modified from a map in "The Origin of the Oceans" by E. Bullard. Copyright © 1969 by Scientific American, Inc. All rights reserved.)

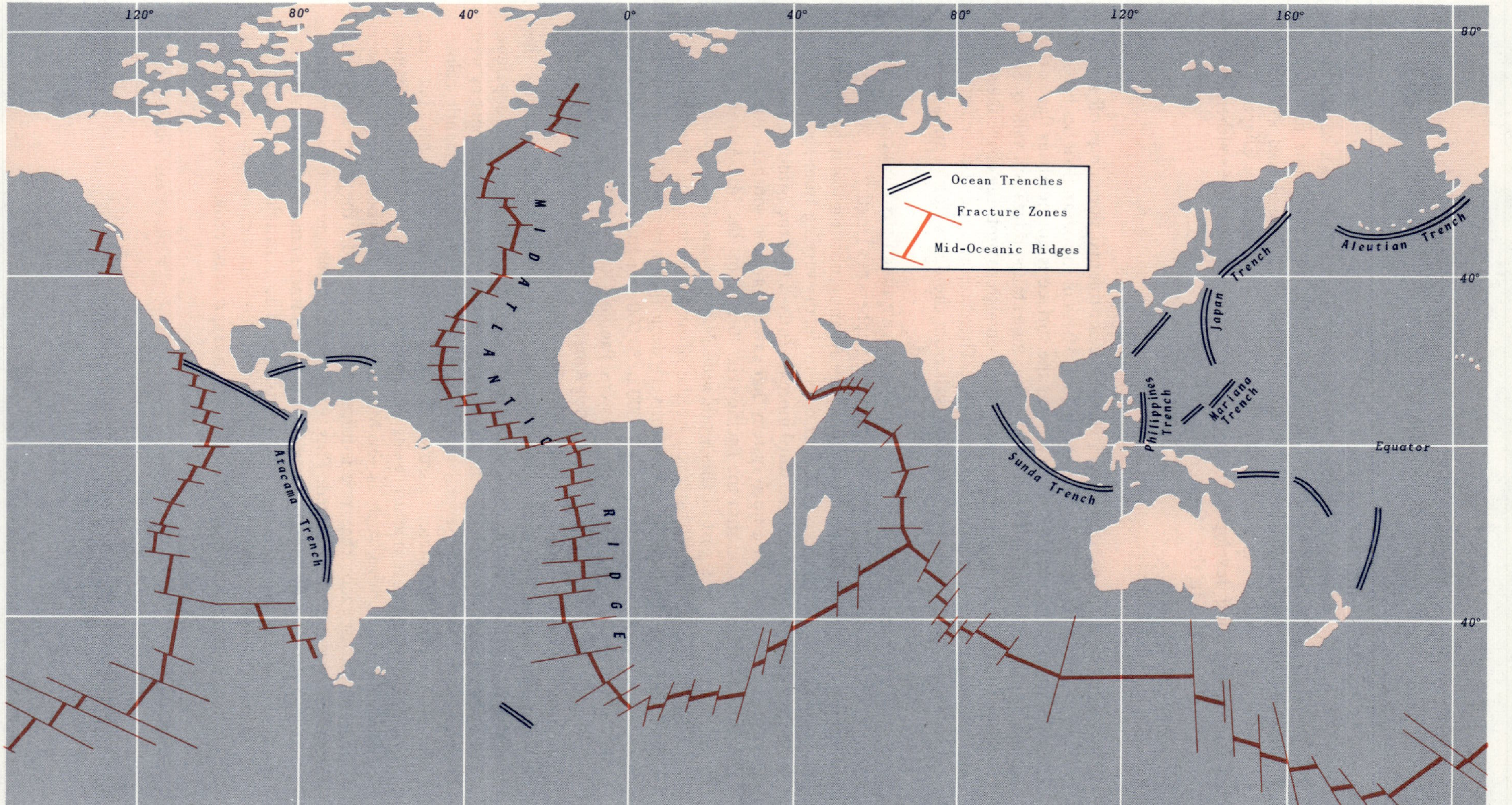

as the Red Sea between Africa and Arabia does today. Gradually the plates spread farther apart (as those of South America and Africa have done), and the ocean widens and takes shape.

From the ocean floors the answers are coming to questions of earth history that have occupied scientists for many years. Again it is modern technology, of cameras and bore holes and floating laboratories, that is making it possible. Exploitive techniques will undoubtedly be developed to secure what is found.

OCEAN WATER

It is possible that the oceans will someday become a principal source of fresh water and, with their vast hydrogen content, a source of fuel as well. Of course, the oceans already *are* the major source of fresh water, but man depends upon nature's hydrologic cycle for the supply. In the future, desalinization plants may stand on the coasts as urban and other water requirements are met in part directly from the oceans. Such plants are already in operation in a few places, but the process is yet too expensive for general use. The technical problems involved in large-scale hydrogen extraction also await solution. Huge desalinization projects and power plants may become part of the twenty-first-century landscape.

Properties

Ocean water has several properties of which we should be aware. First, and most obviously, ocean water is salty. It is a salt solution, a brine, that contains about 35 pounds of salt for every 1000 pounds of water. This salt is mostly sodium chloride, but other chlorides and sodium sulphate are present as well. There is some variation in the salinity of the oceans. The range is between 33 and 37 parts per 1000 generally, but there are places, such as the Red Sea, where there is not much external circulation and a lot of evaporation. There, the salinity may go as high as 40 and even higher. On the other hand, where the fresh water of rivers flows into the ocean, the salinity level is much less than average. If the oceans are filled by fresh-water rivers, how did they come to be saline? The fact is that rivers do carry chemicals in solution, derived from weathered and eroded rocks in their drainage basins. Those chemicals are washed into the seas. Some are deposited on the ocean floor, locked up in marine sediments; others remain in solution. Thus surface evaporation and selective deposition combine to maintain the existing level of salinity.

Second, the water of the oceans has varying temperatures. Looking at it vertically, ocean water generally is warmest at the surface. Because there is constant moving and mixing, a rather thick layer forms in which the temperature is fairly constant. This layer extends down to about 1500 feet, below which there is a rapid drop in temperature. In low-latitude zones, the upper layer may have a temperature of from 70 to 80°F, which drops to the low 40s at the base of the warm upper layer and to the high 30s not far below that. Cold water, like cold air, is denser than warmer water; water that is very salty is also quite dense. In equatorial latitudes, high surface evaporation causes high salinity and thus high density in the water, so that upper portions of the ocean tend to sink down, replacing less dense water below. In near-polar regions, water that has become extremely cold at the surface, and therefore very dense, also slides down to replace water at certain depths. Thus a sort of convection system exists in water just as it does in air.

Ocean Currents

Viewing the distribution of temperature in ocean water horizontally, we note that the warmer temperatures of tropical latitudes are carried poleward and that colder temperatures of polar latitudes come equatorward in *ocean currents.* In Chapter Two we considered the effects of Coriolis force, the force related to the earth's rotation that affects moving objects on the surface. As we noted, atmospheric cells of circulation develop as the air moves in giant circles. The water of the oceans moves, too, not at random, but in huge systems powered by the wind patterns and, to a lesser degree, by salinity and density differences.

The system of ocean currents is perhaps best explained by a close look at the situation in the Atlantic Ocean. Once we understand the system of circulation there, the worldwide mechanics will not be difficult to grasp. Figure 10.4 represents in generalized form the circulation systems of the Atlantic. Note, first, that some of the water movements are called *currents* (such as the Benguela Current), others are identified as *drifts* (the West Wind Drift), and still another is referred to as

---

THE GULF STREAM AS ENERGY SOURCE?

In the continuing search for energy supplies, three scientists recently proposed that it may be possible at some future time to harness the great Gulf Stream to produce electricity. In a 1973 paper entitled "The Florida Current as a Potential Energy Source?" Dr. W. S. von Arx, Dr. H. B. Stewart, Jr., and Dr. J. E. Apel point out that the Gulf Stream carries 30 million cubic meters of water per second past any given point along the Florida south coast, which is over fifty times the total flow of all the rivers of the world. The rate of speed, about 5 miles per hour, is very slow compared with that of rivers and dam outlets where turbines have been placed. But it should be possible to design honeycombs of large, slow-speed machines, rather like water windmills, and position them 100 to 400 feet below the surface across a 12-mile section of the current. The scientists predict that such a system might produce 1 billion watts on a year-round, 24-hours per day basis, which is as much as two nuclear plants could generate. If the pilot plan were to succeed, other coastal locations on several continents might establish similar systems to supplement power supply from other sources.

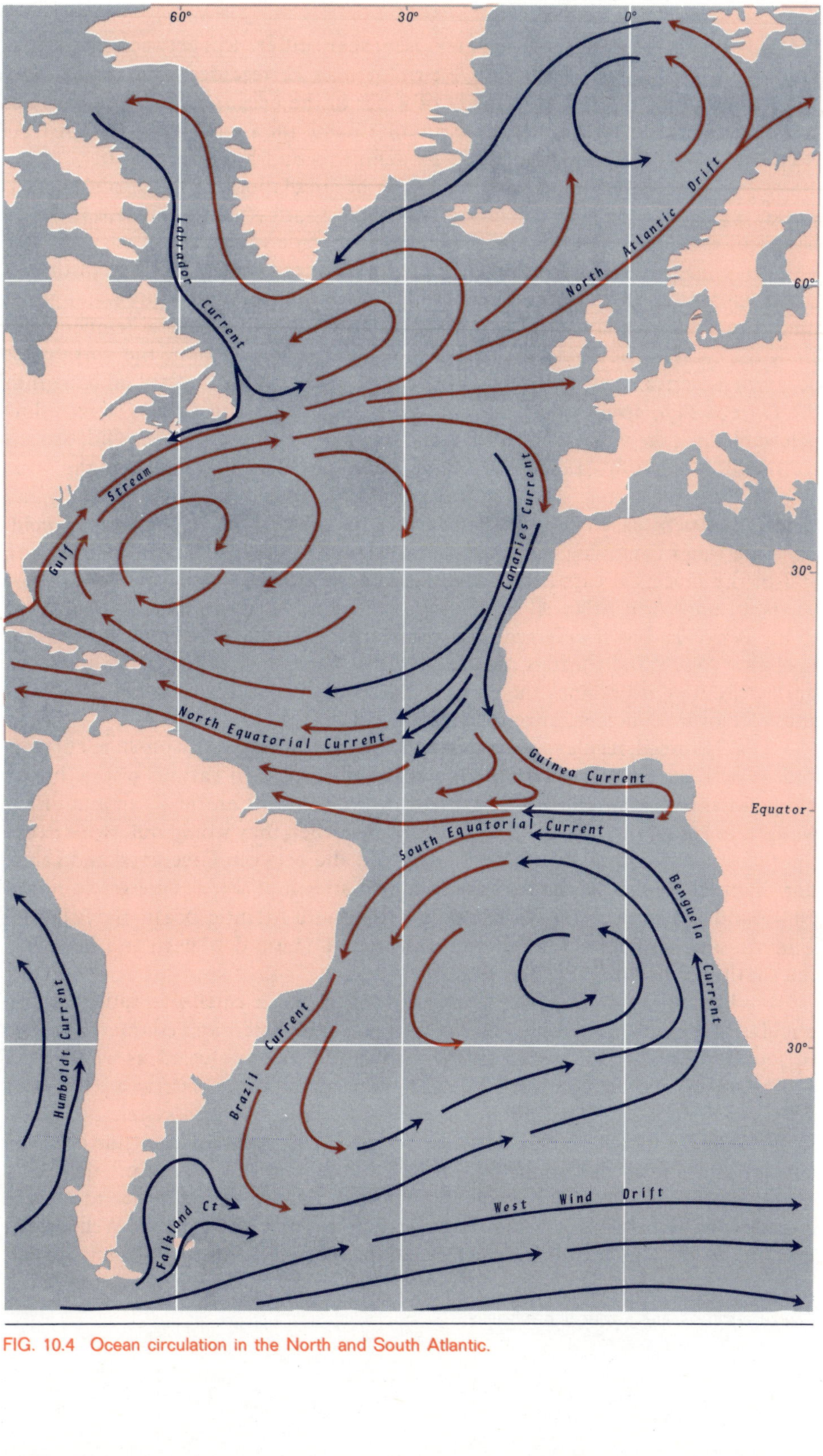

FIG. 10.4 Ocean circulation in the North and South Atlantic.

the Gulf *Stream.* These terms do not represent sharp differences between the various kinds of water bodies in motion. A current may be regarded as a rather more defined, fairly rapidly moving "river" of water in the ocean; a drift is slower, less confined. The Gulf Stream is simply another name for a current in that area. As it crosses the North Atlantic, it loses its definition and becomes a drift.

Also note that the dominant system of circulation in the North Atlantic Ocean is clockwise; it is counterclockwise in the South Atlantic. Perhaps it would be useful to compare Fig. 10.4 with Fig. 2.6: the trade-wind pattern suggests how water in equatorial latitudes would be driven from east to west in the North and South *Equatorial Current.* Upon reaching South America, some of the water piling up in the western Atlantic turns southward, carrying warm tropical temperatures along the coast in the *Brazil Current.* Other water moves on along the north coast of North America, eventually entering the Caribbean Sea or directly merging into the *Gulf Stream,* also a warm current. The Gulf Stream and Brazil Current flow poleward along the North and South American coasts respectively, the Gulf Stream fading into the *North Atlantic Drift* and the Brazil Current turning into the *West Wind Drift.* The water is cooling now, especially in the South Atlantic; by the time the current called the *Benguela Current* develops, the water is quite cold. Now the water is moving from higher, near-polar latitudes toward tropical zones, and in the process it carries the coldness of high latitudes along the continental coasts. In the North Atlantic, the *Canaries Current* is something of an equivalent to the Benguela, but it emerges from the North Atlantic Drift, still a lot warmer than the West Wind Drift is. As the water nears the equatorial zones it warms up; by the time it becomes the Equatorial Current, it has reached its between 70 and 80°F or even higher range once again.

This broad, regular system, whereby east coasts have warm offshore currents and west coasts have cold currents nearby, is disrupted in various places. In the North Atlantic Ocean, the North Atlantic drift crosses the entire ocean and brings warmth to European ports, keeping them ice-free when they would otherwise freeze up; the water then turns back along Greenland and between Greenland and Labrador, from where the *Labrador Current,* a cold stream of water, emerges to round Newfoundland and cool the Gulf Stream. In the South Atlantic Ocean, the *Falkland Current,* another cold-water body, moves around Tierra del Fuego and heads off the southward-flowing Brazil Current.

Broadly, this pattern, a westward movement in equatorial latitudes and circular systems of circulation, clockwise in northern waters and counterclockwise in southern oceans, is repeated in the Pacific and Indian Oceans as well. These oceans are less confined than the Atlantic, though, and so the patterns are less clear and tend to duplicate. The Indian Ocean has a dual circulation system according to the stated principle, but the Pacific has two cells of circulation north of the equator and one to the south. In both the Pacific and Indian Oceans there is a feature that is less well developed in the Atlantic, namely an eastward-flowing countercurrent between the North and South Equatorial Currents. Well-known currents in the Pacific and Indian Oceans include the *Kuroshio Current,* which brings warmth to Japan, the *California Current,* which cools the waters off North America's west coast, the *Humboldt,* or *Peru, Current* (Fig. 10.4), which brings Antarctic coolness to the Chilean and Peruvian Pacific, and the *Agulhas Current,* whose warm waters squeeze between Madagascar and mainland Africa.

Tides

We already are aware that ocean currents have important effects on human life in several areas, by defrosting North European ports, for example, and have economic implications by producing conditions favorable to the development of large fish populations in particular zones. But in what way could tides affect us, other than by causing problems for ships tied up in ports? The answer lies neither in environment nor in economic geography, but in political geography. The term *tide* denotes the periodic rise and fall of the oceans' level. When the tide is high, low-lying coastal flatlands are inundated. When the tide goes down, such areas are exposed. Anyone who has spent a day on the beach has experienced this change, however slight. It might not matter, except that countries measure their territorial waters from their coasts. There are places where the difference between measuring from high-tide limits and from low-tide marks involves not just a few yards, but several miles. Conversely, someone who owns property along a coast may want to know if his or her rights extend to the low-tide or the high-tide line. Imagine that oil were to be found under that beach!

It is not simply a matter of deciding upon high-tide or low-tide lines for these purposes. Different kinds of high tide exist, and different kinds of low tide as well. There is no need for us to go into the details of how tides are generated; this involves the centrifugal force of the earth's rotation and the gravitational attraction of the moon and the sun. When the moon, earth, and sun are lined up, *spring* tide occurs, during which high water is unusually high and low water unusually low. During *neap* tides, the range is much less, because the moon and sun tend to work against each other. Thus any legal language relating to boundaries must specify whether the high- or low-tide limits used for measurement are those of spring tide, neap tide, or some other period.

Tides travel across the world's oceans, but the places where they are felt most strongly and where they have their fullest impact on human affairs are along the coasts. In nearly enclosed bodies of water such as the Mediterranean Sea, tidal ranges are very slight. In protected waters such as the Gulf of Mexico and Hudson Bay, the range may amount to only 2 or 3 feet. On exposed coasts, the tidal range is larger, often as much as 7 to 10 feet, and it can present problems in ports, where ships must be able to move with the rise and fall along their quays. Still greater tidal ranges occur in some estuaries and funnel-shaped bays, where the water piles up when the tide rises and flows out in a rush when the tide goes down. The result is an enormous difference, amounting to as much as 30 to 40 feet and even more. Nova Scotia's Bay of Fundy is known for its great tidal ranges, measuring 40 to 50 feet.

Waves

When the wind sweeps over a body of water, whether a lake, sea, or ocean, it generates waves in the water. The waves travel unimpeded across open water, and anyone who has taken a flight across a large water body will have noticed wave patterns in the water below, often resembling giant arcs. Sometimes whitecaps dot the waves as the winds whip water loose in a foamy spray. One after another the waves roll across the sea, displaying three features: *length,* that is, the distance between the top of one wave and the next; *height,* or the vertical distance between the crest and the trough of each wave; and *velocity,* which is the speed with which successive wave crests pass a marker. When the wind becomes stronger, the height of the waves increases. When the wind is between 5 and 10 miles per hour, wave height will be only 1 to 2 feet. But a 40 to 50 mile-per-hour storm can throw the ocean surface into incredible convulsions, with wave heights reaching as much as 60 feet. Many a steel ship has been battered to pieces in such weather by the force of water and wind.

The development of waves depends to a considerable extent on the distance the wind has been able to travel over open water. This is called the *fetch;* also important is the time the wind has had to do its work in throwing the water into oscillation. If you have been on the open sea in a ship or boat and have seen wavelike undulations in the water without any noticeable wind prevailing, it means

A severe storm can do great damage to structures along the shoreline. This is what Point O'Woods, Fire Island, New York, looked like following a storm on March 6, 1962.

that the wind had already died out and you experienced its aftermath in the water, or *swell.* It might be that a forceful wind was whipping up waves somewhere else and you felt waves that were dying out far away from that locale.

In what way are waves important to us? If you have done any surfing you will know something of the force a wave can exert on an object and the speed with which it may approach the land. Thus waves mean very little as long as they traverse open water (except that large ones can endanger boats and ships), but when they smash into coastlines, they take on a very different character. Newspaper photographs following a severe storm can tell you what the waves have done in one night: destroyed are dunes, levees, dikes, homes, roads, and ports. Waves are constantly modifying shorelines. They add sandy sediment by washing it up on beaches; they break through and destroy what they are building in one stormy attack. They are almost constantly at it, altering coasts, building sandbars, driving back dunes. They do so just where man would like to establish some permanence for so many reasons, recreational, economic, and political.

COASTLINES

Let us look more closely at the coastlines that the currents pass by, the tides inundate periodically, and the waves attack so often. It is not necessary to consult a large-scale map to be able to see the differences between coasts. Consider the world base map we have been using throughout this book (page 378). Note the difference between the coastline of southern Chile and that, say, of eastern Florida. Some coasts are so smooth and straight that, when you fly over them, they look as though they were laid out along a ruler. Florida's east coast exemplifies this. Other coastlines are full of bays, curves, promontories, cliffs, and other irregularities, as is that of southern Chile. There is almost every variety between these extremes.

In most parts of the world, the water immediately adjacent to the shoreline is shallow and increases gently in depth away from the beach. This has an important effect upon the waves that come rolling in from the ocean, for the increasing shallowness slows them down and causes the crests to come close together (thus, the wave length decreases rapidly). The bottom of the wave is slowed down as it scrapes the floor, but the top continues and falls over in a *breaker.* During a storm, pieces of rock, large stones, and pebbles are hurled up the beach and against structures that are normally out of reach of the waves. So it is not only the water, but also the projectiles carried by the water that do the damage along coastlines. In fact, during periods of relative quiescence there is little erosion, and the waves actually build beaches by bringing fine-grained materials ashore.

Where there is no shallow zone offshore and the surface below the water descends rapidly, nothing slows the waves down as they come in from the ocean, and they hurl themselves against a coastline that reflects this situation: cliffs form, caves are etched out of the rock faces, and especially resistant sections of rock may survive for a while out in the water as *stacks.* Such landforms reflect circumstances under which the sea is rising relative to the land (the land itself may be sinking),

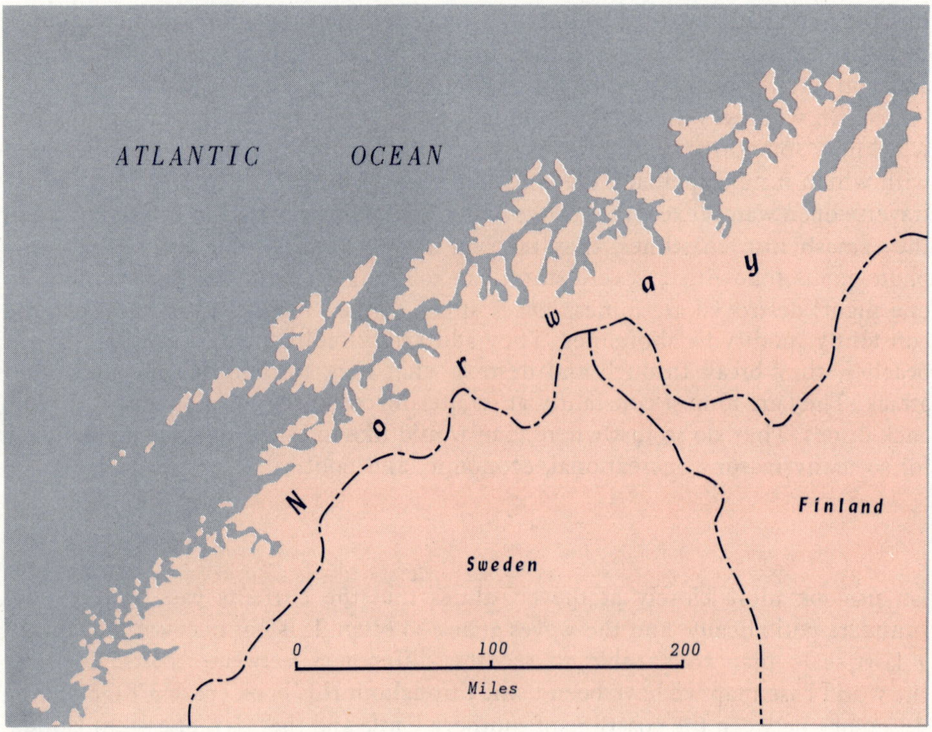

FIG. 10.5 Detail of Norway's north coast.

and there has not been time for the waves to wear down the coastal rocks and to create beaches.

There are still other ways through which highly irregular, deeply indented coastlines may originate. This kind of coastline has also resulted from the activity of glaciers, which at one time occupied deep valleys that extended into the sea. Today those valleys are no longer occupied by ice, but by sea water, which invaded them when the ice melted. This situation is reflected by the coast of Norway (Fig. 10.5), a *fjord* coastline. If the relative position of sea and land does not change over the next thousands of years, the headlands separating those fjords will gradually be worn away, and the whole coastline will take on a smoother appearance. Just as nature's rivers seek to smooth out and wear down topographic irregularities on land, so the waves are constantly reducing the irregularities of coastlines.

Even currents have an effect on coastlines. As these streams of water flow parallel to the coast, they carry along sedimentary material that is sometimes dropped, for example, where the current is forced to take a turn or where its velocity is reduced. Thus elongated ridges of sediment, or *bars*, can form, sometimes across the mouth of a river or an estuary. Along shallow coastlines, waves and currents may combine to build *offshore bars*. In due course these offshore bars become so stable and large that they cut the water between them and the coastline off from the ocean, and a lagoon forms. Miami Beach began as such an offshore bar!

No matter what the morphology of any coastline, it will always be undergoing change. As we noted briefly in Chapter One, physical geographers have tried to recognize order in these changes by classifying coastlines according to whether the land is rising relative to sea level (*emergent* shorelines), whether it appears to be dropping relative to sea level (shorelines of *submergence*), or whether it displays evidence of both processes having occurred in recent times (*compound* shorelines). No shoreline classification is completely satisfactory, however, because the varieties are so many and circumstances vary so much along the tens of thousands of miles of continental margins.

Practical Problems

It is important to be aware of these details relating to tides, currents, and coasts, for shorelines are zones where numerous problems are emerging and intensifying. For example, the increased exploitation of resources in the subsoil has raised questions about the exact location of the landward boundary as well as the seaward boundary along coastlines. Where do a state's rights end and the federal government's rights begin? And the changeable nature of coastlines makes it difficult to establish with precision the maritime boundary of a state, which must be measured, as we shall see, from some legally defined limit along the shoreline. Pollution problems also have worsened, as several times a year we read of oil spills from drilling stations on the continental shelf or from ships, which damage beaches and destroy wildlife. Again there is the question of responsibility and authority. The delimitation of land boundaries between states has always presented problems; the seaward boundaries are now proving to be no less problematic.

RESOURCES In passing, in our previous chapters, we have referred to the actual and potential resources of the continental shelf, the overlying waters, and the ocean itself. In a sense, as we get to know more of the contents of this area of the world,

Drilling platforms stand on the continental shelf, miles from the coast. New resources are constantly being located in continental shelf areas. Other comparatively shallow marine areas (the Mid-Atlantic Ridge, for example) may also yield significant minerals in the future. All this causes jurisdictional problems.

it is as though we are discovering a new continent. It is being explored more intensively; formerly unattainable regions of it are becoming accessible to man's equipment. Nations are realizing that their self-interest is involved in their participation in this maritime venture, just as colonial powers during the nineteenth century secured spheres of interest as much to obstruct their rivals' designs as to obtain wealth. Only this time, weak states have a chance as well as powerful ones. Peru and Sierra Leone, for example, have claimed not only the whole continental shelf adjacent to their territories, but, the shelf being quite narrow, up to 200 miles of territorial waters as well. In Peru's case, this is to protect that country's fishing industry; other countries unable to exploit their maritime resources themselves can at least deal with states whose technological advantage does make it possible. Only the weakest legal restraints exist, and the decades since the Second World War have witnessed an unprecedented invasion of the oceans by countries seeking to extend their sovereignty over sea and shelf.

THE OCEAN HIGHWAY As we noted in earlier chapters, accessibility is a resource, albeit an intangible one. For centuries the oceans constituted such a resource to hundreds of cities, localities, and countries. They were the open, free highways between producer and market, between colonial power and dependency, between émigrés' homeland and a better future. The European countries surrounding the North Sea recognized a need for a zone of coastal waters over which the adjacent states would have sovereignty, but they also wanted that zone to remain quite narrow in order to present a minimal impediment to their shipping. To the states of western and northern Europe, the sea's avenues were routes to power and wealth, and although they were at times tempted to try to claim fishing rights over certain shallow parts of the North Sea (fishing also played an important role in their economies), their desire for open, free waters always won the day. Thus their territorial seas remained 3 to 4 nautical miles in width, and narrow straits were internationalized, as were some rivers.

The situation remained relatively stable for centuries, until, in fact, the last 25 years. Now, as maritime claims reach dozens and even hundreds of miles, straits and narrows are being closed off as international maritime highways, bays and gulfs are appropriated by states, and the scramble for the seas is on in full force. Again the analogy with nineteenth-century colonialism presents itself as maps are making their appearance proposing the partitioning of water bodies such as the North Sea and the Gulf of Mexico among "eligible" countries (Fig. 10.6). We are witnessing the end of the concept of the universal high seas.

FISHING The mushrooming food needs of man's growing numbers has strained the fish resources of the oceans. There was a time when fewer mouths to be fed and lesser technological capacities combined to ward off any danger of overfishing, but those days are past. As we noted in Chapter Eight, the oceans' fish fauna tend to concentrate in certain areas favorable to the growth of the foods that fish need. Involved are the temperature of the ocean water and other factors, but the combined

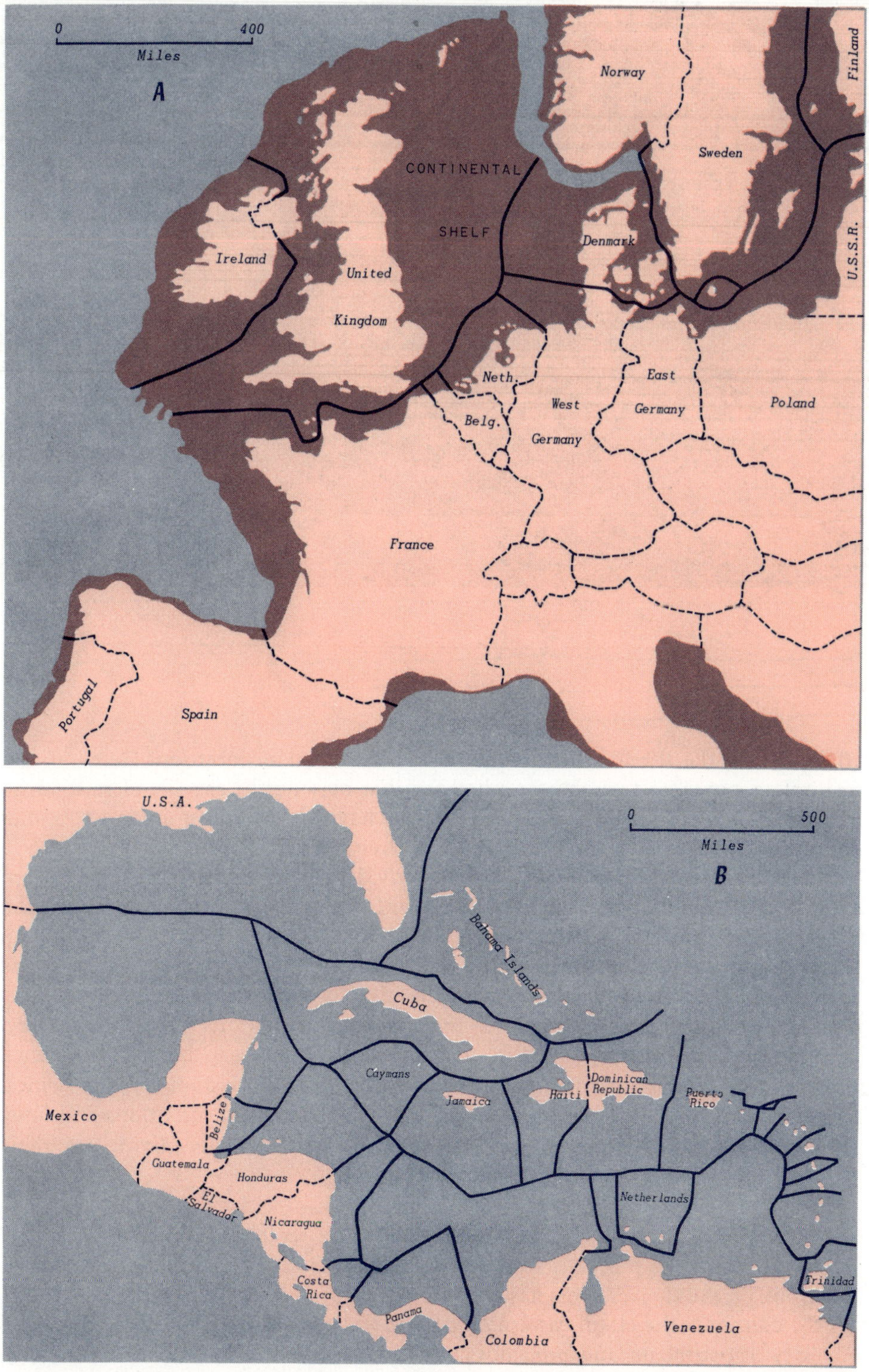

FIG. 10.6 The North Sea and the Gulf of Mexico allocated. (From a map entitled "The Gulf and Caribbean Allocated," in D. L. Capone and A. F. Ryan, *The Regional Sea: a Theoretical Division of the Gulf of Mexico and the Caribbean Sea*, Transactions of the Miami Geographical Society, 3, 1 (March, 1973), pp. 7–8.)

Tuna fishing off Sicily. The tuna catch still requires the work of many fishermen, but more modern methods of harvesting tuna could endanger the species. A sad by-line to the tuna catch relates to the dolphin, which tends to cohabit with the tuna and gets caught in tuna fishermen's nets. The fish at right is a tuna, but at left a dolphin is being dragged aboard.

effect is to make the fish harvestable because of its concentration. Demand and technology now join to endanger a balance that may not be capable of restoration once it has been disrupted; some fish species have already been rendered extinct and other ocean life, whale species among them, are in danger.

Only international agreements, clearly defined and strictly adhered to, can save the situation, but cheating is easy and enforcement (for example, through the International Court of Justice) is difficult. If one country fails to sign such an agreement it would benefit from the voluntary restraint of the others; a single nonparticipant can destroy the anticipated effect of the treaty. The assignment of designated fishing zones in the oceans to certain fishing nations is similarly ineffective: fish migrate from one zone into the next, and all it takes to destroy the balance is overfishing of one of the zones. Agreements may involve a quota system, which is also difficult to police. Again, governments may have good intentions and long-range conservationist objectives, but fishermen are there to make a living, and their short-range rewards increase as their volume does.

MINERALS The continental shelves, we know, are geologically part of the continents. The array of minerals found on the landmasses also will be represented, therefore, on the shelves. This has already been proved in numerous places, including the continental shelf off eastern North America, which is yielding an increasing quantity of petroleum and probably contains more of this fuel than is presently known. Coal seams in Britain are known to extend under the sea; natural gas is being taken from the floor of the North Sea. When the demand warrants it, other minerals will be exploited as well. This makes it rather understandable that countries with substantial continental shelves would wish to assert their primacy and privilege over the "seabed and subsoil," as the legal treaties describe the sea floor and what lies beneath. As long as they are not certain of what riches their continental shelves might contain, coastal states want to be sure that some foreign power does not come along to build installations and exploit them.

Of course, the international community must pay a price for such action. It is but a short step from the seabed and subsoil to the overlying superjacent waters, and states that claim their continental shelves and actively exploit them will not permit the unchecked use of the waters in which their installations stand. So Fig. 10.2 may very well become a politico-geographical map as well as a physiographic one.

The United States was the first country to announce its claim of sovereignty over the continental shelf. In 1945, President Truman issued a proclamation to that effect, a document that became known as the Truman Declaration. This proclamation specifically included the seabed and subsoil, but also stipulated that the overlying waters of the Atlantic Ocean were to remain high seas, free and open. In those days the United States claimed only 3 nautical miles of territorial sea (under normal conditions), but the effect of the Truman Declaration was to encourage other states to extend their claims, not only over the continental shelf but *also* over the waters above, and even waters beyond. Chile (in 1947), and Peru and Ecuador (1952), announced their claims over their narrow continental shelves and 200 miles of territorial waters; other countries in various parts of the world followed suit in the next two decades. The Pandora's box had been opened.

THE TERRITORIAL SEA

It is likely that coastal peoples have had a proprietary view of the sea near them as long as organized human communities have existed. Certainly the ancient Greeks and Romans thought this way about the Mediterranean Sea; later, when Western Europe's countries began to develop powerful fleets, the North Sea came into contention. As early as the thirteenth century, Norway's rulers decreed that no foreign vessel could sail north of the latitude of Bergen without a Norwegian royal license. Denmark, which subsequently gained control over Norway, maintained this practice until the 1590s, when a novel rule was instituted. Since the objective of the Bergen rule was to stop foreign fishing boats from plying the rich waters near Iceland (also a Danish possession at the time), the Danes now announced that a continuous belt of water, 8 miles in width and surrounding Iceland, would henceforth constitute Danish territorial waters. Not long thereafter the Danish Crown established similar belts along Scandinavian coasts they controlled. Soon the belts became wider, and during the mid-seventeenth century the Danes claimed as much as 24 miles of sovereign sea.

The First Legal Issue

While the Danes were closing off extensive waters, the English were busy establishing *baselines* along the coast of Britain, connecting capes and promontories and thus closing off bays and estuaries. They then demanded that Dutch fishing boats obtain permits to fish within those lines, some of them 50 miles in length. Since the Dutch did not have indented, embayed coastlines similar to those of Britain, their position was to argue that the British (and Danish) closures were violations of "customary" international law. And so, in the seventeenth century, a legal issue came under debate that has yet to be resolved. A Dutch jurist, Grotius, published a treatise entitled *The Free (Open) Sea.* The British legal expert Selden soon had an answer prepared, and his book, predictably, came out under the title *The Closed Sea.* One interesting concept emerged during this exchange, namely, the idea that the width of any zone of territorial waters should be determined by the reach of weapons positioned on the shore line. In other words, if a cannon standing on a beach dune could shoot a cannonball 3 miles, then that state would have jurisdiction over 3 miles of territorial sea.

Time was on the side of the Dutch. As the English fleet grew in size and capacity, the British realized that a general application of the closed-sea concept would adversely affect their interests. In due course they shifted their position toward that of the Dutch and agreed that a narrow belt of maritime sovereignty would be better than massive closures. Now another set of issues arose. Should such a belt be continuous, as the Danes had it, or should the belt in fact consist of a set of noncontiguous semicircles and take on reality only where shore-based cannon could enforce it? The Dutch jurist Van Bynkershoek suggested just such a solution, arguing in 1702 that no state should be permitted to claim adjacent waters over which it had no control by force of arms. Then he introduced another

---

NAUTICAL MILE

Measurements at sea are normally given as *nautical* miles, which differ from the *statute* mile used on land. The statute mile equals 5280 feet. Its origins lie in Roman times, a measure amounting to 1000 paces, each 5 feet in length. Around the beginning of the sixteenth century, the division into 5280 feet was adopted, but European countries eventually abandoned the measure and went to the metric system and the *kilometer* (about $\frac{5}{8}$ of a mile).

The nautical mile is calculated by dividing the circumference of the earth into 360 degrees and each degree into 60 minutes. One nautical mile, then, equals the distance of one such minute. There is just one problem: the circumference of the globe is not the same in all directions, because the earth is not a perfect sphere. Thus international agreement had to be reached on the exact length to be used, and it is 6076.12 feet. This means that the nautical mile is about 1.15 statute miles.

---

principle, one that is still being debated today, that of "intended and continuous occupation." Van Bynkershoek felt that any state should be allowed to claim permanently any part of the sea or ocean where it could establish permanent control. In his day, that would have meant the assignment of a fleet, but in modern times, intended and continuous occupation can be achieved by the construction of permanent steel structures such as offshore drilling rigs and platforms.

One result of the discussion was the limitation of claims. The Danes, who had consistently increased their territorial sea width, agreed to observe a continuous zone of just 1 league, equal to 4 nautical miles. This was in 1745; the 4-mile limit became the general rule in Scandinavia for the next two centuries. The Dutch and British observed a 3-mile limit, and countries of the Mediterranean generally declared a 6-mile limit to their territorial waters. With Europe's colonial expansion, these rather modest claims were applied also in Europe's overseas dependencies. During the nineteenth century, the United States and Canada adhered to a 3-mile limit, as did Australia. As late as 1951, the self-interest of maritime powers in such modest claims could be read from the statistics: in that year, 80 percent of the merchant-shipping tonnage of the world was registered in countries that subscribed to the 3-mile limit, and another 10 percent belonged to states that adhered to the 4-mile zone. As we shall see, the situation today is very different.

Offshore Zones

From the very beginning, coastal states were not satisfied merely with a territorial sea. Eighteenth-century England, Denmark, and Holland announced that they would reserve the right to search foreign vessels in a "customs zone" lying outside the territorial waters. A "zone of neutrality" also appeared in the legal exchanges. Today it is possible to recognize five zones, each with a greater or lesser degree of general acceptance among interested states (Fig. 10.7).

INTERNAL WATERS When a state draws a baseline across an estuary or a bay, it thereby makes that estuary or bay an *internal* water body, not a part of the territorial waters, but of the same wholly sovereign status as a lake or a river. As we shall see shortly, the delimitation of such internal waters is important, for from the same baselines that cut off the estuary or bay, the state's maritime boundary at sea must be measured.

TERRITORIAL WATERS The zone of *territorial waters* is the critical zone, that belt of sea or ocean water whose outer edge legally constitutes the maritime boundary of the state. No international agreement yet exists to define a standard width for this belt, and so this varies from state to state. There are still countries adhering to the 3- and 4-mile limit, but, as we have noted, there are others claiming 200 miles.

Within the territorial waters a state possesses essentially the same sovereign rights it holds over its land territory. Foreign vessels do have the right of innocent

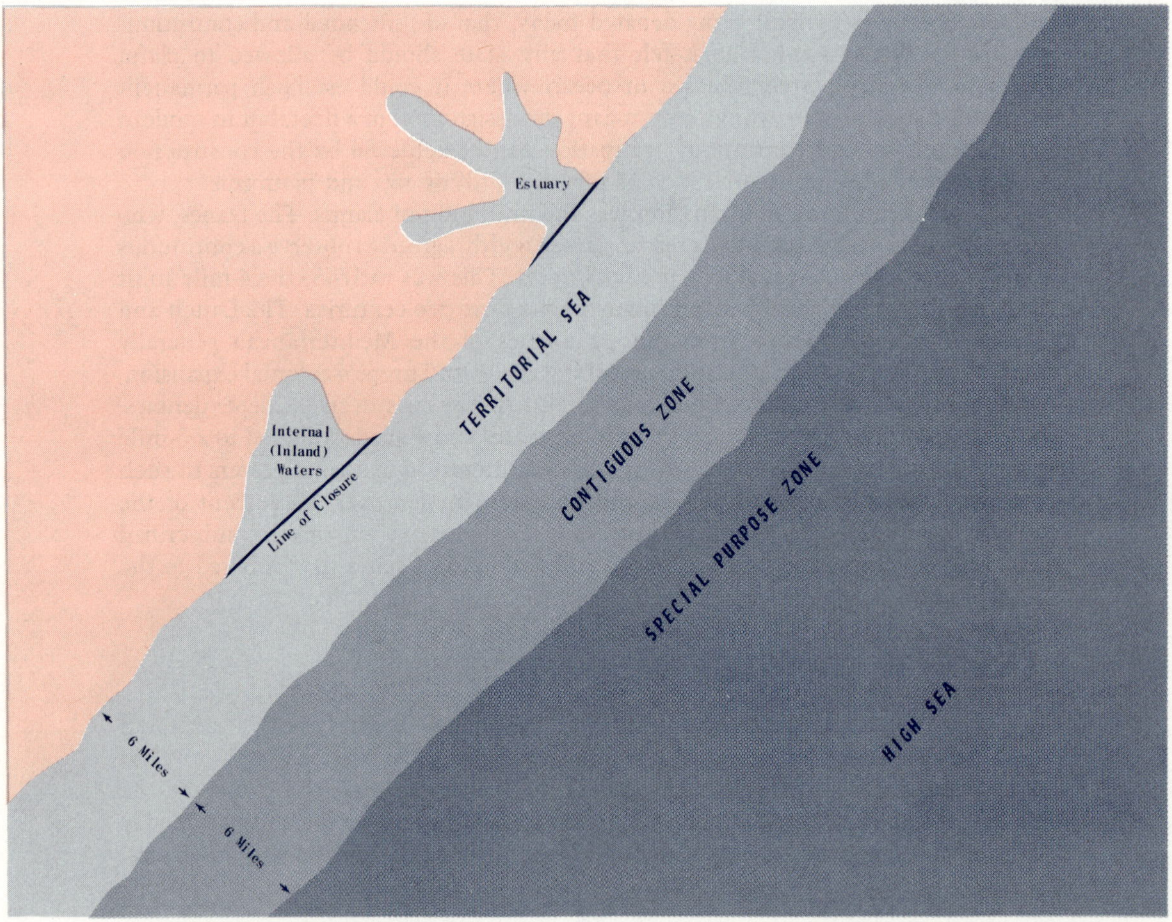

FIG. 10.7 Offshore zones.

passage, but unauthorized entry into the territorial waters is the same as the violation of a land boundary. Contraband found inside the territorial waters has already been smuggled into the state; a crime committed aboard a vessel inside the territorial sea may be treated as though it had been carried out on land, since national laws prevail. Countries that have claimed wide territorial seas have proved that they intend to exercise their powers there, as Peru has proved repeatedly in arresting and towing to port the fishing boats of foreign countries including the United States. The reason for the arrest of the naval vessel *Pueblo* off the North Korean coast also was that the *Pueblo* allegedly violated North Korean territorial waters.

CONTIGUOUS ZONE States find it necessary to exercise certain powers even beyond the territorial waters. For example, if a state claims a 6-mile territorial sea, it could not permit foreign vessels to discharge waste, oil, or other effluents just outside that 6-mile boundary. Winds would quickly drive those pollutants into the territorial waters and, quite possibly, onto beaches and into ports. Thus when states began to confer about the problem of territorial waters, nearly all of them insisted that they should be empowered to control a zone beyond the territorial sea. In this way the idea of a *contiguous zone* gained acceptance. This zone, too, has a designated width, but again no agreement exists on its definition. The consensus among states has been that the territorial sea *plus* the contiguous zone should not exceed 12 miles in width (so that a state with a 6-mile territorial sea could insist on a 6-mile contiguous zone), but that consensus was destroyed by the extension of claims during the 1960s and 1970s. When there are states claiming 200-mile territorial waters, it seems fruitless to negotiate about a 6-mile contiguous zone.

Oil pours onto the ocean from a well at the bottom of the Santa Barbara Channel, off California, February, 1969. Before the leak was stopped, huge amounts of oil floated onto beaches, killing wildlife and soaking the sand. Other oil spills have come from tankers that have broken up at sea. Governments want jurisdiction in adjacent waters so that responsibility may be assigned in such cases.

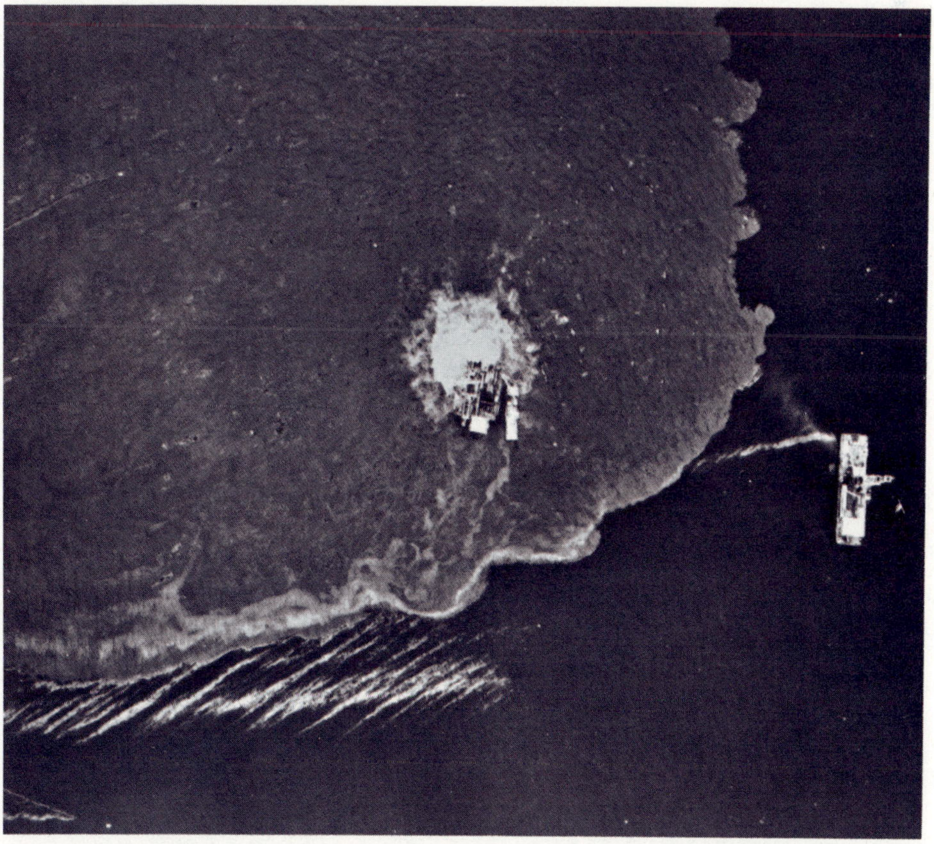

SPECIAL-PURPOSE ZONES A fourth zone that has received some sanction is still less clearly defined than the contiguous zone. This is an area of waters in which a powerful state may frequently carry on naval maneuvers, where it may test new weaponry, or where it is engaged in some other activity in the national interest. In the case of the United States, the Anti-Smuggling Act of 1935 helped develop this idea, for it is desirable, naturally, to catch smugglers before they come as close to the coast as the territorial waters would permit. The *special-purpose zone* has no standard width, and an attempt is not even made to formalize such a width. As the term implies, this is a region of waters where the state's interests and activities are specialized and occasional, not permanent and intensive. Perhaps we would do well to visualize the waters overlying the continental shelf in this context. There is no standard width to the continental shelf either, but obviously a state cannot have an interest in the shelf below without in some way affecting the waters that lie above it.

HIGH SEAS Finally, the *high seas* are those waters so far unclaimed by coastal states. Unlike the other zones we have mentioned here, the area of the high seas is dwindling under the onslaught of political invasion. Figure 10.6 may be a precursor of things to come: it is not a very large step from a subdivision of the North Sea to an allocation of the North Atlantic. High-sea passages that formerly led through straits and narrows are being cut off by encroachment of national claims. True, such "nationalization" is not necessarily an immediate impediment to international shipping, but it has dangerous potential.

THE MARITIME BOUNDARY

The definition, delimitation, and demarcation of the seaward boundaries of states involve a series of problems, some of which we have referred to in passing and should now look at in some detail. When a coastal state decides upon a width for its territorial sea, it must next establish the points along the shoreline from which that width should be measured. Ships at sea must be able to determine, in a fairly simple triangulation, whether they are within or outside the territorial sea or the contiguous zone. Charts must be published on which the baselines drawn across bays and estuaries are indicated as well as the shoreline points of measurement.

Now the problem of the tidal range comes into play. Along some stretches of coastline, low tides expose miles of shallows that are inundated during high tide. At a conference held in 1930 in The Hague, the coastal states agreed that the state should be allowed to measure its seaward boundary from the point of maximum land exposure, that is, when low tide is at its lowest. As we mentioned earlier, the tidal range is greatest during spring-tide conditions, so the rule became that measurement should be from the limit of low water during spring tide. But in 1958, a conference in Geneva changed this practice, and *mean low water* was adopted as the legal line of delimitation. This conforms to the practice used in publishing marine charts, which generally give depths based on mean low water.

LOW-TIDE ELEVATION When ocean waves and currents build an offshore bar, a problem arises. When it is high tide, the offshore bar is inundated. But under low-tide conditions, it is exposed. Still, there may be water between the mainland's shoreline and the offshore bar, even at low tide. The question is: can a state measure its maritime boundary from the outer edge of the offshore bar at low tide rather than from the shoreline? The answer is affirmative, but conditionally. The offshore bar must lie close enough to the shoreline that it would be within the territorial waters if measured from the mainland. In other words, a state cannot claim an offshore bar 20 miles from its coastline and argue that this is its property. Such an offshore bar, used for purposes of seaward boundary delimitation, is called a *low-tide elevation.*

BAYS AND BASELINES Another kind of problem arises in connection with the definition of the term *bay.* Is a bay simply any indentation, any bend in the coastline? In that case, a state could draw baselines across all concave sections of its coast and measure its seaward boundary from those. Again, several conferences were necessary to solve this problem. Under present rules, states should not close off a bay whose mouth is wider than 24 miles or twice the 12-mile limit of territorial waters. A bay less than 24 miles at its mouth is treated as follows: a semicircle is drawn across the bay, with its radius one-half the length of the line of closure (Fig. 10.8). If the resulting semicircle is larger than the waters closed off behind that line, this is not a bay and it cannot be closed off. If the water area behind the line of closure is larger than the semicircle, it is indeed a bay and can be cut off from the territorial sea; it becomes an area of internal waters.

Bays and estuaries frequently are sensitive areas, and many states are disinclined to apply the stated rules. The U.S.S.R., for example, takes this position in Peter the Great Bay, on whose shore Vladivostok is located. Ports, military installations, and other attributes of the state may be concentrated along embayed coastlines, and it can be inimical to the country's national interest to allow the high seas to penetrate such areas. For states that feel this way, there is a legal ground upon which to claim waters that might otherwise be classified as high seas. International law recognizes *historic* waters, estuaries and bays that have played special roles in the evolution of the nation and that still form an integral part of the land area, and *strategic* waters, bays where naval facilities and other military structures are concentrated, where equipment is tested, and where ocean traffic should stay away.

Boundary Making

With all these (and still other) stipulations, how should the seaward boundary of a coastal state be established? This is no simple matter either. It would be easiest,

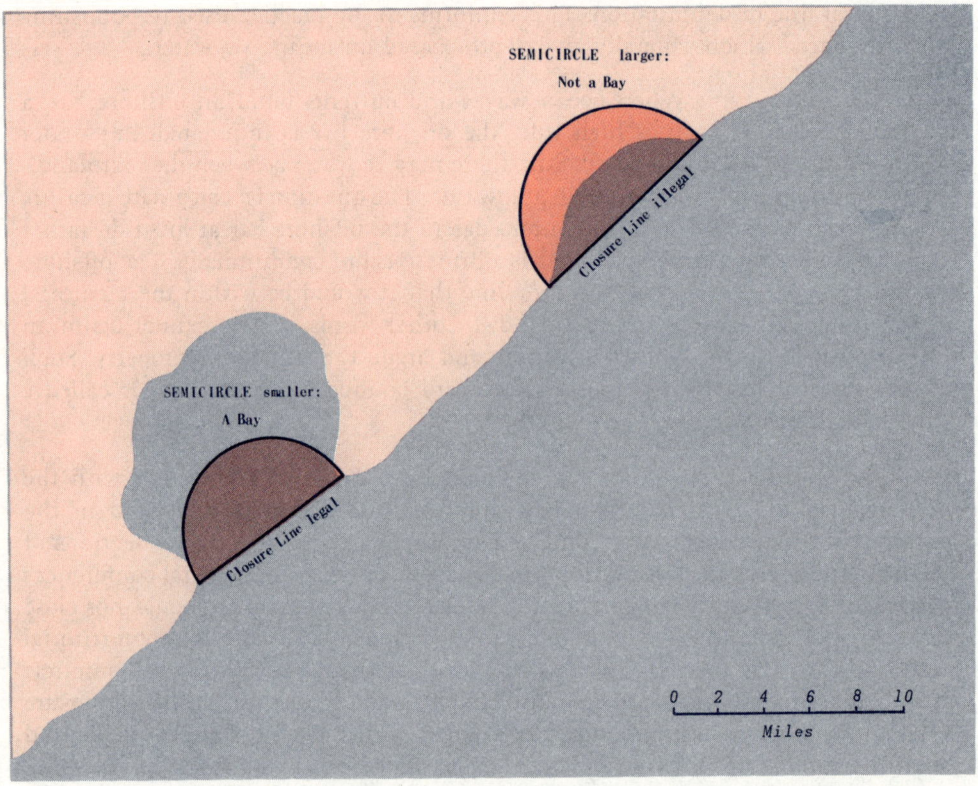

FIG. 10.8 Legal and illegal closure of bays and estuaries.

of course, to straighten shorelines out by drawing baselines all along them, and then to draw those baselines on marine charts, a given distance away. But baselines cannot be established along all parts of coasts, as we have just noted. Indentations in the coast that do not qualify as bays may not be cut off this way. So the *baseline method* cannot be universally applied.

Another possibility might be to reproduce the coastline a certain distance (3, 4, 12 miles) away. This is referred to as the *replica* method. For example, to draw the maritime boundaries of Madagascar (the Malagasy Republic), you would simply enlarge the map by, say, a 12-mile belt and draw the outlines of the island again. The trouble is that the replica method does not produce equidistance from all points on the coasts, that is, the belt of territorial waters does not have a constant width. You can easily prove this to yourself by taking a coastline section, say of western France, and trying to draw a replica of it some distance away. It will not work satisfactorily.

The method in general use today is the *envelope method* (Fig. 10.9). This method was selected at The Hague Conference (1930) mentioned earlier, and it stipulates that "the seaward limit of territorial waters is the envelope of all arcs and circles drawn from all points on the coast . . . or from the seaward limit of those (internal) waters which are contiguous with the territorial waters." Thus the curves and bends of the coastline are reproduced by a series of arcs, a rather simple process; it is also a matter of simple triangulation for a ship to calculate whether it is within one of those arcs or not. Charts are needed for baseline positions, of course, but as long as a navigator knows the width of territorial sea claimed by the coastal state, he can calculate his position with reference to the seaward boundary—when the coast is in sight. This simplicity disappears where states claim hundreds of miles of territorial sea.

MARITIME CLAIMS

This brings us to the thorniest of problems today: the diversity and inconsistency of national claims to territorial waters. Over the past 25 years, despite efforts to

FIG. 10.9 The envelope method of offshore boundary construction.

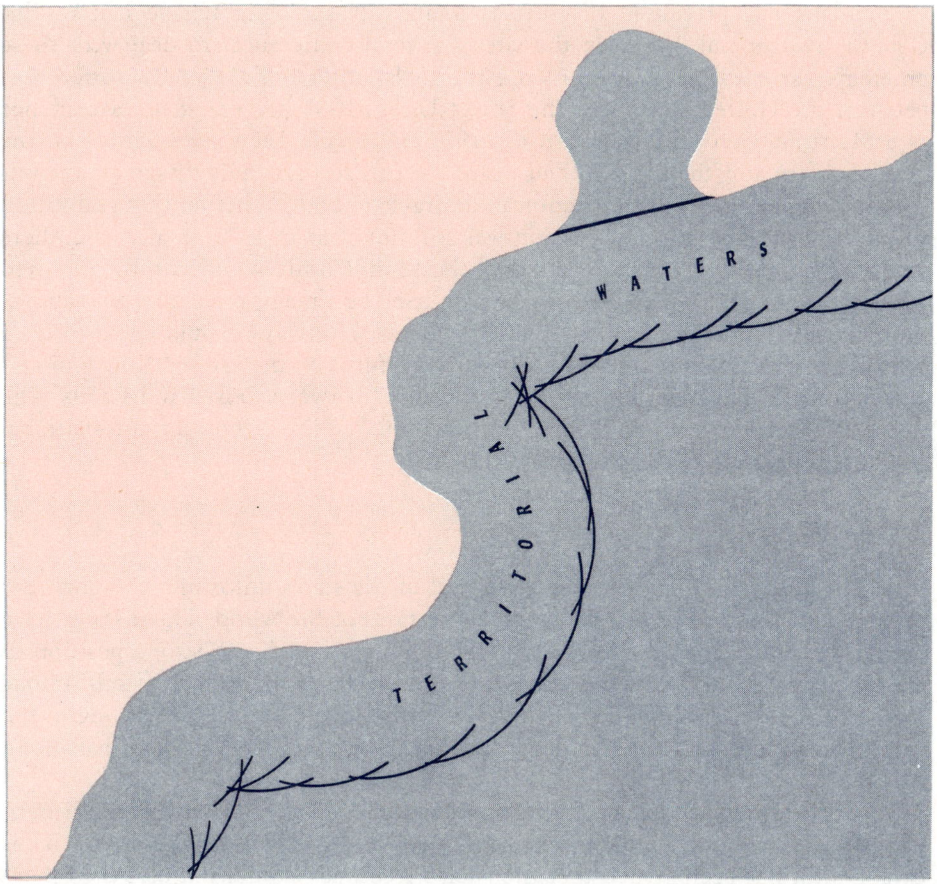

bring order to the chaos, the scramble for the oceans has proceeded at an unprecedented pace. As recently as 1950, the 3-mile limit was still claimed by 40 states. The 4-mile limit was observed by 4 states, and a 6-mile zone was claimed by another 11 states. By 1970, only 25 states still adhered to the 3-mile limit, but no less than 44 states claimed a 12-mile territorial sea. And nearly a dozen Middle and South American states claimed a 200-mile zone, either as territorial sea or as exclusive fishing preserve. The idea that territorial waters should be thought of in terms of hundreds of miles rather than just a few spread from Latin America to other parts of the world. In Africa, Guinea and Senegal, both former French dependencies, decided to demand a 130-mile territorial sea; Sierra Leone, formerly administered by Britain, appropriated 200 miles. India announced a 100-mile exclusive fishing zone. Iceland pushed its maritime boundary 50 miles into the North Atlantic. Elsewhere, states still adhering to the old 3-mile territorial sea announced that they would also claim a 9-mile contiguous zone to protect their fisheries.

Certainly efforts were made to stabilize the situation, but the national self-interest (and ideological objectives) of the powerful states from whom leadership must come prevented the sort of unanimity that might have had results. Geneva, Switzerland, has been the site of several conferences to deal with these problems, the latest in 1971. Ever since the 1958 conference at Geneva, differences between the United States and the Soviet Union have prevented the acceptance of a Maritime Convention, but in the 1971 conference there were signs that the United States and the U.S.S.R. had begun to realize that their disagreement was opening the door to unilateral action undesirable to both. Whether that realization would be translated into joint action was still not clear in 1973. In any case, there seems little hope that anything less than the 12-mile limit would gain the required majority of votes, and such a limit would in itself be a major problem. For example, the English Channel, one of the world's busiest waterways, would cease to have a high-seas corridor and would be shared henceforth by the United Kingdom and France, whose maritime boundaries would meet between Dover and Calais. The universal acceptance of a 12-mile limit would require a host of international guarantees of passage.

Landlocked States

In all this discussion, nothing has been said of the states unfortunate enough not to have any coastline at all, the landlocked states of the world. These states have been represented at recent conferences, but they are not in a strong position to lay claim to the maritime resources the coastal states are parceling out among themselves. None of the landlocked states of the world is especially strong in the international arena; of the two dozen landlocked countries, more than half lie in Africa.

The landlocked states may not be in a position to claim territorial waters, but should they not get a share of the resources of the continental shelf now being exploited? It has been argued that the shelf's resources should not be left to those

One hope for the protection of the marine environment and the reasonable allocation and exploitation of resources existing there lies in international conventions accepted by all countries. The United Nations has sponsored several successive Law of the Sea conferences, attended by all member countries with maritime interests. The photograph shows the conference at Geneva in 1960, attended by representatives of 88 states.

powers capable of claiming them and exploiting them, but should be mined and utilized for the benefit of all states. In other words, those resources should be allocated to all states, coastal or landlocked. Another suggestion has been that the continental shelves should be broken down into units, say, squares of 100 miles each, and these units should be assigned to states, irrespective of their location on land. These proposals have not been formalized in a convention, though, and meanwhile the landlocked states are locked out of the scramble for the oceans.

Until recently it was possible to discern a spatial pattern in the nations' claims to territorial seas. The states claiming the 3-mile limit were concentrated in the region where the concept had its modern origins, namely, Western Europe, and in the non-European states influenced or directed by these powers. The 4-mile

LANDLOCKED STATES OF THE WORLD

AMERICA	EUROPE	AFRICA	ASIA
Bolivia	(Andorra)*	Botswana	Afghanistan
Paraguay	Austria	Burundi	Bhutan
	Czechoslovakia	Central African Republic	Laos
	Hungary	Chad	Mongolia
	(Liechtenstein)	Lesotho	Nepal
	Luxembourg	Malawi	Sikkim
	Switzerland	Mali	
	(San Marino)	Niger	
	(Vatican City)	Rhodesia	
		Rwanda	
		Swaziland	
		Uganda	
		Upper Volta	
		Zambia	

*Microstates are in parentheses.

limit always was a characteristic of northern European states, and the 6-mile limit was long claimed by states with coasts on the Mediterranean, including Greece, Italy, and Turkey. The independent directions taken by former colonial territories have led many of them to renounce the claims with which they became independent, to go to the 12-mile limit or even farther. And the inexorable push outward has begun to capture even the old Western European countries still hanging on in principle to the 3-mile limit. The existence of one of the potentially richest continental shelves off the European mainland has as much to do with this as the desire of several countries (Iceland, Norway, the United Kingdom) to protect their fishing industries.

DISPUTES

The period following the Second World War has been marked by the frequent eruption of disputes over fishing rights and related maritime issues. As the national maritime claims have expanded, so these disputes have grown in number and intensity, underscoring the need for international regulation and control. But what kind of control could be acceptable to all the interested parties, or even most of them? States whose fleets ply all the oceans tend to favor regulation by quota, for the quotas would then be based on their recently reported total catches and not on the quantity of fish resources on their own continental shelves. States whose own continental shelves are especially rich wish to assert their primacy in such areas, and they would prefer regulation by the allocation of exclusive fishing zones, based on proximity. An example of the latter is Iceland, a country whose national economy is enormously dependent on fish taken from adjacent waters. Iceland does not have world-ranging fleets, but it does have some of the world's richest fishing grounds right off its coasts, on the extensive continental shelf that surrounds the island. In the 1960s Iceland and the United Kingdom repeatedly were in conflict over Iceland's demand that a 12-mile zone, measured from lengthy baselines, be recognized as its exclusive fishing preserve. British trawlers entered the zone, and when Icelandic gunboats confronted them, the British sent warships in support of their fishermen.

Such disputes are sometimes taken to the International Court of Justice, but the problem is that the court's decisions are meaningless unless the states involved agree to abide by the arbitration. In Iceland's case, the court had ruled that foreign trawlers should be allowed to operate in Icelandic waters, despite Iceland's argument that these were territorial seas and historic waters, and notwithstanding the fact that the fishing industry is far more crucial to Iceland's overall economy than British fishing is to the United Kingdom's economy. Iceland decided not to obey the International Court's ruling and, taking note of other countries' actions, extended its territorial sea from 12 to 58 miles from its coasts. In the latter part of 1972, British and West German fishing boats were again harassed by Icelandic gunboats, and the Icelanders gave notice that they had no intention of adhering to any court ruling they considered damaging to their interests. Nothing, other than the power of another state, can compel any state to withdraw from a position it has taken relative to its maritime claims.

This is not to suggest that disputes over fishing rights are confined to the North Sea or the North Atlantic. The United States shrimp fleet was threatened during its operations off South American coasts; Soviet lobster fishers dragged lines through United States traps off the northeast coast, precipitating a quarrel there. As we noted, Peru (and Ecuador and Brazil) have arrested foreign fishing vessels within their 200-mile territorial seas. As long as there is no effective international rule governing these operations, there will be conflicts of this kind.

How shall international law be established to solve the growing problem of international maritime rights? A conference sponsored by the United Nations, scheduled to be held in Santiago in 1974, will seek the answer to this question, but without the certainty even that an overwhelming unanimity will in fact constitute a solution. Any reasonable proposition would have to involve the reduction of some of the national claims now existing, and the chances that states would be willing to do this seem very slight. If states find that they need not adhere to International Court of Justice rulings, what are the chances that they will conform to stipulations arrived at by a UN conference that has no authority?

The search for a viable solution must be carried on, however, because the continued absence of regulations and control can endanger the oceans' life and raises a specter of even more serious trouble as the exploitation of the continental shelves gains momentum. Earlier in this book we viewed the deteriorating resource situation on the world's mainland areas, a circumstance that will increase the pressure upon the newly open possibilities in and beneath the oceans. Sometimes expressions of concern over the future in this context cause adverse reactions, criticism of doomsday viewpoints, and the answer, "there will always be alternatives even if and when the oceans cease to yield as they have in the past." That may be so, but in the meantime there were, in mid-1973, only about 100 blue whales left in all the oceans, the humpback whale was about to become extinct, and the finback whale did not appear likely to survive the decade. The United States had banned the importing of all whale products, but it was too late, and other countries had refused to follow suit in any case. For these whale species, the necessary controls could never be implemented, and it is now too late. For countless other marine species, the same fate undoubtedly lies ahead, unless mankind at some moment, and very soon, transforms the mistakes of the past into a conservation program for the future.

As we know too well, these words apply also to the resources of our lands—to the soils, the vegetation, the waters, the minerals, and to tigers and cheetahs, eagles and polar bears. Or will geography books in the future commence something like this:

Not long ago, from the viewpoint of man's lifespan on this earth, people shared with animals and plants a worldly habitat of amazing variety and color. Although great urban centers already flourished, magnificent herds of elephant and buffalo still held their own in East Africa, and wildlife survived in large numbers in many other areas. The oceans teemed with fish and were open for all to traverse, unbounded and free. Vegetation of infinite variety graced even the streets of the cities, and "parks" provided the urbanites a sample of the world beyond their daily sphere. Food was grown mostly on vast fields, where herds of "livestock" were also raised. Fish further varied diets that must have been quite unlike our laboratory-produced meals of today. When, early during the second half of the twentieth century, the first timid space flights were undertaken, the returning astronauts made much of the magnificent colors their planet earth generated. They must not have known that they were gazing into the twilight of a habitat that had sustained human progress for millenia, but whose destruction was already well under way . . .

INDEX

Information found in maps, diagrams, or photographs is indicated by **boldface** page numbers.